TRACK GEOTECHNOLOGY
and
SUBSTRUCTURE MANAGEMENT

Ernest T. Selig

Professor of Civil Engineering
University of Massachusetts
Amherst, Massachusetts, U.S.A.

and

John M. Waters

Railway Geotechnical Consultant
Derby, England

Acknowledgements

Many persons, too numerous to mention individually, contributed to the development of technical material presented in this book. However, an effort has been made to recognize their contributions through the frequent use of references cited in the text.

Continuing support over a number of years of research in railway geotechnology by the first author has been responsible for much of the information incorporated in this book. This research has been sponsored by the Federal Railroad Administration of the United States Department of Transportation, by the Association of American Railroads, and by individual North American railroads. A key person in encouraging this work is Dr. A. J. Reinschmidt of AAR.

Years of experience in railway geotechnology gained working for British Railways has provided the foundation for the contribution to the book by the second author. British Railways also made available previously unpublished material for inclusion in the book. In addition Dr. D. L. Cope, Permanent Way Engineer, British Railways, Intercity, provided encouragement and valuable editorial help. Plasser and Theurer and Plasser South Africa also provided support and encouragement for the work of the second author.

Invitations to give short courses on railway geotechnology by the Universidade Federal do Rio de Janeiro, Brazil, by the University of Pretoria, South Africa, and by Railways of Australia have provided a major incentive for writing this book. Significant factors in completion of the first edition were the sabbatical leave support by the University of Massachusetts, together with the support from the University of Pretoria and SPOORNET during a six-month visit to Pretoria in 1992 by the first author. Key persons in making this happen were Dr. Willem Ebersöhn, Senior Lecturer, Civil Engineering Department, University of Pretoria, Professor A. Rohde, Civil Engineering Department Head at the University of Pretoria, and C. E. Malan, Chief Engineer, Infrastructure, SPOORNET. The encouragement and assistance provided by Dr. Ebersöhn were particularly important.

The authors' wives, Rae Selig and Sally Waters, provided much encouragement during the writing as well as typing and editorial help. In addition Rae Selig prepared the final camera-ready manuscript. Pamela Stefan, Technical Illustrator in the College of Engineering at the University of Massachusetts, prepared many of the illustrations. The help of other friends and colleagues with figures and proof reading is also appreciated.

E. T. Selig

J. M. Waters

TRACK GEOTECHNOLOGY

AND

SUBSTRUCTURE MANAGEMENT

ERNEST T. SELIG

AND

JOHN M. WATERS

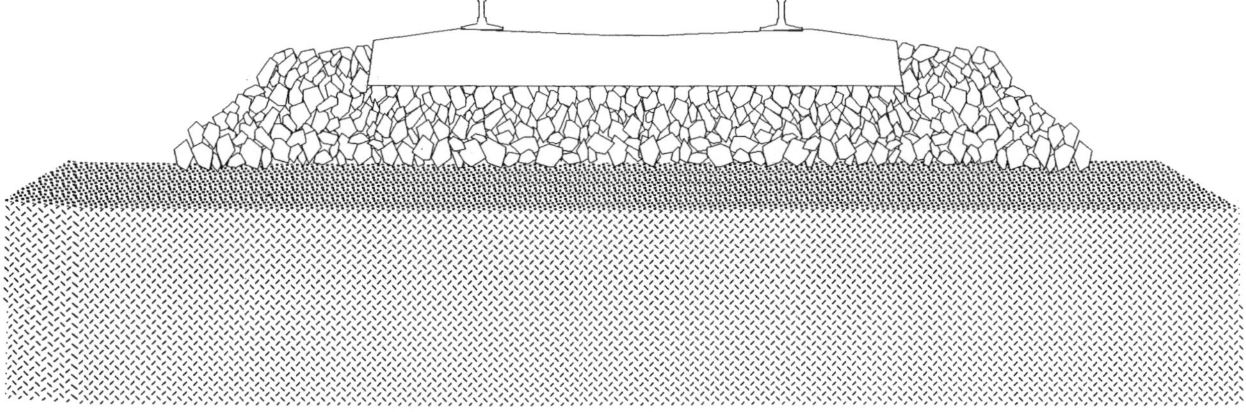

Thomas Telford

Published by Thomas Telford Services Ltd, Thomas Telford House, 1 Heron Quay, London E14 4JD

First published 1994
Reprinted 2000, 2002, 2007

Distributors for Thomas Telford books are
USA: American Society of Civil Engineers, Publications Sales Department, 345 East 47th Street, New York, NY 10017-2398
Japan: Maruzen Co. Ltd, Book Department, 3—10 Nihonbashi 2-chome, Chuo-ku, Tokyo 103
Australia: DA Books and Journals, 648 Whitehorse Road, Mitcham 3132, Victoria

A catalogue record for this book is available from the British Library

Classification
Availability: Unrestricted
Content: Current best practice
Status: Established knowledge
User: Civil and railway engineers

ISBN: 978 0 7277 2013 9

Printed and bound in Great Britain by
CPI Antony Rowe, Chippenham and Eastbourne

Table of Contents

1. Introduction

The railway track system is an important part of the transportation infrastructure of a country, and plays a significant role in sustaining a healthy economy. The annual investment of funds to construct and maintain a viable track system is enormous. The optimum use of these funds is a challenge which demands the best technology available. Unfortunately the pressures on a railroad to reduce operating costs usually result in cutting or eliminating investment in technology because it often does not generate a short term return on investment.

The performance of a railway track system results from a complex interaction of the system components in response to train loading. The superstructure consisting of the rails, fasteners and sleepers has received the most attention in the past. The substructure consisting of the ballast, the subballast and subgrade has been given much less consideration, especially the subballast and subgrade components, even though it has a major influence on the cost of track maintenance. The properties of the substructure components are much more variable and difficult to define than those of the superstructure. This presents a challenge to the railway track engineer.

The state-of-the-art of railway track design and maintenance planning , especially with regard to the substructure, typically has been very empirical, with trial and error usually the basis for the decisions. However this approach has two major limitations:

1) Ample opportunity for error exists for reasons such as the wide variety of field conditions, the length of time needed to experience the results of trials, the lack of historical maintenance and performance records, and the changing track service demands over the years.

2) This process does not lend itself to optimizing use of financial resources.

Fortunately a body of knowledge has evolved which permits a more scientific approach to the practice of track geotechnology and substructure management.

This body of knowledge has not been readily available to the practicing railway engineer because it exists in unwritten form and in a diverse assortment of reports, theses, informal records, and publications which often are not accessible or are not suitable for application in maintenance practice. Furthermore they represent fragments of the total subject like pieces of an unassembled puzzle. Nowhere, to the writers' knowledge, has the practicing railway engineer been able to turn for a comprehensive treatment of the topics which together define the subject of track geotechnology. The goal of this book is to help remedy this situation.

The book provides a written record of the subject as it is understood by the writers. It does not contain a complete literature review on each topic. Instead, the examples are taken from material readily available to the writers. Omission of any other material should not be taken to imply that such material is unsuitable or unimportant. It was simply not possible to collect and summarize all existing material that is within the scope of railway geotechnology.

The writers hope that this edition of the book will stimulate interest in the subject of railway geotechnology, and will generate questions and comments which will result in continuing improvements. The ultimate goal is to provide the basis for design, construction, maintenance and renewal of railway track, as they relate to geotechnology, that will result in tracks which have the desired quality, achieved with the minimum maintenance effort and optimum use of resources.

Chapter 2 describes the components of track and forces to which the track is subjected. Chapter 3 summarizes stress-strain and strength properties of soils under both static and cyclic (repeated) loading. Laboratory tests for measuring these soil properties are defined. Chapter 4 describes field (in situ) tests for measuring soil properties that are useful for substructure investigations. Analytical models for representing the elastic behavior of the track structure, including the interaction of the superstructure and substructure, are described in Chapter 5. Chapter 6 illustrates the residual stresses and plastic deformations that develop in granular layers, such as ballast, under repeated traffic loading. Chapter 7 summarizes tests for defining ballast particle characteristics, discusses the effects of these characteristics on ballast behavior, and then illustrates repeated load behavior of ballast. Chapter 8 describes various aspects of ballast performance over its life cycle from the time that it is placed in track until it is cleaned or replaced. Chapter 9 discusses the requirements of subballast. Chapter 10 describes the role of subgrade and means of insuring satisfactory performance. Chapter 11 summarizes methods for subgrade improvement and applications of asphalt concrete in the track substructure. The types and uses of geosynthetic materials in track, such as filter fabrics and geogrids, are given in Chapter 12. Principles of track drainage are explained in Chapter 13. Machines and methods for track construction and maintenance are described in Chapter 14. Track geometry and track quality are the subject of Chapter 15. Chapter 16 describes a mechanics-based computer model for cost evaluation of alternative maintenance plans.

2. Track Components and Loading

The purpose of a railway track structure is to provide safe and economical train transportation. This requires the track to serve as a stable guideway with appropriate vertical and horizontal alignment. To achieve this role each component of the system must perform its specific functions satisfactorily in response to the traffic loads and environmental factors imposed on the system.

This chapter will describe the components of conventional track structures and their functions, and then discuss the loading characteristics.

2.1. Component Descriptions and Functions

Figure 2.1 shows the main components of ballasted track structures. These may be grouped into two main categories: 1) superstructure, and 2) substructure. The superstructure consists of the rails, the fastening system, and the sleepers (ties). The substructure consists of the ballast, the subballast and the subgrade. Thus the superstructure and substructure are separated by the sleeper-ballast interface.

2.1.1. Rails

Rails are the longitudinal steel members that directly guide the train wheels evenly and continuously. They must have sufficient stiffness to serve as beams which transfer the concentrated wheel loads to the spaced sleeper supports without excessive deflection between supports. The rails also may serve as electrical conductors for the signal circuit (each rail being a separate conductor connected by the train axles), and as the ground line for the electric locomotive power circuit.

The profile of the rail surface together with the wheel profile influences the guidance of the vehicles as they roll. Also rail and wheel surface defects can cause large dynamic loads which are detrimental to the track structure.

Steel rail sections may be connected either by bolted joints or by welding. Bolted rails are most commonly used on curves to provide stress relief from thermally induced length changes or on secondary lines. The bolted rail joints have been one of the major locations of maintenance problems. Discontinuity of the track running surface produces dynamic impact loads battering the rail surface and the joint ends. This creates rough riding track and undesirable train vibration. The combination of the impact load and the reduced rail stiffness at the joints causes greater stress on the ballast and subgrade. This in turn, increases the permanent settlement which produces uneven track. The pumping action at the joints also accelerates rail failure, sleeper wear, and fouling of ballast at the joint. Hence joints generally increase track deterioration.

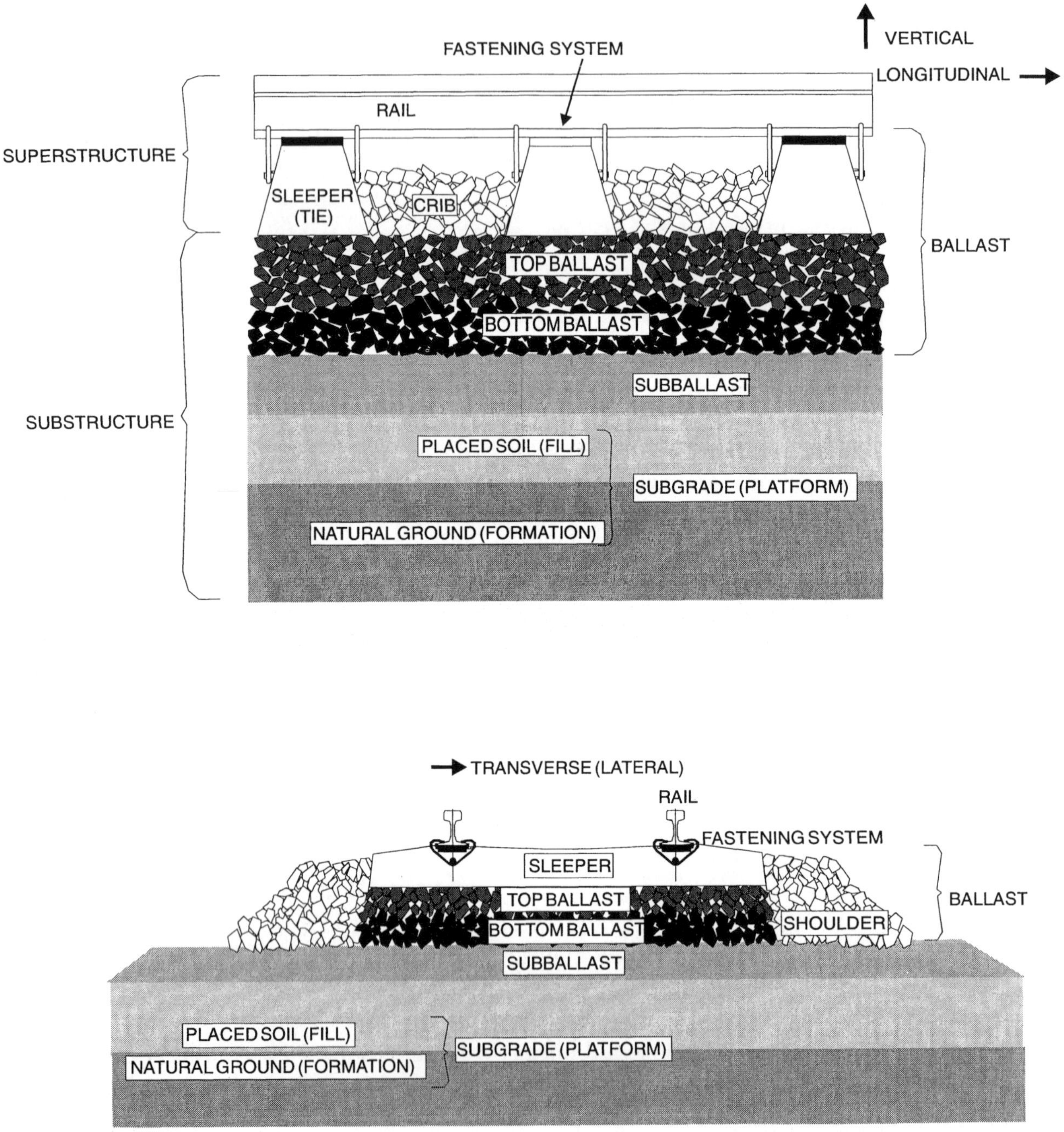

Fig. 2.1 Track structure components

Although much progress has been made in improving joints, a better solution has been to eliminate the joint entirely by the use of continuous welded rail (CWR). This approach is preferred on lines with high speed, with high axle loads, or with high traffic density. Various advantages of CWR include substantial savings through reduced maintenance costs as a result of extended rail life from the elimination of joint wear and batter, improved riding quality, reduced wear and tear on rolling stock, and less damage to the substructure. Disadvantages are 1) breakage of rail or buckling of track from temperature induced changes in rail stress, 2) difficulties in changing worn and defective rails, or in sleeper renewal, and 3) higher initial costs of welding, transporting, and laying longer rails.

Rail gage is the spacing between the inside faces of the rail head. Example values are given in Table 2.1.

Table 2.1 Rail gage values

Location	Gage (mm)	Gage (in.)
North America	1435	56.5
Europe	1435 - 1668	56.5 - 65.7
South Africa	1065	41.9
Australia	1524 - 1676	42 - 63
China	1435	56.5

Typical properties of rails are given in Table 2.2 (Note: Moment of Inertia, I_x, is about horizontal axis). For North America the two lightest rails are used on light and medium traffic density lines. The three heaviest rails are used on high density and heavy haul lines.

2.1.2. Fastening System

The connections between the sleeper and the rails have many variations. These details are beyond the scope of this subsection. Only the main functional features will be mentioned, including the differences between wood and concrete systems. For convenience the resilient pads used with concrete sleepers will be included in this subsection.

The purpose of the fastening system is to retain the rails against the sleepers and resist vertical, lateral, longitudinal, and overturning movements of the rail. The force systems causing these movements are from the wheels and from temperature change in the rails.

Wood sleepers require steel plates under the rail to distribute the rail force over the

Table 2.2 Typical rail sections

A) North America (AREA) (Ref. 2.1)

Rail Size	Mass		Area		Moment of Inertia, I_x	
	lb/yd	kg/m	in.^{2}	cm^2	in.^{4}	cm^4
100 RE	101.5	50.4	9.95	64.19	49.0	2039
115 RE	114.7	56.9	11.25	72.58	65.6	2730
132 RE	132.1	65.5	12.95	83.55	88.2	3671
136 RE	136.2	67.6	13.35	86.13	94.9	3950
140 RE	140.6	69.8	13.80	89.03	96.8	4029

B) South African Railway (Ref. 2.2)

Rail Size	Mass	Area	Moment of Inertia, I_x
	kg/m	cm^2	cm^4
48	48	60.18	1822
57	57	73.24	2651
60	60	76.13	2703

C) Peoples Republic of China (Ref. 2.8)

Rail Size	Mass	Area	Moment of Inertia, I_x
	kg/m	cm^4	cm^4
43	44.7	57	1489
45	45.1	57.6	1606
50	51.5	65.8	2037
60	60.6	77.5	3217
75	74.1	95.1	4490

wood surface. This provides suitable bearing pressure for the wood and protects the wood from mechanical wear. In addition these plates 1) assist the fasteners in restraining lateral rail movement through friction, and 2) provide a canted surface to help develop proper wheel/rail contact. The size of the sleeper plate is an important factor in sleeper performance. If the sleeper plate size is inadequate for the rail loading, the resulting presssure may exceed the wood fiber compressive strength and thus cause accelerated plate cutting and premature deterioration of the sleeper. In North America the sizes are on the order of 200 mm (8 in.) wide by 300 to 360 mm (12 to 14 in.) long.

Driven cut spike fasteners, commonly used in North America, restrain the sleeper plates horizontally. The plates have shoulders to hold the rail laterally when the plate is securely fastened to the sleeper. However, longitudinal rail movement must be restrained by separate anchors clipped to the rails and placed against the sides of the sleepers in the cribs. Driven spikes provide little rail uplift restraint. In fact the head of the spike is often not in contact with the base of the rail so that separation of the rail from the sleeper plate followed by impact can occur as the train wheels move along the track. Screw-type steel spikes are an alternative to driven spikes for wood sleepers. When tightened down against the base of the rail these provide some uplift and longitudinal restraint as well as lateral restraint to the rails. Resilient pads are not used with wood sleepers because the wood itself provides resiliency.

Concrete sleepers have spring fasteners which provide vertical and longitudinal restraint as well as lateral. The fastener is insulated electrically from the sleeper or the rail to minimize signal circuit current leakage. Pads are required between the rail seat and the concrete sleeper surface to fulfill the following functions:

1) Provide resiliency for the rail/sleeper system,

2) Provide damping of wheel induced vibrations,

3)Prevent or reduce rail/sleeper contact attrition, and

4)Provide electrical insulation for the track signal circuits.

2.1.3. Sleeper (Tie)

Sleepers or ties (more specifically cross ties) have several important functions:

1)Receive the load from the rail and distribute it over the supporting ballast at an acceptable ballast pressure level,

2)Hold the fastening system to maintain the proper track gage, and

3)Restrain the lateral, longitudinal and vertical rail movement by anchorage of the superstructure in the ballast.

In addition, concrete sleepers provide a cant to the rails to help develop proper rail/wheel contact.

The two most common types of sleepers are wood (timber) and prestressed/ reinforced concrete. Concrete sleepers generally have a much more secure fastening system than wood sleepers. Concrete sleepers also are heavier than wood sleepers. Concrete is potentially more durable than wood. The combination of these factors provides better rail restraint with concrete sleepers. However, concrete sleepers are more difficult to handle than wood sleepers, and concrete sleepers require pads to provide sufficient resiliency.

Examples of sleeper base dimensions and spacings are given in Table 2.3.

Table 2.3 Typical sleeper dimensions

Location	Material	Width (mm)	Length (mm)	Spacing (mm)
Australia	Wood	210-260	2000-2743	610-760
	Concrete			600-685
China	Wood	190-220	2500	543-568
	Concrete	240-290	2500	568
Europe	Wood	250	2600	630-700
	Concrete	250-300	2300-2600	692
North America	Wood	229	2590	495
	Concrete	286	2629	610
South Africa	Wood	250	2100	700
	Concrete	203-254	2057	700
		230-300	2200	600

2.2. Ballast

Ballast is the select crushed granular material placed as the top layer of the substructure in which the sleepers are embedded.

Traditionally, angular, crushed, hard stones and rocks, uniformly graded, free of dust and dirt, and not prone to cementing action have been considered good ballast materials. However, at present no universal agreement exists concerning the proper specifications for the ballast material index characteristics such as size, shape, hardness, abrasion resistance, and composition that will provide the best track performance. This is a complex subject that is still being researched. Availability and economic considerations have been the prime factors considered in the selection of ballast materials. Thus, a wide variety of materials have been used for ballast such as crushed granite, basalt, limestone, slag and gravel.

Ballast performs many functions. The most important are:

1) Resist vertical (including uplift), lateral and longitudinal forces applied to the sleepers to retain track in its required position.

2) Provide some of the resiliency and energy absorption for the track.

3) Provide large voids for storage of fouling material in the ballast, and movement of particles through the ballast.

4) Facilitate maintenance surfacing and lining operations (to adjust track geometry) by the ability to rearrange ballast particles with tamping.

5) Provide immediate drainage of water falling onto the track.

6) Reduce pressures from the sleeper bearing area to acceptable stress levels for the underlying material.

Note that although the average stress will be reduced by increasing the ballast layer thickness, high contact stresses from the ballast particles will require durable material in the layer supporting the ballast.

Other functions are:

7) Alleviate frost problems by not being frost susceptible and by providing an insulating layer to protect the underlying layers.

8) Inhibit vegetation growth by providing a cover layer that is not suitable for vegetation.

9) Absorb airborne noise.

10) Provide adequate electrical resistance between rails.

11) Facilitate redesign/reconstruction of track.

As shown in Fig. 2.1 ballast may be subdivided into the following four zones:

1) Crib -- material between the sleepers.

2) Shoulder -- material beyond the sleeper ends down to the bottom of the ballast layer.

3) Top ballast -- upper portion of supporting ballast layer which is disturbed by tamping.

4) Bottom ballast -- lower portion of supporting ballast layer which is not disturbed by tamping and which generally is the more fouled portion.

In addition the term boxing may be used to designate all the ballast around the sleeper which is above the bottom of the sleeper, i.e., the upper shoulders and the cribs.

The mechanical properties of ballast result from a combination of the physical properties of the individual ballast material and its in-situ (i.e., in-place) physical state. Physical state can be defined by the in-place density, while the physical properties of the material can be described by various indices such as particle size, shape, angularity, hardness, surface texture and durability. The in-place unit weight of ballast is a result of compaction processes. The initial unit weight is usually created by maintenance tamping, together with mechanical compaction means discussed in Chapter 14, if used. Subsequent compaction results from train traffic combined with environmental factors.

In service the ballast gradation changes as a result of: 1) mechanical particle degradation during construction and maintenance work, and under traffic loading, 2) chemical and mechanical weathering degradation from environmental changes, and 3) migration of fine particles from the surface and the underlying layers. Thus the ballast becomes fouled and loses its open-graded characteristics so that the ability of ballast to perform its important functions decreases and ultimately may be lost.

2.2.1. Subballast

The layer between the ballast and the subgrade is the subballast. It fulfills two functions which are also on the ballast list. These are:

1) Reduce the traffic induced stress at the bottom of the ballast layer to a tolerable level for the top of subgrade.

2) Extend the subgrade frost protection.

In fulfilling these functions the subballast reduces the otherwise required greater thickness of the more expensive ballast material. However, the subballast has some other important functions that can not be fulfilled by ballast. These are:

3) Prevent interpenetration of subgrade and ballast (separation function).

4) Prevent upward migration of fine material emanating from the subgrade (separation function related to item 3).

5) Prevent subgrade attrition by ballast, which in the presence of water, leads to slurry formation, and hence prevent this source of pumping. This is a particular problem if the subgrade is hard.

6) Shed water, i.e., intercept water coming from the ballast and direct it away from the subgrade to ditches at the sides of the track.

7) Permit drainage of water that might be flowing upward from the subgrade.

These are very important functions for satisfactory track performance. Hence in the absence of a subballast layer a high maintenance effort can be expected unless these functions are fulfilled in some other manner.

Functions 3, 4 and 5 form a subset of the subballast functions which represent what is sometimes known as a blanket layer.

The most common and most suitable subballast materials are broadly-graded naturally occurring or processed sand-gravel mixtures, or broadly-graded crushed natural aggregates or slags. They must have durable particles and satisfy the filter/separation requirements for ballast and subgrade. These requirements will be discussed in Chapter 9.

Some of the functions of subballast may be provided by:

1) Cement, lime, or asphalt stabilized local soils,

2) Asphalt concrete layers, or

3) Geosynthetic materials like membranes, grids and filter fabrics (geotextiles).

These items will be discussed in Chapters 11 and 12.

2.2.2. Subgrade

The subgrade is the platform upon which the track structure is constructed. Its main function is to provide a stable foundation for the subballast and ballast layers. The influence of the traffic induced stresses extends downward as much as five meters below the bottom of the sleepers. This is considerably beyond the depth of the ballast and subballast. Hence the subgrade is a very important substructure component which has a significant influence on track performance and maintenance. For example subgrade is a major component of the superstructure support resiliency, and hence contributes substantially to the elastic deflection of the rail under wheel loading. In addition, the subgrade stiffness magnitude is believed to influence ballast, rail and sleeper deterioration. Subgrade also is a source of rail differential settlement.

The subgrade may be divided into two categories (Fig. 2.1): 1) natural ground (formation), and 2) placed soil (fill). Anything other than soils existing locally is generally uneconomical to use for the subgrade. Existing ground will be used without disturbance as much as possible. However, techniques are available to improve soil formations in place if they are inadequate. These techniques will be discussed in a later section. Often some of the formation must be removed to construct the track at its required elevation. Placed fill is used either to replace the upper portion of unsuitable existing ground or to raise the platform to the required elevation for the rest of the track structure.

To serve as a stable platform, the following subgrade failure modes must be avoided:

1) Excessive progressive settlement from repeated traffic loading.

2) Consolidation settlement and massive shear failure under the combined weights of the train, track structure, and earth.

3) Progressive shear failure (cess heave) from repeated wheel loading.

4) Significant volume change (swelling and shrinking) from moisture change.

5) Frost heave and thaw softening.

6) Subgrade attrition.

In addition to its other functions, the subgrade must provide a suitable base for construction of the subballast and ballast.

2.2.3. Drainage

The description of the track components would not be complete without mentioning one of the most important items--the drainage system. The drainage system has several functions: 1) intercepting subsurface water entering the area of the track substructure, 2) intercepting surface water approaching the track structure from the sides, and 3) removing water draining out of the ballast and subballast. Methods for drainage will be discussed in Chapter 13.

2.3. Track Forces

Understanding the type and magnitude of forces that the track substructure must support is basic to track substructure design. In this section the loading environment, load limitations, and dynamic forces generated and transmitted to the track structure are discussed. Types of forces imposed on the track structure are classified as mechanical, both static and dynamic, and thermal. The track structure must restrain repeated vertical, lateral and longitudinal forces resulting from traffic and changing temperature. The combined vertical, lateral, and longitudinal live forces exerted by a train and transferred through the track superstructure to the substructure determine the dynamic loading environment that must be supported by the substructure.

The dynamic interactions between rail vehicle wheels and the rails are a function of track, vehicle, and train characteristics, operating conditions, and environmental conditions. Forces applied to the track by moving rail vehicles (known variously as cars, wagons, trucks or coaches) are a combination of a static load and a dynamic component superimposed on the static load. Maximum stresses and strains in the track system occur under this dynamic loading which is often expressed by a factor which increases the static load. The difference between the dynamic load and the static load is known as the dynamic increment.

High frequency vibrations also result from dynamic loading. Vibrations can significantly affect track superstructure and substructure component performance, particularly at high speeds.

Temperature changes induce thermal stresses in the rail which cause expansion or contraction of the steel. Any restraint to the change in length, as in continuous welded rail, will set up internal stresses generally represented by a force acting in a longitudinal direction in the rail. Without sufficient resistance, track buckling can occur in a vertical or lateral direction due to the longitudinal compression forces in the rails, or rail breaks can occur from tension forces.

Additional information on track forces may be found in Refs. 2.3, 2.4 and 2.5.

2.3.1. Vertical Forces

Vertical forces are considered those that are perpendicular to the plane of the rails. As such, the actual direction is a function of the track cross-level and grade. The types of vertical force are:

1) Vertical wheel force,

2) Uplift force.

In reaction to the vertical downward force on the rail at the wheel contact point, the rail tends to lift up, away from the wheel, as shown in Fig. 2.2. If the uplift force is not compensated by the rail and sleeper weight together with any frictional forces from the ballast, the sleeper will lift momentarily. With the advancing of the wheel the particular sleeper is forced down. This movement causes a pumping action which can cause deterioration of track structure components.

The vertical wheel force is often considered as having a static component equal to the vehicle weight divided by the number of wheels plus a dynamic variation about the static value. The static values range from about 12,000 lb (53 kN) for light rail passenger service to as high as 39,000 lb (174 kN) for heavy haul trains in North America. Sources of vertical dynamic variation are 1) track geometry induced vehicle-rail interaction including rock and roll, and bounce, and 2) wheel impact forces from causes such as wheel flats, rail burns, rail corrugations and rail joints. The impact forces not only can be very large, but they produce vibration in the track structure because of their high frequency component. These vibrations contribute to track settlement and to component deterioration. For example, vibrations can cause powdering of ballast and also ballast shoulder flow.

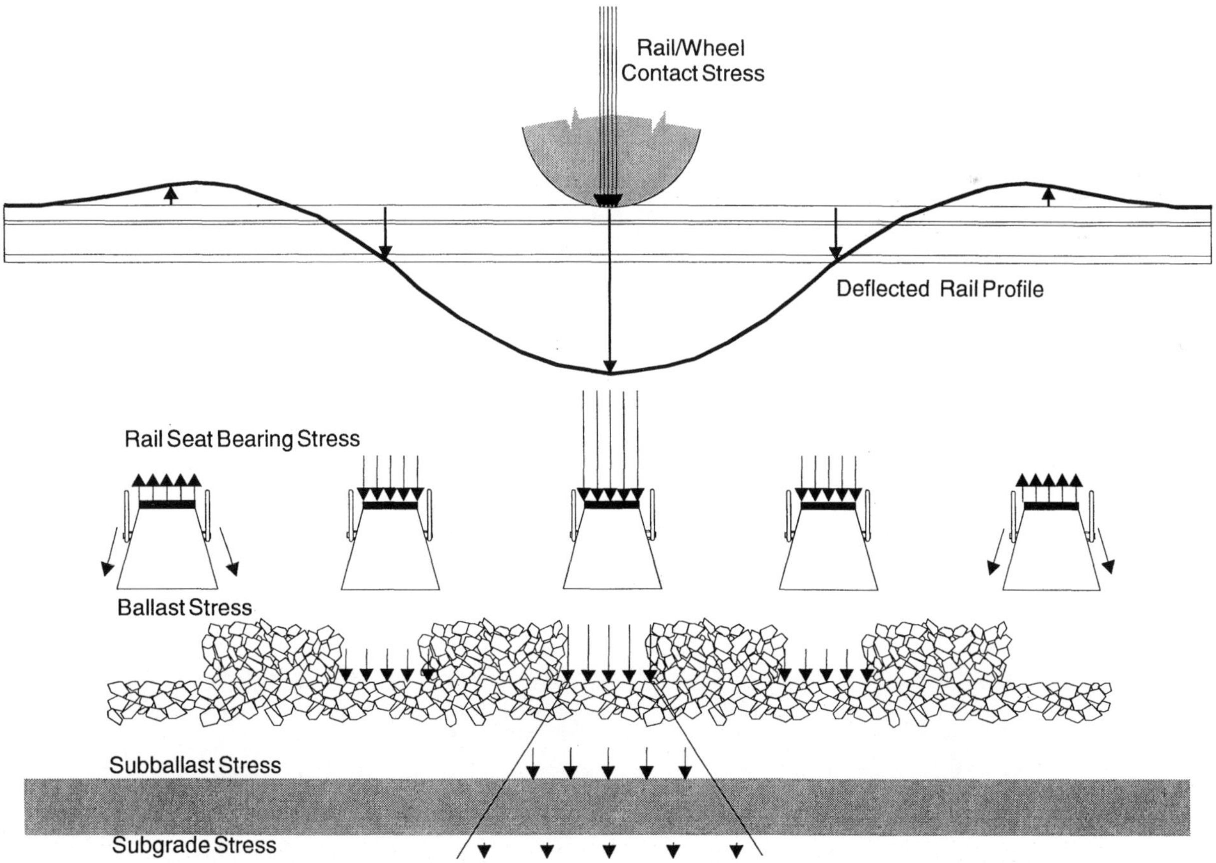

Fig. 2.2 Typical wheel load distribution into the track structure

Vertical wheel load measurements from several concrete sleeper sites in the United States will be shown to illustrate the differences between static and dynamic forces (Ref. 2.6). The loads were measured with strain gages attached to the rail and calibrated for vertical force. A minimum of 20,000 axles were recorded for statistical representation of the dynamic force. The static wheel load distribution was obtained by dividing known individual gross car weights by the corresponding number of wheels.

The results are often plotted in the form of cumulative frequency distribution curves as shown in Figs. 2.3 and 2.4. The vertical axis uses a statistical scale and represents the

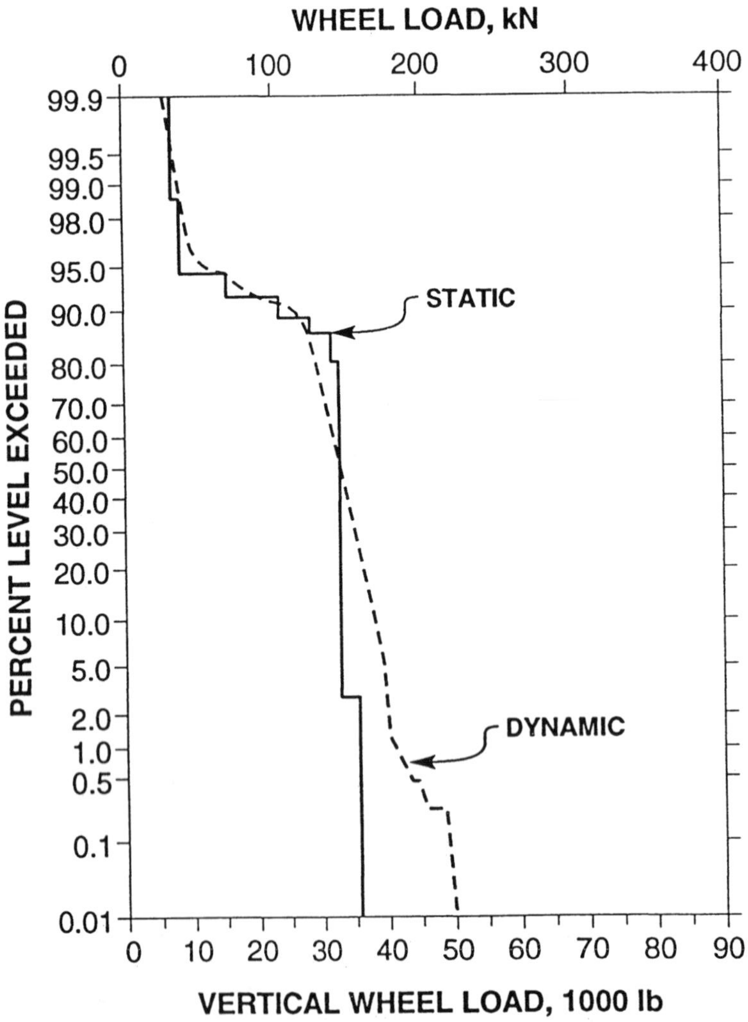

Fig. 2.3 Static and dynamic wheel loads at Colorado test track

percent of the total number of wheel loads which exceed the corresponding wheel load given on the horizontal axis.

The first example is for the Federal Railroad Administration research track (FAST) in Pueblo, Colorado (Fig. 2.3). The dynamic wheel load distribution is not much different from the static wheel load distribution, indicating that the FAST train has almost no wheel tread irregularities. The FAST train primarily consists of 100-ton (890 kN) cars loaded to capacity, with a few selected cars switched in and out for specific tests. The break in the curve at the 90 percent exceeded level indicates that approximately 10 percent of the vehicles carry less than the rated maximum loads, including a few empty cars.

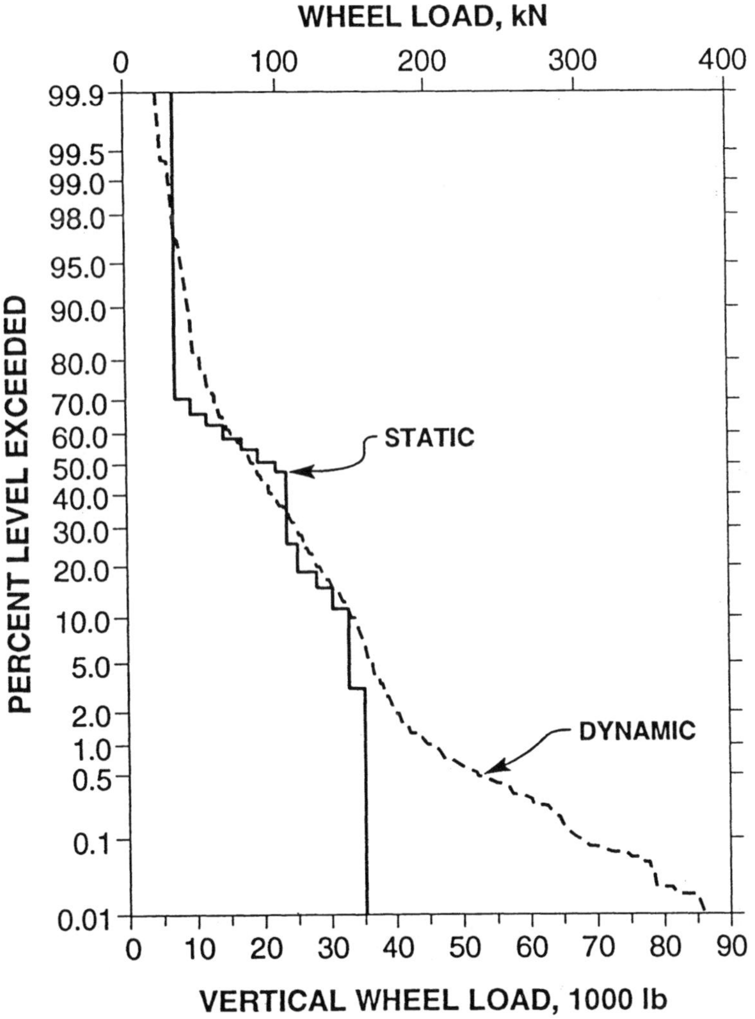

Fig. 2.4 Static and dynamic wheel loads from combined passenger and freight traffic

The second example is for the mainline track between New York and Washington which carries a mix of passenger and freight trains (Fig. 2.4). The dynamic wheel loads were in close agreement with the static loads except for the high end of the range. The large dynamic loads above the 2% level exceeded were caused almost entirely by the passenger cars even though their static wheel load was only about 23,000 lb (102 kN) . The reason for the high dynamic load from these cars was a combination of wheel flats and high speeds (80 - 110 mph or 128 - 176 kph).

2.3.2. Lateral Forces

Lateral forces are parallel to the long axis of the sleepers. These are from two principal sources:

1)Lateral wheel force, and

2) Buckling reaction force.

The lateral wheel force comes from the lateral component of friction force between the wheel and rail, and by the lateral force applied by the wheel flange against the rail. Sources of lateral wheel force are train reaction to geometry deviations (such as sway), self-excited hunting (weaving) motions which result from bogie instability at high speeds, and centrifugal force in curves. Lateral wheel forces are very complex and much harder to predict than vertical forces.

Buckling usually occurs in the lateral direction and is caused by high longitudinal rail compressive stress from rail temperature increase.

2.3.3. Longitudinal Forces

Longitudinal forces are parallel to the rails. These are from several sources:

1) Locomotive traction force including force required to accelerate the train,

2) Braking force from the locomotive and cars,

3)Thermal expansion and contraction of rails, and

4) Rail wave action.

As each set of vehicle wheels passes a point on the track, a depression curve between vehicle wheels and forerunning rail wave will occur and thereby subject the track superstructure to longitudinal forces in the direction of traffic (Ref. 2.7).

REFERENCES

2.1 AREA. *Manual for railway engineering.* American Railway Engineering Association, Washington, D. C.

2.2 South African Transport Services (1983). *South African railway track.* Track Development Section, Chief Civil Engineers Office, Johannesburg.

2.3 DiPilato, M. A., Levergood, A. V., Steinberg, E. I., and Simon, R. M. (1983). *Railroad track substructure-design and performance evaluation practices.* Goldberg-Zoino and Associates, Inc., Newton Upper Falls, Massachusetts, Final Report for U. S. DOT Transportation Systems Center, Cambridge, Massachusetts, Report No. FRA/ORD-83/04.2, June.

2.4 Hay, W. W. (1982). *Railroad engineering.* Second Ed., Wiley.

2.5 Esveld, Coenraad (1989). *Modern railway track.* MRT Productions, Germany.

2.6 Harrison, H. D., Selig, E. T., Dean, F. E., and Stewart, H. E. (1984). *Correlation of concrete tie track performance in revenue service and at the facility for accelerated service testing--Vol. I, a detailed summary.* Battelle Columbus Laboratories and University of Massachusetts, Final Report, Report No. DOT/FRA/ORD-84/02.1, August.

2.7 Clarke, C. W. (1957). **Track loading fundamentals, Part 1.** *Railway Gazette,* 26 pp., January.

2.8 Tung, T. H. (1989). *Railroad Track.* Shanghai Institute of Railway Technology Press.

3. Soil Properties

Many books are available that describe soil properties and conventional methods of evaluating them. However, a general understanding of the basic characteristics of soil deformation and strength properties, especially under repeated loading, is necessary background for the topics in the other chapters. Also an appreciation of methods of measuring the soil properties in the laboratory is necessary for those engineers who may require the results of such tests for solving their railway geotechnology problems. Thus a summary of these topics will be provided in this chapter.

3.1. Classification of Rocks

Geological deposits may be broadly classified as either soil or rock, the distinction being the hardness of the deposit. Rocks, for example, have a high load bearing capacity and require ripping or blasting for removal from their in situ state. Soils consist of assemblies of particles formed by mechanical or chemical breakdown of rock.

The strict geological definition of a rock is "any mass of mineral matter, whether consolidated or not, which forms part of the Earth's crust" (Ref. 3.1). A more widely accepted definition, often used by civil engineers, states that a rock is "something hard, consolidated, and/or load-bearing, which, where necessary, has to be removed by blasting" (Ref. 3.1).

Rocks may consist of only one mineral, but generally consist of an aggregate of mineral species. A mineral is a "naturally occurring, homogeneous solid with a definite, but generally not fixed, chemical composition and an ordered atomic arrangement" (Ref. 3.1). For example, the minerals calcite and aragonite are both composed of calcium carbonate but they have different atomic arrangements, which result in the two minerals having different physical properties of cleavage (tendency of a mineral to break along planes of weak bonding), hardness and specific gravity.

Minerals, when large enough, can be identified in hand sample by such characteristics as their scratch hardness (Mohs' hardness scale), cleavage, color, luster (appearance or quality of light reflected from the surface of a mineral), tenacity (resistance a mineral offers to breaking, crushing, bending or tearing), streak (color when powdered), specific gravity and magnetism. Identification of the mineral composition of a rock is accomplished through petrographic analysis, which consists of hand sample and thin section examinations. A thin section is a representative section of a rock that has been cut, mounted on a glass slide and then ground to the standard thickness of 0.03 mm. At this thickness most minerals will transmit light. The thin section can then be examined using a transmitting light, polarizing microscope. Minerals can be identified in thin section by such characteristics as crystal form or grain shape, grain size, cleavage, color (including changes with orientation in polarized light), and refractive indices. A thin section can also be stained to distinguish the members of the carbonate and feldspar mineral groups.

Rocks are generally classified by texture and mineral composition. The texture of a rock is the relationship between the grains of minerals which form the rock. A description of a rock's texture generally includes grain size, grain shape, degree of crystallinity (ratio of amorphous glass to crystals) and the contact relationships of the grains. Two rocks with similar mineral compositions but different textures would have different names.

Initially, rocks can be grouped into three main categories -- igneous, sedimentary, and metamorphic -- based on the mode of formation. Each category is further divided, with the rocks being classified based on their texture and composition as previously described.

3.1.1. Igneous Rocks

Igneous rocks are those rocks which were formed by the cooling of a magma or lava. Magma refers to molten rock within the Earth and lava refers to molten rock that has reached the surface. As the liquid cools, minerals begin to form (crystallize). The types of minerals which form are a function of the chemical composition of the liquid, as well as pressure and temperature conditions. Also, the longer the cooling period, the larger the crystals (grains) of the resultant rock will be; quickly chilling a lava (e.g. by immersion in water) will result in amorphous (noncrystalline) glass or a very fine-grained groundmass (finer-grained material constituting the main body of a rock, in which larger units are set).

Intrusive (or plutonic) igneous rocks are formed from a magma that has been intruded into the rocks that form the Earth's surface. Since the magma never reaches the surface and is insulated by rock it cools very slowly, resulting in a rock that has large crystals. Granite is an example of an intrusive igneous rock.

Extrusive (or volcanic) igneous rocks are formed from lava that has been extruded on to the Earth's surface, e.g. erupted from a volcano. Since the liquid is on the surface and not insulated, heat loss is rapid and the resultant rock will have smaller mineral grains than an intrusive igneous rock. As with the intrusive rocks, the type of minerals that will form from the lava is a function of the chemical composition of the liquid. Basalt is an example of an extrusive igneous rock.

When classifying igneous rocks, the two most important criteria are mineral composition and texture. Texture in an igneous rock is a function of: 1) grain shape, 2) grain size, and 3) degree of crystallinity. Examples of classification systems are found in Refs. 3.2 and 3.3.

3.1.2. Sedimentary Rocks

Sedimentary rocks are rocks which have been formed at the Earth's surface from material (sediments) derived from the mechanical and chemical weathering (degradation) of pre-existing rocks. Sedimentary rocks can be grouped into two main categories: clastic or chemical. A clastic sedimentary rock, such as a sandstone or shale, is formed by consolidation, through compaction and/or cementing, of an unconsolidated deposit of sediment. A chemical sedimentary rock, such as a limestone, forms from an accumulation of material that is precipitated from seawater or groundwater and organisms (i.e. shell material).

Clastic sedimentary rocks are described and classified based on their mineral composition, grain size, grain shape, and presence and type of cement. Chemical sedimentary rocks, such as limestones, are classified based on the chemical composition of the matrix material, grain size, detrital grains (such as grains of quartz) and cement, as well as presence and type of fossil fragments. Several classification schemes are commonly used for limestones, for example Folk (Ref. 3.4).

3.1.3. Metamorphic Rocks

Metamorphic rocks are those rocks which have undergone a change in mineral composition and fabric as a result of increased pressures and/or temperatures. These increases of temperature and pressure can be caused by the intrusion of a magma (contact metamorphism) and by large scale actions associated with mountain building events and plate tectonics (regional metamorphism).

When a rock, either igneous, metamorphic or sedimentary, is subjected to increased temperatures and pressures its equilibrium is disturbed. The rock will respond, in the solid state, by trying to reach a more stable equilibrium. This results in a change in the mineral composition and fabric of the rock. Fabric is a general term which encompasses both texture (as described previously) and structure, which includes orientation of minerals, and development of planar features such as foliation and schistosity (rock cleavage).

Examples of metamorphic rocks are quartzite and gneiss. Metamorphic rocks are classified based on their mineral composition, texture and fabric. Examples of classification schemes are found in Refs. 3.5 and 3.6.

3.2. Identification and Classification of Soils

Based on its method of formation, a soil deposit may be classified as sedimentary (or transported), residual or fill. For sedimentary soils, the particles are formed in one location, transported by such means as water, air or glacial action, and deposited in another location. Residual soil results when the products of weathering are not transported, but remain in place. Fill is a man-made soil deposit.

Soils contain widely varying amounts of gravel, sand, silt and clay particles. Gravels and sands are distinguished by size. The definitions are given in Table 3.1 based on ASTM D2488 (Ref. 3.7). Silts and clays are also distinguished by size in textural classification systems, but a better way is by their consistency characteristics when combined with water.

Gravels, sands and silts are formed by mechanical breakdown of rock through crushing, grinding, temperature change and erosion. The particles are generally equidimensional and the behavior of soils in which these particles predominate are largely influenced by such factors as particle shape, surface texture, angularity, size, size distribution and packing.

Clays are formed by chemical disintegration of rock with some alteration of the mineral structure. The result is generally plate or tube-shaped particles which are too small to see with an optical microscope. These particles have a very high surface area to volume ratio and the surfaces have unsatisfied electrical charges. This causes electrical forces of attraction and repulsion, in the presence of absorbed water and dissolved ions, to control behavior of soils in which these particles predominate. This feature is responsible for the plasticity characteristic of clays.

Table 3.1 Soil particle size

Component	Size Range	
	U.S. Sieve No.	mm
1. GRAVEL	3" - No. 4	76 - 4.75
a) Coarse	3" - 3/4"	76 - 19
b) Fine	3/4" - No. 4	19 - 4.75
2. SAND	No. 4 - No. 200	4.75 - 0.075
a) Coarse	No. 4 - No. 10	4.75 - 2
b) Medium	No. 10 - No. 40	2 - 0.425
c) Fine	No. 40 - No. 200	0.425 - 0.075
3. SILT and CLAY	< No. 200	< 0.075

One of the most important geotechnical engineering tasks is to properly name and describe the soils involved in a project. In practice this task is often performed poorly. Furthermore the lack of standardized practice can cause misinterpretation of the results of others. One available system is ASTM D2488, a visual/manual soil identification procedure (Ref. 3.7). The soil names used with this system are given in Figs. 3.1, 3.2 and 3.3. The group symbols are from the unified soil classification system. They are defined in Table 3.2. The names are selected by working from left to right in Figs. 3.1 - 3.3, following the correct branch based on information in the boxes. Both learning and using this procedure are simplified and made more efficient by employing a personal computer program SID (Ref. 3.8).

Other methods are available. For example, widely used in South Africa is the work by Jennings et al. (Ref. 3.9).

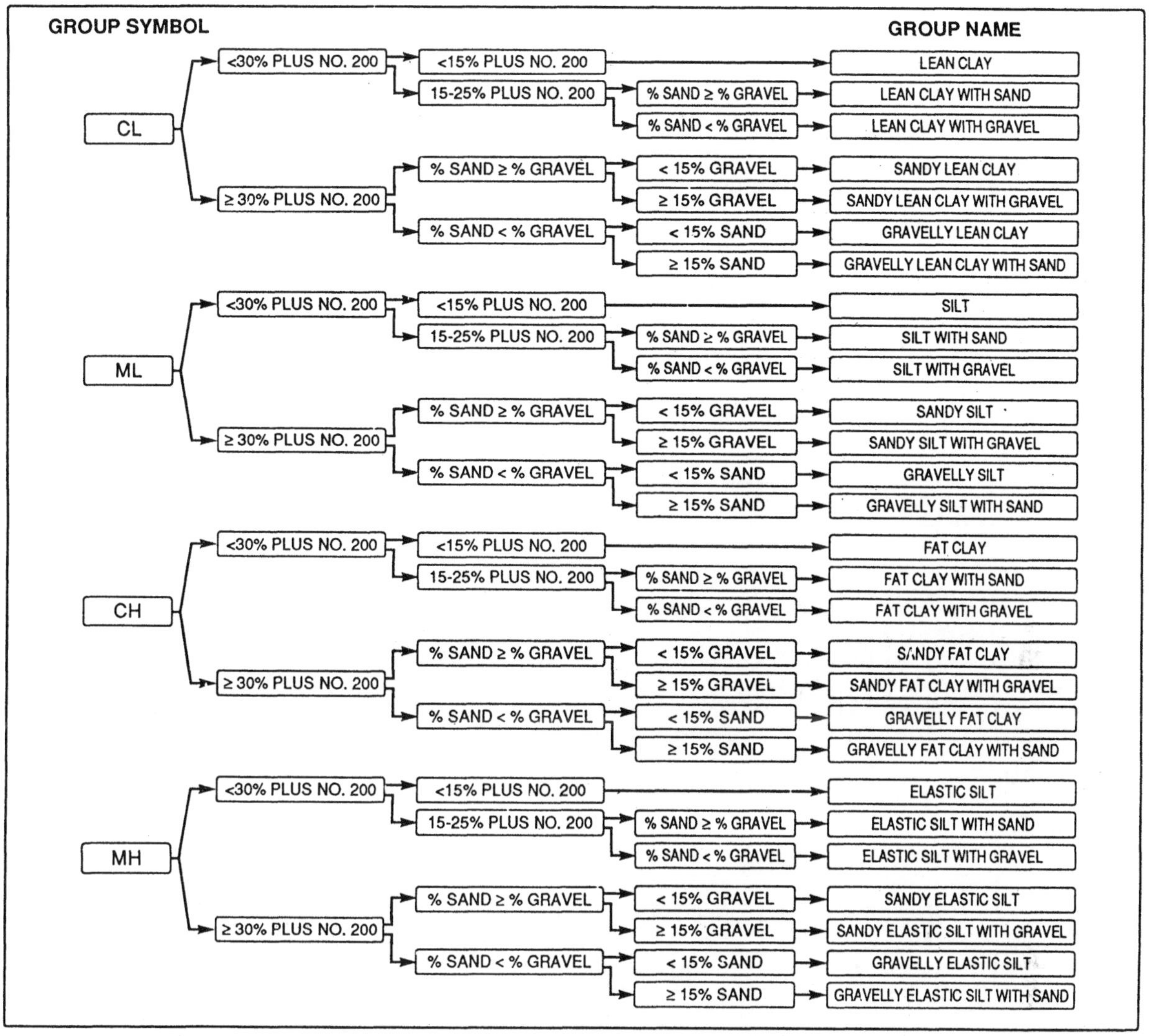

Fig. 3.1 Flow chart for identifying inorganic fine-grained soil (50% or more fines)

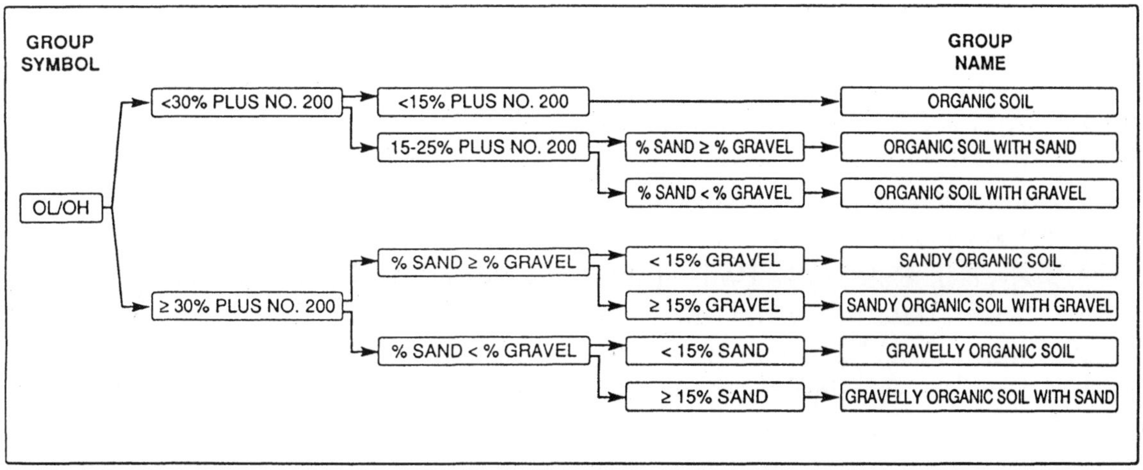

Fig. 3.2 Flow chart for identifying organic fine-grained soil (50% or more fines)

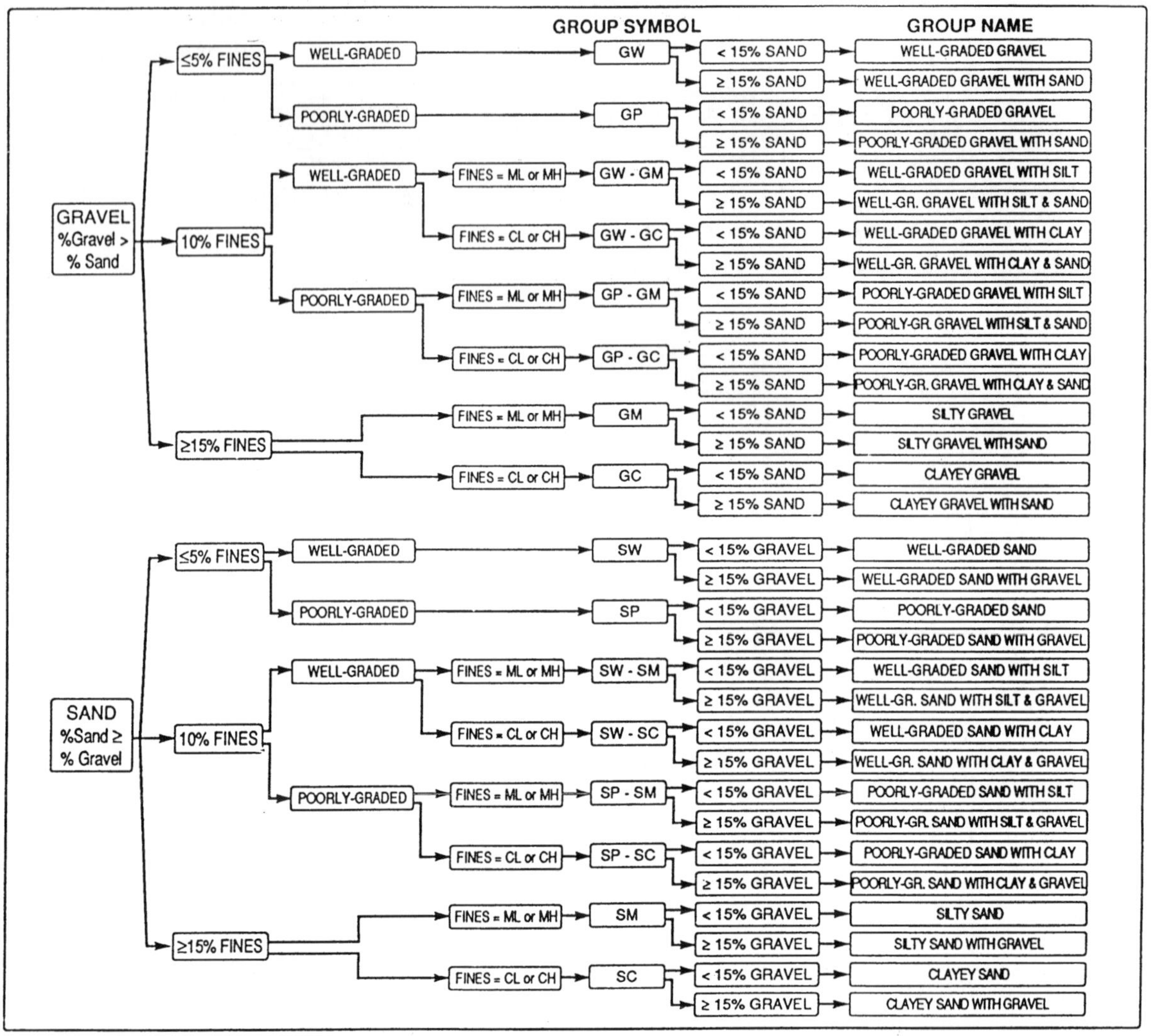

Fig. 3.3 Flow chart for identifying coarse-grained soils (less than 50% fines)

In addition to identification, soils are commonly classified (grouped) according to their characteristics for engineering purposes. The most widely used systems in North America are the Unified Soil Classification System (USCS) as shown in Table 3.2 and Fig. 3.4 (Ref. 3.10), and the American Association of State Highway and Transportation Officials (AASHTO) as shown in Table 3.3 and Fig. 3.5 (Ref. 3.11). A description of both of these systems is given in Ref. 3.12. Both of these systems use soil particle size and consistency characteristics for determining the soil class. The consistency characteristics are represented by liquid limit (Ref. 3.20) and plasticity index, where plasticity index is equal to the difference between the liquid limit and the plastic limit (Ref. 3.21). The ASTM D2488 estimates the USCS classification by visual/manual means. These classifications provide no information on the physical state of the soil (moisture content, strength, stiffness) in situ, because they are based on tests that disturb the soil.

Table 3.2 USCS Soil Classification Chart

Criteria for Assigning Group Symbols and Group Names Using Laboratory Tests				Group Symbol
Coarse-Grained Soils (More than 50% larger than 0.075 mm particle size)	Gravels (More than 50% of coarse fraction larger than 4.75 mm particle size)	Clean Gravels (Less than 5% fines a)	$Cu \geq 4$ and $1 \leq Cc \leq 3$ c	GW
			$Cu < 4$ and/or $1 > Cc > 3$ c	GP
		Gravels with Fines (More than 12% fines a)	Fines classify as ML or MH	GM
			Fines classify as CL or CH	GC
	Sands (50% or more of coarse fraction smaller than 4.75 mm particle size)	Clean Sands (Less than 5% fines b)	$Cu \geq 6$ and/or $1 \leq Cc \leq 3$ c	SW
			$Cu < 6$ and/or $1 > Cc > 3$ c	SP
		Sands with Fines (More than 12% fines b)	Fines classify as ML or MH	SM
			Fines classify as CL or CH	SC
Fine-Grained Soils (50% or more smaller than 0.075 mm particle size)	Silts and Clays (Liquid limit less than 50)	Inorganic	PI > 7 and plots on or above "A" line d	CL
			PI < 4 or plots below "A" line d	ML
		Organic	$\frac{\text{Liquid limit} - \text{oven dried}}{\text{Liquid limit} - \text{not dried}} < 0.75$	OL
	Silts and Clays (Liquid limit 50 or more)	Inorganic	PI plots on or above "A" line	CH
			PI plots below "A" line	MH
		Organic	$\frac{\text{Liquid limit} - \text{oven dried}}{\text{Liquid limit} - \text{not dried}} < 0.75$	OH
Highly organic soils	(Primarily organic matter, dark in color and organic odor)			PT

a) Gravels with 5 to 12% fines require dual symbols: GW-GM, well graded gravel with silt; GW-GC, well graded gravel with clay; GP-GM, poorly graded gravel with silt; GP-GC, poorly graded gravel with clay.

b) Sands with 5 to 12% fines require dual symbols: SW-SM, well graded sand with silt; SW-SC, well graded sand with clay; SP-SM, poorly graded sand with silt; SP-SC, poorly graded sand with clay.

c) $Cu = \frac{D_{60}}{D_{10}}$; $Cc = \frac{(D_{30})^2}{(D_{10} \times D_{60})}$.

d) If Atterberg limits plot in hatched area, soil is a CL-ML, silty clay.

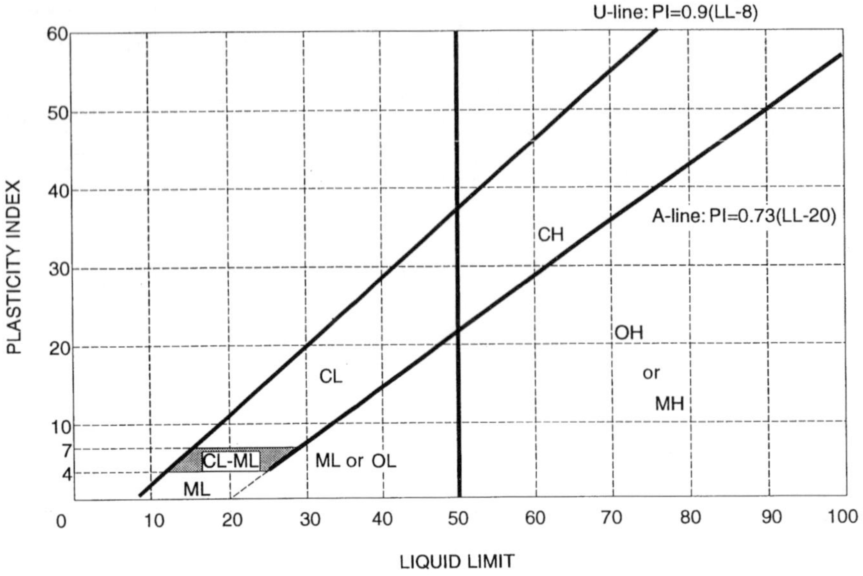

Fig. 3.4 Casagrande's plasticity chart for the USCS system

Table 3.3 AASHTO Soil Classification System

General Classification	Granular Materials (35% or less passing 0.075 mm)							Silt-Clay Materials (More than 35% passing 0.075 mm)			
Group classification	A-1			A-2							A-7
	A-1-a	A-1-b	A-3	A-2-4	A-2-5	A-2-6	A-2-7	A-4	A-5	A-6	A-7-5 A-7-6
Sieve analysis, percent passing:											
2.00 mm (No. 10)	50 max.	—	—	—	—	—	—	—	—	—	—
0.425 mm (No. 40)	30 max.	50 max.	51 min.	—	—	—	—	—	—	—	—
0.075 mm (No. 200)	15 max.	25 max.	10 max.	35 max.	35 max.	35 max.	35 max.	36 min.	36 min.	36 min.	36 min.
Characteristics of fraction passing 0.425 mm (No. 40):											
Liquid limit	—		—	40 max.	41 min.	40 max.	41 min.	40 max.	41min.	40 max.	41 min.
Plasticity index	6 max.		NP	10 max.	10 max.	11 min.	11 min.	10 max.	10 max.	11 min.	11 min.
Usual types of significant constituent materials	Stone fragments, gravel, and sand		Fine sand	Silty or clayey gravel and sand				Silty soils		Clayey soils	
General rating as subgrade	Excellent to good							Fair to Poor			

Natural soil deposits frequently contain a mixture of more than one component (gravel, sand, silt or clay). Hence a wide range of properties can result. This is what makes geotechnical engineering both interesting and challenging. A complication which frequently causes difficulty in practice is that the majority component does not always control behavior. For example, a soil with 60% gravel but with the voids filled with clay will behave more like clay than like gravel. The USCS classifies soil according to whether more than 50% by weight of particles is coarse-grained (sand and gravel) or fine-grained (silt and clay). An experienced geotechnical engineer will understand that with sufficient clay content, a clayey gravel (GC) will not behave like a coarse-grained soil. In recognition of this situation the AASHTO system designates soils behaving as granular materials, as only those containing 35% or less silt and clay particles by weight.

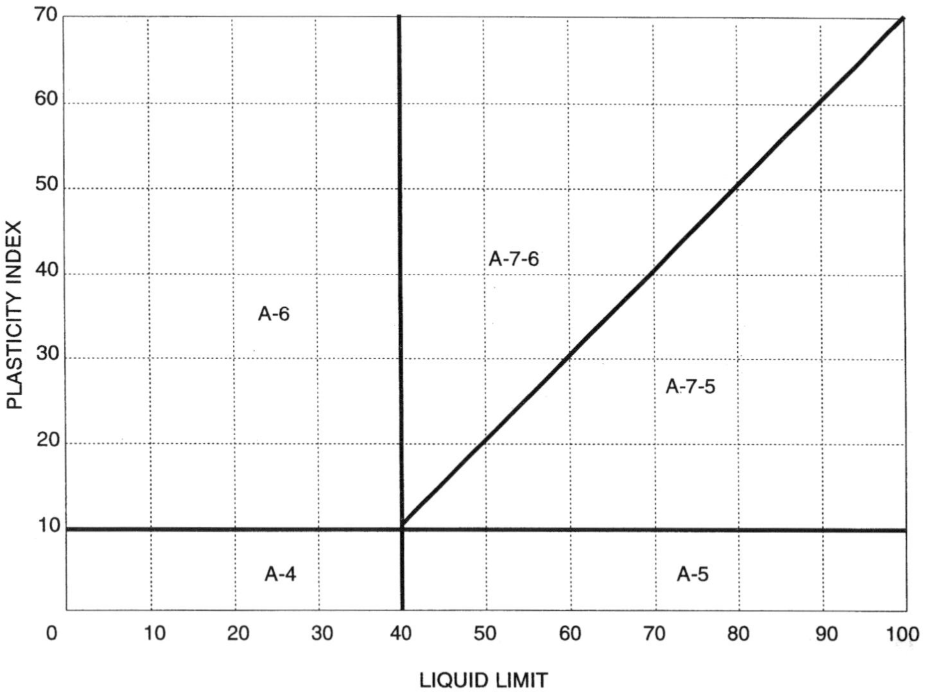

Fig. 3.5 Chart for identifying fine-grained subgroups in the AASHTO system

3.3. Laboratory Tests for Stiffness and Strength

Many laboratory tests have been developed for measuring soil stress-strain and strength properties. The most common are confined compression (also known as one-dimensional compression, consolidation, and oedometer test), direct shear, and triaxial tests. The unconfined compression test may be considered a special case of triaxial test. Two additional tests are useful for measuring basic soil properties. These are the direct simple shear test and the isotropic (hydrostatic) compression test. A description of these six tests will be given together with their uses.

The stresses in these tests may be represented as total or effective, or both, depending on the intended application. Recall that

$$\sigma = \sigma' + u \,, \tag{3.1}$$

where σ = total stress ,

σ' = effective stress, and

u = pore fluid pressure.

Note that a prime (′) is used for designating effective stress.

3.3.1. Confined Compression Test

The confined compression test is commonly performed in an oedometer (Fig. 3.6). The compression stress is applied to the soil in the vertical direction causing vertical compression strain while the strain in the horizontal directions is kept at zero by the rigid confining ring. Thus the vertical strain equals the volumetric strain. Vertical strain can occur in unsaturated specimens by compression of the air voids and in saturated samples by allowing sufficient time for pore water to squeeze out and allow water void reduction. The stress-strain relationship is highly nonlinear, being curved as illustrated in Fig. 3.7. Because of the lateral constraint on the soil specimen, failure will not occur. Thus this test is only suitable for measuring soil stiffness and can not be used to determine strength. The slope of the confined stress-strain curve is constrained modulus, D, and the reciprocal is coefficient of volume change, m_v. Note that D increases with increasing stress. This parameter directly represents the vertical compression of unsaturated soil deposits (such as compacted fill) in situations where lateral movement is restricted.

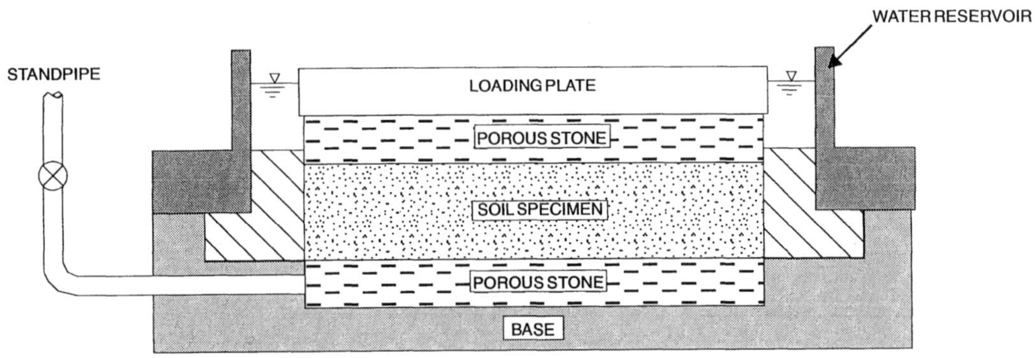

Fig. 3.6 Fixed ring oedometer

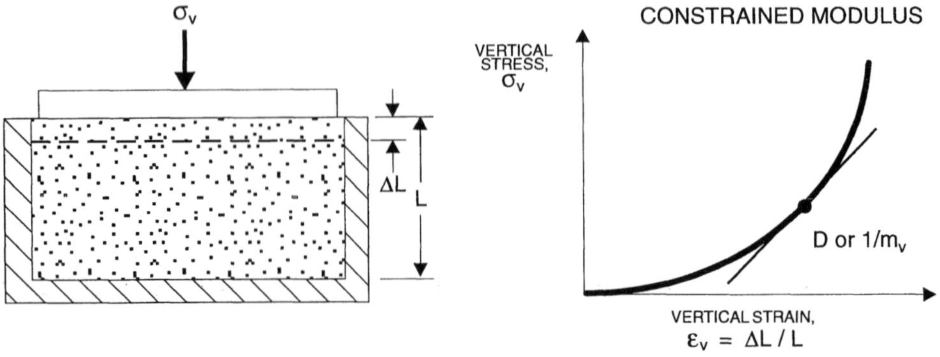

Fig. 3.7 One dimensional (confined) compression test

The most common use of this confined compression test is for measuring the properties of saturated soils during consolidation. Because the specimen is saturated (no air voids) when a compression stress increase is first applied, no significant vertical strain occurs. Instead the pore water pressure increases by an amount equal to the compression stress increment. Gradually this excess pore pressure dissipates as water drains out of the specimen, allowing volume change and hence vertical strain. Given sufficient time, the stress increment will be transferred fully from the pore water to the soil skeleton. This process is known as consolidation. Neglecting secondary compression, the sequence of stress-strain points representing the end of primary consolidation generate a curve like that shown in Fig. 3.7. However it is traditional to plot the data for saturated soils on axes of void ratio vs log of vertical effective stress, σ'_v, in which case the slope is known as compression index, C_c (Fig. 3.8a). The advantage is that for many soils this semilog plot gives a relatively linear relationship. The disadvantage is that void ratio, e, is used to represent strain, ε_v. However,

$$\varepsilon_v = \frac{\Delta e}{1 + e_0}, \tag{3.2}$$

where e_0 is the initial void ratio. Hence void ratio may easily be converted to strain.

As an alternative, the void ratio is sometimes plotted as a function of vertical effective stress on a linear scale (Fig. 3.8b), in which case the slope is termed coefficient of compressibility, a_v.

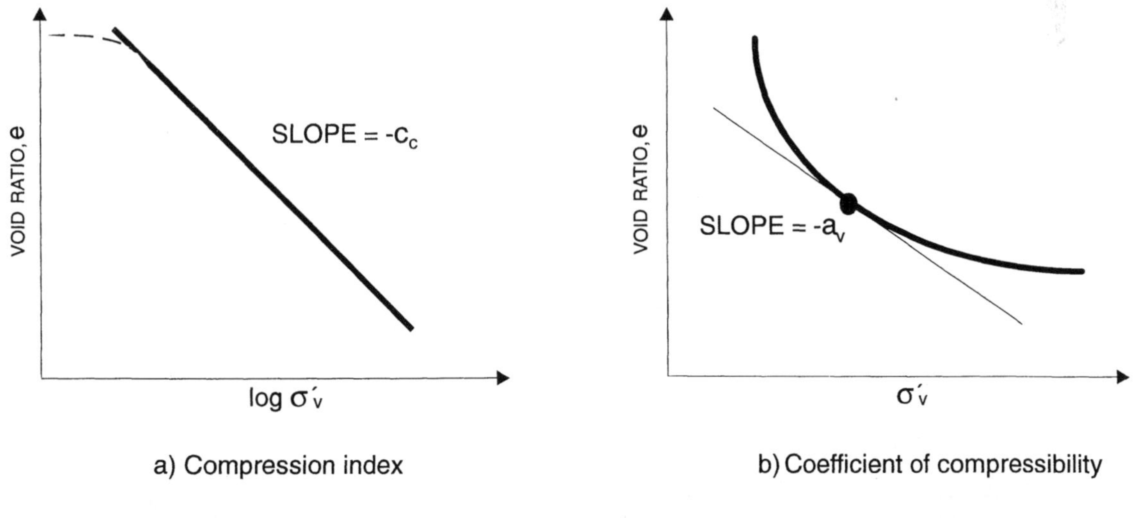

Fig. 3.8 Alternate forms of plotting confined compression test data

The relationships among D, m_v, C_c and a_v are given in Table 3.4 (Ref. 3.13). Although the variety of terms and form of plot tend to be confusing, the thing to keep in mind is that they are all variations of the basic confined stress-strain relationship in Fig. 3.7.

Table 3.4 Relations between various stress-strain parameters for confined compression

	Constrained Modulus	Coefficient of Volume Change	Coefficient of Compressibility	Compression Index
Constrained modulus	$D = \frac{\Delta\sigma_v}{\Delta\varepsilon_v}$	$D = \frac{1}{m_v}$	$D = \frac{1 + e_0}{a_v}$	$D = \frac{(1 + e_0)\sigma_{va}}{0.435\ C_c}$
Coefficient of volume change	$m_v = \frac{1}{D}$	$m_v = \frac{\Delta\varepsilon_v}{\Delta\sigma_v}$	$m_v = \frac{a_v}{1 + e_0}$	$m_v = \frac{0.435\ C_c}{(1 + e_0)\sigma_{va}}$
Coefficient of compressibility	$a_v = \frac{1 + e_0}{D}$	$a_v = (1 + e_0)\, m_v$	$a_v = -\frac{\Delta e}{\Delta\sigma_v}$	$a_v = \frac{0.435\ C_c}{\sigma_{va}}$
Compression index	$C_c = \frac{(1 + e_0)\sigma_{va}}{0.435\ D}$	$C_c = \frac{(1 + e_0)\sigma_{va}\, m_v}{0.435}$	$C_c = \frac{a_v\, \sigma_{va}}{0.435}$	$C_c = -\frac{\Delta e}{\Delta \log \sigma_v}$

Note: e_0 denotes the initial void ratio. σ_{va} denotes the average of the initial and final stresses.
(Table is reprinted from Ref. 3.13 with permission of the publisher.)

3.3.2. Isotropic Compression Test

In the isotropic compression test, the soil is subjected to a uniform all around pressure increase (hence hydrostatic or isotropic loading), and the resulting volumetric strain is measured (Fig. 3.9). In the same way as the confined compression test, volumetric strain will occur in unsaturated soils by compressing the air voids and in saturated soils by allowing consolidation. The slope of the isotropic compression stress-strain curve is known as bulk modulus, B. When shearing stresses are applied to soil specimens, either unsaturated or

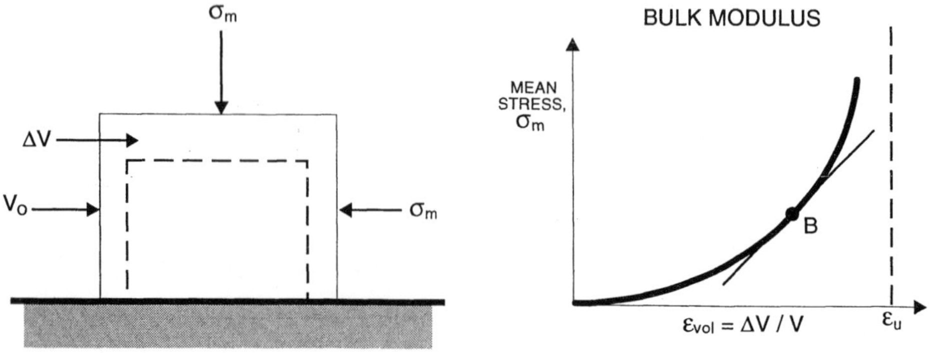

Fig. 3.9 Isotropic compression test

saturated but allowing consolidation to occur, part of the volume change will occur because of the tendency for either dilation or contraction of the soil skeleton during shear. The volumetric strain in the isotropic compression test does not include this effect.

3.3.3. Triaxial Compression Test

One of the most versatile and useful laboratory tests for soil stress-strain and strength properties is the triaxial compression test shown in Fig. 3.10. The soil specimen has the shape of a vertical cylinder and is placed in a chamber capable of applying a hydrostatic confining pressure to the specimen. The specimen is isolated from the confining fluid by a rubber membrane attached to the top cap and base. The base, and sometimes also the top cap, has a porous filter to allow the application of a back pressure to the pore water, or to allow flow of pore water into or out of the specimen through drainage lines during volume change. Also the drainage lines may be closed to prevent water movement during a test. Axial stress is superimposed on the hydrostatic stress by the top cap when it is loaded by a piston. In spite of the test name, the stress state is axisymmetric rather than three-dimensional because the horizontal stress is the same in all directions. Axial extension as well as axial compression stress increments may be applied, while the confining pressure is either constant or varied.

Specimens may be tested in either a saturated or an unsaturated state. Saturated specimens may be tested with the pore water drainage line open (drained) or with it closed (undrained).

The triaxial apparatus permits the specimens to be tested under a variety of stress states, but to simplify the discussion only two will be described here. The first is isotropic compression, explained in 3.3.2. The second, termed compression loading, occurs when the specimen is subjected to increasing axial compression stress with constant confining stress.

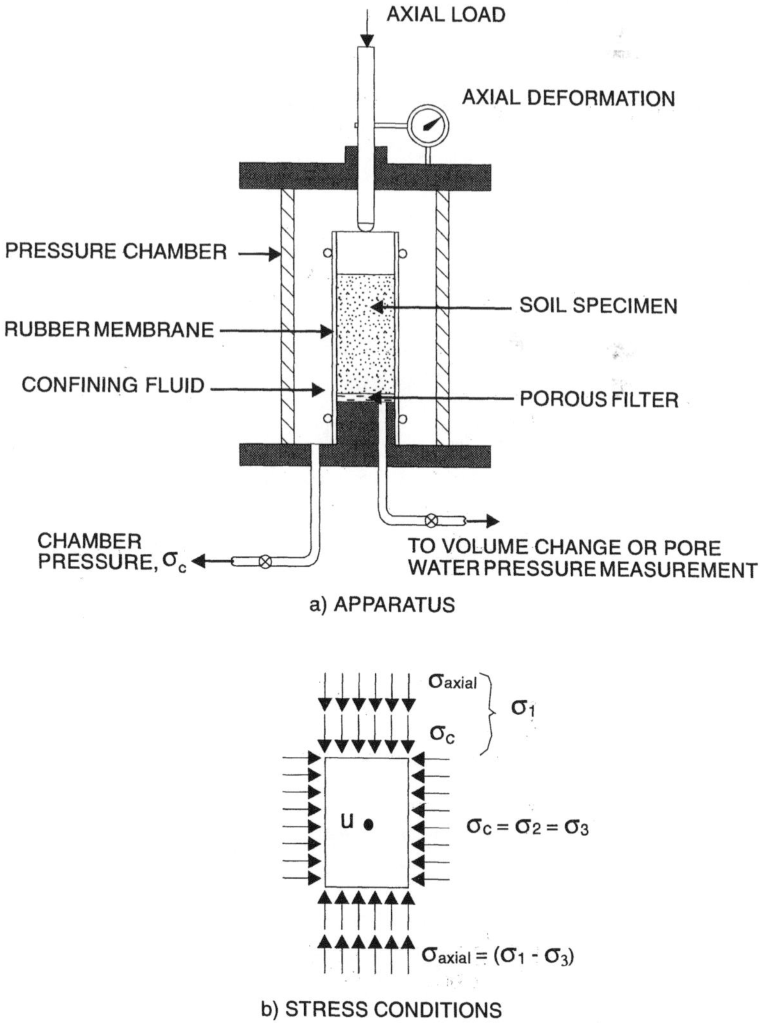

Fig. 3.10 Illustration of triaxial test

For the isotropic compression test, the confining pressure, σ_c, is increased without any superimposed axial load and the specimen volume change measured. For unsaturated specimens this measurement requires external axial and lateral strain measurement on the specimen. For saturated specimens the volume of pore water squeezed out of the specimen with drainage line open is taken as the volume change. Undrained isotropic compression tests (drainage line closed) are not appropriate for saturated specimens because the specimens are essentially incompressible.

For the compression loading test, the confining stress, σ_c, is first applied, either with the drainage line open (termed consolidated) or with the drainage line closed (termed unconsolidated). Then the axial stress is increased, either with the drainage line open (termed drained shear) or with the drainage line closed (termed undrained shear). Thus the three possible test combinations are consolidated-drained (CD), consolidated-undrained (CU) or unconsolidated-undrained (UU). The reasons for these different choices will be explained subsequently. The loading may be increased until the sample fails, in which case the strength is measured as well as compression stress-strain behavior.

Typical trends for triaxial compression tests are illustrated in Fig. 3.11. The axial stress is the major principal stress, σ_1. The confining stress is the minor principal stress, σ_3. The

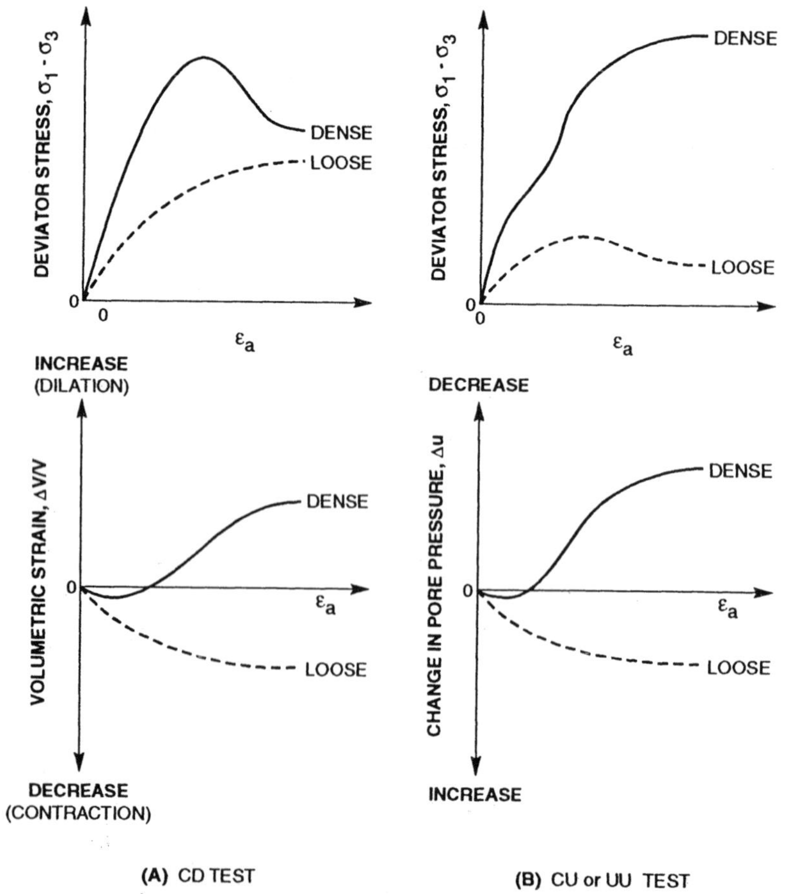

Fig. 3.11 Triaxial test results for saturated sand with constant confining pressure

vertical axis in the plot is the principal stress difference, $(\sigma_1 - \sigma_3)$. Note the trends for volume change during shear in drained tests (no pore pressure change), and for pore pressure change during shear in undrained tests (no volume change). The tendency to change volume during drained shear induces a similar trend for pore pressure change in saturated specimens when drainage is prevented. The slope of the axial stress-strain curve is Young's modulus, E, as long as the lateral stress, σ_3, is constant (Fig. 3.12). Under the same condition Poisson's ratio, ν, is the negative ratio of lateral to axial strain.

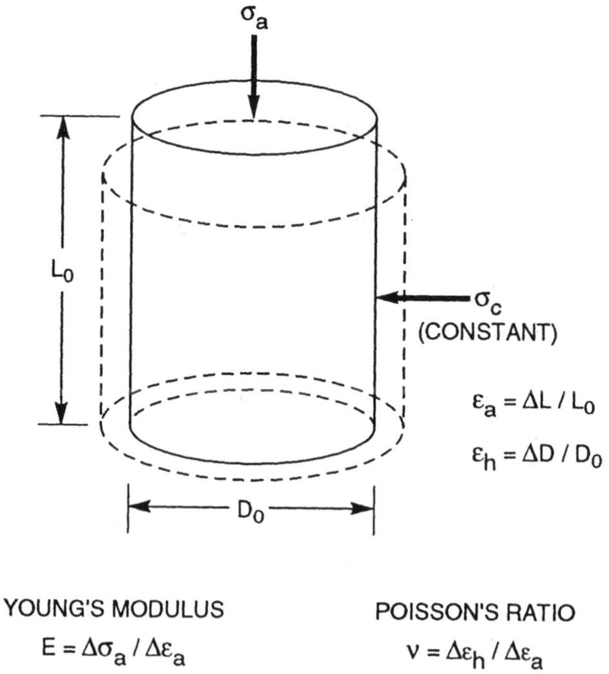

Fig. 3.12 Triaxial test parameters

3.3.4. Unconfined Compression Test

The unconfined compression test may be considered a special case of the triaxial compression test in which $\sigma_c = \sigma_3 = 0$. To determine the strength of the soil under zero confining pressure, the axial load is increased until the specimen fails. The test is only applicable to fine-grained soils which can develop enough negative pore water pressure to stand as a cylinder without an external confining pressure. The soil may or may not be saturated. However, the loading is rapidly applied so that drainage of pore water does not occur.

3.3.5. Direct Shear Test

Perhaps the oldest shear test for soil strength is the direct shear test (Fig. 3.13). The device, in essence, includes a split box which confines the specimen laterally as in the confined compression test, but which permits shear to take place on the horizontal plane of the split. The normal stress on the shear plane can be varied by changing the amount of

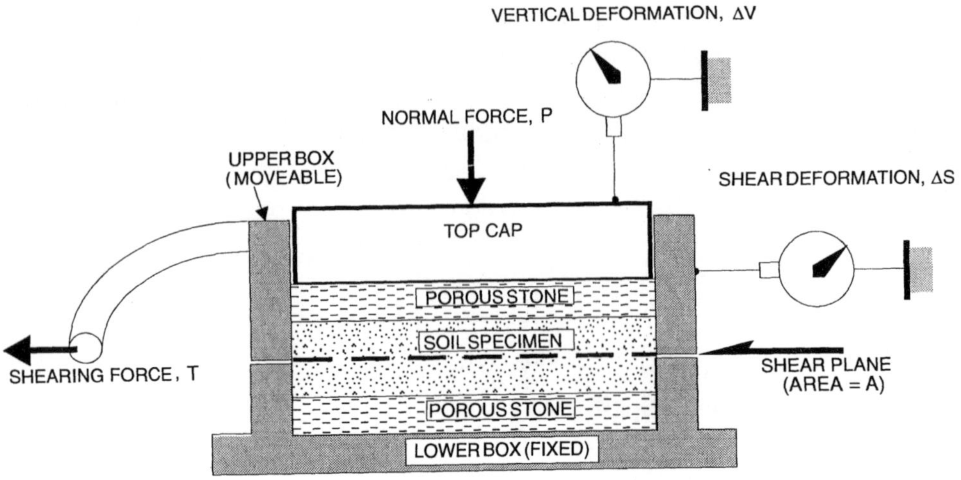

Fig. 3.13 Direct shear apparatus

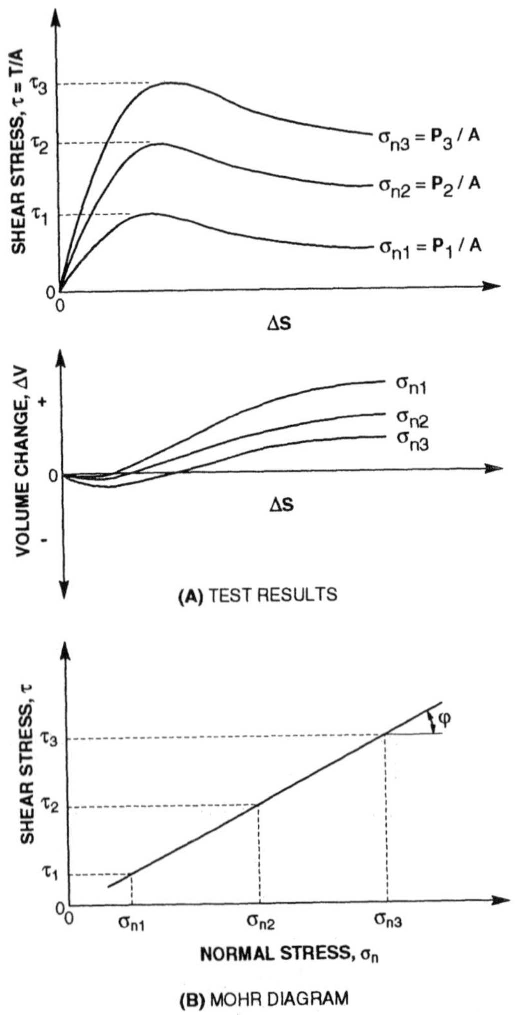

Fig. 3.14 Direct shear results for dense sand (adapted from Ref. 3.12 with permssion of the publisher)

vertical force applied to the top cap. Under a given normal stress the shear stress is measured as a function of shear displacement. Specimen dilation or contraction can also be determined by measuring the top cap movement. Typical results are illustrated in Fig. 3.14. This test gives soil strength but not stiffness, since soil strain can not be determined. The soil specimen may be saturated or unsaturated, but generally the test is only suitable for drained conditions.

3.3.6. Direct Simple Shear Test

In the direct simple shear (DSS) test the entire specimen is deformed in shear while under a controlled vertical stress (Fig. 3.15). The severe stress concentrations along the failure plane in a direct shear test are avoided in the DSS test and, in addition, shear strain is measured. Thus, specimen shear stiffness is determined as well as shear strength. The slope of the resulting shear stress vs. shear strain curve is the shear modulus, G.

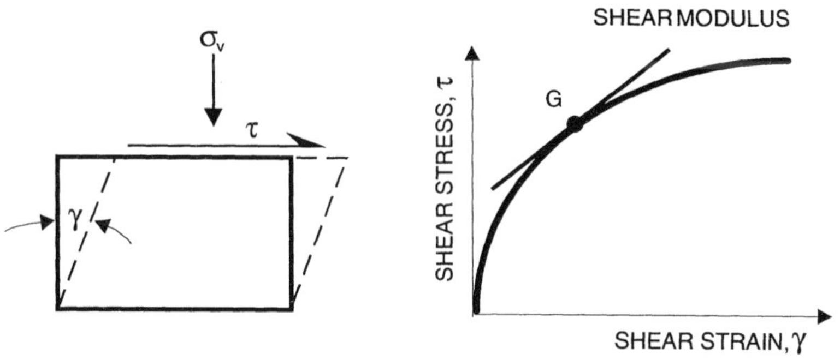

Fig. 3.15 Simple shear test

3.3.7. Repeated Load Tests

Repeated (cyclic) load tests are tests in which the stress applied to the soil varies regularly in magnitude with time. Although the term "cyclic" is often used for designating more general stress state variations than the term "repeated", for application to railway substructure subjected to traffic loading, the terms will be considered interchangeable, and "repeated" will generally be used. The purpose of this test is to represent stiffness and strength behavior of soils subjected to repeated stresses, as from such sources as traffic. Conceptually any of the previously described laboratory tests may incorporate repeated load. However, the triaxial test and the DSS test are most commonly used for this purpose. Further information on cyclic load testing of soils may be found in Refs. 3.14 and 3.15.

An example of test results is given in Fig. 3.16 which shows the stress-strain behavior in a repeated-load triaxial test in which the principal stress difference ($\sigma_1 - \sigma_3$) is cycled between zero (i.e., $\sigma_1 = \sigma_3$), and a maximum value much less than the strength of the specimen. Of primary interest in this test are resilient (elastic) Young's modulus and rate of accumulation of plastic axial strain. Generally for drained or unsaturated specimens, the incremental plastic strain decreases with each repeated cycle. However, for saturated specimens, or for unsaturated specimens with the maximum stress near failure, the incremental plastic strain can increase with each cycle.

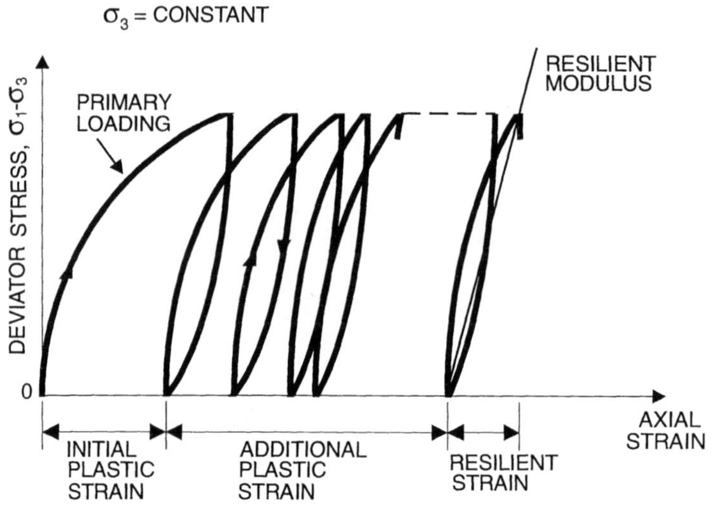

Fig. 3.16 Representation of stress-strain curve from a drained triaxial test under repeated load

3.4. Relationships Among Stiffness Parameters

Soils do not have linear stress-strain relationships; hence they do not have unique values for the stiffness parameters. However as an approximation, particularly for small stress changes, linear elastic behavior is assumed. In this case the following stress-strain relationships apply:

1) Normal strain --

$$\varepsilon_x = \frac{1}{E}\,[\sigma_x - \nu\,(\sigma_y + \sigma_z)]\,, \tag{3.3a}$$

$$\varepsilon_y = \frac{1}{E}\,[\sigma_y - \nu\,(\sigma_z + \sigma_x)]\,, \tag{3.3b}$$

$$\varepsilon_z = \frac{1}{E}\,[\sigma_z - \nu\,(\sigma_x + \sigma_y)]\,, \tag{3.3c}$$

2) Shear strain --

$$\gamma_{xy} = \frac{\tau_{xy}}{G}, \tag{3.4a}$$

$$\gamma_{yz} = \frac{\tau_{yz}}{G}, \tag{3.4b}$$

$$\gamma_{zx} = \frac{\tau_{zx}}{G}, \tag{3.4c}$$

3) Volumetric strain --

$$\frac{\Delta V}{V} = \varepsilon_x + \varepsilon_y + \varepsilon_z . \tag{3.5}$$

In these equations

ε = normal strain,
γ = shear strain,
σ = normal stress,
τ = shear stress,
E = Young's modulus,
ν = Poisson's ratio,
G = shear modulus,

$\frac{\Delta V}{V}$ = volumetric strain,

and the subscripts designate the three orthogonal directions, x, y and z.

There are only two independent elastic constants. In section 3.3 the following parameters were defined from the laboratory tests: D, B, E, ν, and G. In Eq. 3.3 the use of E and ν was most appropriate, and in Eq. 3.4 the use of G was most appropriate. However any two parameters can be used to define all the others. For example some useful relationships are:

$$B = \frac{E}{3(1 - 2\nu)}, \tag{3.6}$$

$$D = \frac{E(1 - \nu)}{(1 + \nu)(1 - 2\nu)}, \tag{3.7}$$

$$G = \frac{E}{2(1 + \nu)}. \tag{3.8}$$

Note from Eq. 3.5 that when $\nu = 0.5$ the volumetric strain is zero, i.e. the material is incompressible. Then from Eqs. 3.6 and 3.7 B = D = infinite, and from Eq. 3.8 G = E/3.

3.5. Soil Stress-Strain and Strength Properties

An understanding of the basic nature of soil stress-strain and strength properties is essential to the practice of railway geotechnology. The details of this subject are readily available in many references (for example Refs. 3.12 and 3.13), and do not need to be repeated here. The purpose of this section will be to summarize the main concepts.

A good starting point is a description of geostatic stresses in the ground prior to any change in loading as illustrated for the soil element in Fig. 3.17. Both total and effective horizontal and vertical stresses are defined. In this case these are principal stresses, but if the ground surface were sloped they would not be. The effective stress is the average stress applied to the soil skeleton. The sum of the effective stress and the pore water pressure is the total stress. The parameter K_0 is the ratio of the horizontal to vertical effective stress under the condition of zero horizontal strain as in the confined compression test (Section 3.3.1). Unfortunately, although K_0 is an effective stress ratio, traditionally this is not designated with a prime ($'$). The symbol, γ, with appropriate subscripts, is used for unit weights (gravity force per unit volume) of the soil and water.

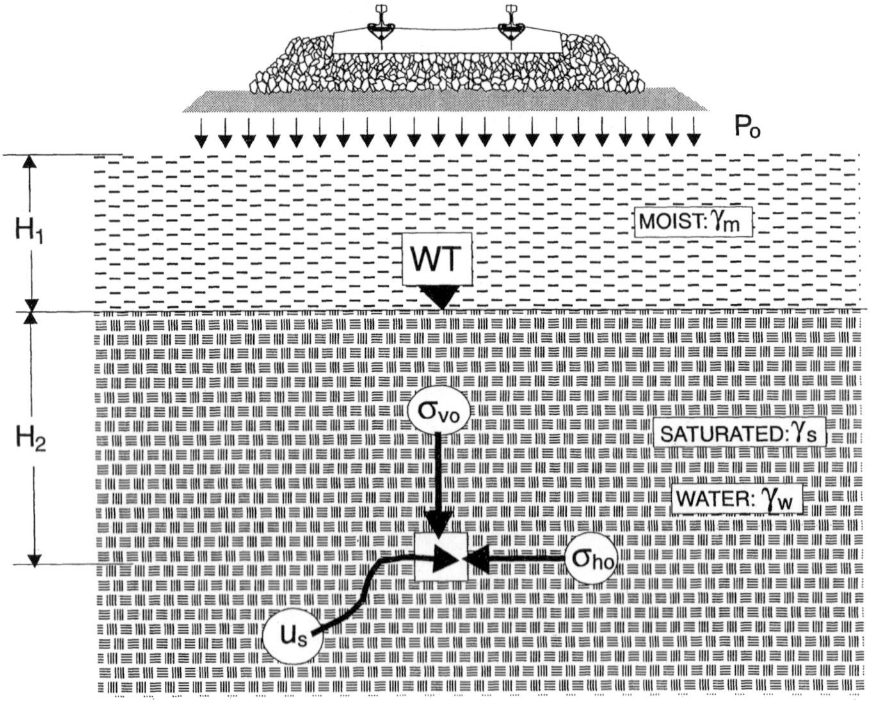

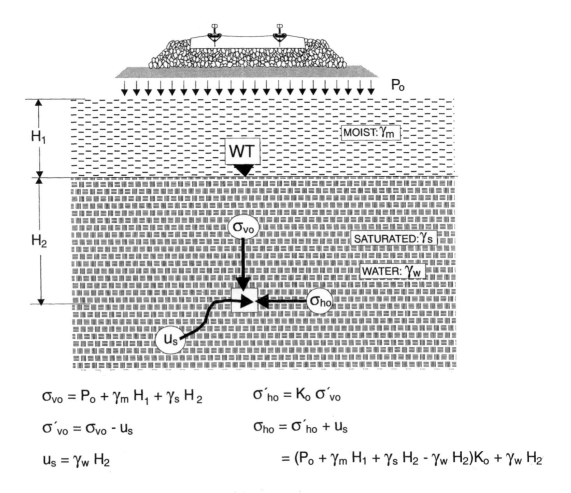

$$\sigma_{vo} = P_o + \gamma_m H_1 + \gamma_s H_2 \qquad \sigma'_{ho} = K_o \sigma'_{vo}$$

$$\sigma'_{vo} = \sigma_{vo} - u_s \qquad \sigma_{ho} = \sigma'_{ho} + u_s$$

$$u_s = \gamma_w H_2 \qquad = (P_o + \gamma_m H_1 + \gamma_s H_2 - \gamma_w H_2)K_o + \gamma_w H_2$$

Fig. 3. 17 Geostatic stresses

After the soil deposit has remained in the stress condition shown in Fig. 3.17 for a long enough time, it will be consolidated under these effective stresses with a static pore pressure, u_s , defined by the ground water conditions. In Fig. 3.17 saturated conditions are assumed below the water table. For partially saturated conditions the pore fluid pressure is a more complicated parameter because both air and water phases exist in the voids, each at a different pressure. A simplified representation of equivalent pore fluid pressure for partially saturated soils is u_e as given by

$$u_e = u_w A_w + u_a (1 - A_w) , \tag{3.9}$$

where u_w is the pore water pressure, u_a is the pore air pressure, and A_w is a factor essentially representing the portion of the voids occupied by water (Ref. 3.13). In this case effective stress is defined as

$$\sigma' = \sigma - u_e . \tag{3.10}$$

If the element of soil has never been subjected to a vertical effective stress greater than that presently existing, σ'_{vo} , as indicated in Fig. 3.17, then the soil is said to be normally consolidated. If, however, the vertical effective stress has been higher, say σ'_{vm}, then the soil is said to be overconsolidated. The overconsolidation ratio, OCR, is defined as

$$OCR = \frac{\sigma'_{vm}}{\sigma'_{vo}} . \tag{3.11}$$

In general the stress-strain and strength trends are the same for all soils, when tested under the same conditions. The main difference among soils is the magnitude of the parameters. This is a big difference, of course, but so is the range of values within one soil type as well.

To illustrate the trends assume that the soils exist at a stress state represented by Fig. 3.17 with $K_0 = 1$. Samples of the soils are taken and saturated specimens prepared for triaxial tests in which the initial conditions are set equal to the stress state in Fig. 3.17. One set of tests is run with the drainage line closed during loading (CU test). Another set is run with the drainage line open during loading which is performed slowly enough to permit water flow into or out of the specimen while maintaining pore water pressure at a value of u_s (CD test). The corresponding stress-strain curves have been illustrated in Fig. 3.11. Highly overconsolidated clays and dense silts, sands and gravels show similar trends. Likewise normally consolidated clays and loose silts, sands and gravels show similar trends. As indicated in Section 3.3.3, a volume change tendency in a CD test results in a similar trend for pore pressure change during a CU test, keeping in mind that a volume increase converts to a pore pressure decrease, and vice versa.

In Fig. 3.11 the CD results are the same as the UU test results would be when the initial specimen consolidation conditions for sample formation are the same as the consolidation conditions for the CD test.

Soil strength will be represented as a simplification by the commonly used Mohr-Coulomb equation (Fig. 3.18), which in terms of total stresses is

$$\tau_f = c + \sigma_{nf} \tan\varphi , \tag{3.12a}$$

and in terms of effective stresses is

$$\tau_f = c' + \sigma'_{nf} \tan\varphi' , \tag{3.12b}$$

where τ_f is the shearing resistance on the assumed plane of failure at failure (i.e. shear strength), σ_{nf} or σ'_{nf} is the normal stress on the plane of failure at failure, c or c′ is the line intercept, and φ or φ' is the slope angle of the line. The equation has been given in both total and effective stress, the choice depending on which method of analysis is used.

It is best to think of c, φ or c′, φ' as parameters of a straight line failure equation, rather than cohesion, and internal friction angle. They really represent stress independent (c, c′) and stress dependent (φ, φ') strength parameters. As will be shown in this section *c* or c′ is not an intrinsic parameter of clay, and φ or φ' is not an intrinsic parameter of sands and gravels. Both apply to all soil categories, the values depending mainly on the type of test and the degree of specimen saturation, as well as soil type.

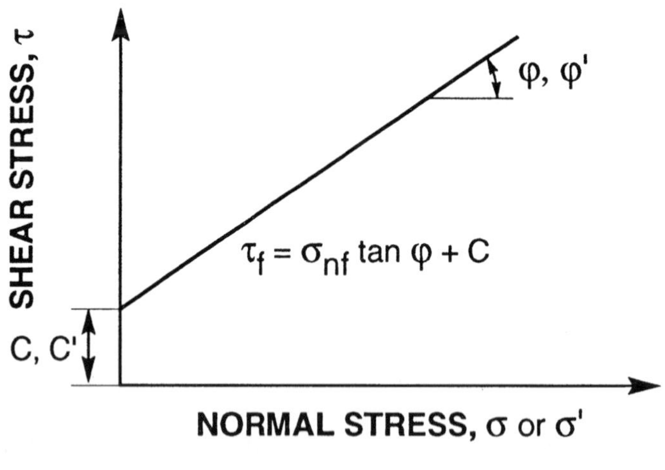

Fig. 3.18 Mohr-Coulomb shear strength relationship

The following discussion will compare total and effective strength behavior for both drained and undrained conditions. The examples apply to all soil types. Deviation from this idealization may be found in the literature for use in situations requiring more precise strength representation.

Consider first the CD tests shown in Fig. 3.11a. The failure stress states can be represented as Mohr circles from which the failure envelopes can be derived (Fig. 3.19). For wording simplification the two cases will be referred to as loose and dense sand, keeping in

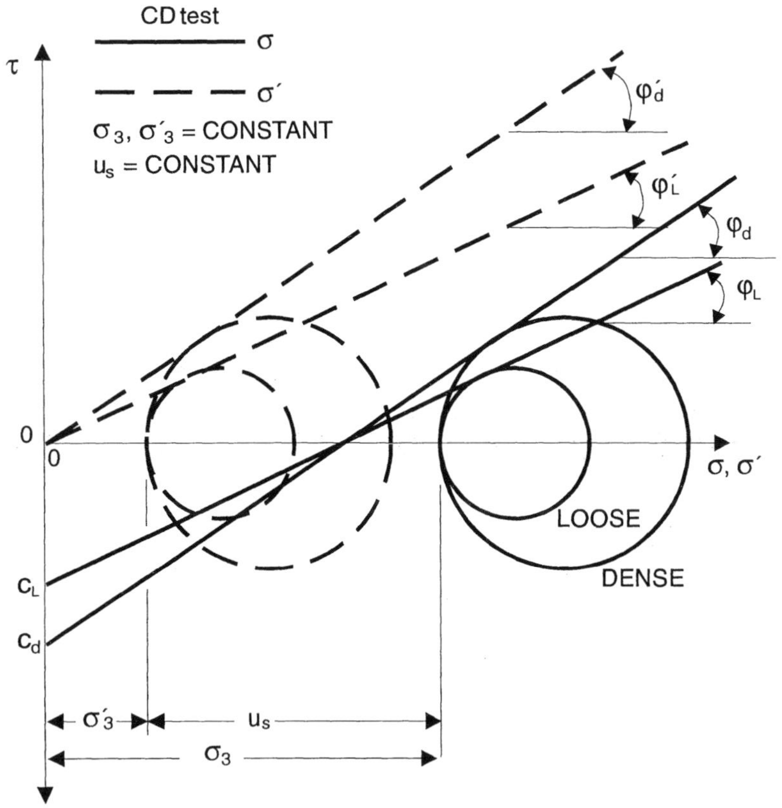

Fig. 3.19 Failure states for loose and dense sands in CD triaxial tests

mind the corresponding cases for other soils. The values of σ_3 , or σ_3' and u_s are constant throughout the test and equal to the initial values in Fig. 3.17 (with $K_0 = 1$). The specimens are assumed to be saturated. The $c' = 0$ for both the loose and dense cases. The φ' values are independent of u_s and are unique strength parameters for the given specimen conditions. Assuming u_s is constant for all CD tests on these specimens, the total stress envelope is parallel to the effective stress envelope so that $\varphi = \varphi'$ and $c = u \tan \varphi'$. Because c is not a unique strength parameter, it is preferable to use effective stress analysis for fully drained loading cases. This is possible because pore pressures are known.

Whether sand or clay, in CD tests using effective stress parameters the soils appear as "cohesionless" materials. However, highly overconsolidated clays will have a small value of c'. Also for all soils over a wide range of consolidation stresses the envelope will actually be curved. The forced fit of a straight line envelope (Eq. 3.12) will also result in non zero values of c' when the strength envelope is actually curved.

Now consider the consolidated undrained tests in Fig. 3.11b. The Mohr circles and effective strength envelopes are shown in Fig. 3.20. Because u will change during shear even though σ_3 is constant, σ_3' is not constant as it was for the CD tests. The pore pressure at failure, u_f, is given by

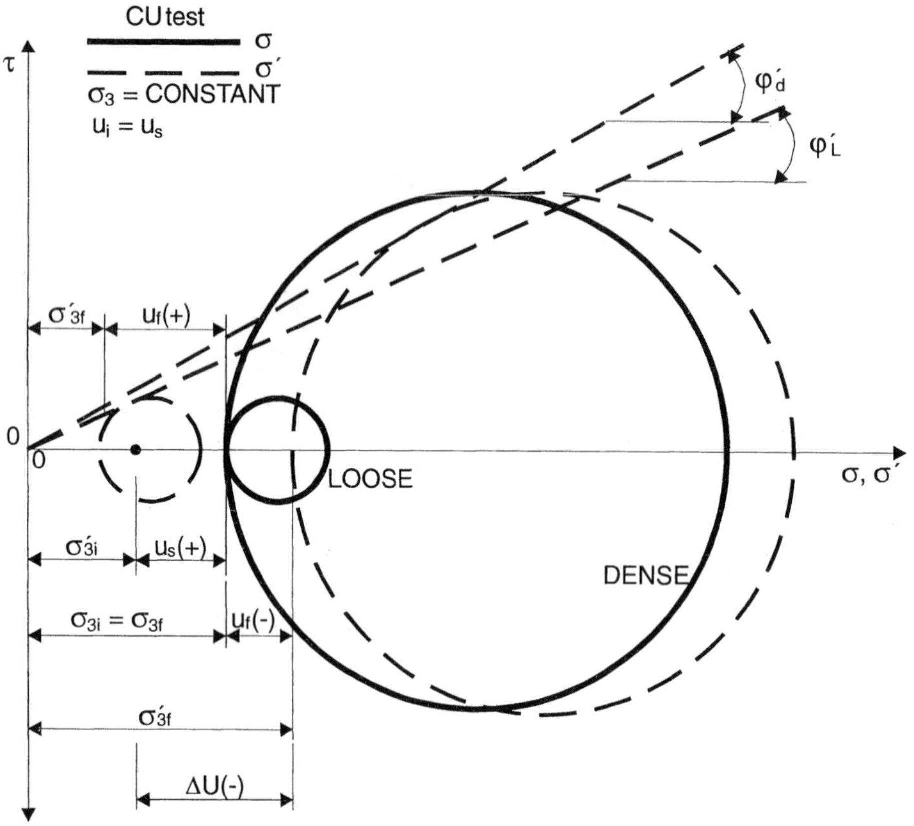

Fig. 3.20 Failures states for loose and dense sands in CU triaxial tests

$$u_f = u_s + \Delta u \; . \tag{3.13}$$

For the dense sand Δu is negative and in fact in this example $\Delta u > u_s$ so that u_f is negative. Thus the effective confining pressure increases during shear. The problem with sand is that the pore pressure can not be very negative before cavitation occurs (water forms a gas phase), at which point the specimen is no longer incompressible and behaves more like a drained case. For the loose sand Δu is positive, hence reducing the effective confining pressure. The strength parameters φ'_d and φ'_l are the same as for the CD tests. However the shear strengths are not, judging from the size of the Mohr circles. In fact for the dense sand

$$(\tau_{ff})_{CU} > (\tau_{ff})_{CD} \; , \tag{3.14a}$$

while for the loose sand

$$(\tau_{ff})_{CU} < (\tau_{ff})_{CD} \; . \tag{3.14b}$$

In the CU test $c \neq c'$ and $\varphi \neq \varphi'$. However, tests at other confining pressures are needed to define c, φ because of the combined effects of u_s and Δu.

The main benefit of the CU test illustration is to explain the UU test results. Specimens for the UU test are always consolidated to some effective stress state in the formation process. Let that state be as shown in Fig. 3.17. Theoretically if a specimen is taken from the ground and the same total stresses are reapplied in the triaxial chamber, then u_s and σ_{vo}' and σ_{ho}' automatically will be reestablished. Because of sample disturbance this will not be exactly true, hence to start UU tests the specimens are often reconsolidated to the field initial effective stresses and pore pressure. In either event the initial σ, σ' stress states for the UU test will be as shown in Fig. 3.20 for the CU test. Also the pore pressure change during shear will be the same for the UU test as for the CU test.

Now take three specimens set up at the initial stress state (σ'_{3i} and u_s) in Fig. 3.20. Then with the drainage line closed change σ_3 of two specimens to values other than in Fig. 3.20. For this change $\Delta u = \Delta\sigma_3$ so that $\Delta\sigma_3' = 0$, i.e. $\sigma_3' =$ constant. Then load the specimens to failure. The effective stresses at failure will be the same for all three tests. Thus the effective stress failure circle and shear strength will be identical for all three tests. The total stress circles will all be the same size but will be shifted by the amount of pore pressure at failure (Fig. 3.21). In this figure the number subscript after f subscript designates the total stress circle number. The consequence is that φ'_d and φ'_l are the same as for the CD and CU tests, but for the UU test $\varphi = 0$, and $\tau_f = c$.

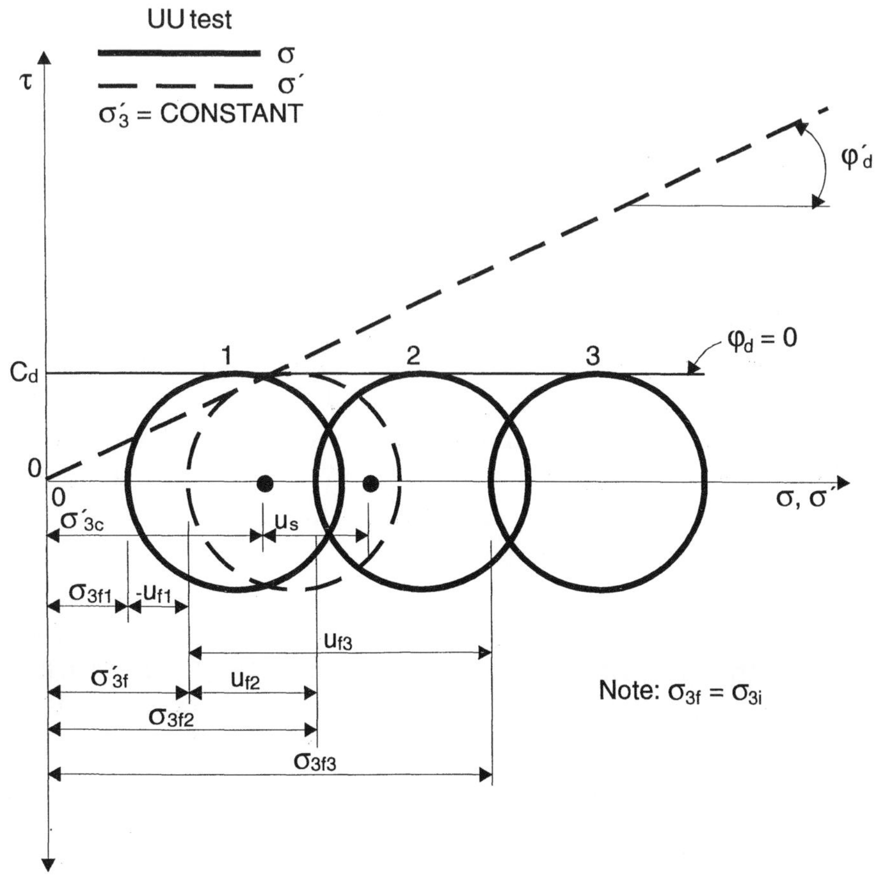

Fig. 3.21 Failure states for dense sand in UU triaxial tests

The unconfined compression test on a saturated specimen ideally would be represented in Fig. 3.21 by a circle of the same size as shown but with $\sigma_{3f} = 0$. Because of sample disturbance and effects of unloading the soil the unconfined compression test circle will generally be smaller than those shown in Fig. 3.21 with the consequence that c_d will be underestimated.

It is most convenient for undrained problems to use total stress analysis with $\varphi = 0$ because the difficulty in estimating pore pressure for effective stress analysis is bypassed. Also the strength parameter c is only a function of initial conditions before loading and is theoretically independent of the loading path.

The confusion in practice is that textbooks often give sand the strength parameters $c = 0$, $\varphi \neq 0$. This is only true for $u = 0$ and drained behavior. The former is not generally true and the latter may not be true. The same books give clay the strength parameters $c \neq 0$, $\varphi = 0$. This is only true for undrained loading of saturated specimens, which is often not appropriate. This practice leads to the misinterpretation of the physical meaning of c, φ as well as numerical mistakes in analysis. For saturated specimens the values of c, φ depend on whether the specimens are tested drained or undrained, not whether they are sand or clay.

Partially saturated soils have more complicated strength behavior than saturated soils. Generally pore pressure is not known so that total stress analysis must be used. If percent saturation is low enough, the soils may behave essentially as drained whether or not a CU, CD or UU test is performed. For high percent saturation the choice of test will have to consider the problem being analyzed. In any of these partially saturated cases non zero values of both c and φ will be obtained.

3.6. Repeated Load Behavior

In general, soil loading can be broadly categorized as either monotonic (increasing or decreasing without reversal) or cyclic. Monotonic loading is often called static loading even though "static" implies nothing is changing. Soil behavior under monotonic loading has been extensively investigated for many years and is better understood than behavior under cyclic (repeated) loading. However, an appreciable amount of data on cyclic loading behavior of soils has been generated from geotechnical studies relating to earthquakes, roads, railways and underwater foundations subject to wave action.

The type of loading of primary interest to railway track performance is cyclic. Failure under cyclic load is progressive and occurs at stress levels below those causing failure under monotonic loading. There are significant differences in behavior trends with these two categories of loading. Nevertheless an understanding of soil behavior under monotonic loading is important for understanding behavior under cyclic loading.

A summary of basic soil behavior under cyclic loading will be given in this section. Specific property data for ballast, subballast and subgrade soil will be provided in later sections. The actual behavior of soil under cyclic loading is complex in part because it depends on stress conditions and stress history and in part because of the wide variety of soil conditions. Only simplified approximations to the real three-dimensional stress conditions can be represented by available test apparatus, and only a limited range of soil conditions can be tested. Thus the results to be presented here are only an indication of the type of behavior that can occur in the field. Applications to railway track are described in Ref. 3.16.

3.6.1. Monotonic Trends

Behavior under monotonic loading increase for loose and dense saturated sand in both drained and undrained conditions was illustrated in Fig. 3.11. The example for CD tests is

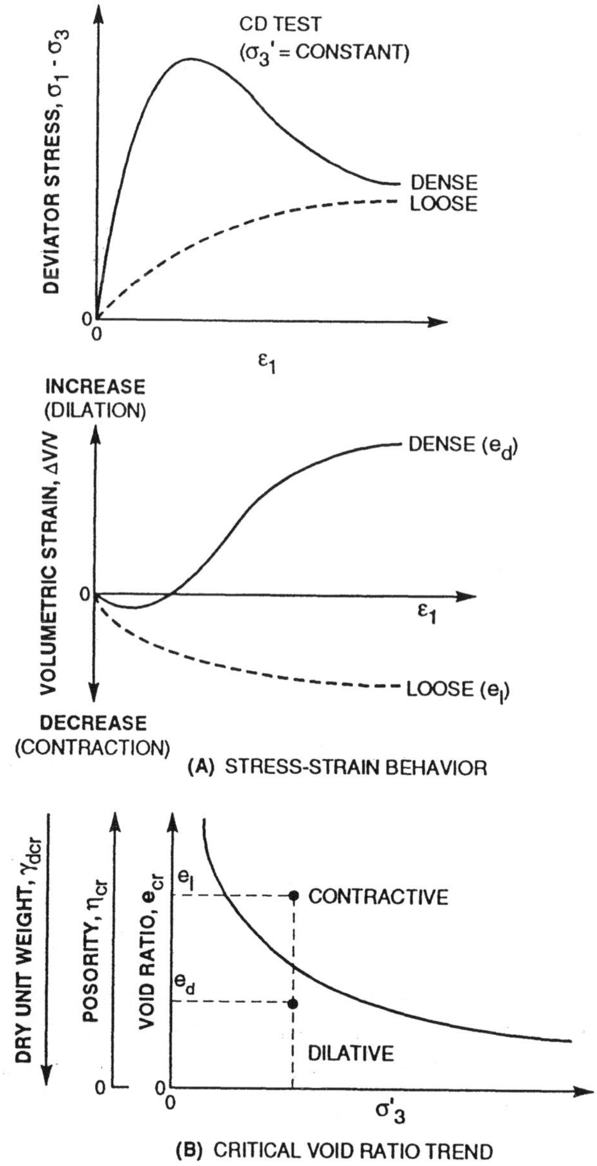

Fig. 3.22 Relationship between critical void ratio and effective confining stress for CD tests on saturated sand

repeated in Fig. 3.22 to show the relationship between void ratio, effective confining stress and volume change tendencies. In Fig. 3.22 void ratio, e, is defined as

$$e = \frac{\text{volume of voids}}{\text{volume of solids}}, \tag{3.15}$$

whereas porosity, n, is defined as

$$n = \frac{\text{volume of voids}}{\text{total volume}}. \tag{3.16}$$

These two parameters are related by

$$n = \frac{e}{(1 + e)}. \tag{3.17}$$

Dry unit weight, γ_d, is also related to void ratio by

$$\gamma_d = \frac{G}{1 + e}\gamma_\omega, \tag{3.18}$$

where G = soil particle specific gravity, and

γ_w = unit weight of water.

Figure 3.22a shows that for a dense sand (low initial e) the sample dilates when sheared to failure, whereas for a loose sand (high initial e) the sample contracts. Obviously at an intermediate initial void ratio, no net change in volume occurs. This void ratio is termed the critical void ratio, e_{cr}. It is a function of effective confining stress, σ'_3, as shown in Fig. 3.22b.

An example of CU tests, to supplement Fig. 3.11 is given in Fig. 3.23 (adapted from Refs. 3.22 and 3.23 based on Ref. 3.12). For specimen C dilation tendency during shear causes a pore pressure reduction above 2% strain, hence increasing effective confining stress and thus shearing resistance. For specimen A contraction tendency during shear causes pore pressure increase up to a constant value above 5% strain. At this point the effective confining pressure is very small (σ'_{3f} = 15 kPa compared to an initial value of σ'_{3c} = 400 kPa) with the result that the shearing resistance is nearly lost. Further strain occurs with constant shearing resistance.

The Mohr circles for specimen A are shown in Fig. 3.24 (adapted from Ref. 3.23 based on Ref. 3.12). An additional increase in pore pressure of 15 kPa would cause σ'_3 to become zero, in which case the Mohr circles would become a point and the sand would have no shearing resistance or stiffness. This condition is termed complete liquefaction.

Examples of complete or nearly complete liquefaction are soil flow slides and quicksand. Flow slides occur in loose sands and sensitive clays when disturbance from some cause (seepage pressure, vibration, or increase in shearing stress) causes pore pressure increase. Flow slides do not occur in soils that tend to dilate during shear. Quicksand can

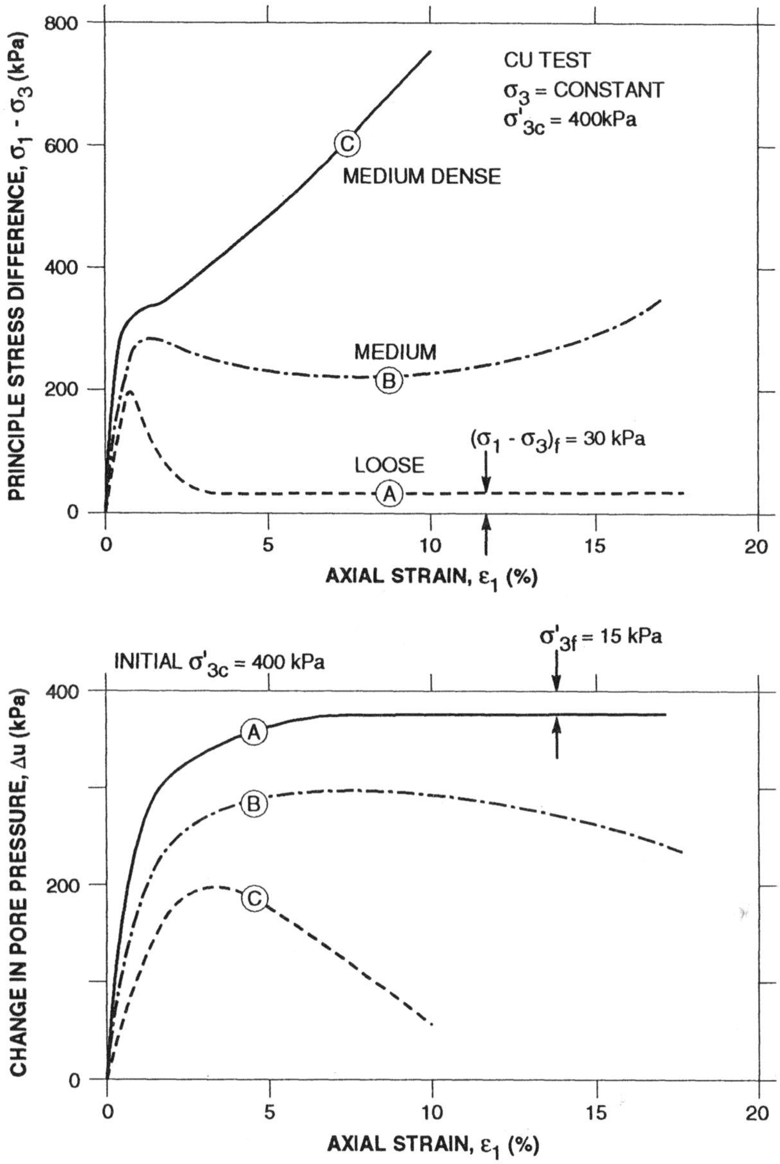

Fig. 3.23 Comparison of three hydrostatically consolidated CU tests on saturated sand

occur over a wider range of void ratios because upward seepage pressure causes the increase in pore pressure and resulting decrease in effective stress.

Data from CU tests such as shown in Fig. 3.23 are used to determine the relationship between void ratio and initial effective confining pressure at which flow will occur (for example specimen A). This relationship is known as the steady state line (Ref. 3.17). It is similar to the critical void ratio line in Fig. 3.22. Specimens of contractive soils, such as C, have a reduction in effective stress during shear, reaching the flow state at point A. Specimens of dilative soils, such as D, have an increase in effective stress during shear and

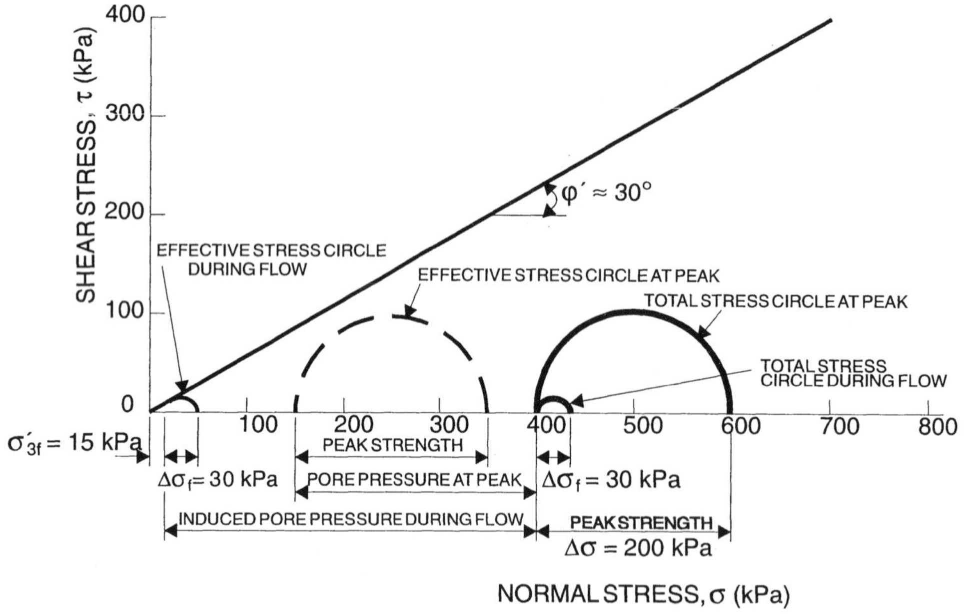

Fig. 3.24 Total and effective stress Mohr circles for liquefaction failure

hence require a high shearing resistance to produce flow. At low void ratios, flow conditions are not reached. Figure 3.25 (Ref. 3.17) shows that the lower the effective consolidation stress (σ'_{3c}) the higher the void ratio (looser specimens) needed to achieve flow.

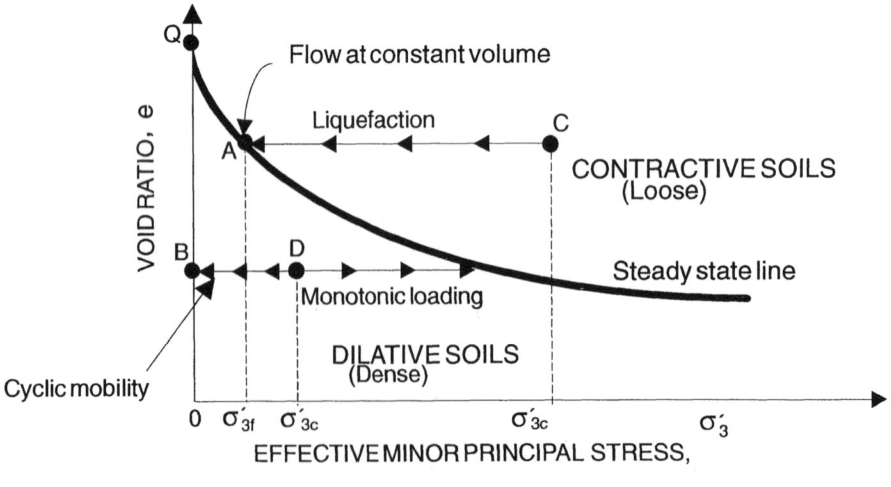

Fig. 3.25 Diagram showing liquefaction potential based on undrained tests of saturated sands (adapted from Ref. 3.17 with permission of publisher)

3.6.2. Cyclic Triaxial Test Results

Results from cyclic triaxial tests will be used to illustrate the behavior of soils under cyclic loading. In these tests the confining pressure will be constant and designated σ_c. A positive or negative piston load may be applied to the specimen so that initially σ_1 does not equal σ_3. This initial principal stress difference will be designated σ_{di}. In addition a cyclic piston load may be applied to create a cyclic principal stress difference, $\Delta\sigma_d$. Thus during cycling the lateral stress is a constant σ_c, and the axial stress, σ_a, is

$$\sigma_a = \sigma_c + \sigma_{di} \pm \Delta\sigma_d \,. \tag{3.19}$$

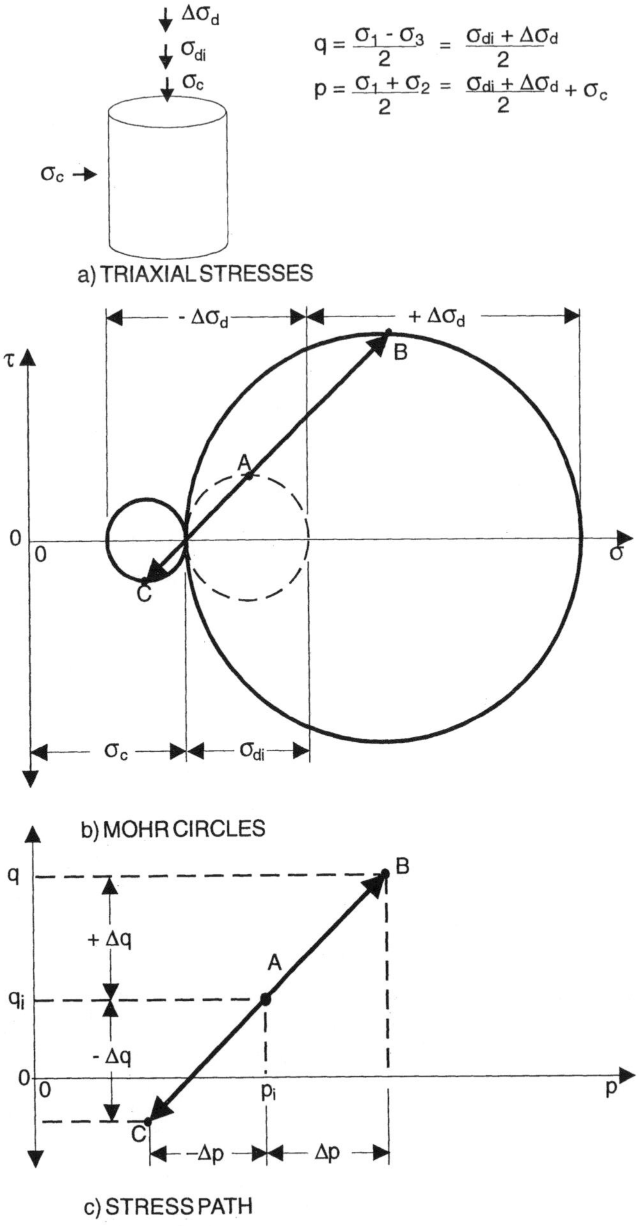

Fig. 3.26 Cyclic triaxial test conditions with constant confining pressure

The minimum principal stress, σ_3, is the smaller of σ_a and σ_c. The maximum principal stress is the larger of σ_a and σ_c. These test conditions and corresponding Mohr circles are shown in Fig. 3.26. The corresponding effective stresses are obtained by subtracting the pore pressure from the total stresses.

In Fig. 3.26b is a line termed "stress path" which connects the peak points on the Mohr circles (Refs. 3.12 and 3.13) for the sequence of stress states during cycling. Point A represents the initial condition, point B represents the maximum compression loading state, and point C represents the minimum compression loading state. Points on the stress path have coordinates p and q, as shown in Fig. 3.26c. The stress path is a convenient way to illustrate the cyclic triaxial stress states.

Note in Fig. 3.26 that the specimen stress state during cycling passes through a hydrostatic state. When this happens, the shearing stresses reverse directions on individual planes. Whether or not shear stress reversal occurs has a significant effect on the mode of specimen deformation and failure in undrained tests. A description of this phenomenon is given in Ref. 3.18.

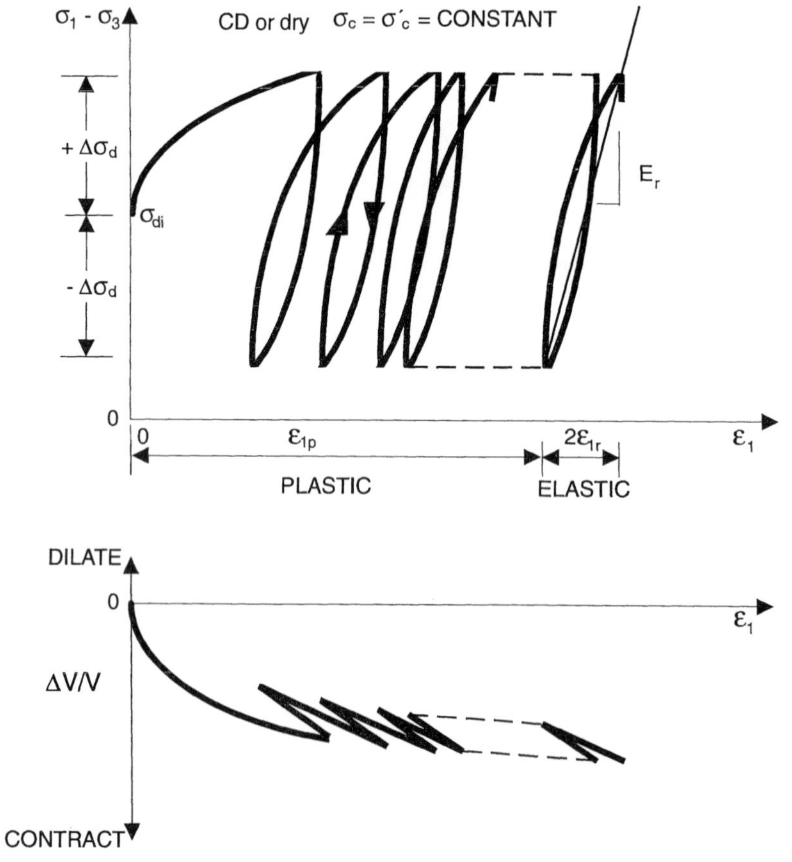

Fig. 3.27 Stress-strain behavior in a CD cyclic triaxial test

3.6.2.1 Consolidated-drained tests

Consolidated-drained cyclic triaxial tests may be done on saturated samples of granular soils (sands, gravels, ballast) if the rate of cycling is low enough to permit flow of pore water into and out of the specimen to accommodate volume changes. This test is also used on dry specimens to study the volume and stiffness changes during cycling.

Typical stress-strain characteristics are illustrated in Fig. 3.27. Regardless of the initial density state, the volume decreases with each cycle at a diminishing rate. Plastic compressive axial strain progressively increases at a diminishing rate. After a sufficient number of cycles of the same stress the plastic strain increment becomes smaller than the elastic strain for the

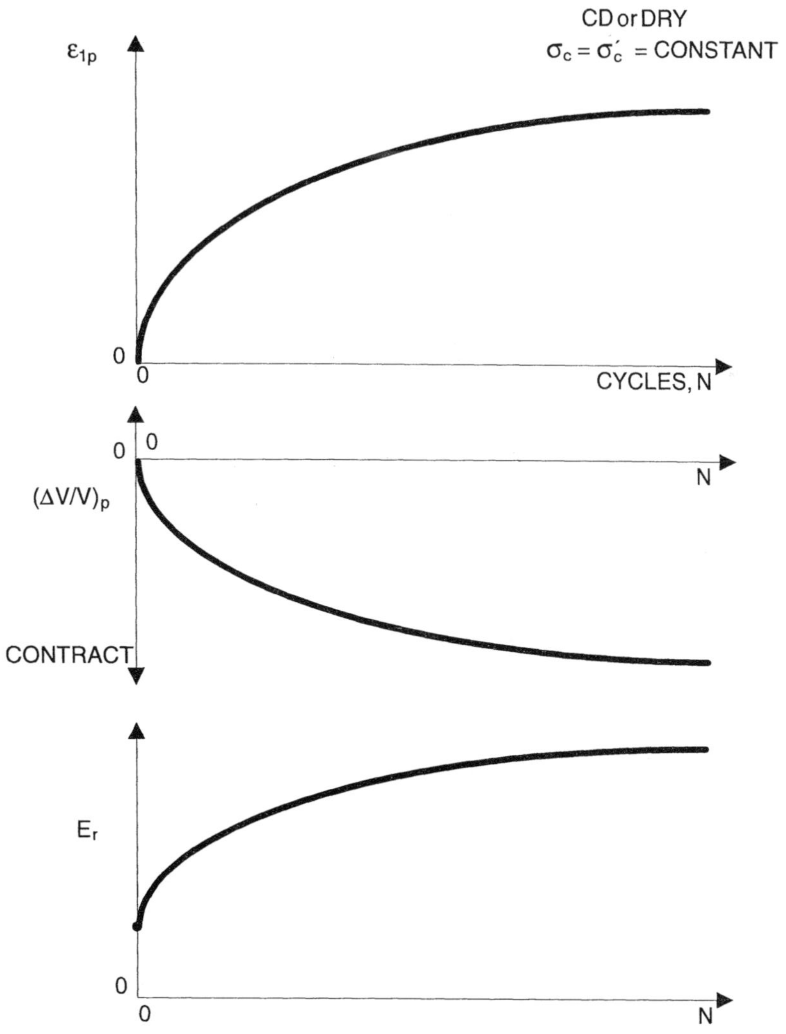

Fig. 3.28 Trends for plastic axial strain, plastic volumetric strain and resilient modulus with successive cycles

same cycle, and the specimen behaves essentially elastically. The resulting ratio of cyclic stress to cyclic elastic strain is termed "resilient Young's modulus", E_r, i.e.

$$E_r = \frac{\Delta\sigma_d}{\varepsilon_{1r}} , \tag{3.20}$$

where ε_{1r} is one half the peak-to-peak elastic strain.

Figure 3.28 shows the trends for plastic axial strain, ε_{1p} , plastic volumetric strain, $(\Delta v/v)_p$, and resilient modulus, E_r , with increasing numbers of cycles. Note that, even though the incremental plastic strain per cycle becomes very small, in cases like railway applications, the cumulative plastic strain still can become very large because millions of cycles may be applied.

Studies of volume change characteristics of dry sands under a variety of stress paths were conducted by Tam (Ref. 3.19) using the constant confining pressure CD cyclic triaxial test with a variety of stress paths. He showed that (Fig. 3.29a), for a given number of cycles

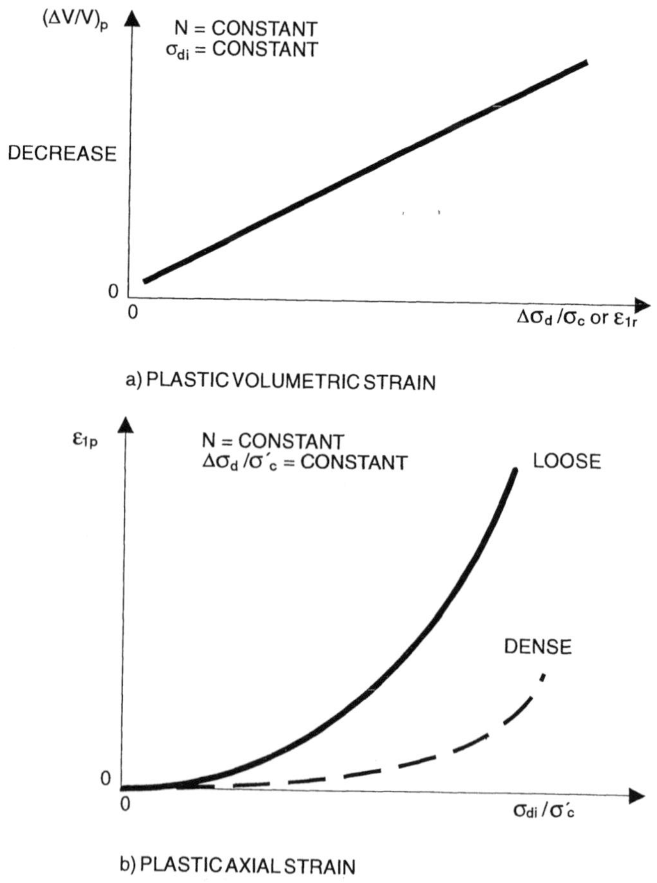

Fig. 3.29 Effect of stress on plastic strains for CD cyclic triaxial tests on sand

with the same initial σ_{di}, the plastic volumetric strain increased in proportion to the cyclic stress ratio, $\Delta\sigma_d/\sigma_c'$ or cyclic strain, ε_{1r}. This indicates that volume change is a strain-controlled response. He also showed that (Fig. 3.29b), for a given number of cycles with the same cyclic stress ratio, the plastic strain increased at an increasing rate with increasing initial compression stress ratio, σ_{di}/σ_c'. This means that as the initial compression stress state increases toward failure, the amount of plastic strain from the same cyclic stress increases rapidly. As expected, the plastic strain is greater for loose sand than for dense sand.

The plastic volumetric strain decreases with cycles as long as the initial stress conditions are not close to failure. However, as failure is approached, Tam (Ref. 3.19) showed that the plastic strain increases in dilation (Fig. 3.30). The transition stress state from contraction to dilation may be identified by static CD triaxial tests as illustrated in Fig. 3.31. Initial stress states resulting in circles below the threshold line will produce volume decrease (contraction) during cycling even when the maximum stress state during cycling crosses the threshold line, whereas initial stress states resulting in circles crossing the threshold line will produce volume increase (dilation) during cycling (Fig. 3.32). The circle shown in Fig. 3.31 will result in no net volumetric strain during cycling.

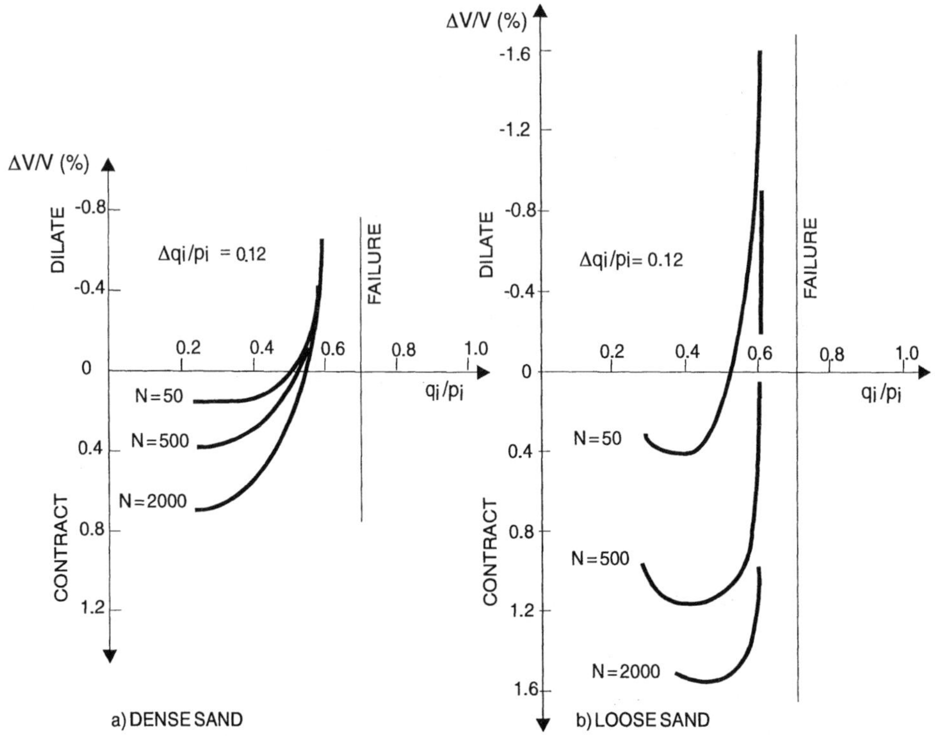

Fig. 3.30 Effect of initial stress on volumetric strain for CD cyclic triaxial tests on dry sand

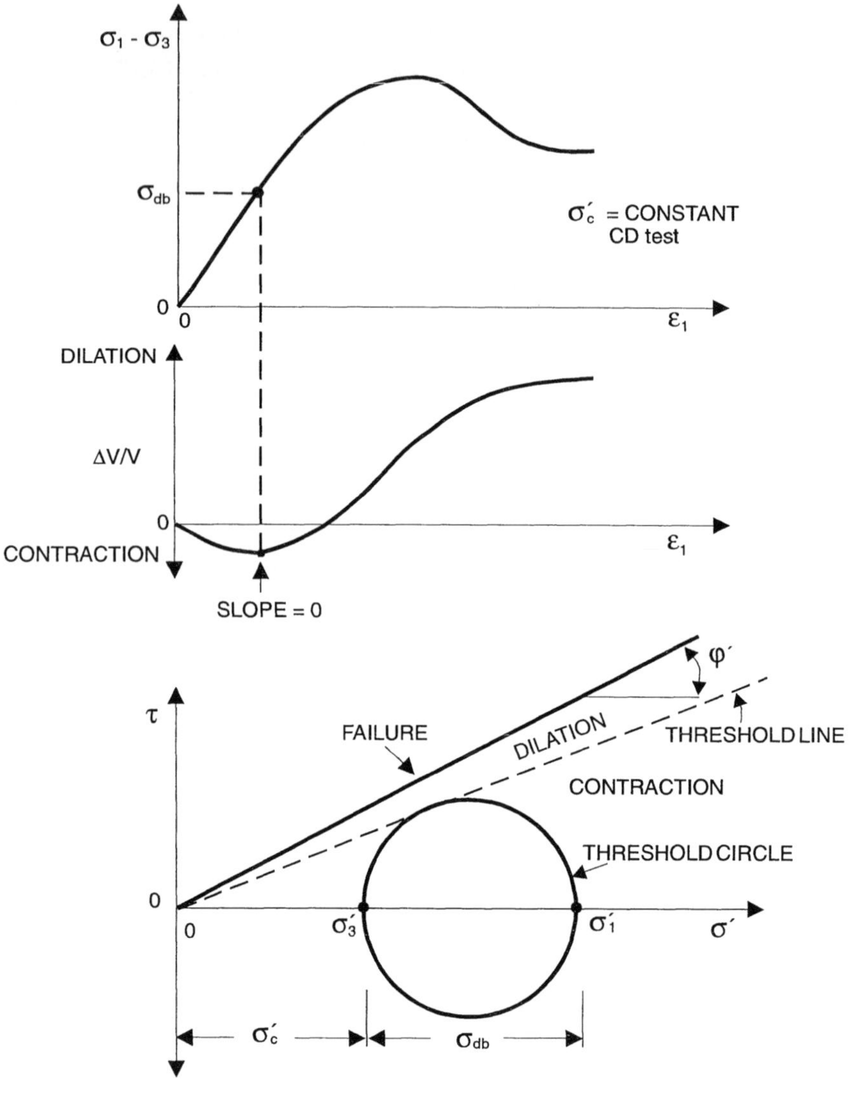

Fig. 3.31 Determination of threshold stress state for cyclic plastic volumetric strain from static CD triaxial tests

3.6.2.2 Consolidated-undrained tests of saturated soils

During undrained cyclic loading of saturated soil samples no volume change occurs. Instead the pore water pressure will change in a manner that may be predicted from the volume change tendencies from drained cyclic tests. A general presentation of the behavior of saturated sands in undrained loading is given in Ref. 3.18. The main points will be summarized in this section and some additional examples given.

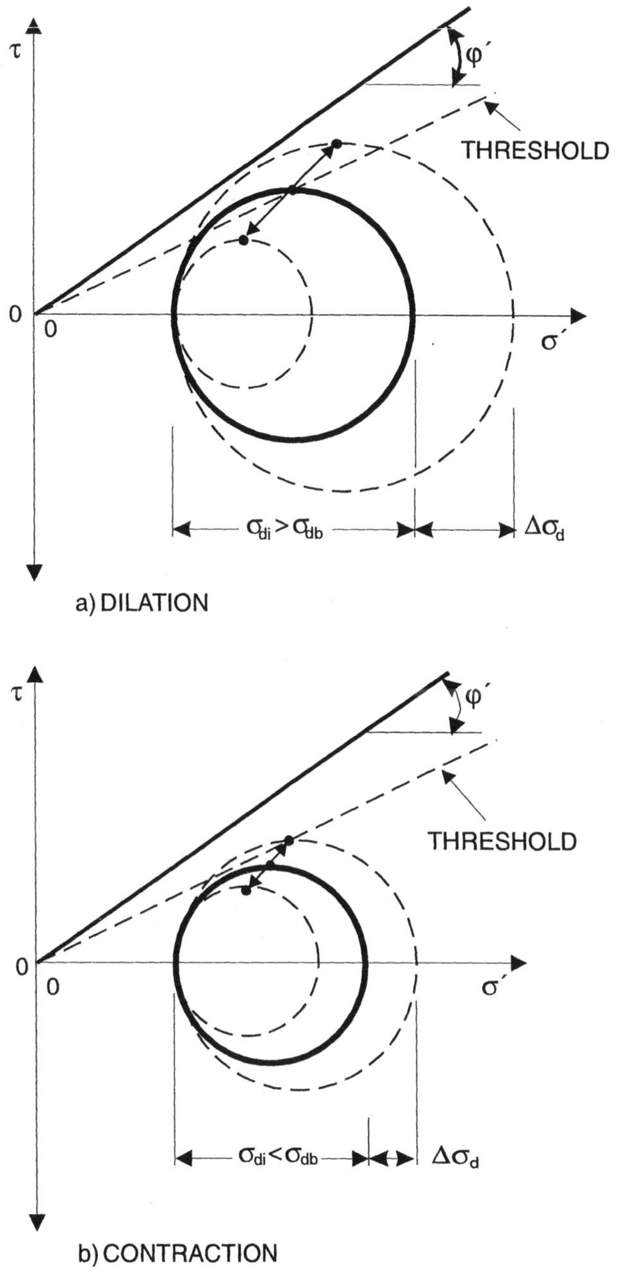

Fig. 3.32 Examples of stress conditions resulting in dilation and contraction during cyclic loading

Except for initial stress states above the threshold line, pore pressure increase will be induced by cyclic loading. Increases in pore pressure will reduce the effective confining stress and may move the stress state closer to failure. In such cases the threshold line may be crossed so that the pore pressure will decrease again. Thus, a complication in undrained cyclic tests is that effective stresses change during cycling, whereas in the drained cyclic tests effective stresses remain constant.

A special case of the stress conditions shown in Fig. 3.26 occurs when $\sigma_{di} = 0$. This is called an isotropic cyclic test. The Mohr circles and corresponding stress paths for the applied total stresses are shown in Fig. 3.33a. The resulting effective stress path, stress-strain and pore pressure behavior are found in Fig. 3.34. Note that the specimen stiffness, defined as the ratio of $\Delta\sigma_d$ to the corresponding cyclic strain amplitude, decreases with each cycle. Note too that the effective confining pressure approaches zero when the stress path crosses the q = 0 axis, but increases again due to dilation tendencies as shear is applied. This maintains the effective stress path within the failure lines. Note that the slope angle of the failure envelope in terms of q and p′ is, due to geometry, less than φ'.

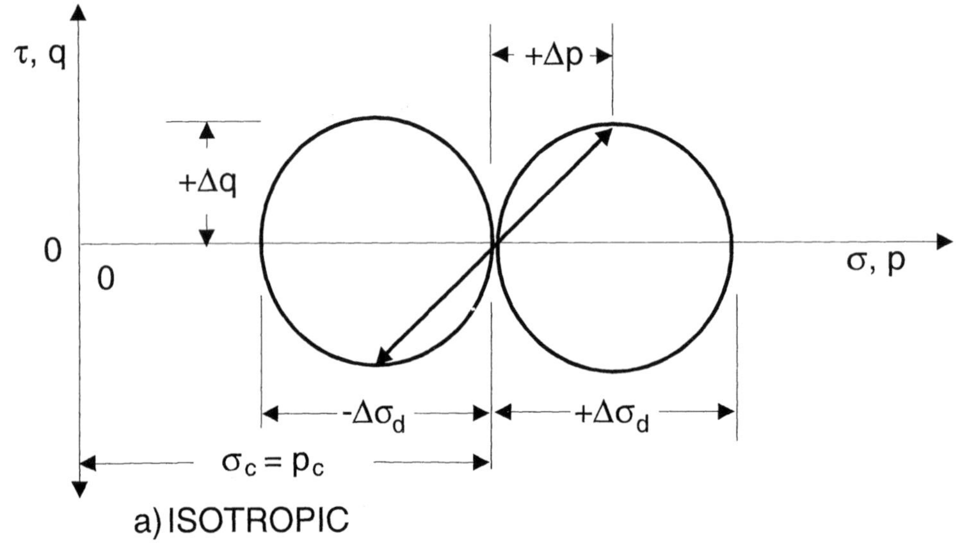

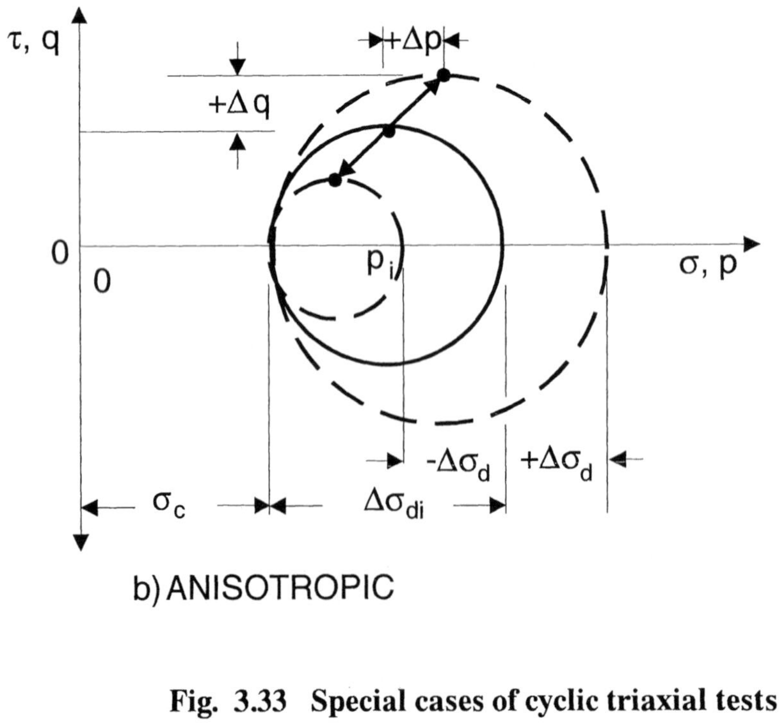

Fig. 3.33 Special cases of cyclic triaxial tests

If the dilation tendency of the specimen is insufficient to keep the effective stresses within the failure lines, then the specimen will not support the applied shearing stresses and instead large deformations will occur. In Fig. 3.34 the increase in axial strain and pore pressure with increasing number of cycles is also shown. The explanation for the pore pressure trend near failure is indicated by Fig. 3.35. The axial strain is symmetrical about zero which means no plastic component (i.e., no change in specimen length).

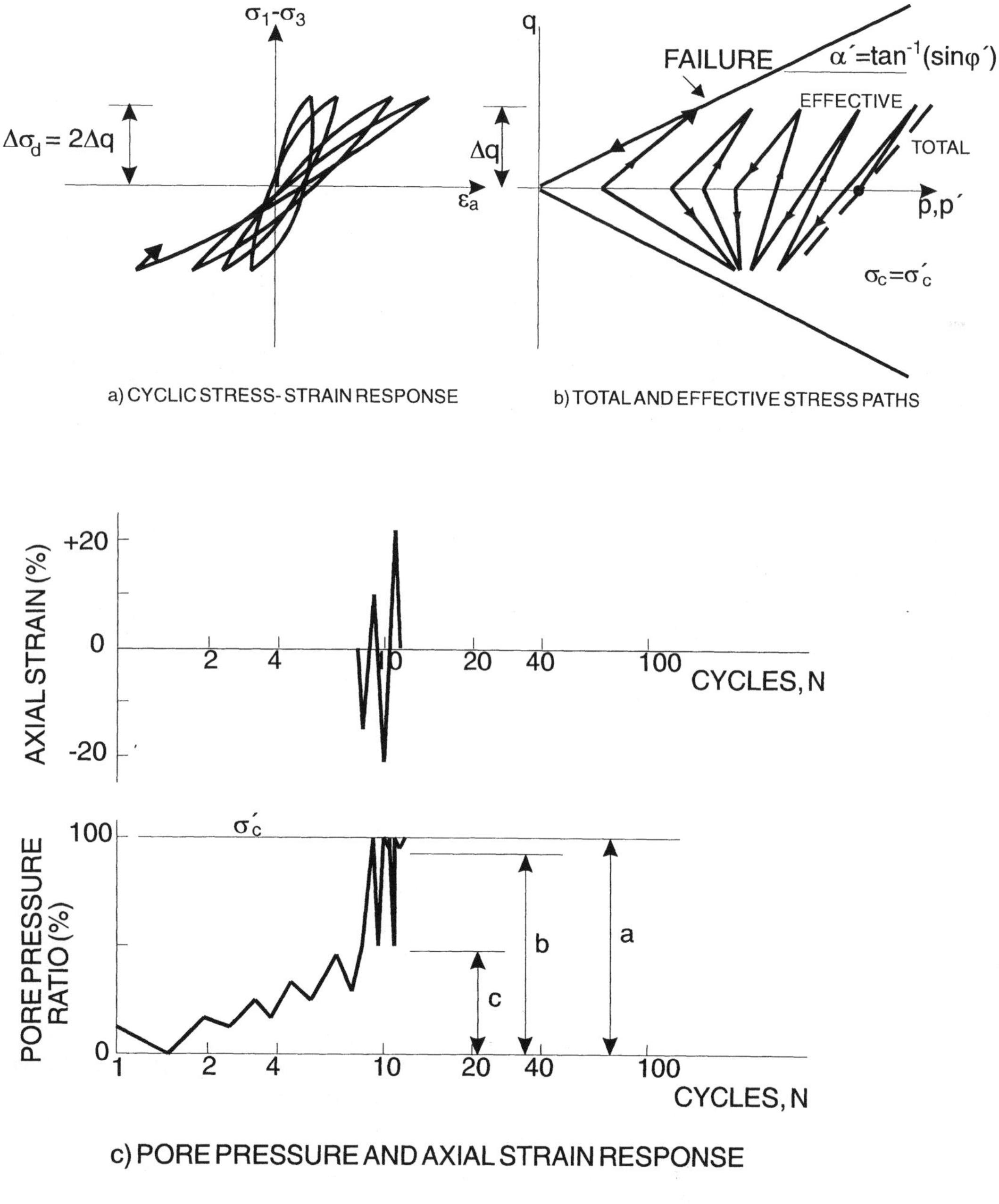

Fig. 3.34 Response of saturated soil sample in CU isotropic cyclic triaxial test

The phenomenon represented by Fig. 3.34 is termed "cyclic mobility" to distinguish it from liquefaction under monotonic loading. In contrast to liquefaction, cyclic mobility will occur at any initial unit weight or void ratio, not just for loose samples. See, for example, the dilative specimen at point D in Fig. 3.25. In monotonic loading the point D moves to the right, whereas for cyclic loading D progressively moves to the left. For the contractive specimen at point C both monotonic and cyclic loading move the point to the left.

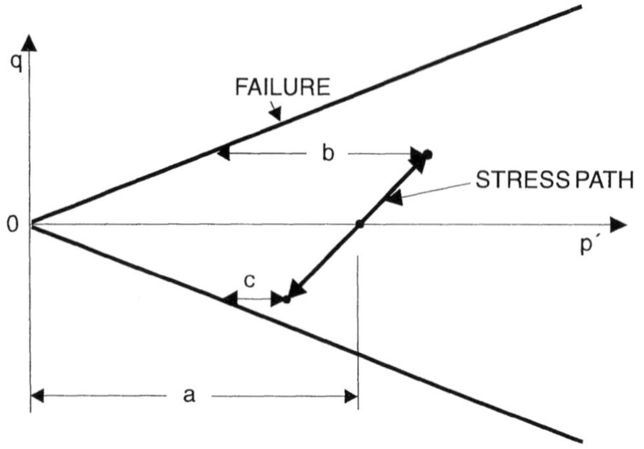

Fig. 3.35 Stress path explaining pore pressure development in Fig. 3.34

The main difference between dense and loose specimens in isotropic cyclic tests is a matter of degree. This is illustrated in Fig. 3.36 (Ref. 3.12) which shows the relationship between magnitude of cyclic stress and number of cycles to cyclic mobility failure (defined by the magnitude of the cyclic strain amplitude) as a function of relative density and effective consolidation stress. Note that increasing relative unit weight, D_r , increases the required cyclic stress to produce failure in a given number of cycles. Increasing the initial effective consolidation pressure also increases the required cyclic stress.

Anisotropic cyclic triaxial tests are those which start from a non-hydrostatic stress state as indicated in Fig. 3.26. A special case of this condition in which no shear stress reversal occurs is shown in Fig. 3.33b. A comparison will be made of the specimen behavior in such an anisotropic cyclic test with the behavior already shown for the isotropic cyclic test. The soil tested was a saturated sand in an intermediate density state. An isotropic cyclic test like that in Fig. 3.33a was performed on specimen A while an anisotropic cyclic test like that in Fig. 3.33b was performed on specimen B. The axial strain and pore pressure response for two tests are given in Fig. 3.37. The pore pressure increased more rapidly in the isotropic test, and for both tests the cyclic component was smaller than the progressive mean value. The pore pressure for the isotropic test rapidly approached the initial effective consolidation stress indicating impending failure. The pore pressure in the anisotropic case increased at a decreasing rate, approaching a limiting value. The plastic strain component for the isotropic case remained small throughout the test (meaning little specimen

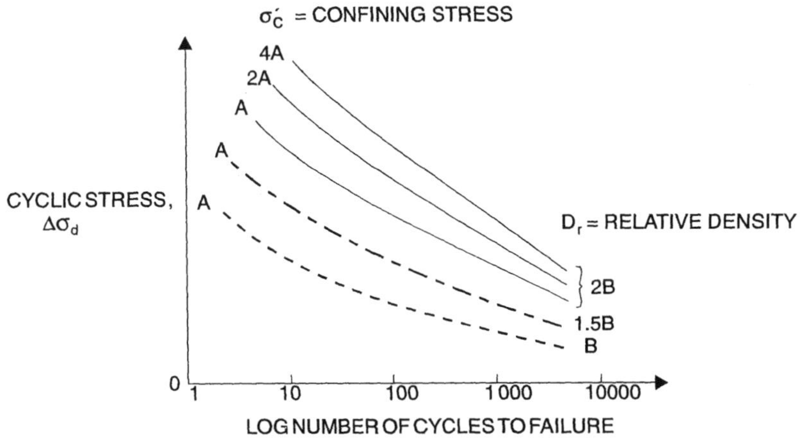

Fig. 3.36 Relationship between cyclic stress and number of cycles to cause cyclic mobility failure

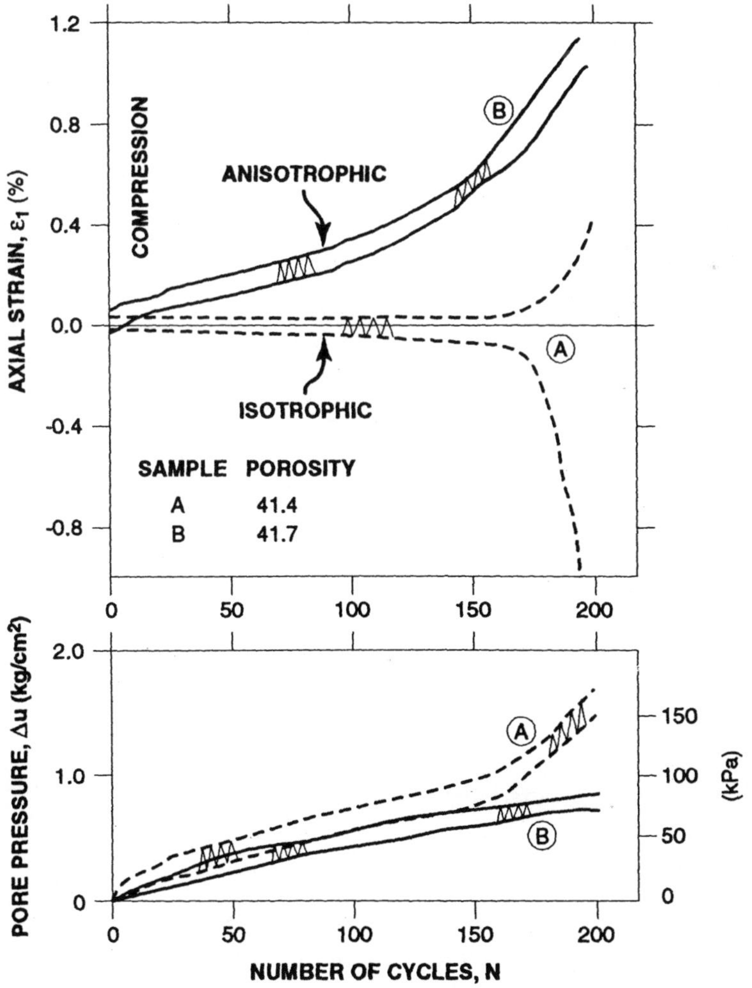

Fig. 3.37 Cyclic and residual behavior for isotropic and anisotropic cyclic triaxial tests

shortening), while the cyclic component increased rapidly near failure. In contrast, the cyclic strain remained small for the anisotropic test while the plastic component increased at an increasing rate. Thus, failure for the isotropic case would be defined by a limiting cyclic strain, while failure for the anisotropic case would be defined by a limiting plastic strain.

The total and effective stress paths for the anisotropic test with no shear stress reversal is shown in Fig. 3.38a for comparison with the isotropic stress paths in Fig. 3.34b. The corresponding case for the anisotropic test with shear stress reversal is shown in Fig. 3.38b. Results of tests with all three cases showed (Ref. 3.18) that cyclic strain governed failure with shear stress reversal while plastic strain governed failure when no shear stress reversal occurred in a cyclic test.

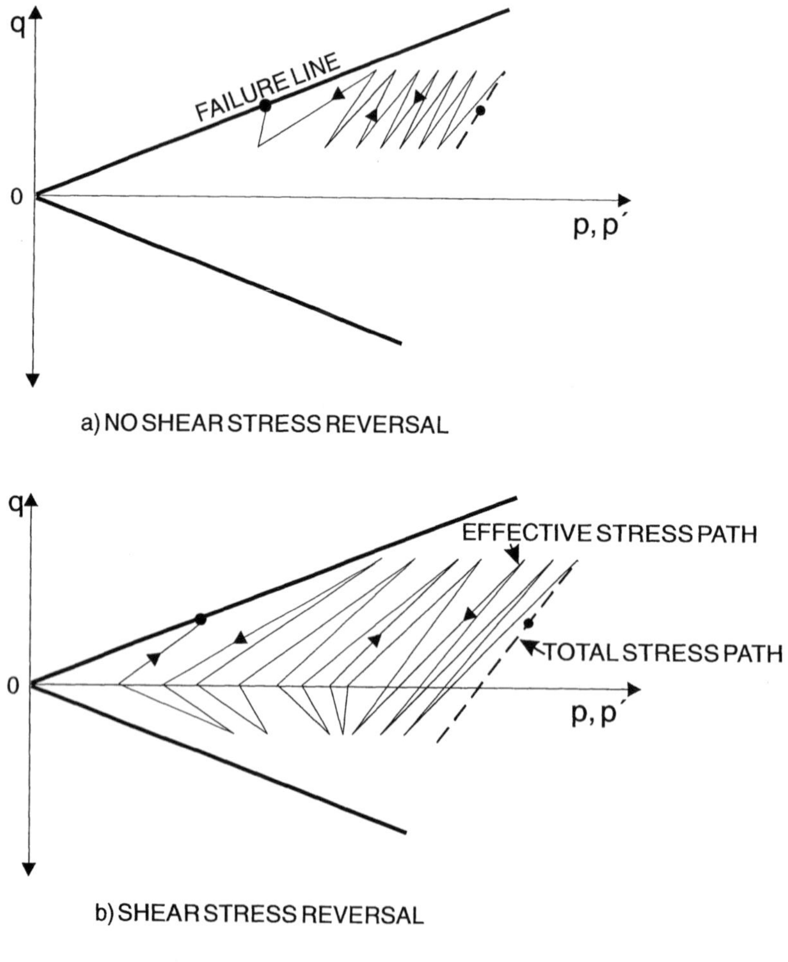

Fig. 3.38 Total and effective stress paths for anisotropic cyclic triaxial test with and without shear stress reversal

3.6.2.3 Partial drainage or partial saturation

Two important cases of cyclic behavior need to be considered for comparison with the examples given. One is for partial drainage and the other is for partial saturation.

For saturated soils, pore water pressure will increase as a result of cycling. Partial drainage of the pore water can occur to reduce the excess pore pressure. Some dissipation will occur simultaneously with build up during cycling because the soil is not completely sealed. The degree will depend on the soil permeability and the drainage path length. In addition the cycling is not continuous, so that dissipation will also occur during periods between cycling, as, for example, between trains. The effect of these mechanisms is to diminish the rate of pore pressure increase indicated in Figs. 3.34 and 3.37.

For partially saturated soils, behavior during cycling will first be similar to that of drained tests in which volume reduction occurs. This will result in an increase in percent saturation. If the soil has a high enough percent saturation when cycling begins and if drainage is limited then, 100% saturation can be approached and the soil will begin to respond as an undrained specimen. The percent saturation, especially of fine-grained soils (silts and clays), will vary with weather conditions, hence seasonal variation in soil behavior under traffic induced cyclic loads is to be expected.

REFERENCES

3.1 Whitten, D. G. A. and Brooks, J. R. V. (1972). *A dictionary of geology*. Penguin Books, N.Y., N.Y.

3.2 IUGS Committee (1973). **Plutonic Rocks: Classification and nomenclature recommended by the IUGS Subcommission on the Systematics of Igneous Rocks.** *Geotimes*, October.

3.3 Streckeisen, Albert (1979). **Classification and nomenclature of volcanic rocks, lamprophyres, carbonatites and metilitic rocks: Recommendations and suggestions of the IUGS subcommission on the systematics of igneous rocks.** *Geology*, Vol. 7, pp. 331-333.

3.4 Folk, R. L. (1959). **Practical petrographic classification of limestones.** *American Association of Petroleum Geologists Bulletin,* Vol. 43, pp. 1-38.

3.5 Best, M. G. (1982). *Igneous and metamorphic petrology*. W. H. Freeman and Co., San Francisco.

3.6 Hurlbut, C. S. and Klein, C. (1977). *Manual of mineralogy*. John Wiley and Sons, N.Y.

3.7 ASTM D2488. **Standard practice for description and identification of soils (visual-manual procedures).** Annual Book of Standards, Section 4, Construction, Vol. 04.08, *Soil and Rock.*

3.8 Selig, E. T. and Selig, Ted (1991). *User manual for SID -visual/manual soil identification procedure, version 3.1.* Amherst, Massachusetts, November.

3.9 Jennings, J. E., Brink, A. B. A. and Williams, A. A. B. (1973). **Revised guide to soil profiling for civil engineering purposes in Southern Africa.** *Die Siviele Ingenieur in Suid-Afrika*, Januarie, pp. 3-12.

3.10 ASTM D2487. **Standard test method for classification of soils for engineering purposes**. Annual Book of Standards, Section 4, Construction, Vol. 04.08, *Soil and Rock.*

3.11 American Association for State Highway and Transportation Officials. *Standard specifications for transportation materials and methods of sampling and testing*, 12th Edition, 1978, and ASTM D3282. **Standard classification of soils and soil-aggregate mixtures for highway construction purposes**. Annual Book of Standards, Section 4, Construction, Vol. 04.08, *Soil and Rock.*

3.12 Holtz, R. D. and Kovacs, W. D. (1981). *An introduction to geotechnical engineering.* Prentice-Hall.

3.13 Lambe, T. W. and Whitman, R. N. (1969). *Soil mechanics.* John Wiley and Sons, New York.

3.14 O'Reilly, M. P. (1991). **Cyclic load testing of soils.** *Cyclic loading of soils: from theory to design,* O'Reilly and Brown Eds., Blackie, London, pp. 70-121.

3.15 Wood, D. M. (1982). **Laboratory investigations of the behavior of soils under cyclic loading: a review.** Chap. 20, *Soil mechanics-- transient and cyclic loads*, edited by G. N. Pande and O. C. Zienkiewicz, Wiley, pp. 513-582.

3.16 Brown, S. F. and Selig, E. T. (1991). **The design of pavement and rail track foundations.** *Cyclic loading of soils: from theory to design*, O'Reilly and Brown Eds., Blackie, London, pp. 249-305.

3.17 Castro, G. and Poulos, S. J. (1977). **Factors affecting liquefaction and cyclic mobility.** *Journal of Geotechnical Engineering Division*, ASCE, Vol. 103, No. GT6, pp. 501-516, June.

3.18 Selig, E. T. and Chang, C. S. (1981). **Soil failure modes in undrained cyclic loading.** *Journal of the Geotechnical Engineering Division*, ASCE, Vol. 107, No. GT5, May, pp. 530-551.

3.19 Tam, Kwok Ho (1986). *Volume change characteristics of dry sand under cyclic loading.* M.S. degree project report, Report No. AAR85-331P, Department of Civil Engineering, University of Massachusetts, Amherst, Massachusetts, January.

3.20 ASTM D423. **Standard test method for liquid limit of soils.** Annual Book of Standards, Section 4, Construction, Vol. 04.08, *Soil and Rock.*

3.21 ASTM D424. **Standard test method for plastic limit and plasticity index of soils.** ASTM Annual Book of Standards, Section 4, Construction, Vol. 04.08, *Soil and Rock.*

3.22 Castro, G. (1969). *Liquefaction of Sands.* PhD Dissertation, Harvard University.

3.23 Casagrande, A. (1975). **Liquefaction and cyclic deformation of sands, a critical review.** *Proceedings of the 5th Panamerican Conference on Soil Mechanics and Foundation Engineering, Buenos Aires.*

4. In Situ Tests for Soil Properties

The soil subgrade is the base upon which a railroad track structure with its ballast and subballast layers is placed. The top of the subgrade is not always clearly delineated, but it is usually within about 1 m of the sleeper bottom. The length and width of effective track bearing area for loads placed on the rail is much greater than just the depth from the sleeper bottom to the subgrade and in fact includes substantial portions of the subgrade as shown in Fig. 4.1. Because of this geometry, the subgrade is well within the zone of influence of the

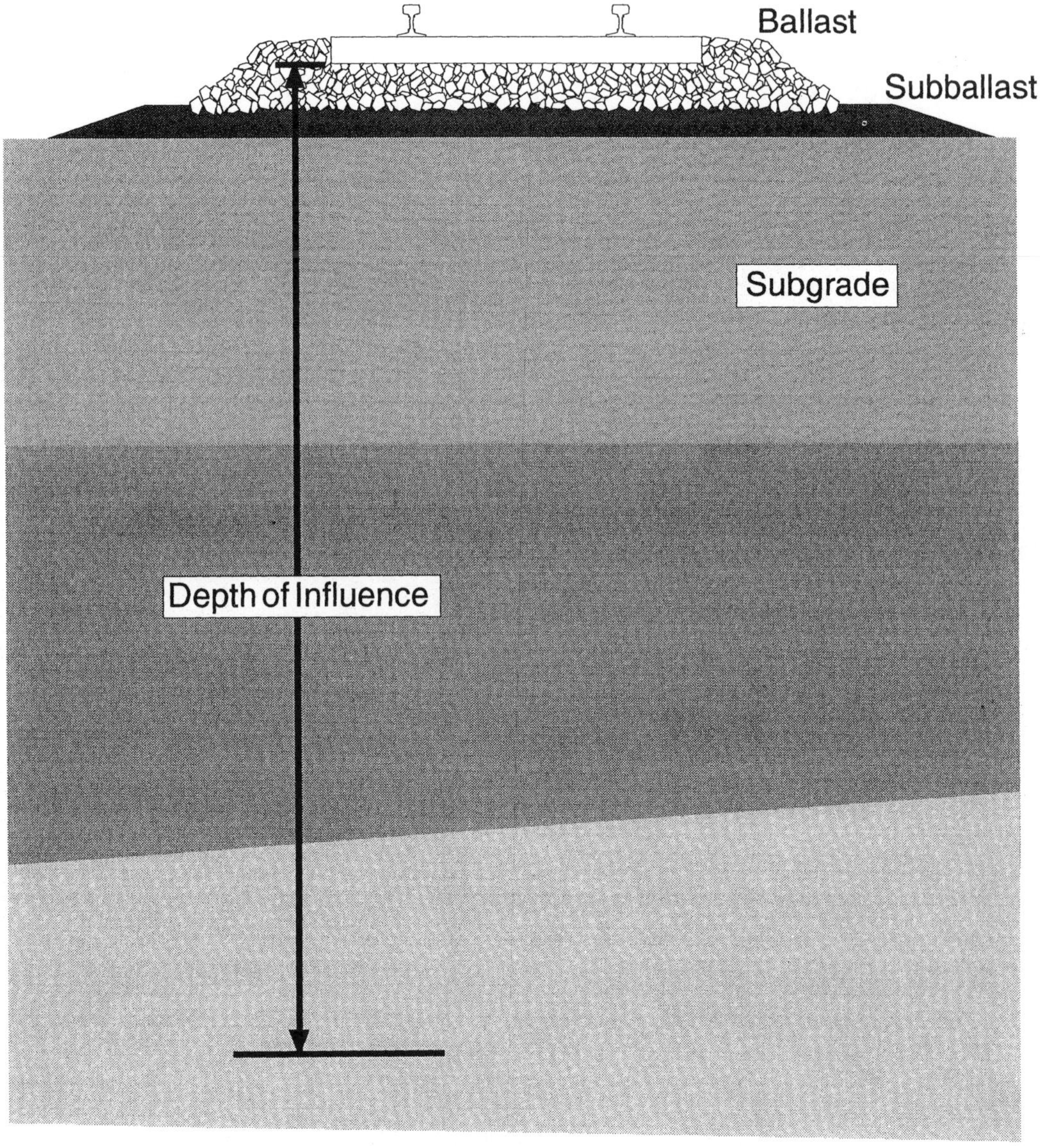

Fig. 4.1 Track substructure profile indicating depth of influence of stress in subgrade from train loading

applied surface load, and as a result the subgrade has a significant effect on both track deformation under load and the maximum load-carrying capacity.

For a given load, the track deformation is a function of the stiffness of individual supporting layers and the thickness of each layer or depth to layer interfaces. The limit-equilibrium loading condition or bearing capacity is a function of individual layer strengths and depths to layer interfaces. Soil stiffness and strength are related to each other in the sense that soils with low strength tend also to have low stiffness. Thus track with a soft subgrade layer at a relatively shallow depth will result in low bearing capacity and high deformation under load, and is likely to experience abnormally high track geometry change under repeated train loading as well as a low track modulus.

Changes in track geometry under repeated traffic load usually result from either cumulative deformation within the ballast/subballast layers above the subgrade or from subgrade movement. Proper field observations and reference to maintenance records at specific locations which require continual maintenance can distinguish which of these factors is the cause. If frequent need for surfacing track by tamping is subgrade related, then a low bearing capacity could also be indicated. Thus maintenance experience can give clues to subgrade conditions, especially on mainline track. This relationship can be explored by means of field in situ tests performed on the subgrade on site and examination of maintenance records.

The purpose of this chapter is to describe in situ tests for assessing track performance. Excluded are tests which require substantial material excavation such as those for CBR and plate bearing resistance. The presentation is based on Ref. 4.1.

4.1. Advantages and Limitations of In Situ Tests

Determination of soil stiffness and strength properties for track subgrade evaluation may be done by one of two general approaches: 1) obtaining good quality field samples and performing laboratory tests, or 2) conducting in situ tests. Laboratory tests have several limitations. First, they are expensive and many are needed to represent soil variability with position along the track and with depth. Second, the results may be inaccurate because of sample disturbance, inadequate representation of in situ stress conditions, or failure to duplicate soil in situ physical state (moisture content and unit weight) which vary over time. Even without the limitations of laboratory tests, in situ tests are needed to identify field variability and define subsurface strata before laboratory samples can be selected. In situ tests have the ability to provide estimates of soil properties rapidly and relatively inexpensively for each test location. At the same time they often allow the engineer to examine a larger portion of the subsurface than would normally be possible. Table 4.1 summarizes the main advantages of using in situ tests in evaluating track subgrade conditions.

Table 4.1 Main advantages of in situ tests for evaluating track subgrade conditions

*Test may be conducted in soils which are difficult to sample.
*Soil properties which may not be investigated easily by laboratory tests may be determined.
*A larger total volume of soil supporting the track system may be tested.
*Difficulties in handling soil samples are avoided.
*Possibility of obtaining near continuous vertical record of subgrade properties.
*Relative ease of deployment at different locations.
*Avoiding problems with handling contaminated samples.
*Significant reduction in time of investigation.
*In situ tests may be able to assess influence of macrofabric on soil behavior.
*Potential for large cost savings.

A limiting characteristic of most in situ tests is that, in general, they do not measure real soil properties; instead they provide some intermediate parameter, such as stress or torque, which is a measured response of soil behavior. This measurement is then used to generate a desired soil property via an empirical, semi-empirical or theoretical transformation. Usually, simplifying assumptions are associated with the transformation and therefore the accuracy of individual test results may be directly related to the assumptions. Therefore, in situ tests are not without limitations and these limitations should be remembered when evaluating the results obtained by different tests. The main limitations of in situ tests in evaluating track subgrade conditions are presented in Table 4.2.

Table 4.2 Main limitations of in situ tests in evaluating track subgrade conditions

*Boundary conditions may be poorly defined.
*Drainage conditions are generally unknown.
*Soil disturbance is usually unknown.
*Mode of deformation or failure may be different than full-scale track system.
*Strain rates are usually higher than laboratory tests or track loading.
*Nature of soil being tested is usually unknown.
*Effects of environmental changes are difficult to assess.
*Repeated loading behavior is difficult to assess.

4.2. Evaluation Approach Using In Situ Tests

There are essentially two separate design approaches that can be taken when using in situ tests in evaluating the behavior of track subgrades: 1) indirect design, and 2) direct design.

The indirect design approach relies on the interpretation of in situ tests to obtain conventional design parameters of soils and subsequent application of those parameters in a more-or-less traditional design methodology. An example would be to use the Field Vane Test to estimate the undrained shear strength of a clay which would then be used in an appropriate bearing capacity equation to predict the undrained bearing capacity of a track section. This approach is one which most engineers would be more comfortable with since it uses procedures with which they are most familiar.

The direct design approach gives the opportunity to pass directly from the in situ measurements to the performance of the subgrade without the need to evaluate intermediate soil parameters. An example of this class of test would be to use the modulus value obtained from a Dilatometer or Cone Penetrometer to predict the load/deflection characteristics of a track directly. Most likely, this approach would involve the use of an empirical correlation obtained from a number of full-scale track modulus measurements.

Soil parameters such as undrained shear strength and soil modulus are not unique soil properties, but are dependent on test conditions such as strain rate, stress path to failure, etc. Therefore, one should not expect every test to provide the same value of a particular parameter. What is of importance is to assess to within what accuracy each test may be used to predict performance. In many cases the accuracy of the prediction is limited to the quality of the assumptions made.

It should be kept in mind that in terms of predicting track subgrade performance, there are essentially only two conditions to be evaluated: 1) subgrade strength (for load bearing capacity), and 2) subgrade stiffness (for deformation analysis). However, in addition to these two soil parameters, it is equally important to delineate subsurface layering and be able to identify changes in stratigraphy that affect the overall performance of the surface track system.

4.3. Description of In Situ Tests

A large number of in situ tests are available to the geotechnical profession for use on a variety of problems and materials. The more common and readily available in situ tests to determine parameters for predicting subgrade performance will be presented in the following sections. The general applicability of these tests in different soil conditions is indicated in Table 4.3. Sufficient references are given for the interested reader to seek more detailed information.

Table 4.3 Application of in situ tests in different soils

Soil Type	rameters		
	Undrained Shear Strength Su	Friction Angle φ'	Modulus E or G
Very soft to soft clay	FVT CPTU CPT DMT SPT	BST	DMT
Medium clay	FVT CPTU CPT DMT SPT PLT	BST	PMT DMT
Stiff to very stiff clay	CPT DMT PMT CPTU SPT PLT	BST	PMT DMT
Loose sand	N/A	CPT SPT DMT BST	DMT
Dense sand	N/A	CPT BST SPT DMT	PMT
Gravelly sand	N/A	SPT CPT DMT	PMT

SPT - Standard Penetration Test
CPT - Cone Penetration Test
CPTU - Piezocone
FVT - Field Vane Test
PMT - Pressuremeter Test
BST - Borehole Shear Test
DMT - Dilatometer Test
PLT - Plate Load Test

4.3.1. Field Vane Shear Test

The field vane shear test (FVT) has been the most traditional test used by geotechnical engineers to assess the undrained shear strength of saturated cohesive soils (clays). The concept of the test is to introduce a multibladed vane into the soil, provide a rotational torque to the vane, and initiate a shear failure in the soil around the perimeter and ends of the soil cylinder (Fig. 4.2). The concept is straightforward and the equipment used to conduct the test is relatively simple.

Determination of the undrained shear strength is made by assuming the soil fails as a right circular cylinder and that the strength is isotropic such that the undrained strength, s_u ,

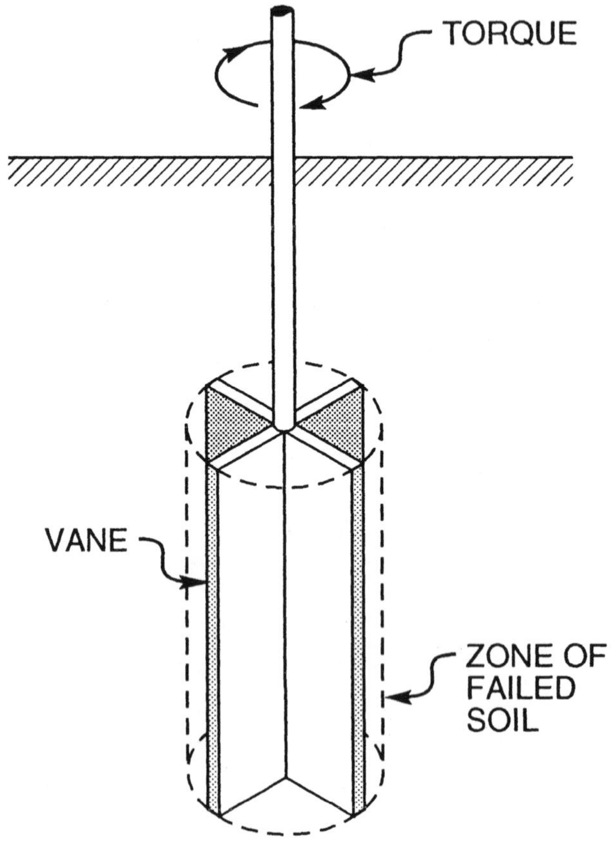

Fig. 4.2 Field vane

may be obtained from the measured torque. For a standard vane with a blade height-to-diameter ratio of 2, the strength is given as:

$$s_u = \frac{6M}{7\rho D^3}, \tag{4.1}$$

where: M = measured torque,

D = diameter of the vane, and

ρ = factor which accounts for shear stress distribution at ends of vane.

Currently available equipment allows for testing either at the bottom of a borehole in stiffer clays or pushing from the ground surface without a borehole in soft soils. Torque measurements may be either manual or automatic, with the torque applied either manually or using a mechanical control. Some form of installation, usually by a hydraulic push, is required to place the vane at the test depth. Tests are normally conducted no closer than about 600 mm (2 ft) in vertical distance. Thus the FVT may provide a relatively small data base over the zone of load influence (about 2 track widths) unless multiple test locations with alternating test depths are used. Deployment through the ballast and subballast would require that a temporary casing or borehole be installed which would disturb the track system.

There are a number of factors which influence the results obtained by the FVT, some of which relate to apparatus and some of which relate to test procedures. These include:

1) size of vane (height, diameter, thickness of blades),

2) height/diameter ratio (H/D) of vane,

3) rate of rotation of the vane, and

4) time to initiate rotation after insertion.

The recommended procedure for field vane testing is governed by ASTM D2573 (Ref. 4.2); however, this standard leaves substantial room for deviation which may introduce significant errors. For example, two vanes of the same H/D may not always give the same strength. Similarly, the influence of rotation rate on the strength measured with the FVT varies with soil type. Regardless of the source of potential errors, systematic errors can be reduced by using identical procedures and equipment at all sites.

In the past, the use of field vane test results to predict performance of shallow foundations such as a track system simulates has suggested a correction to measured field vane data to match observed field performance (Ref. 4.3). A number of recent studies have shown that such corrections may not be warranted in all cases and individual sites need to be assessed separately. This is particularly true in cases where the track subgrade is composed of stiff clays or fissured clays.

4.3.2. Cone Penetration Test

The cone penetration test (CPT) allows rapid site assessment by measuring the quasi-static thrust (pushing force) required to advance the tip of a 60 deg apex cone with a diameter of 34.5 mm into the ground. The tip force divided by the projected end area gives a measure of cone tip resistance, q_c . At the same time the frictional force acting on a cylindrical portion of the instrument behind the tip is measured and is denoted as the sleeve friction, f_s . Mechanical cones use hydraulic pressure to measure the tip and frictional forces. The electrical cones, which are newer, use strain gaged load cells to measure the forces as in Fig. 4.3.

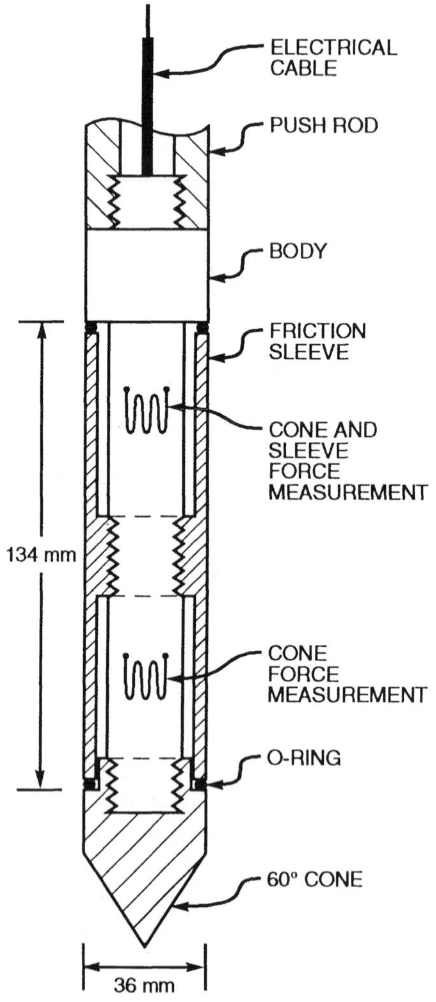

Fig. 4.3 Electric cone penetrometer

This test, which has been widely used in Europe, has gained popularity in the U.S. during the past 10 years and is now considered common place in most parts of the country. Because the rate of advance is rapid, the test is considered to simulate an undrained loading condition. Since the cone diameter is small in comparison to the depth of penetration, the failure of the soil created by the advancing cone models a deep foundation bearing capacity failure. Therefore, a simple theory may be used to estimate the undrained shear strength. Unfortunately, several factors influence the results of cone tests and not all soils respond the same. The equipment and procedure is described in ASTM D3441 (Ref. 4.4) which attempts to reduce substantial variations in the test.

The majority of applications of the CPT in clays makes use of a simple equation to obtain undrained shear strength:

$$s_u = \frac{q_c - \sigma_{vo}}{N_k}, \tag{4.2}$$

where: q_c = cone tip resistance,

σ_{vo} = in situ total vertical stress, and

N_k = an empirical factor.

The value of N_k is not a constant but appears to be influenced by plasticity, stress history, and stiffness, and may reflect the influence of lateral stress on the cone resistance, which is normally neglected. In practice the value of N_k ranges from about 10 to 40 with a

Table 4.4 The coefficient for estimating constrained modulus, D, using CPT data

$$D = \frac{1}{m_v} = (\alpha)\, q_c$$

Range of q_c	Range of α	Corresponding Soil Type
< 7 bars 7 - 20 bars > 20 bars	3 - 8 2 - 5 1 - 2.5	Clay of Low Plasticity (CL)
> 20 bars < 20 bars	3 - 6 1 - 3	Silts of Low Plasticity (ML)
< 20 bars	2 - 6	Highly Plastic Silts and Clays (MH, CH)
< 12 bars	2 - 8	Organic Silts (OL)
< 7 bars, and 50 < w < 100 100 < w < 200 w > 200	 1.5 - 4 1 - 1.5 0.4 - 1	Peat and Organic Clay (P_t, OH)

q_c = cone tip resistance, w = water content

mean value of about 16. The reliability of Eq. 4.2 obviously reflects the quality of the reference value of s_u which is normally taken from field vane tests in softer clays and either plate loading or reference laboratory tests in stiffer clays.

The deformation characteristics of the soil may be obtained through empirical relationships suggested by Mitchell and Gardner (Ref. 4.5) relating q_c to soil modulus as given in Table 4.4.

The CPT may represent the most versatile of available in situ tests in that a borehole is usually not necessary. Thus, deployment of a cone may be easily adapted to contract drillers or may be developed from a special stand-alone vehicle for railroad track applications.

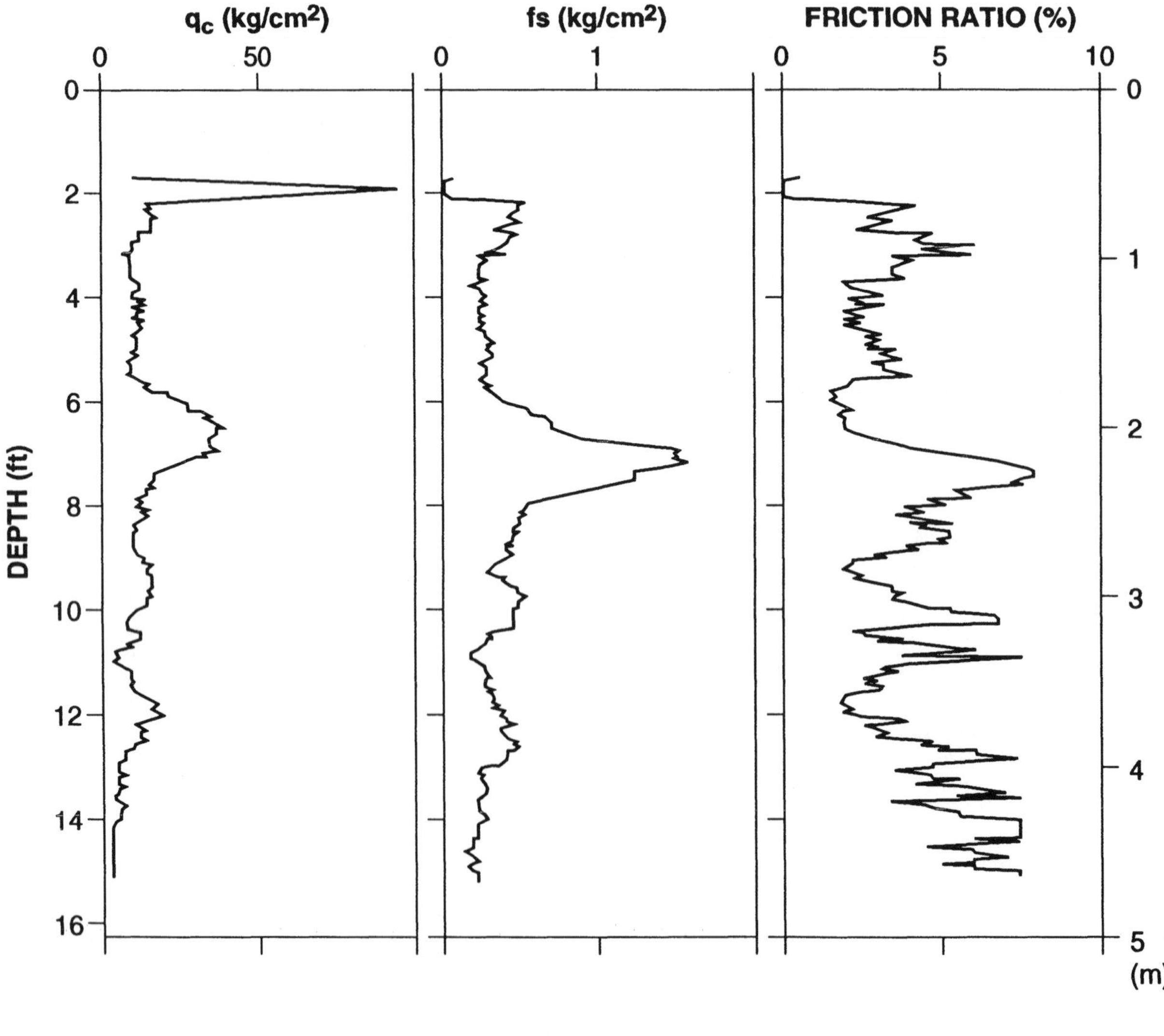

Fig. 4.4 Typical cone penetration test profile

The use of an electric cone also allows rapid data acquisition and evaluation by a portable lap-top personal computer. Therefore variations in site subsurface characteristics may be determined in relatively close vertical spacing, usually every 6 mm (0.25 in.). A typical CPT profile of measured tip resistance (q_c) and sleeve friction (f_s) versus depth are shown in Fig. 4.4 together with a calculated friction ratio (f_s/q_c). This amount of test data may be useful for a statistical assessment of site conditions and may form the basis for a probabilistic design.

The results of the CPT may be used to provide a basis for soil classification by combining the tip and sleeve resistance, for example as shown in Fig. 4.5 (Ref. 4.20). Several schemes have been presented in the literature to identify soil type using the CPT.

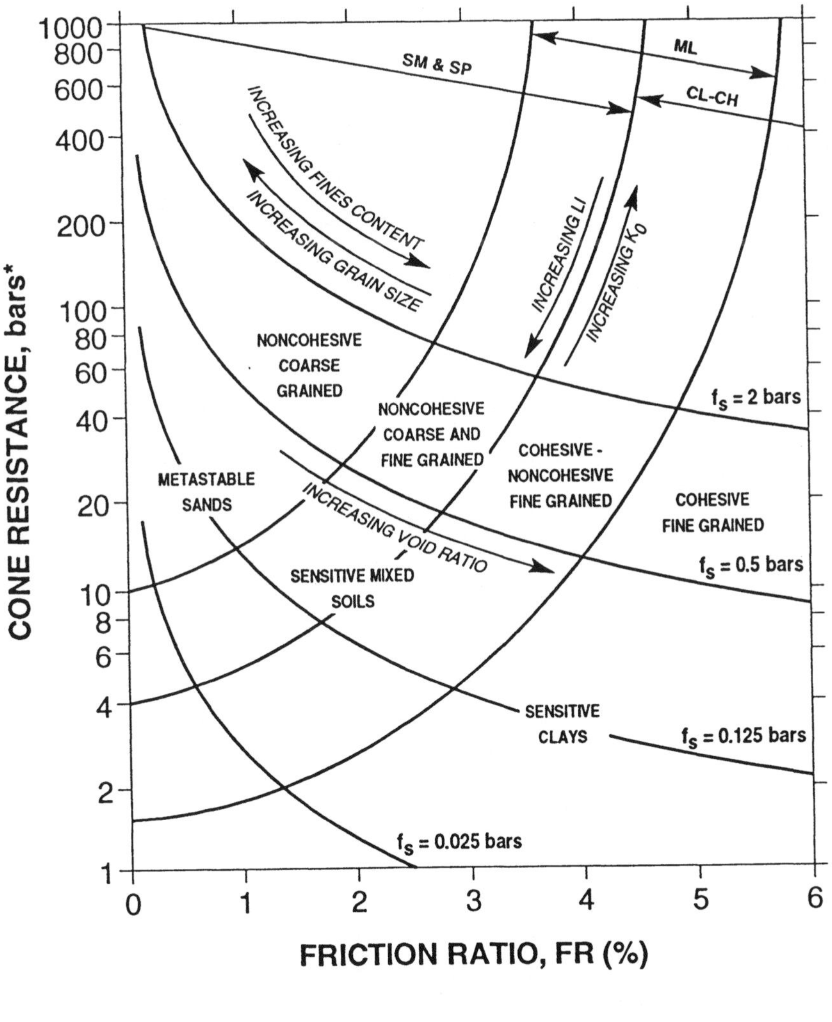

*1 bar = 100 kPa = 1 kg/cm^2 = 14.7 psi

Fig. 4.5 Soil classification chart for CPT (reprinted from Ref. 4.20 with permission of the publisher)

4.3.3. Piezocone

The addition of a porous element and pressure transducer to measure pore water pressure during cone penetration results in a piezocone (CPTU). This device has greatly enhanced the capabilities of the CPT to identify soil stratigraphy and estimate geotechnical properties. While the CPTU is a relatively new addition to in situ testing, the mechanics of test interpretation have advanced rapidly such that there is now an element of familiarity among users.

Since the test only constitutes a design change to the cone, as shown in Fig. 4.6, the installation procedures are nearly identical to those of the CPT with one exception -- the test is intended for saturated soil conditions only and therefore a borehole extending down to the water table and backfilled with water or other drilling fluid is usually necessary. This is essential if the porous element is to remain saturated prior to penetration. Cavitation of the system, allowing air to enter, will give erroneous results.

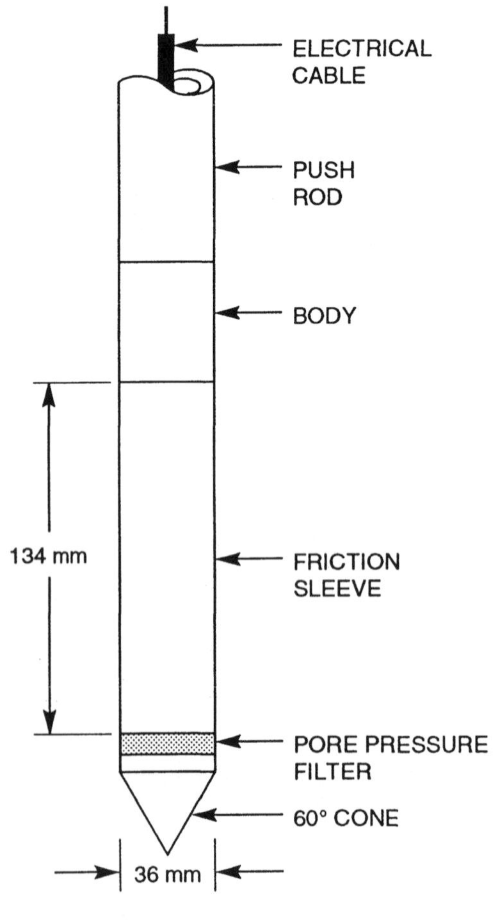

Fig. 4.6 Piezocone

The increased capabilities of the CPTU over the CPT result from the sensitivity of the pore water pressure response of different soils to the penetration of the cone. There is a decidedly different response depending on the location of the porous element, e.g., cone tip, cone face, cone base, or cone shaft. The following comments are for pore pressure measurements made at the cone base.

In normally consolidated soils, high positive pore water pressures are generated by the advance of the CPTU, the magnitude of which is influenced by soil strength, stiffness, and plasticity. The use of simple cylindrical cavity expansion theory to predict the magnitude of undrained shear strength is generally given as:

$$s_u = \frac{u - u_o}{N_u} = \frac{\Delta u}{N_u}, \tag{4.3}$$

where: u = measured pore pressure,

u_o = in situ pore pressure,

N_u = an empirical factor.

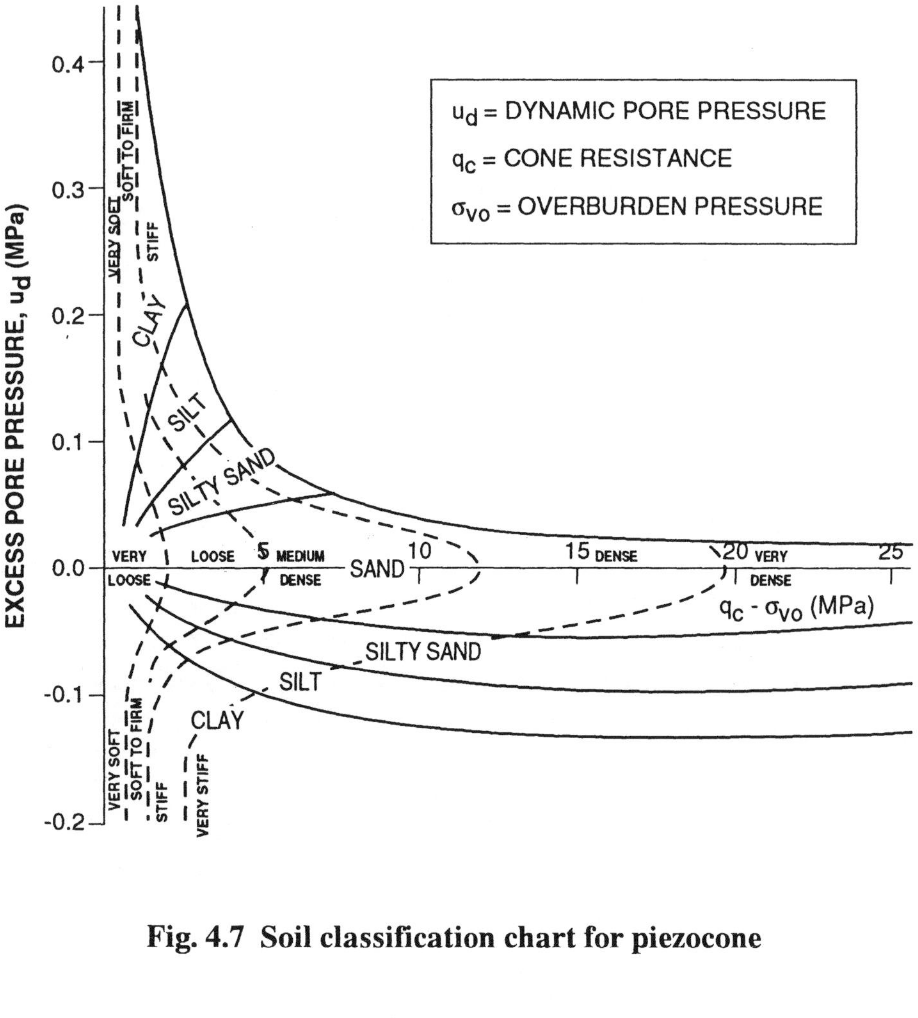

Fig. 4.7 Soil classification chart for piezocone

As with N_k, the value of N_u varies somewhat with soil plasticity, but has a much narrower range than N_k. Typically N_u values range from 2 to 8.

In overconsolidated soils, low or even negative pore water pressures may be measured during CPTU profiling. Estimates of undrained strength may be obtained using normalized soil parameters, i.e., by reference of the measured response to the in situ overburden stress and the normally consolidated value (Ref. 4.6). This is a fairly new approach, but is well founded in the SHANSEP and CAMCLAY models and uses the cone to estimate the shear strength from both the measured pore pressure and tip resistance.

The CPTU offers a redundancy in the estimate of the undrained shear strength which is attractive: s_u from q_c and s_u from Δu. As with the CPT, the CPTU also lends itself to automatic data acquisition which again means that a statistical data base may be developed.

The combination of pore water pressure and cone tip resistance measurements may be used to provide a much more refined technique for soil identification, for example as shown in Fig. 4.7 (Ref. 4.7). While such schemes are subject to revision and modification, e.g., by incorporating the influence of horizontal as opposed to vertical stress, they appear to be much more accurate than those based on just cone sleeve and tip resistance.

4.3.4. Flat Dilatometer

The flat dilatometer (DMT) was introduced by Marchetti (Ref. 4.8) as a penetration tool for obtaining rapid design properties of soil in situ. The test consists of a sharpened flat plate which measures approximately 200 x 95 x14 mm (3.85 x 7.75 x 0.55 in.) and which is equipped with a 60 mm (2.4 in.) diameter flexible diaphragm mounted on one face of the blade (Fig. 4.8). The diaphragm is inflated by nitrogen gas pressure once penetration has reached the desired test depth. In many soils, penetration may begin at the ground surface, and tests are generally conducted at 0.8 to 1.0 ft intervals. Therefore, like the quasi-static cone penetrometers (CPT and CPTU) in many cases a borehole is not required, particularly for shallow ($<$ 7 m or 20 ft) investigations. Pushing thrust should be quasi-static and not by dynamic impact; therefore the blade can usually be advanced from conventional drilling rigs.

In operation, two pressure inflation readings are normally obtained after penetration: 1) the pressure required to lift the membrane off the face of the blade, designated P_o, and 2) the pressure required to force the center of the flexible diaphragm a distance of one millimeter against the soil, designated as P_1. The test takes only about two minutes to complete. These pressure readings are then used along with estimates of the soil unit weight and groundwater location to estimate general soil type and a number of conventional soil parameters (Refs. 4.8 and 4.9). In saturated cohesive soils, the DMT represents an undrained expansion test, while in freely draining materials such as sands and silts, the results generally represent drained conditions (Ref. 4.10). The exact test procedure is described by ASTM subcommittee D18.02 (Ref. 4.19).

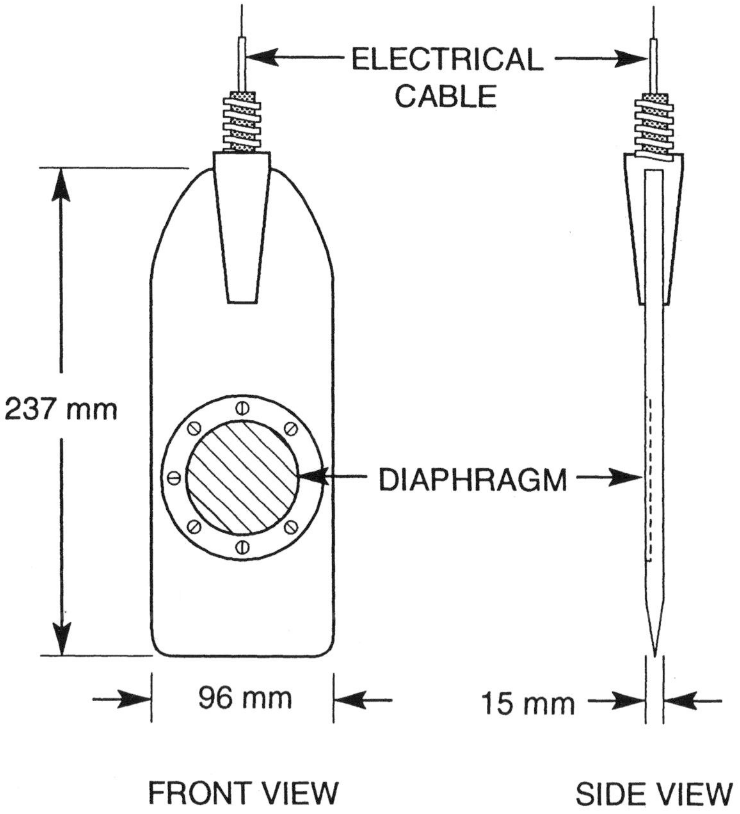

Fig. 4.8 Flat dilatometer

The initial pressure reading, P_0 is a measure of the total stress acting on the face of the blade immediately after penetration and therefore includes the effects of installation pore pressures. This is similar to what occurs in the CPTU and therefore P_0 may be used to obtain an estimate of the undrained shear strength. Since tests are conducted at relatively close vertical spacings considerable data are obtained for analysis.

One of the problems with penetration type tests is that a large amount of unknown soil disturbance occurs as a result of installation. This is true of the FVT, CPT, CPTU, and DMT. The DMT, unlike the other tests described so far, tries to "see through" this disturbed zone with the second, P_1, pressure reading. In this way, a direct measure is taken of the "elastic" response of the soil or response to continued loading after penetration. This is distinctly different than other penetration tests and may be useful for estimating load/deformation behavior. For example, the modulus obtained using this pressure reading has been used to evaluate pavement subgrade response (Ref. 4.11).

The DMT is the newest of the in situ tests presented, but has gained popularity at a very fast rate such that there is now a wide range of experience available in a wide range of materials. The equipment has proven to be very robust and easy to operate. Repair is also easy since there are only minimal moving components.

4.3.5. Full-Displacement Pressuremeter Test

The full-displacement pressuremeter test (FDPMT) is a total stress expansion test which determines the soil response to expansion of an approximately cylindrical cavity. The volume increase of a cylindrical probe is measured as expansion progresses so that the results may be presented as a volume-pressure or cavity strain-pressure diagram (Fig. 4.9). An expandable probe is pushed into the ground using the quasistatic thrust from a drilling rig. After penetration, the rubber membrane of the probe is expanded with a simple screw pump/console using a fluid such as water to produce a cavity strain of about 30%. The entire test apparatus, which was introduced by Briaud and Shields (Ref. 4.12), is shown in Fig. 4.10 and consists of the probe and a simple control console. The rubber membrane of the probe is covered with a metallic sheath to protect the membrane during penetration.

As with the majority of the other penetration tests described, the FDPMT usually doesn't require a borehole and penetration can usually start from the ground surface. The probe is slightly smaller in diameter than a CPT (about 28 mm) and in some cases the probe tip has been equipped with a load cell so that both a CPT and FDPMT are incorporated into the same device.

The utility of a FDPMT over other in situ tests is to allow determination of the load-deformation response of soils under conditions which permit comparison with theoretical analysis for cylindrical cavity expansion. Thus, the results may be analyzed using a well-formulated solution. Accordingly, the elastic response (shear modulus) and plastic response (undrained shear strength) may both be obtained from the same test.

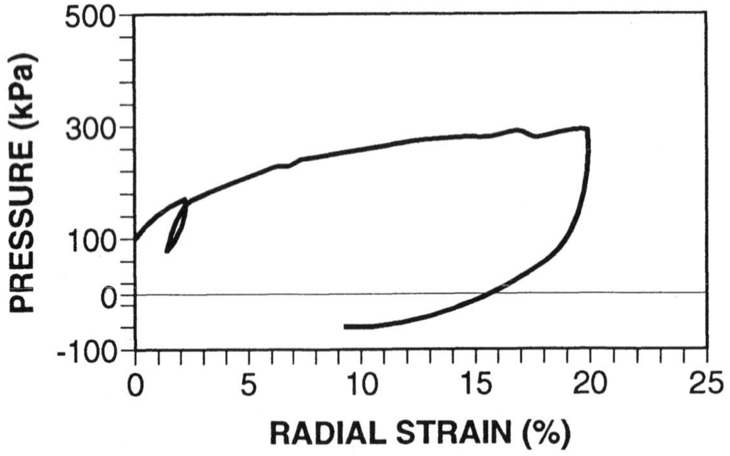

Fig. 4.9 Relationship between cavity strain and cavity pressure obtained from a full-displacement pressuremeter

The drawbacks are that data reduction is time consuming and usually not performed in the field, and tests are usually not performed any closer vertical spacing than 600 mm (2 ft). Again, there appears to be a trade-off, as with all other tests.

4.3.6. Standard Penetration Test

The standard penetration test (SPT) has had a longstanding place in North American geotechnical engineering, primarily in evaluating the in-place density of cohesionless soils, i.e., sands. In the U.S. the test has become so common place that it would be difficult to find a drilling contractor who does not have the equipment readily available on every drill rig.

The SPT is a dynamic penetration test which uses a split barrel sampler to retrieve a soil sample after the barrel is driven into the ground using a fall hammer (Fig. 4.11). The

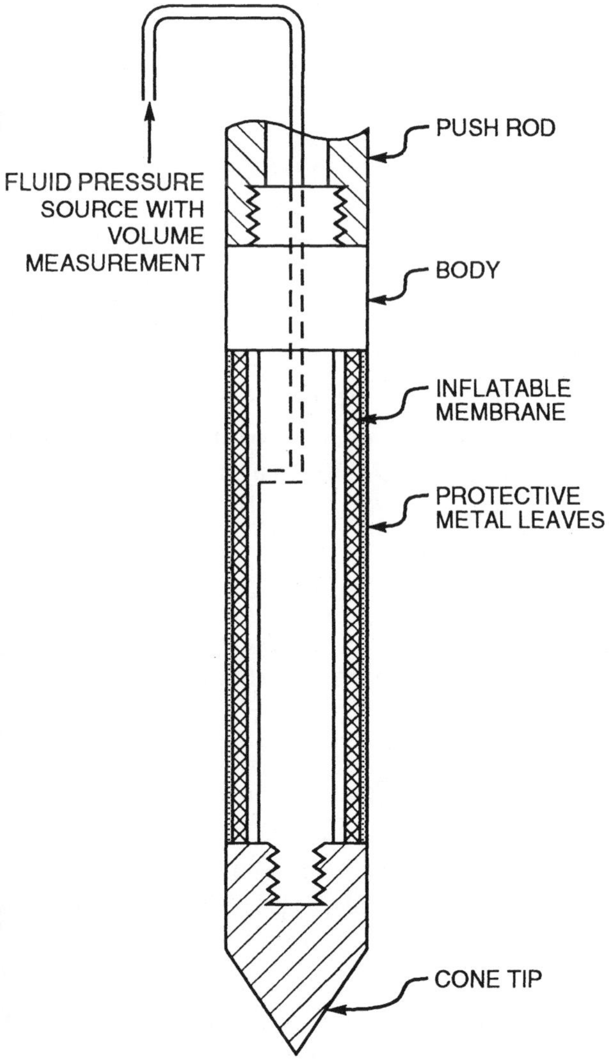

Fig. 4.10 Full-displacement pressuremeter

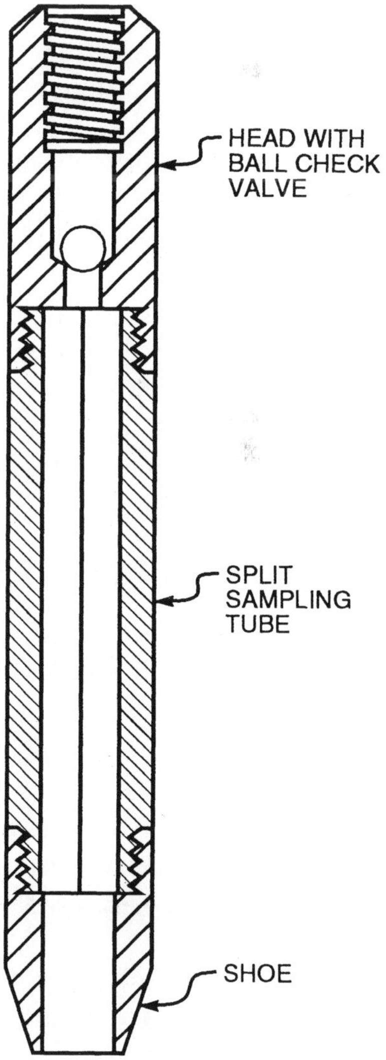

Fig. 4.11 Standard penetration test sampler

test procedure is outlined in some detail in ASTM D1586 (Ref. 4.13); however in practice, there appears to be substantial deviation from the recommended procedures. Because of variations in equipment and test procedure, the results often show considerable scatter, and it is not uncommon for SPT results to vary as much as 100% for the same soil conditions.

In operation, the barrel is driven at the bottom of a drilled hole and the number of hammer drops or blows required to cause penetration is recorded for three consecutive 152-mm (6-in.) increments of penetration. While the full penetration is thus 457 mm (18 in.), the number of hammer blows for the last two 152-mm (6-in.) increments, i.e, from 152 to 457 mm (6 in. to 18 in.), are added together to give the SPT blow count in blows per foot, designated as N.

In saturated cohesive soils the test is conducted rapidly such that undrained conditions are assumed to exist. Therefore, an estimate of the undrained strength can be made from the blow count. Because of the multitude of external (test equipment and procedure) and internal (soil stiffness, soil strength, in situ stress conditions, etc.) variables that may influence the recorded value of N, SPT gives only very approximate values for design parameters in clays. Correlations between N and s_u are often site/soil specific and are sometimes difficult to extrapolate. Still, SPT results may be useful in a qualitative sense for distinguishing soft versus hard zones. Additionally, it is the only in situ penetration test which provides a soil sample for direct classification.

On the other hand, the SPT does require a borehole and the results represent an average soil response to penetration over a depth of 457 mm (18 in.). The number of tests which can be conducted over the depth of interest for most track applications is thus limited. Overall, the SPT in its current form should be viewed as a low technology form of site investigation which perhaps results in a conservative approach to site evaluation because of all the uncertainties involved with the testing.

4.3.7. Dynamic Cone Penetration Test

A simple and rapid site exploration technique which has been used in the past to inexpensively evaluate shallow depth soil conditions is the dynamic cone penetration test (DCT). This test is a variation of the SPT in that the split barrel sampler is replaced with a simple 60 deg apex steel cone attached to the end of the drill rods. The same driving mechanism is used as with the SPT, although a smaller hand operated tool may also be used. In this case, the number of hammer blows to drive the cone each 152 mm (6 in.) is recorded. The results may be reported as either incremental or cumulative blows per 152 mm (6 in.), as shown in Fig. 4.12.

One problem with this test in cohesive soils is that there is a tendency to accumulate frictional resistance or adhesion along the drill rods, thus reducing the ability to distinguish distinct soil layering. This appears to be less of a problem however at shallower depths. Estimates of design parameters may be difficult and are generally highly empirical. Thus, there may be a high degree of uncertainty associated with applying the results.

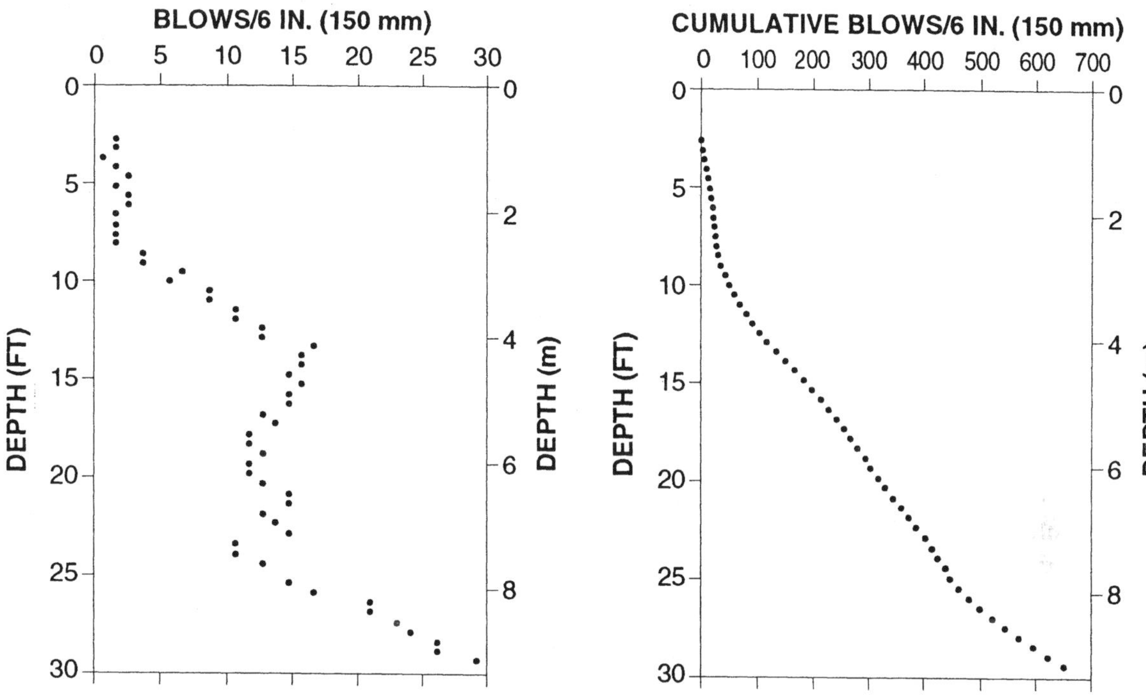

Fig. 4.12 Results from a dynamic cone penetration test

A hand-operated dynamic cone penetrometer (DCP) is described by Ayers and Thompson (Ref. 4.14). It consists of a 78-N (17.6-lb) sliding cylindrical weight, a fixed 1200-mm (48-in.) long calibrated penetration shaft, and a 60 degree drive cone tip (Fig. 4.13). The hand-operated DCT test is conducted by releasing the sliding weight from the top of the shaft and recording the penetration of the calibrated shaft, i.e., the penetration produced by one fall of the weight. The sequence is repeated until the desired depth is reached up to a maximum depth of approximately 1200 mm (48 in.). This hand-operated device is useful for defining approximate depths to ballast and subballast layer boundaries. However only testing of the top portion of the subgrade is feasible.

Advantages of the DCP include:

1) Minimal equipment and personnel required (2 operators).

2) Rapid (approximately 5 minutes per test depending on the material being tested).

3) Can be used for a broad range of material types.

4) Can be used to characterize subsurface layers (i.e. differentiate layers).

5) Correlated with CBR (Refs. 4.15 and 4.16), unconfined compressive strength (Ref. 4.15) , and shear strength (Ref. 4.17).

The primary disadvantage of the DCP is the high degree of test data variability associated with large open-graded granular materials. The variability is attributable to the maximum aggregate size, the amount and nature of fines present, and the relative density (Ref. 4.17).

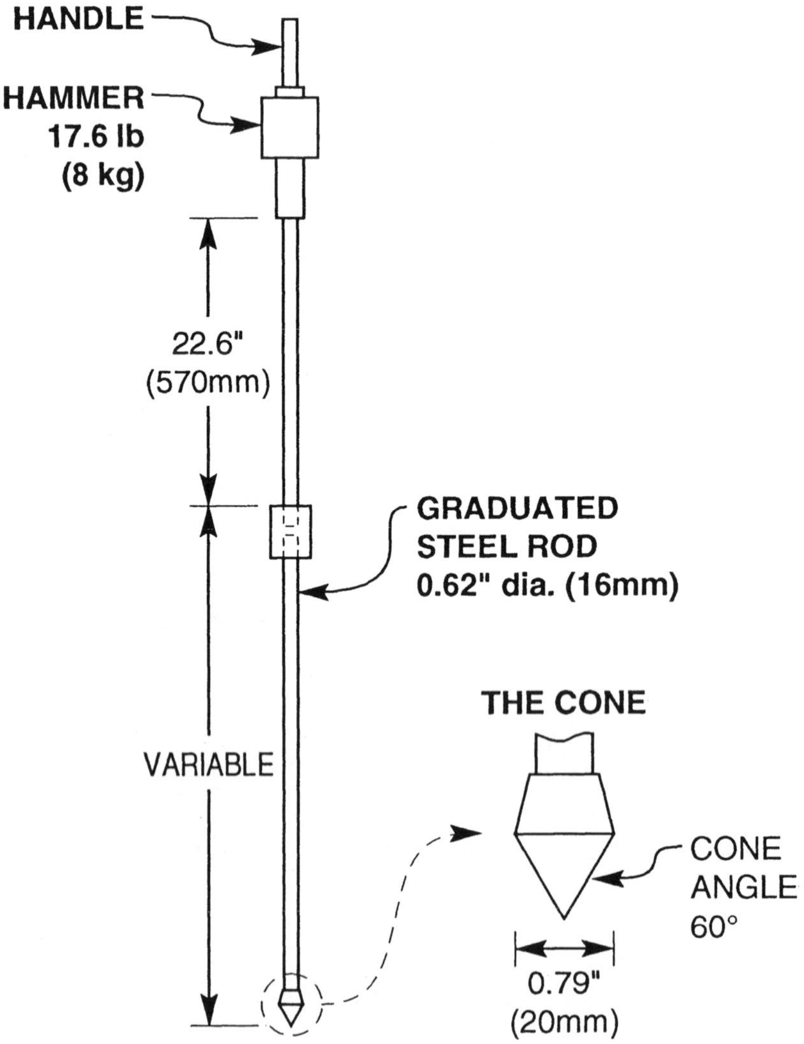

Fig. 4.13 Dynamic cone penetrometer

4.4. Site Investigation Strategy

The selection of in situ tests and the development of a site investigation plan is often a difficult task in view of the large number of tests available and the specific design requirements and soil types for the individual project. However, there are a number of factors which should be considered in selecting a test or tests to be used. These include both technical and nontechnical issues and invariably must also include an economic consideration for most projects. Among the factors which should be considered are:

1) Soil Type

Not all in situ tests are ideally suited to all field conditions. Table 4.5 presents a summary of the relative suitability of the tests described in this report to determine strength and deformation characteristics of cohesive soils.

Table 4.5 Suitability of in situ tests for determining undrained shear strength and load deformation performance in cohesive soils

	Soil Type					
Test	Very Soft Clay	Soft Clay	Medium Clay	Stiff Clay	Silty Clay	Silt
FVT	A	A	B	C	C	D
CPT	A	A	A	B	B	C
CPTU	A	A	A	B	B	C
DMT	A	A	A	A	A	B
PMT	B	B	A	A	B	C
SPT	D	C	B	A	A	A
DCT	A	A	A	B	B	B

A - High suitability C - Limited use
B - Moderate suitability D - Not applicable

2) Availability of Equipment

Many engineering firms may not have immediate access to specific test equipment which means that an outside specialty contractor or consultant may be required.

3) Time

Some tests require more time to complete than others and also require peripheral equipment such as borehole drilling capabilities as opposed to simple pushing thrust.

4) Difficulty of Test

A number of tests are relatively simple to conduct and require little operator instruction prior to obtaining meaningful results, e.g. CPT, VST, DMT. These tests are essentially operator independent. Other tests are more mechanically complex and require more skilled technicians, e.g. FDPMT.

5) Amount of Data Obtained

Depending on the logistics of a project, the amount of data obtained within the overall design depth may become critical. This generally depends on the test interval. For example, within a typical subgrade profile relatively few FDPMT's may be obtained, whereas a CPT may give a much larger data base, which in turn can provide the basis for a probabilistic analysis.

Additionally, tests which provide information relative to site stratigraphy, or which may provide a number of pertinent soil parameters may have some decided advantages, e.g., CPT over FDPMT. Again, this will depend on site and project specifics.

6) Data Interpretation

Quite often tests may require expert and experienced individuals to provide interpretation of the data obtained. This may explain a reluctance to use a certain in situ test, the attitude being that it's too difficult to interpret the results.

7) Test Variability

A problem with all soil tests is that they have a certain amount of intrinsic variability such that the properties measured are probably not true soil properties. The variability of the measured property is a combination of both soil variability and test variability. While nothing can be done to eliminate soil variability apart from recognizing separate soil strata, test variability may be reduced by using standardized equipment and procedures, and frequently checking calibrations.

Handy (Ref. 4.18) has suggested that if soil variability is high, i.e. where subsurface stratigraphy is complex, a less expensive test should be selected so that a large number of tests can be conducted. At sites where the stratigraphy is simple, consisting of one or two layers, and the soil variability is expected to be low, a more expensive test may be used since fewer tests will be required. However, this doesn't necessarily mean that more expensive tests provide better data. In fact, several relatively inexpensive tests provide very good results.

REFERENCES

4.1 Selig, E. T. and Lutenegger, A. J. (1991). *Assessing railroad track subgrade performance using in situ tests.* Geotechnical Report No. AAR91-369F, Department of Civil Engineering, University of Massachusetts, Amherst, Massachusetts, September.

4.2 ASTM D2573. **Standard test method for field vane shear test in cohesive soil.** *Annual Book of ASTM Standards*, Section 4, Construction, Vol. 04.08, Soil and Rock.

4.3 Bjerrum, L. (1972). **Embankments on Soft Ground**. *Proc. Conf. on Earth and Earth-Supported Structures,* ASCE, Vol. 2, pp. 1-54.

4.4 ASTM D3441. **Standard test method for deep quasi-static, cone and friction-cone penetration tests of soil.** *Annual Book of ASTM Standards*, Section 4, Construction, Vol. 04.08, Soil and Rock.

4.5 Mitchell, J. K. and Gardner, W. S. (1975). **In situ measurement of volume change characteristics.** State-of-the-Art Report, *Proceedings of the Conference on In-Situ Measurement of Soil Properties,* Specialty Conference of the Geotechnical Division, North Carolina State University, Raleigh, Vol. II.

4.6 Lutenegger, A. J. and Kabir, M. G. (1988). **Interpretation of piezocone results in overconsolidated clays.** *Proceedings of the Symposium on Penetration Testing in the United Kingdom,* pp. 43-46.

4.7 Jones, G. A. and Rust, E. (1983). **Piezometer probe (CPTU) for subsoil identification.** *International Symposium in In Situ Testing*, Vol. 2, Paris.

4.8 Marchetti, S. (1980). **In situ tests by flat dilatometer**. *Journal of the Geotechnical Engineering Division, ASCE,* Vol. 106, pp. 299-321.

4.9 Lutenegger, A. J. (1988). **Current status of the Marchetti dilatometer test.** *Proceedings of the First International Symposium on Penetration Testing,* Vol. 1, pp. 137-155.

4.10 Lutenegger, A. J. and Kabir, M. G. (1988). **Dilatometer C-Reading to help determine stratigraphy.** *Proceedings of the First International Symposium on Penetration Testing*, Vol. 1, pp. 549-554.

4.11 Borden, R. H., Aziz, C. N., Lowder, W. M. and Khosia, N. P. (1985). **Evaluation of pavement subgrade support characteristics by dilatometer tests.** *Transportation Research Record* No. 1022, pp. 120-127.

4.12 Briaud, J. L. and Shields, D. H. (1979). **A special pressuremeter and pressuremeter test for pavement evaluation and design.** *Geotechnical Testing Journal,* GTJODJ, Vol. 2, pp. 143-151.

4.13 ASTM D1586. **Standard method for penetration test and split-barrel sampling of soils,** *Annual Book of ASTM Standards,* Section 4, Construction, Vol. 04.08, Soil and Rock.

4.14 Ayers, M. E. and Thompson, M. R. (1989). *Evaluation techniques for determining the strength characteristics of ballast and subgrade materials.* Report for Asssociation of American Railroads by University of Illinois, July.

4.15 Harison, J. A. (1983). **Correlation between California bearing ratio and dynamic cone penetrometer strength measurement of soils.** *Proceedings, Institution of Civil Engineers,* Part 2, December.

4.16 Kleyn, E. G. and Savage, P. F. (1982). **The application of the DCP to determine the bearing properties and performance of road pavements.** *Proceedings,* International Symposium on Bearing Capacity of Roads and Airfields, Trondheim, Norway, June.

4.17 Ayers, M. E. and Thompson, M. R. (1988). *Rapid shear strength evaluation of in situ ballast/subballast materials.* Department of Civil Engineering, University of Illinois at Urbana-Champaign, Technical report submitted to USA-CERL, June.

4.18 Handy, R. L. (1980). **Realism in site exploration: Past, present, future and then some all-inclusive.** *Proceedings of the Symposium on Site Exploration in Soft Ground Using In Situ Techniques,* FHWA, pp. 239-248.

4.19 Schmertmann, J. H. (1986). **Suggested method for performing the flat dilatometer test.** *ASTM Geotechnical Testing Journal,* GTJODJ, Vol. 9, pp. 93-101.

4.20 Douglas, Bruce J. and Olsen, Richard S. (1981). **Soil Classification Using Electric Cone Penetrometer.** *Cone Penetration Testing and Experience,* G. M. Norris and R. D. Holtz Eds., ASCE, pp. 209-227.

5. Analytical Track Models

The principal function of the track models considered in this chapter is to interrelate the components of the track superstructure and substructure for properly representing their complex interaction in determining the effect of the traffic loads on the stresses, strains, and deformations in the system. Such modelling provides the basis for predicting track performance, and therefore technical and economical feasibility of track design and maintenance procedures. Accurate analysis is limited by a number of factors including:

1) Uncertainties in the magnitude of loading.

2) Complex ballast properties which change with traffic, maintenance, and environmental conditions.

3) Lack of quantitative information on substructure characteristics.

Furthermore, the track is subjected to loading in three directions: vertical, lateral and longitudinal. However, the available geotechnical models only consider the vertical component. Thus the coupling effects of the three loading directions are not represented.

This chapter will first present the classical beam-on-elastic-foundation model and then describe some more recent computer models which provide a detailed representation of the substructure. These models all represent the actual dynamic wheel load by an equivalent static load. This is a reasonable approximation as long as inertial and wave propagation effects do not substantially alter the effects of the dynamic load.

5.1. Beam on Elastic Foundation

The theoretical formulation of the beam-on-elastic-foundation model is based on the assumption that each rail acts like a continuous beam resting on an elastic support. The track modulus (actually track foundation modulus), u, is defined as the supporting force per unit length of rail per unit vertical deflection of the rail. This track model is shown in Fig. 5.1. The rail foundation, represented by u, includes the effects of the fastener, sleeper, ballast, subballast and subgrade. The differential equation for this model is (Ref. 5.1 and 5.2):

$$EI \frac{d^4y}{dx^4} + u\,y = 0 , \tag{5.1}$$

where E = rail modulus of elasticity,

I = rail moment of inertia,

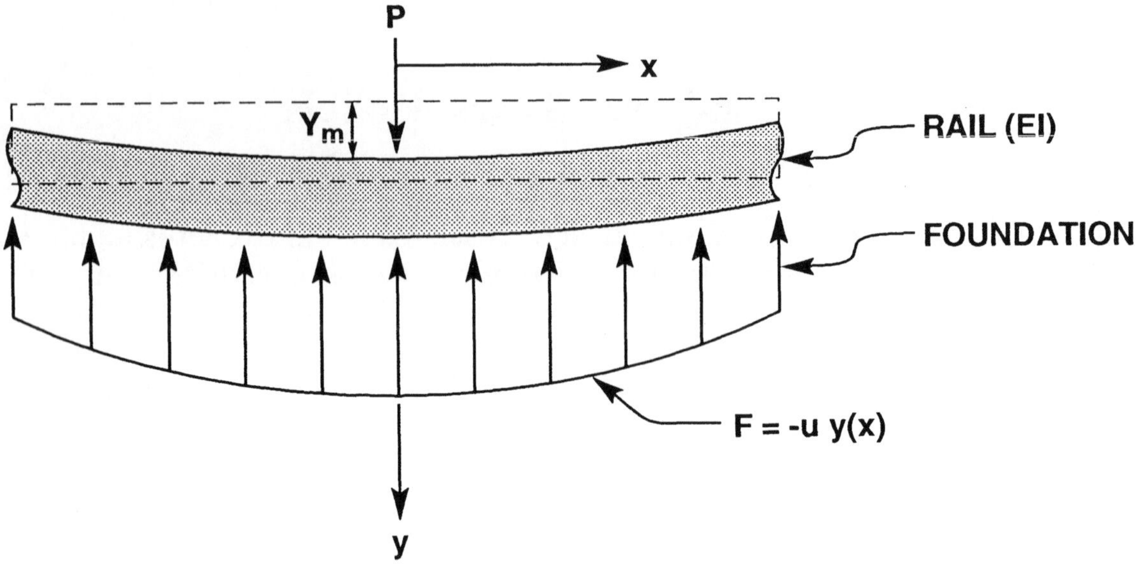

Fig. 5.1 Beam on elastic foundation model

u = modulus of elasticity of the track support or track foundation modulus, and

y = rail deflection.

The solution to Eq. 5.1 for the rail deflection, y(x), at any distance, x, along the rail from the single point load, P, is

$$y(x) = \frac{P\lambda}{2u} e^{-\lambda x}(\cos \lambda x + \sin \lambda x) , \tag{5.2}$$

where

$$\lambda = \left(\frac{u}{4EI}\right)^{\frac{1}{4}} . \tag{5.3}$$

The supporting line force against the rail, F(x), is then obtained from

$$F(x) = - u\ y(x) . \tag{5.4}$$

The successive derivatives of Eq. 5.2 give equations for slope, bending moment, and shear at any distance along the rail from the point load. The slope, $\theta(x)$, is given by

$$\theta(x) = - \frac{P\lambda^2}{u} e^{-\lambda x} (\sin \lambda x) , \tag{5.5}$$

the bending moment, M(x), is given by

$$M(x) = \frac{P}{4\lambda} e^{-\lambda x} (\cos \lambda x - \sin \lambda x) , \tag{5.6}$$

and the shear force, V(x), is given by

$$V(x) = -\frac{P}{2} e^{-\lambda x} (\cos \lambda x) . \tag{5.7}$$

The distributions of y, θ, M and V are illustrated in Fig. 5.2.

The maximum values of deflection, Y_m, moment M_m, and supporting line force, F_m, occur directly beneath P. These are given by

$$Y_m = \frac{P\lambda}{2u} , \tag{5.8}$$

$$M_m = \frac{P}{4\lambda} , \tag{5.9}$$

and

$$F_m = u \; Y_m . \tag{5.10}$$

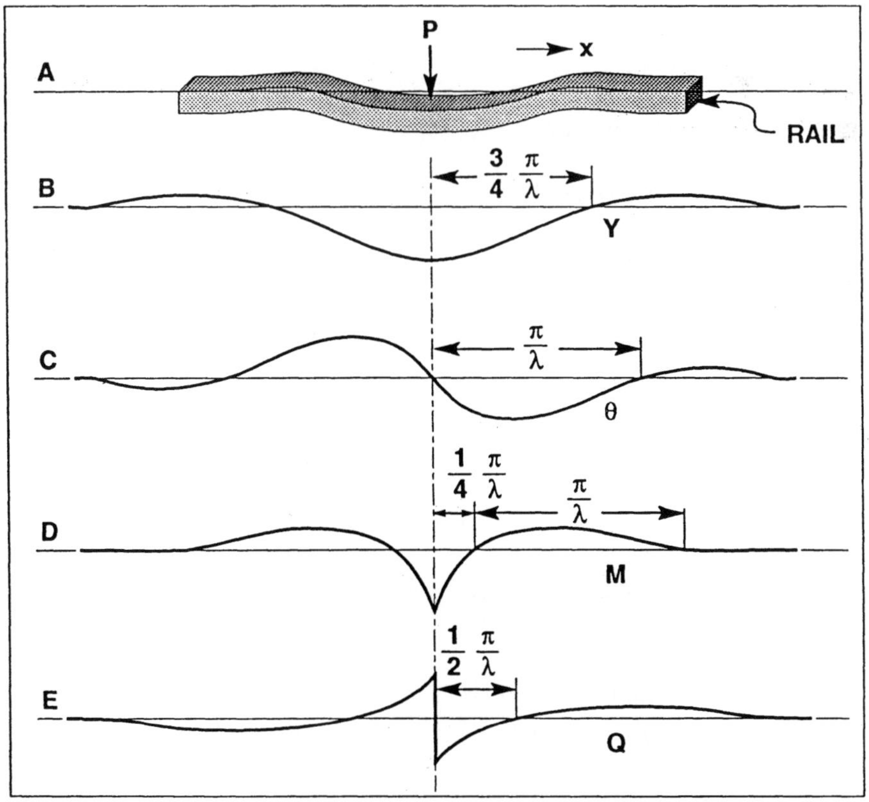

Fig. 5.2 Variation of parameters along rail

An upper bound value of rail seat load, Q_m, can be estimated by

$$Q_m = F_m S , \tag{5.11}$$

where S = sleeper spacing.

The ballast pressure, P_b, on the sleeper bearing area, A_b, can be estimated as

$$P_b = \frac{2 Q_m}{A_b} . \tag{5.12}$$

The vertical stress applied to underlying layers can then be estimated from vertical stress distribution charts.

With this model, dynamic load is considered by increasing the value of P. The deflections and moments in the rail from multiple axles can be obtained by superposition of single axle results.

The track modulus, u, can not be calculated from the properties of the components, i.e. the fastener, sleeper, ballast and underlying layers. Therefore the model can not consider their individual effects. Instead the track modulus must be calculated from field measurements of track deflection under load. There are basically three ways to do this:

1) Single point load test,

2) Deflection basin test, and

3) Multiple axle vehicle load test.

The first method is best. However, it requires apparatus for applying known vertical loads to the top of rail at a single point and measuring the resulting vertical deflection. An example of such an arrangement is shown in Fig. 5.3. The track modulus is then calculated from

$$u = \frac{\left(\frac{P}{y_m}\right)^{\frac{4}{3}}}{(64\,EI)^{\frac{1}{3}}} . \tag{5.13}$$

The second method is based on the fact that for vertical equilibrium of forces with the beam-on-elastic-foundation model the integral of the supporting line force must equal the applied force. Hence,

$$P = \int_{-\infty}^{+\infty} u\, y\, dx . \tag{5.14}$$

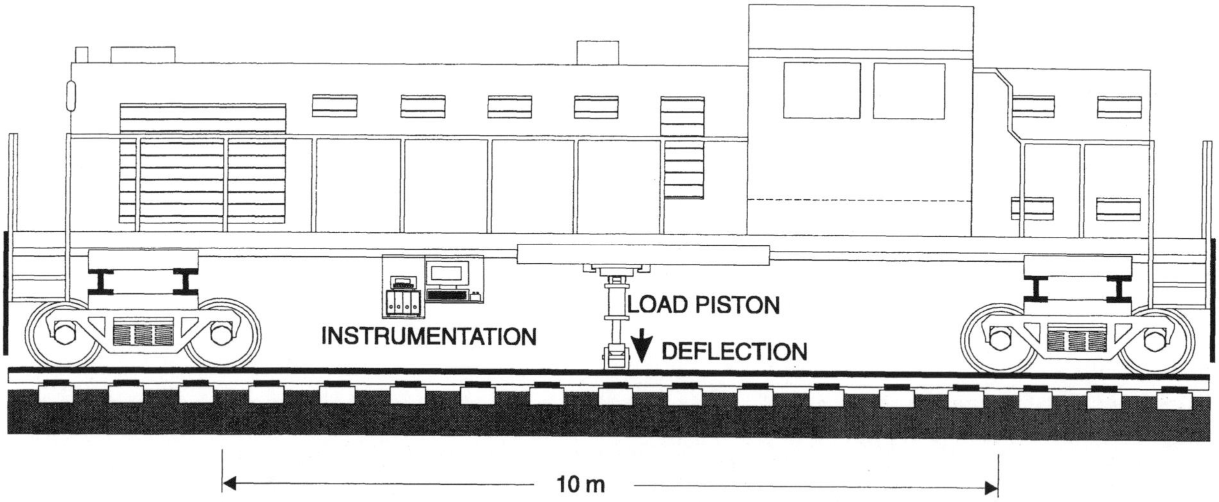

Fig. 5.3 Test arrangement for track modulus

If u is considered constant along the rail, then Eq. 5.14 becomes

$$P = u A_y , \tag{5.15}$$

where A_y is the area of the deflection basin caused by vertical force P. For this method, the deflection basin is measured for a single point load and u calculated from Eq. 5.15. This method was used by Talbot (Ref. 5.3) and further evaluated by Zarembski and Choros (Ref. 5.4). Note that Eq. 5.15 is independent of the rail EI although the shape of the deflection basin will vary with EI. To eliminate the effect of slacks in the track, a light and a heavy vertical force may be used instead of just a single force. Then A_y is the difference in the two deflection basin areas and P is the difference between the heavy and light forces. The main disadvantage of this method is that it requires deflection measurements at many rail positions simultaneously. The advantage is that it tends to average over a length of rail rather than just rely on one point.

The third method applies the load using two or three axle bogies on rail vehicles. This is the most convenient method for applying the load, but the interpretation of the data relies on a potentially involved wheel load superposition analysis. Kerr (Ref. 5.5) proposed a method for this analysis. The disadvantage of his method is that it references the loaded deflection to the unloaded deflection and hence incorporates slack, whereas the first two methods permit referencing a seating load deflection. This method, however, can be extended to use heavy axle loads referenced to light axle loads as shown by El-Sharkawi (Ref. 5.6).

Table 5.1 Comparison of track response model inputs

Input Parameter	BOEF	GEO-TRACK	ILLI-TRACK	KEN- TRACK
GENERAL:				
Maximum number of layers below sleeper		5	6	6
Consider sleeper-ballast separation?	no	optional	NR	NR
Type of analysis: longitudinal or transverse		NR	X	NR
Total section depth		NR	X	NR
MATERIAL DATA REQUIRED FOR EACH LAYER:				
Layer thickness		X	X	X
Young's modulus / Poisson's ratio		X	X	X
Unit weight		X		NL
Coefficient of lateral earth pressure		X		NL
Coefficient for stress dependent modulus calculation		X	X	NL
Are failure criteria used?	no	NR	optional	optional
Angle of distribution for finite element calculations		NR	X	NR
Track modulus	X			
SLEEPER DATA:				
Length / width		X	X	X
Thickness			X	X
Spacing	X	X	X	X
Young's modulus		X	X	X
Poisson's ratio			X	
Moment of inertia		X	X	X
Unit weight		X		NL
Cross sectional area		X		
Effective tie bearing length		NR	X	NR
Number of segments along sleeper		X		X
Sleeper spring constant			X	
Can nonstandard cross-sections be considered?	no	no	no	yes
Location of rail on sleeper		X		X
RAIL DATA:				
Weight per unit length		X		X
Cross-sectional area		X		
Young's modulus	X	X	X	X
Moment of inertia	X	X	X	X
Section modulus				X
Gage		X		
Rail fastener stiffness		X		X
LOADING CONDITIONS:				
Number of loads	unlimited	max. 4	not given	max. 25
Load locations	X	X	X	X
Wheel load magnitude	X	X	X	X
Specified deflection (input instead of load)			optional	

Note: X = Input is required; NR = Input not required - program automatically considers; Optional = Input is optional depending on computations and output desired; NL = Input required only for nonlinear analysis.

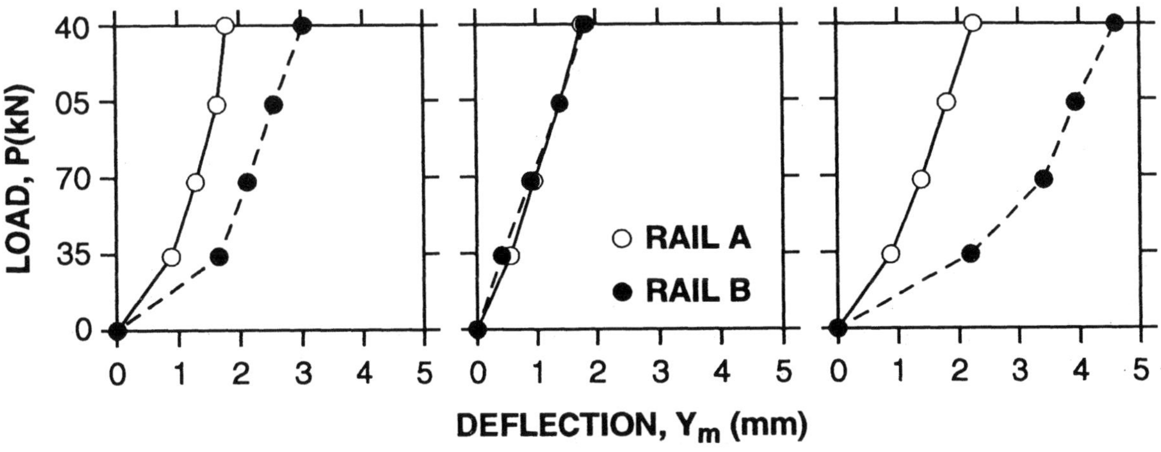

Fig. 5.4 Vertical track load-deflection measurements

Examples of vertical load-deflection measurements using the single point load test are shown in Fig. 5.4. The curves are highly nonlinear, with the degree of nonlinearity generally increasing with track settlement under traffic. This shows the importance of the choice of load levels in determining track modulus.

5.2. Computer Models

A number of computer models have been developed for the track structure under vertical wheel load which include separate components representing the rails, the fasteners, the sleepers and the substructure layers. Thus they overcome a major limitation of the beam-on-elastic- foundation model. Some of these models have been superceded by more recent models. Others are not readily available, or are too expensive to use in general. Three models will be discussed in this section which represent a variety of options for design and analysis of track with special attention to the substructure influence. These models are ILLITRACK, GEOTRACK and KENTRACK. A comparison of the basic features of these three models with those of the beam on elastic foundation (BOEF) model are given in Table 5.1 which is adapted from Ref. 5.10.

5.2.1. ILLITRACK

A finite element model, ILLITRACK, was developed at the University of Illinois (Ref. 5.7). It is not a three-dimensional model. It consists of two two-dimensional models, one transverse and the other longitudinal (Fig. 5.5), employing output from the longitudinal model as input to the transverse model. In this manner, a three-dimensional effect is obtained with much less computer cost than with a three-dimensional model. Non-linear properties for the material are obtained in the laboratory from repeated load triaxial tests. An incremental load technique is employed to effect a solution. Explicit failure criteria were developed for the ballast, subballast and subgrade materials. However, the model does not prevent tension from being developed between the sleeper and the rail.

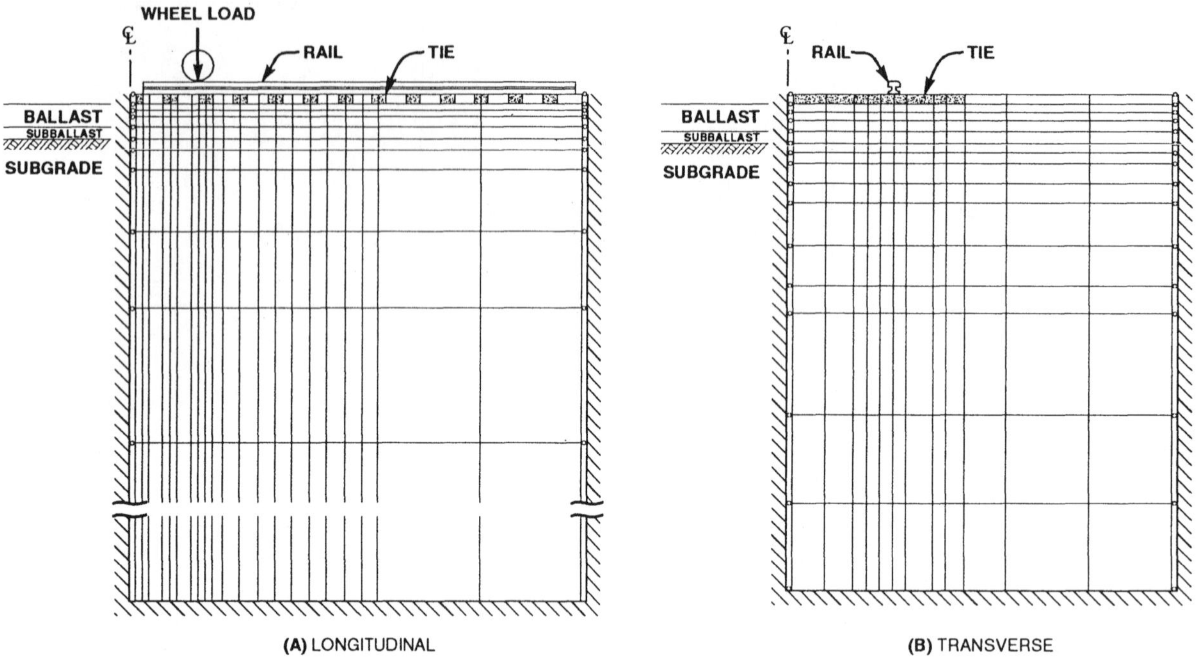

Fig. 5.5 Longitudinal and transverse two-dimensional finite element meshes

The main advantage of this model is its ability to vary the properties of the substructure in the longitudinal and transverse directions. The main disadvantage is that it is only a two-dimensional approximation to the actual three-dimensional problem.

5.2.2. GEOTRACK

Based on an evaluation of previously existing models (Ref. 5.8) a model, GEOTRACK, was developed (Ref. 5.9) which was intended to contain the minimum required features at a reasonable computer cost. This model permits calculation of track deflection and track modulus as a function of 1) axle load, 2) properties of the rails and sleepers, 3) properties of the ballast and underlying layers, and 4) geometry, including sleeper spacing and layer thicknesses. The model also provides an estimate of the stresses and deformations in the ballast, subballast and subgrade layers as a function of the same variables. The values of these parameters are needed for studying the behavior of the ballast and subgrade in track, and for predicting the permanent deformation of track by methods to be described in a later section.

Concurrently with the computer model development, field measurements of resilient response were obtained in the ballast and subgrade at the research track in Pueblo, Colorado, known as the Facility for Accelerated Service Testing (FAST). These measurements have provided insights into track dynamic response, as well as data for evaluating the GEOTRACK model.

All of the existing data of this type for FAST have been compiled and statistically analyzed. The findings are presented in a later section along with the important trends. Also, the field results are compared with calculated values from the GEOTRACK model. Finally the effects of variations in the track structure parameters on track response are illustrated with results from GEOTRACK.

The GEOTRACK computer program is a three-dimensional, multi-layer model for determining the elastic response of the track structure, using stress-dependent properties for the ballast, subballast and subgrade materials. The output of the program provides rail seat load, sleeper-ballast reactions, sleeper and rail deflections, and sleeper and rail bending moments. In addition, the output provides vertical displacement and the complete three-dimensional stress states at selected locations in the ballast, subballast and subgrade.

The components of the track contained in the model structure are shown in Fig. 5.6. The rails are represented as linear elastic beams with a support at each sleeper. The rails span eleven sleepers and are free to rotate at the ends and at each sleeper. Each connection between the rails and sleepers is represented by a linear spring. The sleepers are represented as linear elastic beams supported at ten equally-spaced locations on the underlying ballast. The individual sleeper-ballast reactions are applied to the ballast surface over circular areas, whose sizes are related to the sleeper dimensions.

The ballast, subballast and underlying subgrade soil are represented as a set of up to five linear elastic layers of infinite horizontal extent. The bottom layer extends to infinite

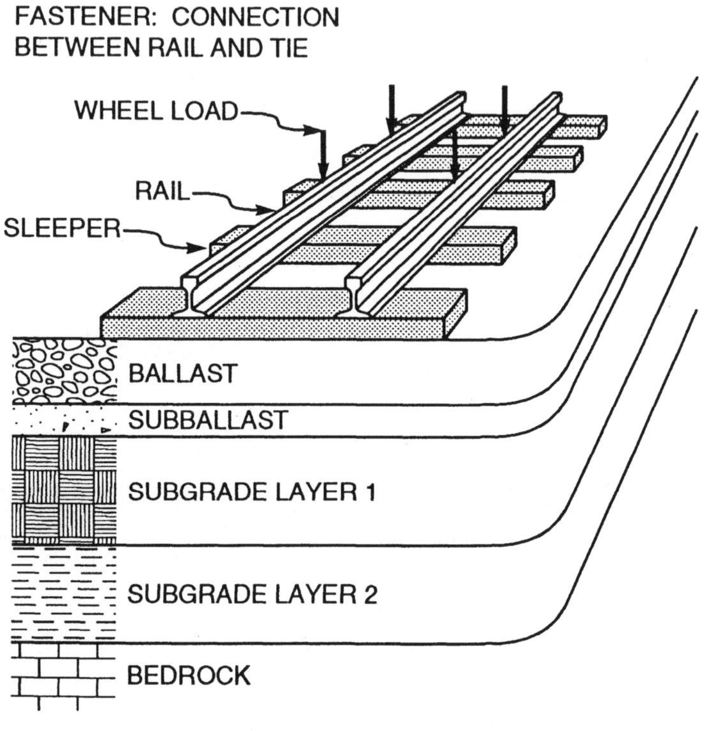

Fig. 5.6 Configuration of GEOTRACK

depth. Each layer has a separate Young's modulus of elasticity and Poisson's ratio. No slip is permitted at the layer interfaces.

The model only allows a vertical component of wheel load. For single axle loading, the wheel load is placed directly over the center sleeper of the eleven-sleeper track structure. For multiple axle loads, the single axle solution is shifted to each additional axle location and combined algebraically with the center sleeper single axle solution.

In general, the axle spacings will not be integer multiples of the sleeper spacing, and thus the axles will not all be located over the center of sleepers. However, as an approximation either the axles are located at the center of the sleepers nearest to the actual locations or an axle load is subdivided into two parts which are assigned to adjacent sleepers. Because the sleeper spacings are small relative to the axle spacings, and the rails are sufficiently stiff, the error caused by this approximation is not of practical importance.

Since the model treats each iteration as a linear elastic problem, unequal axle loads can be accommodated by multiplication of the single axle solution by the appropriate ratio of wheel loads prior to the summations.

The rail and sleeper weights have been included in the GEOTRACK program for two purposes. The first involves the consideration of sleeper-ballast separation. Depending upon the loading and foundation support conditions, the calculated sleeper-ballast reactions acting at locations four to six sleepers away from the loaded sleeper in the single axle case can be negative, indicating an upward incremental tensile force on the ballast which is not physically permissible. Thus, for single axle cases the program contains an option that limits the maximum incremental tensile force allowed on any sleeper segment to the portion of the static weight of sleeper and rail that acts on that sleeper segment. The program version described in Ref. 5.9 limited the tensile force to zero and, hence, neglected the rail and sleeper weights. The consequence was that the rails and sleepers separated from the ballast within a short distance from the loaded sleeper.

In the current GEOTRACK model, the static sleeper weights are distributed equally to each sleeper segment. The rail weights associated with each sleeper are distributed to the sleeper segments in inverse proportion to the distance from the center of the sleeper segment to the rail location.

The computer program solves the structural load distributions and compares the calculated sleeper-ballast reactions with the limiting static weights at each segment. Successive corrections to the loads are made and the calculations repeated until the tensile forces are equal to these limiting values.

The second use of the sleeper and rail weights is in the calculation of the soil stress states. Each layer in the GEOTRACK model is characterized as an elastic material with a stress-state-dependent resilient modulus, E_r. The magnitude of the modulus is defined by the stress state in each layer.

To determine the stress state, the static sleeper and rail weights previously described are converted to surcharge pressures acting at the surface of the ballast layer. The geostatic vertical stress is determined by summing the products of layer unit weights and depths from the bottom of the sleeper to a selected point within each layer, usually in the middle. The geostatic vertical stress is then added to the surcharge pressure. The vertical stress multiplied by K, which is the assumed coefficient of lateral total earth pressure in each layer, gives the unloaded horizontal stress for the layer. These unloaded stresses are then added to the incremental stresses predicted with the GEOTRACK program to get the stresses in the loaded state.

These stresses are then used to calculate the resilient moduli using relationships derived from experiments. The layer moduli are updated after each successive iteration of stress calculation, and the calculations repeated until convergence is achieved between calculated stresses and corresponding resilient moduli. Generally, three iterations are sufficient for reasonable moduli convergence.

5.2.3. KENTRACK

KENTRACK (Ref. 5.11) is similar to GEOTRACK and has been shown (Ref. 5.10) to give the same results when the calculation input conditions are the same. It uses the same multi elastic layer theory for the substructure as GEOTRACK. However, KENTRACK uses a finite element model for the sleepers which permits a variation of the sleeper cross section properties. KENTRACK also permits the use of more vertical rail forces.

Although this model was developed specifically for track structures having a hot-mix asphalt underlayment between the ballast and subgrade, the model is versatile enough to be applied to the design and/or analysis of conventional ballast track or concrete slab track with either wood or concrete sleepers. Two types of failure criteria have been included in KENTRACK. The first is the maximum vertical compressive stress or strain in a specified layer (ballast or subgrade) to control permanent deformation. The second is the maximum horizontal tensile strain in the bottom of the asphalt layer (if present) to control fatigue cracking. A detailed description of the model may be found in Ref. 5.12 which serves as a user's guide for the model.

5.3. Resilient Track Response

5.3.1. Field Conditions

An extensive track substructure response measurement program was conducted at the FAST track in Pueblo, Colorado. Included were strains in the ballast and subballast, vertical stress at the subballast-subgrade interface, and vertical deformation of the subgrade surface relative to an anchor point approximately 3050 mm (10 ft) below this surface. A typical layout is shown in Fig. 5.7. The strain measurement instrumentation is described in detail in Ref. 5.13.

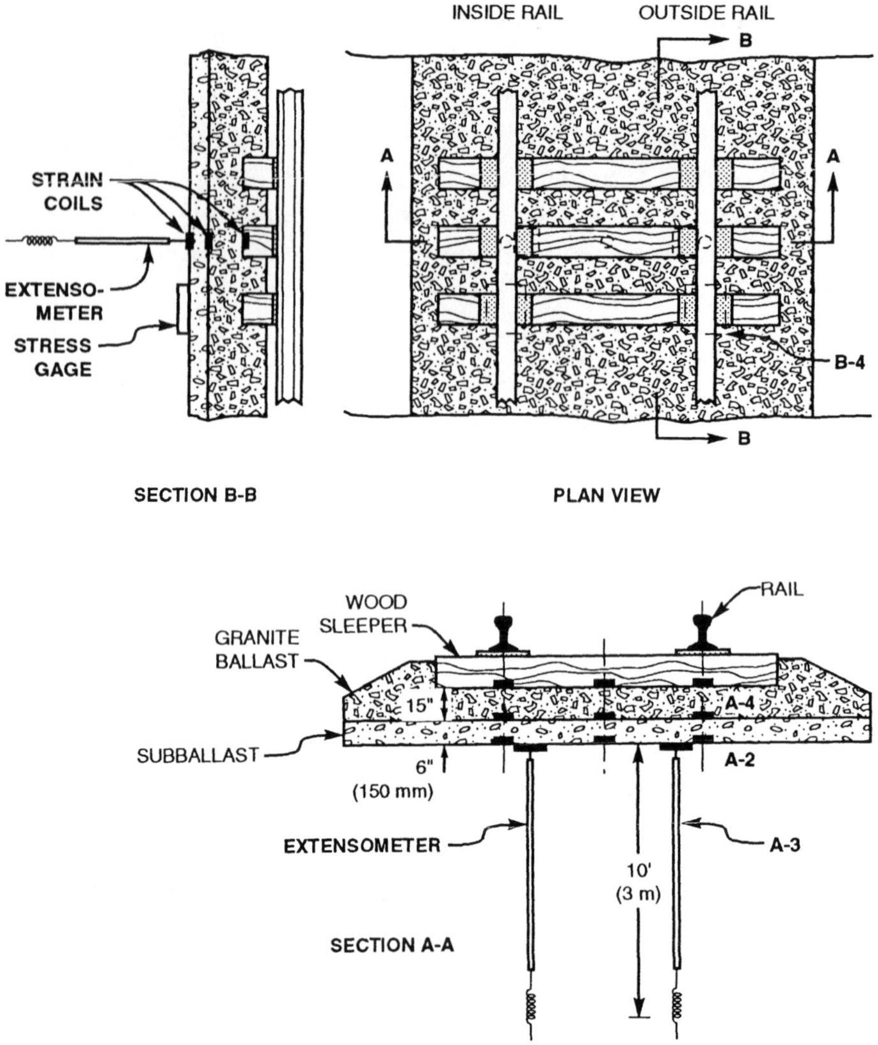

Fig. 5.7 Instrumentation arrangement for FAST track

The sets of different conditions for the instrumented track sections are listed in Table 5.2A. The instrumented sections contained wood and concrete sleepers, ballast nominal depths ranging from 380 to 530 mm (15 to 21 in.), and several types of ballast. The subballast in each set was a 150-mm (6-in.) layer of compacted well-graded gravelly sand. The subgrade was a silty fine to medium sand, designated as SM in the Unified Soil Classification System, and A-1 to A-4 in the AASHTO System.

The measured results for the period from the time of construction through 1558 GN (175 million gross ton or MGT) of accumulated train traffic are given in Refs. 5.14 and 5.15. Subsequently, observations have been obtained at intervals up to 300 MGT (2670 GN) of traffic (Ref. 5.16).

A section of FAST track was reconstructed in 1979, and new instrumentation installed (Refs. 5.17 and 5.18). The track conditions are given in Table 5.2B. For the rebuild, the old ballast and subballast were removed and replaced with new ballast. Thus, a subballast layer was not included in the new section.

Table 5.2 FAST track section conditions

Track Section	Variable Track Parameters						
	Track Geometry	Ballast Type	Ballast Depth		Sleeper Type	Sleeper Spacing	
			in.	mm		in.	mm
A) Original Track							
17C	Curved	Granite	15	380	Concrete	24.0	610
17E	Tangent	Granite	15	380	Concrete	24.0	610
18A	Tangent	Granite	21	530	Wood	19.5	495
18B	Tangent	Granite	15	380	Wood	19.5	495
20B	Tangent	Limestone	15	380	Wood	19.5	495
20G	Tangent	Traprock	15	380	Wood	19.5	495
B) Reconstructed Track							
22A	Tangent	Traprock	15	380	Wood	19.5	495
22B	Tangent	Traprock	15	380	Concrete	24.0	610

5.3.2. Example of Response

A set of initial dynamic measurements obtained after gage installation is shown in Fig. 5.8 to illustrate the elastic response when a three-car train passed slowly over the instrumented wood sleeper section. Among the observations from these records are the following:

1) The permanent strain and deformation from one pass of the train were negligible compared to the elastic components.

2) The 119 metric ton (131-ton) hopper cars produced larger response than the 119 metric ton (131-ton) locomotive, because of the higher axle loads.

3) The variation in stress, strain or deformation as each individual axle in a group passes over the gage is small compared to the group average, indicating that the rail is distributing the axle loads over distances exceeding the axle spacing.

4) The vertical strain in the ballast is mostly negative (extension) beneath the center of the sleeper at the centerline of the track. The extension and compression strains beneath this point in the subballast are approximately equal.

5) The subgrade deflection was always downward relative to the unloaded track position, and the subballast strains beneath the rail were essentially only compressive.

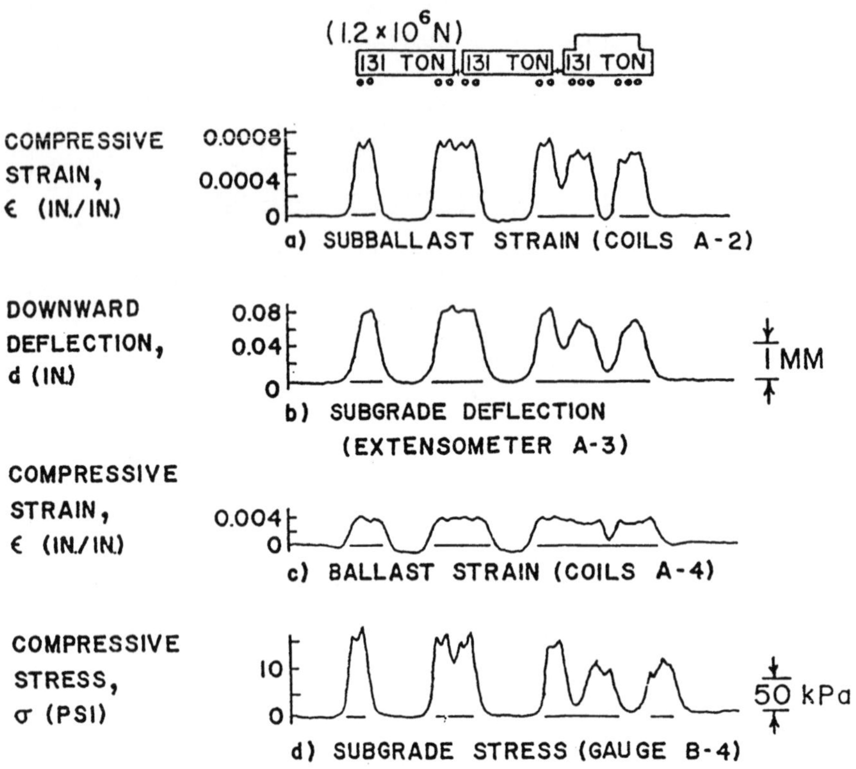

Fig. 5.8 Dynamic measurements at locations in Fig. 5.7

6) The ballast strains were extensional at the midpoint of the cars as a result of spring-up of the rail. However, part of this extension could be a result of lifting of the sleeper from the ballast because the top part of the strain gage was attached to the sleeper rather than to the ballast surface.

5.3.3. Statistical Results

Periodically, as train traffic accumulated on the test track, dynamic response from the moving trains was measured with the installed instruments. The available peak resilient values from the same magnitude wheel load were recorded and averaged at each instrument location. The measurements at individual instrument locations within a particular track section contained little variability, i.e., replicate measurements for any individual instrument were quite consistent. This analysis did show, however, that although the track properties in each section were nominally the same, large variability was present among the instrument responses at different locations within the same track section. When the analysis showed that there were no significant differences in the mean track section responses, the pooled averages and standard deviations of those groups with similar responses were calculated. These pooled values are given in Table 5.3.

Table 5.3 Pooled average resilient track response statistics

Sleeper Type	Wheel Load		Measurement	Mean	Standard Deviation
	lb	kN			
Concrete	32900	146	Ballast strain	0.00107	0.00101
	24000	107		0.00074	0.00071
	14000	62		0.00075	0.00060
Wood	32900	146	Ballast strain	0.00481	0.00178
	24000	107		0.00555	0.00296
	14000	62		0.00468	0.00185
All	32900	146	Subballast strain	0.00058	0.00043
	24000	107		0.00048	0.00018
	14000	62		0.00042	0.00019
Concrete	32900	146	Subgrade stress,	70 (10.1)	4.8 (0.7)
	24000	107	kPa (psi)	56 (8.1)	5.2 (0.8)
	14000	62		47 (6.8)	17.7 (2.6)
Wood	32900	146	Subgrade stress,	39 (5.7)	11.4 (1.7)
	24000	107	kPa (psi)	28 (4.0)	7.6 (1.1)
	14000	62		26 (3.8)	18.9 (2.8)
All	32900	146	Subgrade	0.84 (0.033)	0.17 (0.007)
	24000	107	deflection,	0.73(0.029)	0.19 (0.008)
	14000	62	mm (in.)	0.64 (0.025)	0.30 (0.012)

A comparison of the mean responses of all of the track sections indicated that there were no significant differences between the ballast strains measured in any of the wood sleeper sections independent of ballast type or depth. The ballast strains measured in the concrete sleeper sections were generally an order of magnitude less than those in the wood sleeper sections for all wheel loads. A possible explanation for these large differences between the wood and concrete sleeper sections is the effect of sleeper seating in the wood sleeper section. The upper inductance coils of the ballast strain sensors were attached to the wood sleepers. Any sleeper spring-up when the track was unloaded, which caused a gap between sleepers and ballast, would lead to an apparent strain in the ballast under wheel loading. The coils in the concrete sleeper sections were not attached to the sleepers, but instead rested directly on the ballast surface, thus eliminating the possiblity of this type of systematic measurement error. The analysis of the resilient subballast strains indicated no significant differences between the mean responses from all of the track sections. The same results were found for the resilient subgrade deflections. Although the magnitude of strain and deflection response increased with wheel load increase, the trends betwen track sections did not change.

The subgrade stresses were measured only in track sections 17E (concrete sleepers) and 18B (wood sleepers). A comparison of these results was made to evaluate the effect of sleeper type on resilient stress response. The analysis of the data indicated that there were significant differences between the subgrade stresses measured in the wood and concrete sleeper track sections. The concrete sections produced resilient subgrade stresses which were higher than those in the wood sleeper sections for the range of wheel loads monitored.

5.3.4. GEOTRACK Parameters

The material parameters for the rails, sleepers and fasteners representing the FAST track conditions are given in Table 5.4. The track geometries, ballast layer thicknesses, and sleeper spacings are given in Table 5.2. The ballast, subballast, and subgrade resilient moduli formulations for the FAST track from Refs. 5.19, 5.20 and 5.21, respectively, are given in Table 5.5 together with assumed Poisson's ratios, assumed coefficients of lateral total earth pressure at rest, K, and measured unit weights (Ref. 5.22) for these same materials.

Table 5.4 Sleeper, rail and fastener properties

A) SLEEPER

Variable	Units	Wood	Concrete
Spacing	in. (mm)	19.5 (495)	24 (610)
Length	ft (m)	8.5 (2.6)	8.5 (2.6)
Width	in. (mm)	9.0 (229)	10.8 (274)
Cross-sectional area	$in.^2$ (mm^2)	63.0 (40600)	86.6 (55900)
Weight	lb (N)	250 (1110)	850 (3780)
Young's modulus	psi (kPa)	1.5×10^6 (10.3×10^6)	3.0×10^6 (20.7×10^6)
Moment of inertia, I_x	$in.^4$ (cm^4)	257 (10700)	582 (24200)

B) RAIL (CONTINUOUS WELDED)

Variable	Units	Value
Spacing	in. (mm)	59.3 (1510)
Cross-sectional area	$in.^2$ (mm^2)	13.4 (8650)
Weight per unit length	lb/yd (N/m)	136 (662)
Young's modulus	psi (kPa)	30×10^6 (207×10^6)
Moment of inertia	$in.^4$ (cm^4)	94.9 (3950)

C) FASTENER

Variable	Units	Value
Stiffness	lb/in. (MN/m)	7×10^6 (1230)

Table 5.5 Soil parameters for GEOTRACK predictions

Soil Layer	Resilient Modulus, E_r	Modulus Coefficients K_1	Modulus Coefficients K_2	Poisson's Ratio, ν	Earth Pressure Coefficient, K	Unit weight lb/cu ft	Unit weight Mg/m^3
Ballast	$E_r = K_1 + K_2(\theta)$	22685*	425*	0.3	1.0	106	1.70
Subballast	$E_r = K_1 P_a (\frac{\theta}{P_a})^{K_2}$	940.8	0.69	0.4	0.75	144	2.30
Subgrade	$E_r = K_1 P_a (\frac{\theta}{P_a})^{K_2}$	877	1.1	0.4	0.75	112	1.79

* for E_r in psi Note: θ = bulk stress, P_a = atmospheric pressure

Parametric studies with the GEOTRACK model have shown that the predicted elastic response of the track structure was relatively insensitive to variations in ballast type, i.e., limestone, traprock or granite. The statistical analysis of the FAST field measurements (Ref. 5.23) also showed that differences in response due to ballast type could not be detected. For these reasons, the material properties for granite ballast were used in all of the computer calculations for comparisons of measured and predicted elastic track responses.

Figure 5.9 shows the idealized soil layers used in the GEOTRACK model to represent the FAST track conditions for 380 mm (15 in.) and 530 mm (21 in.) ballast depths.

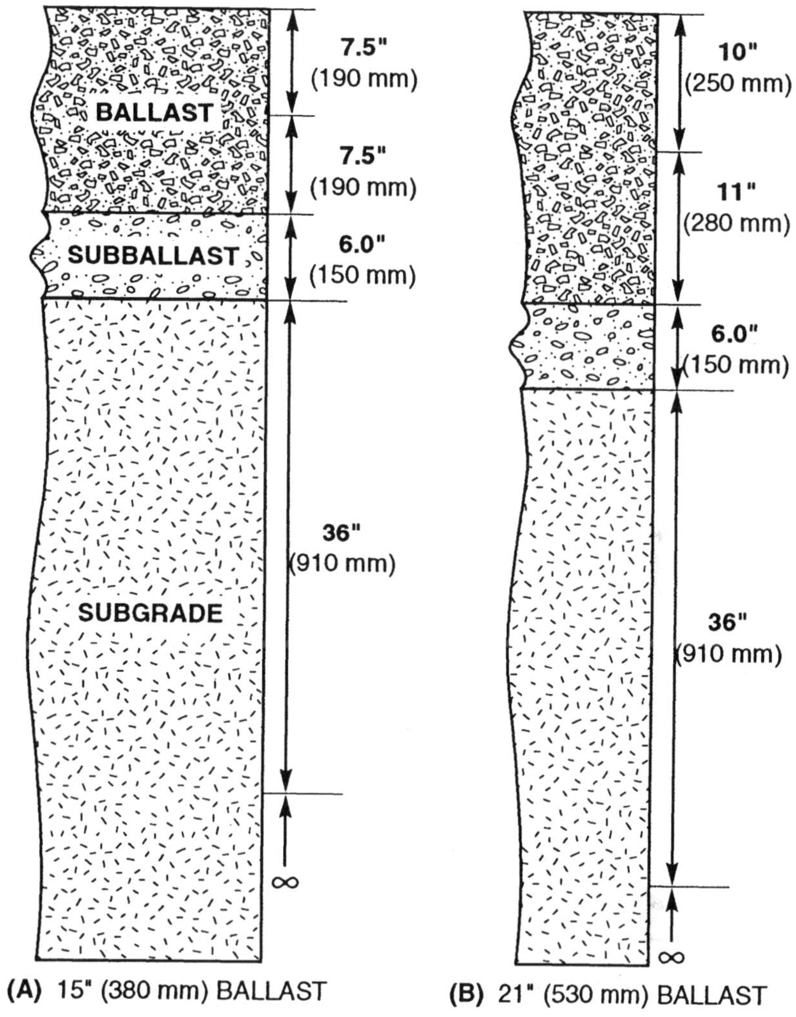

Fig. 5.9 Soil layers for GEOTRACK analysis

The loading conditions analyzed using the GEOTRACK computer model involved four axles, with equal wheel loads of either 22, 45, 89, or 146 kN (5000, 10000, 20000 or 32900 lb). The axle spacings of 1490 to 1980 mm (59 to 78 in.) are representative of the wheel trucks and wheel truck coupling distances found on typical rolling stock.

To indicate the relative layer stiffnesses, the resilient moduli values determined from the final iteration of GEOTRACK are given in Tables 5.6, 5.7, and 5.8 for the various track conditons, soil layers, and wheel loads used for the comparisons between measured and predicted resilient responses. These moduli were based on the weighted average bulk stresses at the midpoints of the corresponding layers or sublayers. For the underlying infinite depth subgrade layer, the bulk stresses were arbitrarily selected at about 0.6 m (2 feet) below the top of this layer.

5.3.5. Comparison of Measurements and Predictions

The measured average values are shown in Figs. 5.10 and 5.11 as triangles or circles. The 95% confidence interval above and below each average is shown as a vertical line. The results from the GEOTRACK model are shown as solid curved lines. The measured strains for sections with wood sleepers were much larger than the corresponding strains in concrete sleeper sections for all wheel loads. A possible explanation for the differences in measured ballast strains in concrete and wood sleeper sections is the sleeper seating effect in the wood sleeper sections, as previously discussed. The magnitude of the measured strains for both track sections generally increased as wheel load increased, roughly parallel to the predicted strain curve shown on Fig. 5.10 a. The single predicted ballast strain curve shown represents wood and concrete sleeper sections having a 380 mm (15 in.) ballast depth and wood sleeper sections having a 530 mm (21 in.) ballast depth. Although the GEOTRACK model did show differences in ballast strains for these three cases, the differences were very small and much less than the 95% confidence limits of the measured response. The ballast strains measured in the concrete sleeper sections tend to be in general agreement with those

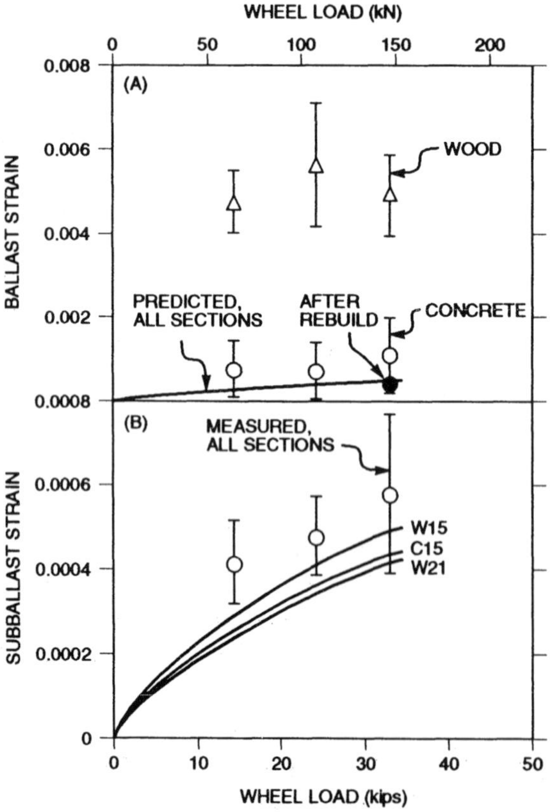

Fig. 5.10 Ballast and subballast strains

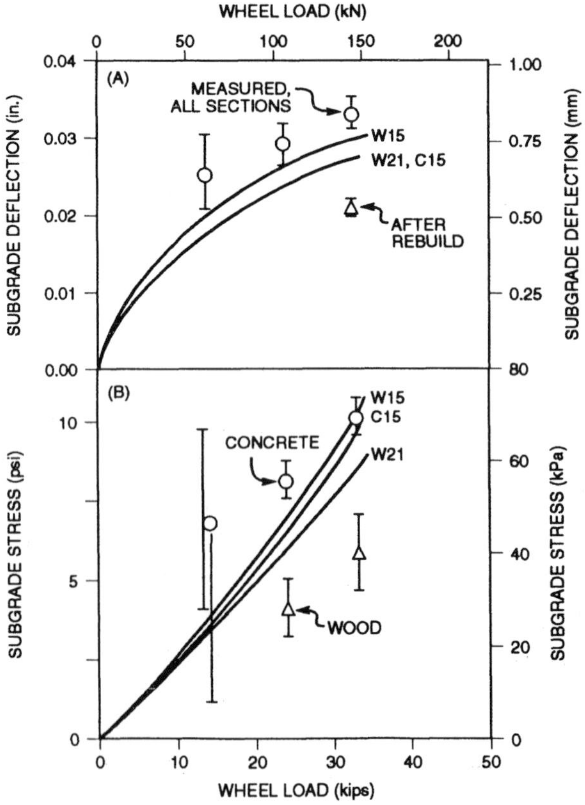

Fig. 5.11 Subgrade stress and deflection

predicted by the computer model, while the measured wood sleeper strains are much higher than predicted.

Although the measured subballast strains (Fig. 5.10b) were generally larger than those predicted for all wheel loads and track conditions, the magnitudes are in general agreement. The predictions show that the subballast strains increase at a decreasing rate as the wheel load increases. The predictions also show that as ballast depth increases, the subballast strains decrease, and that wood sleeper sections produce larger subballast strains than concrete sleeper sections. Although the statistical analysis showed that there were no significant differences in the measured subballast strain caused by varying track conditions compared to the variability of the field measurements, the measurements actually show the same trend as that predicted by the model. For example, the average subballast strains for the wood sleeper sections with 380 mm (15 in.) ballast (W15) were actually larger than those measured in concrete sleeper sections with 380 mm (15 in.) ballast (C15). Also, the average measured subballast strains in track sections having wood sleepers with 530 mm (21 in.) ballast depth (W21) were the smallest, just as predicted by GEOTRACK. This suggests that there may actually be differences in subballast strains between the track sections, but that detection of such small differences is limited by existing field variability.

The predicted values of subgrade deflection (Fig. 5.11a) were obtained by taking the differences between the calculated vertical deflections at the subgrade surface and the vertical deflections 3.05 m (10 feet) below the subgrade surface. This corresponds to the locations of the reference anchors on the extensometers used for the field measurements. The predicted subgrade deflections increased at a decreasing rate as the wheel load increased. The measured results are consistent with this trend. The subgrade deflections at FAST would be expected to show a marked decrease in magnitude for wheel loads below 44.5 kN (10000 lb), but field data were not available for these low service loads. The measured subgrade deflections were generally 10 to 15 percent larger than those predicted by the GEOTRACK model. The predictions show that subgrade deflections decrease as ballast depth increases for wood sleeper sections, and that wood sleeper sections having a 530 m (21 in.) ballast depth result in subgrade deflections of the same magnitude as concrete sleeper sections with a 380 mm (15 in.) ballast depth.

The measured values of resilient subgrade vertical stresses (Fig. 5.11b) were obtained only for one concrete and wood sleeper section, each having 380 mm (15 in.) ballast depth. Unlike the strains and deflections, the subgrade stresses increase at a slightly increasing rate as the wheel load increases. The reason could be a result of the ballast stiffness increase with total wheel load (Tables 5.6, 5.7 and 5.8).

Although the stress measurements at any particular MGT level were quite variable in both wood and concrete sleeper sections, the average of these field measurements over traffic levels from 27 to 2670 GN (3 to 300 MGT) did show that the concrete sleeper sections resulted in higher vertical stresses than the wood sleeper sections. However, the GEOTRACK model results do not show this trend. The difference between the concrete sleeper and wood sleeper measurements may have been caused by differences in sleeper support conditions which would be amplified by the differences in sleeper stiffnesses. Since the GEOTRACK model considers homogeneous isotropic layers supporting the track bed, variations in the sleeper support conditions such as centerbinding cannot be taken into account.

Of the factors considered for the GEOTRACK predictions, ballast depth has the largest effect upon subgrade stresses. As shown in Fig. 5.11b for wood sleeper sections, the subgrade stresses tend to decrease as ballast depth increases. Field data were not available from FAST to validate this conclusion, but Ref. 10.16 confirms this trend.

Table 5.6 GEOTRACK parameters and soil moduli for FAST comparisons-- Part A

Conditions: wood sleeper with 19.5 in. (495 mm) spacing; 15 in. (380 mm) ballast depth; 4-axle superposition.

Soil Layer	Subdivided Layer Depth		Wheel Load		Soil Modulus, E_r	
	in.	mm	lb	kN	psi	MPa
Ballast	7.5	190	5000	22.3	25900	178
			10000	44.5	28250	195
			20000	89.0	33200	229
			32090	142.8	39750	274
	7.5	190	5000	22.3	24850	171
			10000	44.5	25850	178
			20000	89.0	28150	194
			32090	142.8	31500	217
Subballast	6.0	152	5000	22.3	7650	53
			10000	44.5	9550	66
			20000	89.0	13250	91
			32090	142.8	17850	123
Subgrade	36.0	914	5000	22.3	4500	31
			10000	44.5	6100	42
			20000	89.0	9850	68
			32090	142.8	15350	106
	∞		5000	22.3	10200	70
			10000	44.5	11300	78
			20000	89.0	13550	93
			32090	142.8	16600	114

Table 5.7 GEOTRACK parameters and soil moduli for FAST comparisons -Part B						
Conditions: wood sleeper with 19.5 in. (495 mm) spacing; 21 in. (533 mm) ballast depth; 4-axle superposition.						
Soil Layer	Subdivided Layer Depth		Wheel Load		Soil Modulus, Er	
	in.	mm	lb	kN	psi	MPa
Ballast	10.0	254	5000	22.3	25750	177
			10000	44.5	27900	192
			20000	89.0	32200	222
			32090	142.8	37850	261
	11.0	279	5000	22.3	25100	173
			10000	44.5	25950	179
			20000	89.0	27900	192
			32090	142.8	30650	211
Subballast	6.0	152	5000	22.3	8200	56
			10000	44.5	9850	68
			20000	89.0	13100	90
			32090	142.8	17100	118
Subgrade	36.0	914	5000	22.3	5150	35
			10000	44.5	6700	46
			20000	89.0	10200	70
			32090	142.8	15250	105
	∞		5000	22.3	11200	77
			10000	44.5	12300	85
			20000	89.0	14450	100
			32090	142.8	17350	120

Table 5.8 GEOTRACK parameters and soil moduli for FAST comparisons - Part C						
Conditions: concrete sleeper with 24 in. (610 mm) spacing; 15 in. (380 mm) ballast depth; 4-axle superposition						
Soil Layer	Subdivided Layer Depth		Wheel Load		Soil Modulus, E_r	
	in.	mm	lb	kN	psi	MPa
Ballast	7.5	190	32090	142.8	40400	278
	7.5	190			31900	220
Subballast	6.0	152			18250	126
Subgrade	36.0	914			17000	117
	∞				19050	131

FAST test section 22 was rebuilt and instrumented in 1979, after approximately 3800 GN (425 MGT) of traffic. Details of the rebuilding procedure and instrumentation plan can be found in Refs. 5.17 and 5.18. The rebuild involved removal of the track structure, followed by excavation of the ballast and subballast layers and the top of the subgrade. The exposed subgrade surface was wetted and compacted with a vibratory roller, using sufficient roller coverages to obtain field dry densities equivalent to 95% of the maximum dry densities obtained from modified compaction test ASTM D1557. No subballast was added; instead, the ballast was placed directly on the subgrade.

Prior to placing the ballast, extensometers were installed in holes drilled into the prepared subgrade at locations to be subsequently below rail seats. The extensometer casings were 3.05 m (10 feet) long sections of continuous PVC pipe, grouted into the subgrade at the bottom anchor. The top of the extensometer was located at the subgrade surface.

Inductance coils like those used in sections 17, 18, and 20 were installed in pairs to measure the total vertical ballast strains beneath the rail seats. The bottom coil in each pair was placed on the subgrade surface before the ballast was added. The top coil was placed on the ballast surface beneath the sleeper, but not attached to the sleeper.

Stress cells installed in the subgrade surface provided measurement of resilient vertical stress under the rail seats and under the sleeper centers.

The predictions in Figs. 5.10 and 5.11 were based on sections 17, 18, and 20 which had 380 to 530 m (15 to 21 in.) of ballast with 15 cm (6 in.) of subballast. The new section 22 had only 380 mm (15 in.) of ballast. Nevertheless, a comparison of the resilient measurements from section 22 with the predictions will help assess the validity of the computer model.

Measurements of resilient ballast strains in both wood and concrete sleeper sections following the FAST rebuild have been obtained for up to 1200 GN (135 MGT) of additional traffic. The new data from both wood and concrete sleeper sections were not significantly different. Hence the averaged resilient ballast strain value for both sections for heavy car wheel loads is shown as a solid circle on Fig. 5.10a. The rebuild data showed much less variability between the measured responses, resulting in a very narrow confidence interval (not shown) for the mean response. The close agreement between the rebuild data and the predicted response supports the contention that sleeper lift-up and sleeper-ballast gap closure was the cause of the large ballast strains measured in the wood sleeper sections during the previous FAST experiments, where the upper coils were attached to the sleepers. The stress gages at the subgrade surface were too few and showed too large a variability for quantitative interpretation. However, the wood and concrete sections appeared to have about the same magnitude of subgrade stress, with values that tended to be equal to, or higher than, those in Fig. 5.11b. A higher subgrade stress would be expected because section 22 had a 380 mm (15 in.) depth from sleeper to subgrade, whereas sections 17, 18, and 20 had 530 to 690 mm (21 to 27 in.) depths, which included subballast layers.

The resilient subgrade deflection readings for the rebuilt section were analyzed for 146 kN (32900 lb) wheel loadings. There were no significant differences between the responses of the wood and concrete sleeper sections. These findings are consistent with the measurements obtained in sections 17, 18, and 20. The pooled average and 95% confidence limits for the dynamic subgrade deflections in section 22 after the rebuild are shown in

Fig. 5.11a. The rebuild readings were on the order of 60 to 70% of the previous values. Field borings indicated that the subgrade in the rebuilt sections was considerably stiffer than the initial conditions in FAST sections 17, 18, and 20.

5.4. GEOTRACK Parametric Comparisons

In order to investigate the sensitivity of the GEOTRACK model to the main track variables, a parametric study was done. Table 5.9 gives the fixed and variable track properties that were used for the comparisons. The nominal case is representative of a mainline wood sleeper track. No stress-dependent moduli formulations and soil properties were used for these comparative cases, since program iterations with stress-dependent moduli would have led to changes in soil moduli as the other variables were changed. For all cases, a single wheel load of 146 kN (32900 lb) was used. The main responses studied were rail seat load, vertical displacements of sleepers and soil layers, vertical stresses at the ballast and subgrade surfaces, and track modulus.

Table 5.9 Fixed and variable track properties for single axle parametric study

A) FIXED PARAMETERS	
Parameter	Value
Number of layers	2
Subgrade modulus, MPa (psi)	55 (8000)
Subgrade Poisson's ratio	0.4
Rail cross-sectional area, mm^2 ($in.^2$)	86.5 (13.4)
Rail Young's modulus, kPa (psi)	207×10^6 (30×10^6)
Sleeper length, mm (in.)	2590 (102)
Sleeper width mm (in.)	229 (9)
Wheel load, kN (lb)	142 (32000)

B) VARIABLE PARAMETERS				
Parameter	Nominal Value	Values used holding all other parameters at nominal value		
Ballast Young's modulus, MPa (psi)	310 (45000)	55 (8000)	689 (100000)	
Ballast depth, mm (in.)	305 (12)	152 (6)	610 (24)	
Ballast Poisson's ratio	0.3	0.1	0.49	
Sleeper spacing, mm (in.)	495 (19.5)	254 (10)	610 (24)	914 (36)
Sleeper moment of inertia, cm^4 ($in.^4$)	10700 (257)	24200 (582)		
Sleeper Young's modulus, kPA (psi)	10.3×10^6 (1.5×10^6)	3.4×10^6 (0.5×10^6)	20.7×10^6 (3×10^6)	
Rail moment of inertia, cm^4 ($in.^4$)	3950 (94.9)	1610 (38.7)	2080 (50)	6240 (150)
Rail fastener stiffness, MN/m (lb/in.)	1230 (7×10^6)	18 (0.1×10^6)	879 (0.5×10^6)	176 (1×10^6)

The vertical soil displacement predicted at locations beneath the rail seat, using the nominal track parameters, is shown in Fig. 5.12. The displacement decreases with increasing depth, but even at 2.03 m (80 in.), the vertical deformation accounts for 50% of the total surface deformation.

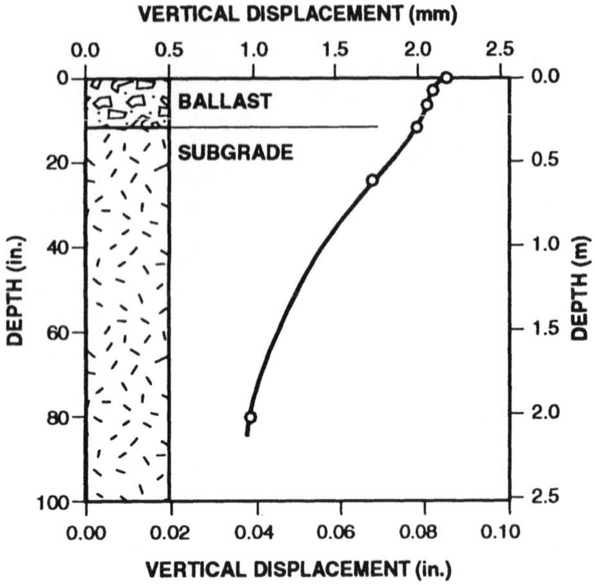

Fig. 5.12 Vertical displacement with depth

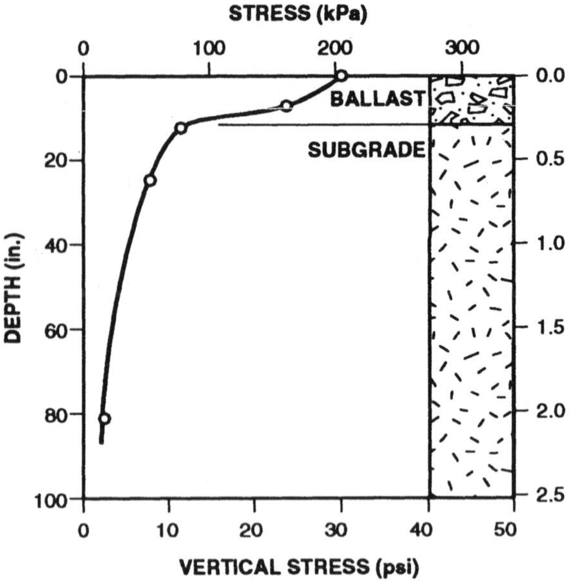

Fig. 5.13 Vertical stress with depth

The vertical stress from the wheel load directly below the rail seat is shown in Fig. 5.13 for the nominal case. The stress attenuation with depth is much more rapid than the displacement. The GEOTRACK model indicates a sharp reduction in vertical stress at the subgrade interface, attributable to the stiffer ballast layer overlying a softer subgrade.

The horizontal variations of vertical stresses at the ballast and subgrade surfaces beneath the loaded sleeper are shown in Fig. 5.14a for the nominal case. These stresses are highest beneath the rail seat. At the ballast-subgrade interface, the stresses are small but more uniform than at the ballast surface.

Vertical ballast strains for the nominal case with a 300 mm (12 in.) ballast depth are shown in Fig. 5.14b. The strain pattern under the loaded sleeper follows the stress distribution pattern, with the largest strains occurring beneath the rail seat.

The distribution of the applied wheel load to the track structure rail seats is shown in Fig. 5.15a for the nominal case. The ballast tension release option was used for the parametric study computer runs, which resulted in the applied load being distributed to only the loaded sleeper and two sleepers on each side. The rail seat at the loaded sleeper has approximately 36% of the applied load, while the second rail seat on each side has 22% and the third rail seat has 10% of the applied load.

The vertical sleeper displacements and the ballast surface displacements beneath the rail seats for the nominal case are shown in Fig. 5.15b. Although the rail seat loads on

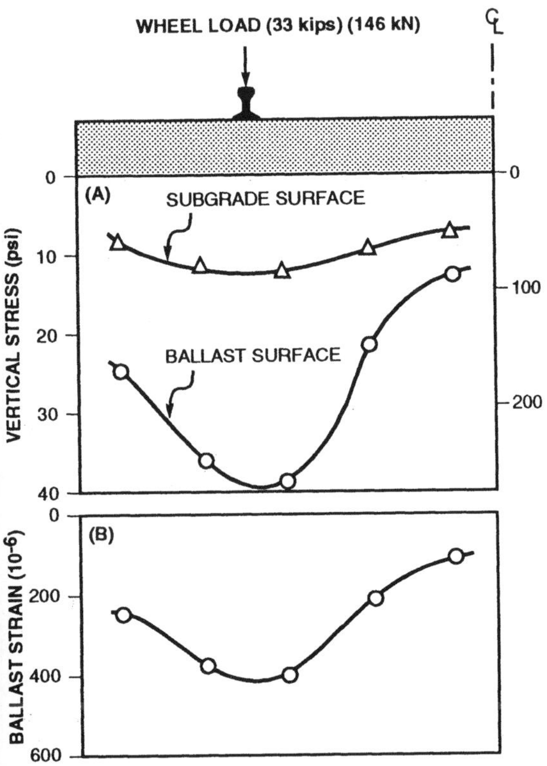

Fig. 5.14 **Vertical stresses and ballast strain along sleeper**

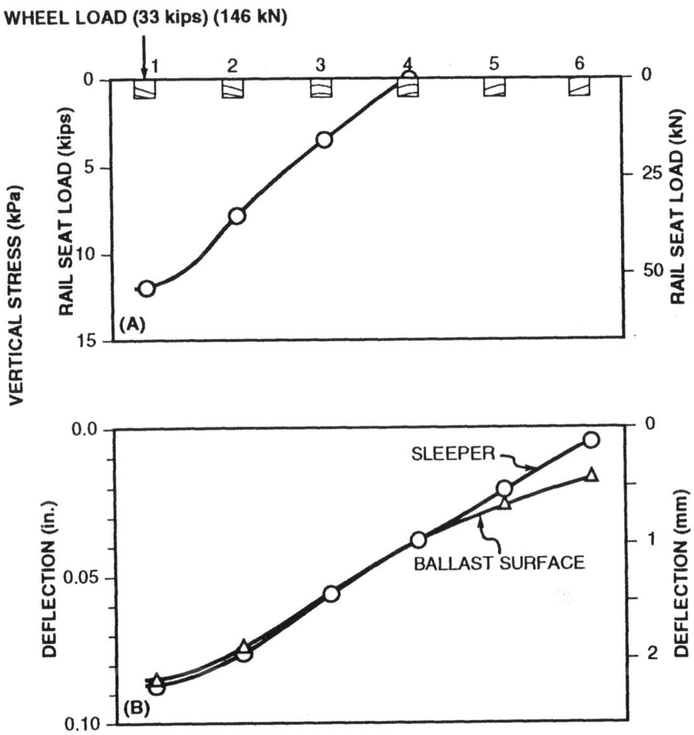

Fig. 5.15 Rail seat load and vertical deflections along rail

sleepers four through six are negligible, the sleeper deflections at these points are not. The effect of the tension release can be seen by the loss of contact between sleepers five and six and the ballast surface.

The predicted track response beneath the applied wheel load for the eighteen variations in conditions in Table 5.9 are compared in Figs. 5.16 and 5.17 with results for the nominal case. In these figures, the horizontal lines represent the nominal results. The vertical lines, labeled with the values of the track parameters used in the sensitivity study, show the range of predicted effect of each individual parameter. These comparisons were made for locations directly under the applied wheel load. The effects of these conditions on track response, both parallel to the rail and parallel to the loaded central sleeper, can generally be inferred by making corresponding adjustments to the trends presented for the nominal case in Figs. 5.12, 5.13 and 5.14.

The rail seat load was affected most by the sleeper spacing (Fig. 5.16a). As the sleeper spacing increased from 250 to 910 mm (10 to 36 in.), the load applied to the sleeper beneath the wheel increased by a factor of about 3. The rail moment of inertia was the next most influencing factor on rail seat load. An increase in this factor from 1610 to 6240 cm^4 (38.7 to 150 $in.^4$) decreased the rail seat load by 40%. Sleeper properties (E and I) and ballast Poisson's ratio had a negligible effect on the rail seat load.

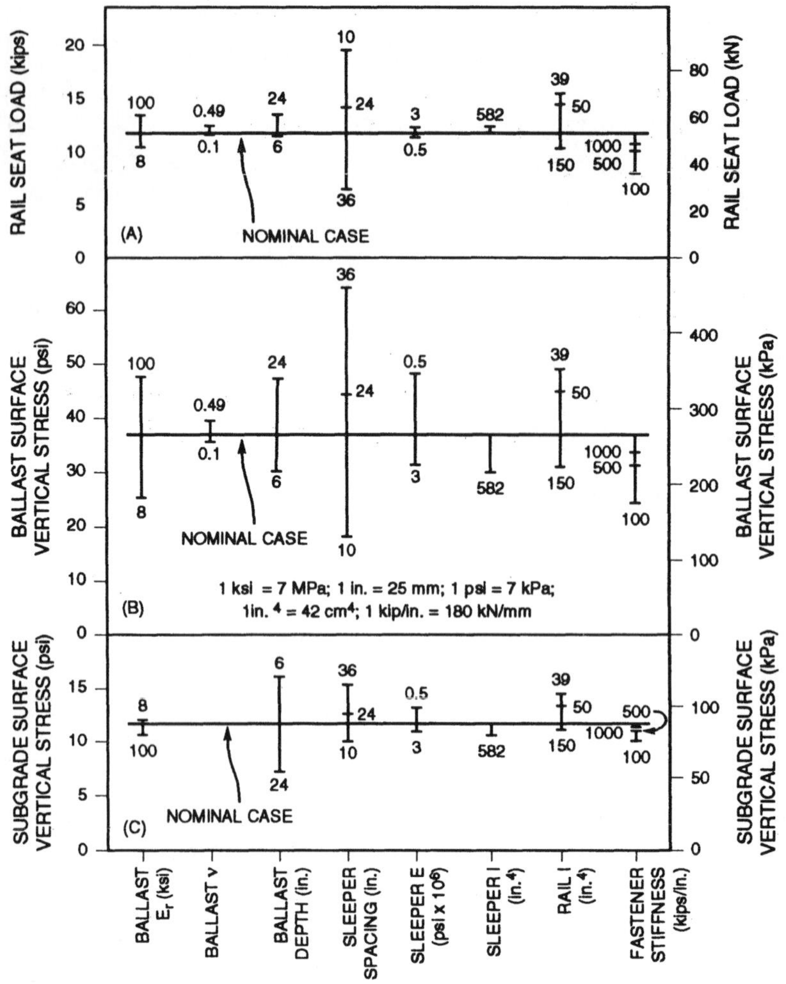

Fig. 5.16 Effect of variable parameters on stress and rail seat load

The vertical stress at the sleeper-ballast interface beneath the point of wheel load application was affected most by sleeper spacing (Fig. 5.16b). As the sleeper spacing increased from 250 to 910 mm (10 to 36 in.), the load increased by a factor of about 4. Significant increases in ballast stress were also caused by an increase in ballast Young's modulus and ballast depth, and by a decrease in sleeper Young's modulus and rail moment of inertia. Ballast Poisson's ratio and sleeper moment of inertia had much smaller effects.

The subgrade surface vertical stress beneath the wheel was affected most by the ballast depth (Fig. 5.16c). A decrease in ballast layer thickness from 610 to 150 mm (24 to 6 in.) more than doubled the subgrade stress. An increase in sleeper spacing from 250 to 910 mm (10 to 36 in.) increased the subgrade stress by 50%. Sleeper Young's modulus and rail moment of inertia had a smaller, but still significant, effect on subgrade stress. Ballast Young's modulus, ballast Poisson's ratio, and sleeper moment of inertia had negligible effects on the subgrade stress.

A decrease in fastener stiffness by a factor of 70 did significantly decrease the rail seat load, ballast surface stress and subgrade surface stress (Fig. 5.16). However, even with this very large change in fastener stiffness, this parameter was not among those causing the largest changes in the track response. In fact, a change in fastener stiffness from 1226 MN/m (7 x 10^6 lb/in.) to 175 MN/m (1 x 10^6 lb/in.) resulted in insignificant changes in the above track responses.

The sleeper vertical deflection beneath the wheel load, which is equal to the ballast surface deflection at the same location, was not affected more than 20% by the changes in any of the track parameters (Fig. 5.17a). The track variables having the largest effect were ballast Young's modulus, ballast depth, sleeper spacing, rail moment of inertia, and fastener stiffness. Negligible effects were caused by the changes in ballast Poisson's ratio and sleeper moment of inertia.

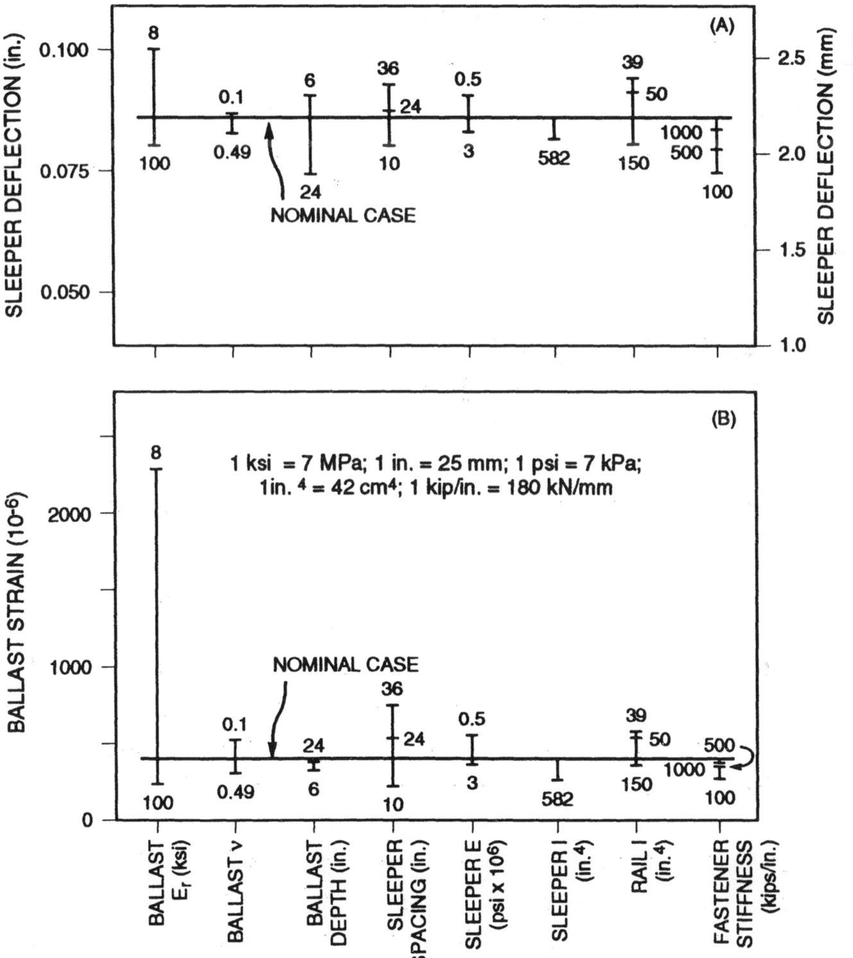

Fig. 5.17 Effect of variable parameters on deflection and ballast strain

The average ballast vertical strain was calculated by dividing the differential displacement between the upper and lower ballast surfaces by the initial layer thickness. By far the greatest effect on ballast vertical strain was caused by the change in ballast Young's modulus (Fig. 5.17b). A decrease in modulus from 689 to 55 MPa (100000 to 8000 psi) caused an increase in ballast strain by a factor of about 9. However, sleeper spacing also had a large influence on ballast strain. An increase in sleeper spacing from 250 to 910 mm (10 to 36 in.) increased ballast strain by a factor of 3. Ballast Poisson's ratio, sleeper Young's modulus and rail moment of inertia had approximately a factor of 2 influence on ballast strain. Ballast layer depth had the smallest influence on the ballast strain.

The total ballast strains for the 150 and 610 mm (6 and 24 in.) layer depths were smaller than the strain for the nominal 300 mm (12 in.) depth. The reason for this trend is that the strain distributions through the layers were very different for the various ballast depths (Fig. 5.18). Analysis of the strain distributions for various layer thicknesses can be very important when developing methods for track settlement predictions, based on representative soil elements and stresses for use in laboratory stress path approaches. For constant layer thickness, variations in other track parameters did not result in such large differences in ballast strain distribution, so the bar graphs shown in Fig. 5.17b can be interpreted more easily.

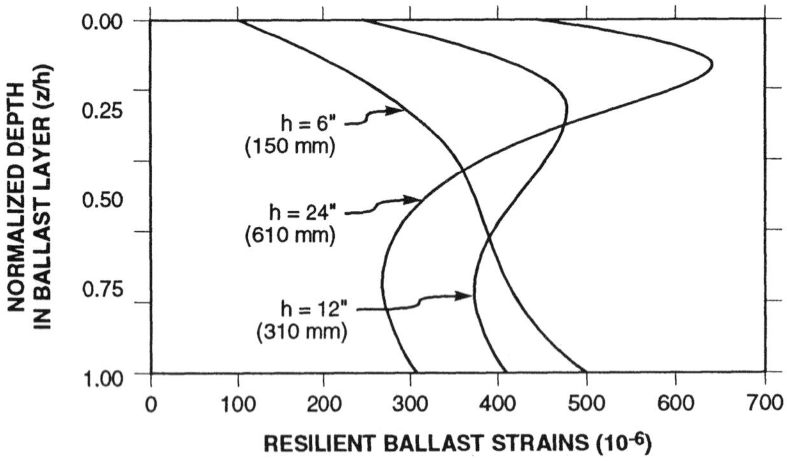

Fig. 5.18 Vertical ballast strain distribution

Track modulus was also predicted for each case using the rail deflections calculated by the GEOTRACK program. The effects of variations in track conditions on track modulus are shown in Fig. 5.19.

The parameter influencing the track modulus most was fastener stiffness (Fig. 5.19). The decrease in fastener stiffness decreased track modulus by approximately 50%. However, actual variations in the fastener stiffness resulting from different types of rail

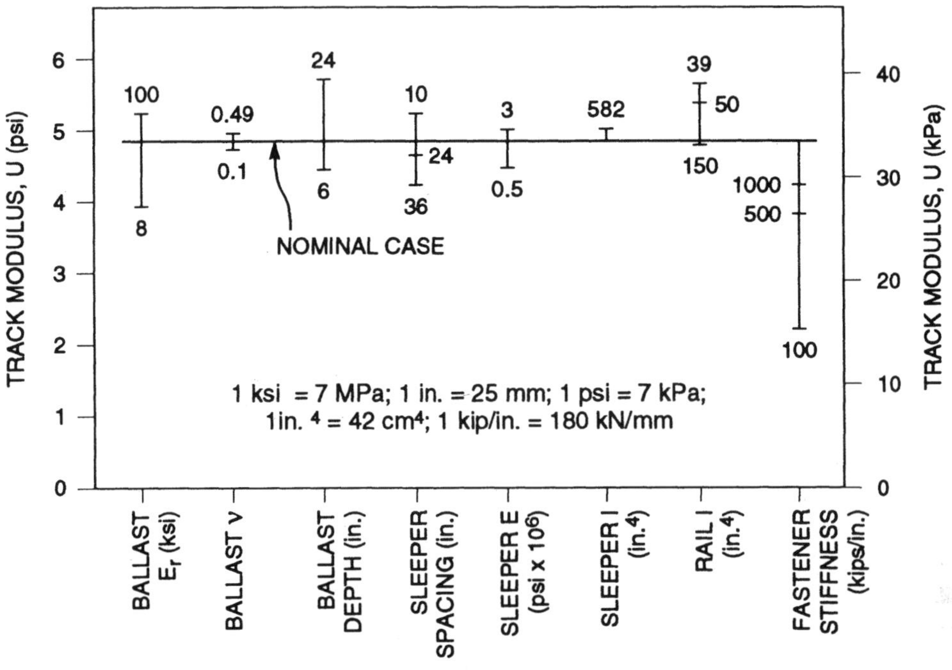

Fig. 5.19 Effect of variable parameters on track modulus

fasteners and pads currently used in the industry would not have as large a range as used in this study. A variation of 10% from the nominal value of 1226 MN/m (7 x 10^6 lb/in.) would be reasonable for the types of rail fasteners commonly used. Track modulus was increased by about 10 to 20% for the increase in ballast Young's modulus and ballast depth, and decrease in sleeper spacing and rail moment of inertia. The ballast Poisson's ratio and sleeper moment of inertia had negligible influence on the track modulus.

Figure 5.20 shows the vertical stress distributions beneath the sleeper at a) the ballast surface, and b) the ballast-subgrade interface. The curves in Fig. 5.20 are for sleepers having stiffnesses of two times and one-third the nominal case stiffness. As sleeper stiffness increased from the nominal case, the sleeper approximated the behavior of a rigid footing supported by an elastic foundation. The maximum vertical stress at the ballast surface occurred at the sleeper ends and generally decreased towards the sleeper centerline. As sleeper stiffness decreased from the nominal case, the maximum ballast surface vertical stress beneath the rail seat increased and the trend was analogous to that of a flexible footing on a stiff foundation. The maximum stress occurred beneath the load point and decreased toward the sleeper end and sleeper centerline. As sleeper stiffnesses increased, the maximum vertical stress at the subgrade surface decreased and was generally more uniform across the sleeper. The magnitudes and distributions of the subgrade vertical stresses were, however, quite similar for all of the sleeper bending stiffnesses.

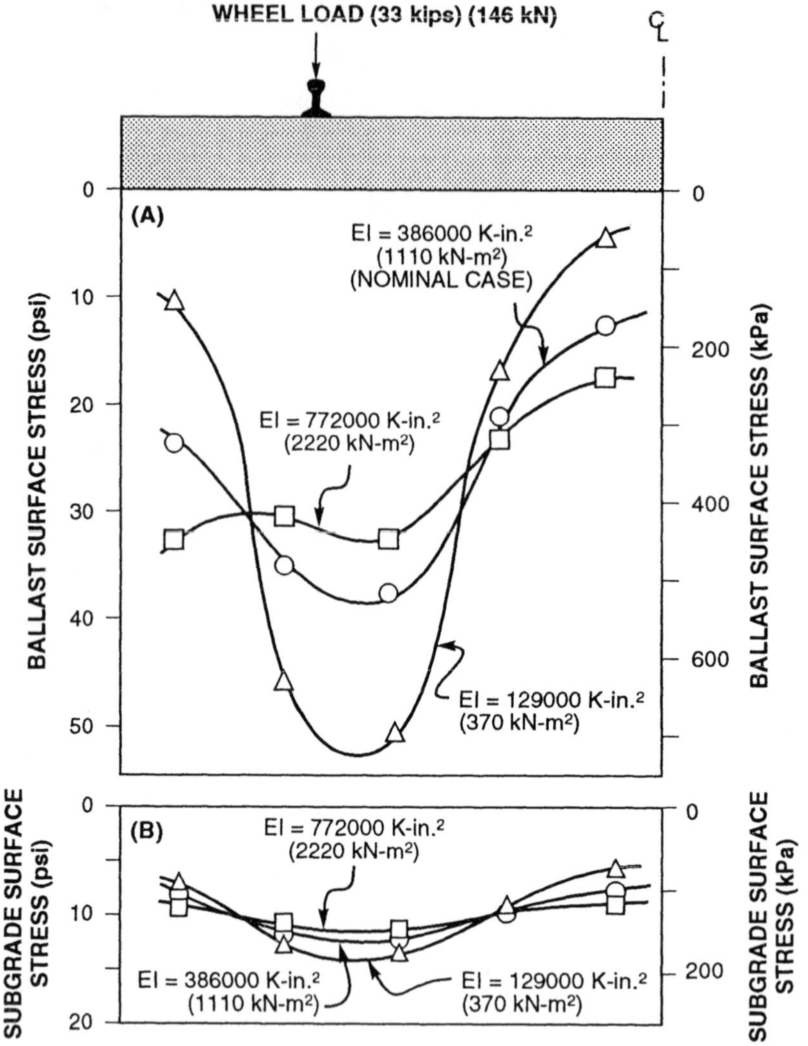

Fig. 5.20 Effect of sleeper stiffness on vertical stress along sleeper

REFERENCES

5.1 Hetenyi, M. (1946). *Beams on elastic foundation*. University of Michigan Press.

5.2 Hay, William W. (1982). *Railroad Engineering*. Second Edition, John Wiley and Sons, Inc., New York, 1982.

5.3 Talbot, A. N. (chairman) (1918). **First progress report of the special committee on stresses in railroad track.** *Bulletin of the American Railway Engineering Association*, Vol. 19, No. 205, March, pp. 875-1058.

5.4 Zarembski, A. M., and Choros, J. (1979). **On the measurement and calculation of vertical track modulus.** *AREA Bulletin* 675, November-December, pp. 156-173.

5.5 Kerr, A. D. (1983). **A method for determining the track modulus using a locomotive or car on multi-axle trucks.** *AREA Bulletin* 692, pp. 269-286, May.

5.6 El-Sharkawi, A. E. (1991). *Correlation of railroad subgrade resilient modulus with cone penetration test data.* Geotechnical Report No. AAR91-389P, University of Massachusetts, Department of Civil Engineering, Amherst, Massachusetts, September .

5.7 Robnett, Q. L., Thompson, M. R., Knutson, R. M., and Tayabji, S. D. (1975). *Development of a structural model and materials evaluation procedures.* Ballast and Foundation Materials Research Program, University of Illinois, report to FRA of US/DOT, Report No. DOT-FR-30038, May .

5..8 Chang, C. S., Adegoke, C. W. and Selig, E. T. (1979). **A study of analytical models for track support systems.** *Transportation Research Board Record* 733, Washington, D.C., pp. 12-19.

5.9 Chang, C. S., Adegoke, C. W. and Selig, E. T. (1980). **The GEOTRACK model for railroad track performance.** *Journal of Geotechnical Engineering Division*, ASCE, Vol. 106, No. GT11, November , pp. 1201- 1218.

5.10 Coleman, D. M. (1985). *Synthesis of railroad design methods, track response models, and evaluation methods for military railroads.* Miscellaneous Paper GL-85-3, Final Report, Waterways Experiment Station, Corps of Engineers, Vicksburg, Mississippi, March .

5.11 Huang, Y. H., Lin, C., and Deng, X. (1984). **Hot mix asphalt for railroad trackbeds-structural analysis and design.** Technical Sessions, Association of Asphalt Paving Technologists, Scottsdale, Arizona, April.

5.12 Huang, Y. H., Lin, C., and Deng, X. (1984). *KENTRACK, a computer program for hot mix asphalt and conventional ballast railway trackbeds.* Research Report RR-84-1, The Asphalt Institute, College Park, Maryland, April .

5.13 Selig, E. T. (1975). **Soil strain measurement using inductance coil method, ASTM-STP584.** *Performance Monitoring for Geotechnical Construction,* August, pp. 141-158.

5.14 Yoo, T. S. and Selig, E. T. (1979). **Field observations of ballast and subgrade deformations in track.** *Transportation Research Board Record* 733, Washington, D.C., pp. 6-12.

5.15 Selig, E. T., Yoo, T. S., Adegoke, C. W. and Stewart, H. E. (1979). *Status report - ballast experiments, intermediate (175 mgt) substructure stress strain data,* Interim Report No. FRA/TTC/Dr-10(IR), Prepared for FAST Program, Transportation Test Center, Pueblo, Colorado, submitted September .

5.16 Selig, E. T. (1980). *FAST substructure dynamic stresses, strains and deflections from 3 to 300 MGT.* Report No. TSC80-270I, Department of Civil Engineering, University of Massachuetts, Amherst, Massachusetts prepared for U.S. DOT, Transportation Systems Center, Cambridge, Massachusetts, December .

5.17 *Documentation report for ballast and subgrade instrumentation, 1979 rebuild, FAST section 22.* Report prepared by Haley and Aldrich, Inc., Cambridge, Massachusetts, for DOT/TSC, Cambridge, Massachusetts, March 1980.

5.18 *Documentation of ballast and subgrade aspects of the 1979 rebuild of FAST sections 3, 17 and 22.* Report prepared by Haley and Aldrich, Inc., Cambridge, Massachusetts for DOT/TSC, Cambridge, Massachusetts, June 1980.

5.19 Alva-Hurtado, J. E. (1980). *A methodology to predict the elastic and inelastic behavior of railroad ballast.* PhD Dissertation, Report No. OUR80-240D, Department of Civil Engineering, University of Massachusetts, Amherst, Massachusetts, May .

5.20 Thompson, M. R. (1977). *FAST ballast and subgrade material evaluation.* Ballast and Foundation Materials Research Program, University of Illinois, Report to FRA, U.S. DOT, Report No. FRA/ORD-77/32, December.

5.21 Stewart, H. E. (1980). *Index and property testing of FAST subgrade.* Report No. OUR80-240I, Department of Civil Engineering, University of Massachusetts, Amherst, Massachusetts, March.

5.22 Panuccio, C. M., Yoo, T. S., and Selig, E. T. (1982). *Mechanics of ballast compaction-volume 3: field test results for ballast physical state measurements.* Final Report to the U.S. DOT, Transportation Systems Center, Cambridge, Massachusetts, DOT-TSC-FRA-81-3, III, March .

5.23 Stewart, H. E. (1980). *One-way analysis of variance applied to dynamic and static track foundation reponses at FAST.* Report No. FRA80-224I, Department of Civil Engineering, University of Massachusetts, Amherst, Massachusetts, May .

6. Residual Stresses and Plastic Deformations in Granular Layers

Chapter 3 demonstrated that soils accumulate plastic strain under repeated loading. Two sources were involved:

1) progressive volume change, and

2) progressive shear strain (specimen distortion).

For a given material and initial density state, Chapter 3 showed that the amount of plastic strain from each of these two sources is a function of the particular cyclic stress path.

In a granular layer, such as ballast and subballast, plastic deformations will occur under repeated traffic loads because of these plastic strains. The vertical component of the plastic deformation results in track settlement. Because the material within the layer is confined, the stresses in the unloaded state will also progressively change under the effects of repeated load. These residual stresses have an important influence on the granular layer plastic strain and resilient modulus. In this chapter the mechanisms for this settlement will first be described. Then examples of the layer system behavior will be given from laboratory tests. Included is a demonstration of residual stress development.

6.1. Mechanisms of Vertical Plastic Deformation

Progressive shear strain under repeated wheel loading causing vertical deformation occurs because the vertical stress in the layers in the loaded state is greater than the horizontal stress. Shear strain in granular materials requires particle rearrangement. The shear strain produces a tendency for horizontal spreading with accompanying vertical shortening of the layer, independent of whether volume change also occurs. The horizontal spreading is resisted in the ballast by confinement from the cribs and shoulders and from friction at the sleeper and underlying layer interfaces. In the subballast, transverse confinement is provided by the boundaries of the layer. Longitudinal confinement in the subballast and the lower portions of the ballast is provided by the material adjacent to the loaded area. No net longitudinal strain should occur in these zones.

Progressive volume change under repeated wheel loading causing vertical deformation occurs for two reasons:

1) particle rearrangement to a more dense packing, and

2) particle breakage with the smaller particles moving into the voids of the larger particles.

Particle breakage is caused not only by the particle contact forces created by the wheel loads, but also from cycles of freezing and thawing, and from chemical breakdown due to the environmental factors.

Any vibration accompanying the wheel loading will increase settlement because of its influence on particle rearrangement and lateral flow of the ballast. Also it appears likely that vibration can cause a reduction (release) of horizontal residual stresses which confine the layers.

Over a period of time the ballast voids become progressively filled with fine particles (fouled). The effect on volume change due to fouling from ballast particle breakage has already been mentioned. If the particles come from the underlying layer, then a volume reduction will occur because of interpenetration. Settlement will occur as the ballast particles migrate downward. This mechanism results from an inadequate subballast layer and is aggravated by poor water drainage. If the fouling comes from sources external to the substructure then volume reduction will not accompany the fouling.

As a result of all three sources of fouling, the ballast tendency for volume change and plastic shear strain will be progressively altered. The fouling can produce either an increase or a decrease in the rate of volume change and plastic strain accumulation, depending on the nature of the fouling materials. Water and mud will lubricate the particles and make rearrangement easier. Fine granular particles will increase ballast interlocking and make rearrangement more difficult.

Perhaps the biggest effect of fouling on plastic deformations occurs when ballast is disturbed by tamping. The loosening effect of tamping is much greater with fouled ballast. As a result, the plastic deformations under subsequent traffic, caused by the mechanisms described, will be much greater for fouled ballast of any type than for clean ballast. Thus the plastic strain in the bottom ballast will be affected differently due to fouling than that in the top ballast.

All of the above mechanisms of plastic deformation involve changes within the granular layers. Displacement of the subgrade also causes settlement through such mechanisms as subgrade consolidation, compaction under traffic, frost action, swelling and shrinking from moisture change, shear displacement, and subgrade squeeze. This category will be discussed in Chapter 10.

Plastic deformation leads to differential settlement and hence track geometry loss, which eventually results in raising and tamping to restore geometry. The tamping process principally rearranges the ballast particles to fill the voids under the sleeper caused by the raise. However, tamping has a number of adverse effects. It causes particle breakage. Also, the top ballast is loosened which results in a reduction of bearing stiffness and lateral track stability. The particle rearrangement causes new wear and breakage surfaces which will accelerate the ballast degradation when traffic resumes.

Finally, the vibration from tamping will cause the finer particles in the voids to settle to the bottom of the ballast bed and thus eventually impede drainage. Subsequent traffic will cause the same effects as before, as the cycle repeats itself. As will be shown in later chapters, this cycle can be broken if consideration is given to the geotechnical principles presented in this book.

6.2. Box Test Demonstration

The volume change and shear strain mechanisms for vertical plastic strain development were illustrated by cyclic triaxial tests in Chapter 3 (Figs. 3.27 and 3.28). The

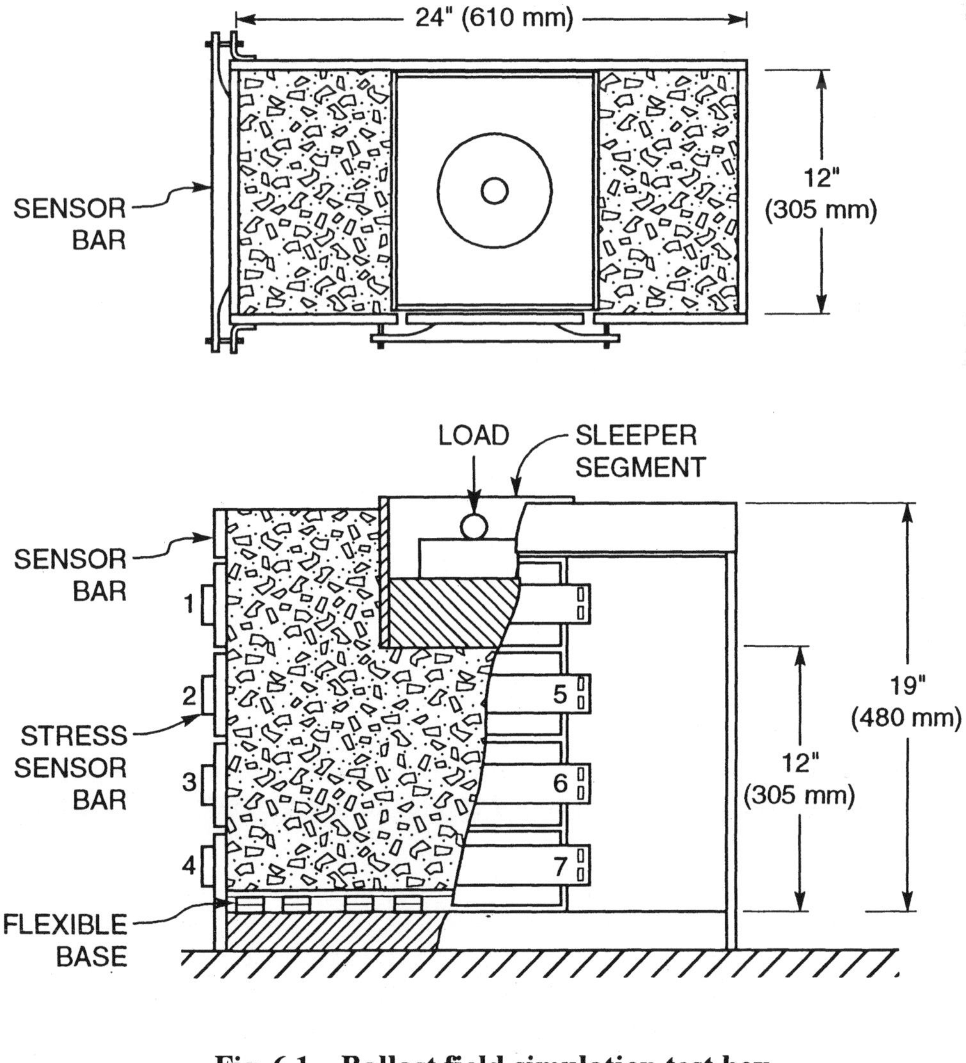

Fig. 6.1 Ballast field simulation test box

effects of these mechanisms together with a contribution from ballast breakage will be demonstrated with a special box test developed at the University of Massachusetts to simulate field loading conditions (Ref. 6.1).

The configuration of the apparatus for this test is shown in Figure 6.1. The bottom of the box is flexible to represent the effect of subgrade. Either a wood or concrete sleeper segment can be used. Instrumented side and end panels were installed to measure the lateral ballast pressure.

To set up a test, a 300-mm-thick (12-in.) layer of ballast is first placed in the box and tamped. Then a 300-mm-long (12-in.) segment of sleeper is placed on top and surrounded by crib ballast. This results in eight ballast zones as indicated in Fig. 6.2. The two zones immediately beneath the sleeper segment (zones 7 and 8) contain ballast that has been dyed to aid in measuring and observing ballast degradation under repeated loading.

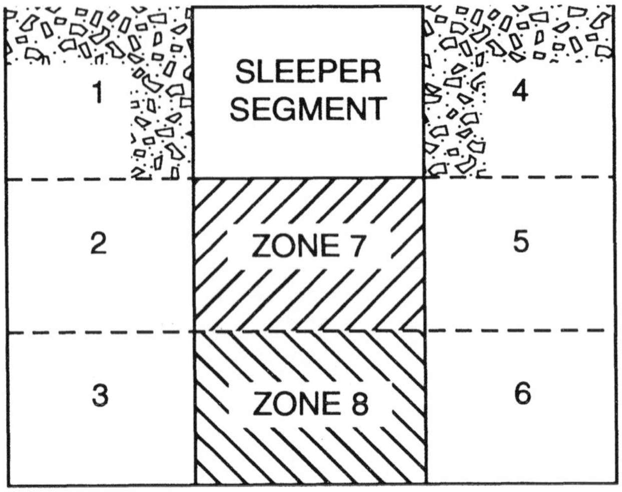

Fig. 6.2 Ballast box sampling zones

When the box has been prepared, it is placed in a load frame to provide vertical repeated loading from a servo-hydraulic testing machine. Although horizontal loading is believed to be important as well, only vertical loading is simulated in this test. The amount of load applied to the sleeper segment is determined from GEOTRACK analysis (Chapter 4) which relates the wheel loading to the ballast pressure at the bottom of the sleeper beneath the rail seat, taking into consideration the stiffness of the rails, the sleepers, and the underlying layers. The number of cycles of a given wheel loading can be directly related to traffic in millions of gross tons. With this device a mix of wheel loads can be applied and the corresponding effects determined. Also, the ballast can be loosened and rearranged periodically to simulate one of the effects of maintenance.

The following effects can be measured in this test: sleeper settlement, ballast breakage, ballast abrasion, ballast density and stiffness change, and horizontal residual stresses in the ballast.

In the interpretation of the box test results it is important to consider the wheel load characteristics observed in the field, and how these are represented in the box. The nominal static wheel load is one-eighth of the gross car weight. With continuously welded rail and good maintenance, the actual or dynamic wheel load will probably not exceed the static load by more than ten to twenty percent. However, wheel and rail defects can cause very high

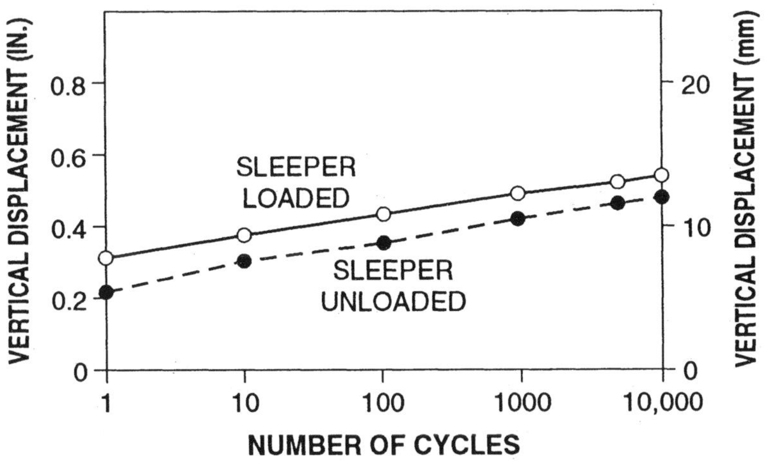

Fig. 6.3 Cumulative sleeper settlement from repeated load

loads, as much as three times the nominal static wheel load. Furthermore, these high loads are associated with high frequency vibration, even exceeding one hundred Hertz. In the field simulation box test dynamic loads are considered by applying a greater equivalent, static load to the rail in the GEOTRACK computer model.

In the simulation tests, sleeper settlement increased typically in proportion to the log of the number of load cycles (Fig. 6.3) for up to about one million cycles. From the initial tamped state at which the ballast is placed in the box, typically 13 to 25 mm (0.5 to 1 in.) of settlement develops in a 300-mm-thick (12-in.) layer of ballast with five hundred thousand load cycles. Settlement was also observed to increase with increasing load.

One of the important observations with this apparatus was the occurrence of high horizontal residual stresses in the ballast as observed with the instrumented side panels (Fig. 6.4). The results show that the horizontal stress in the loaded state decreases with increasing number of load cycles. On the other hand, the horizontal stress in the unloaded state increases with number of cycles and eventually becomes approximately equal to the horizontal stress in the loaded state. The unloaded horizontal stress may even exceed the loaded horizontal stress. This horizontal stress in the

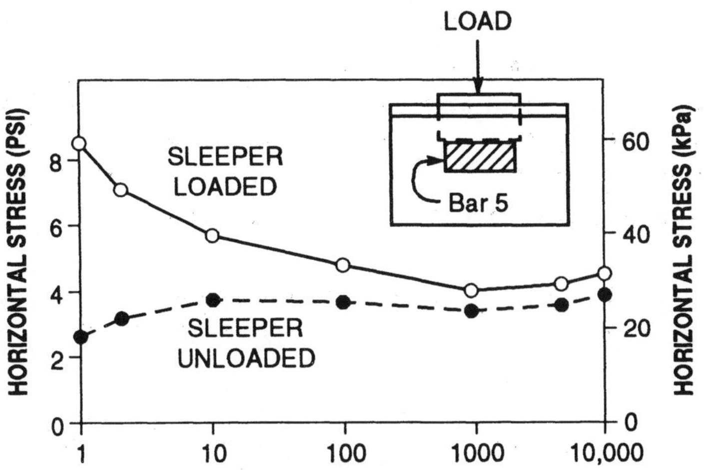

Fig. 6.4 Effect of repeated load on horizontal stress at sensor bar 5 in Fig. 6.1

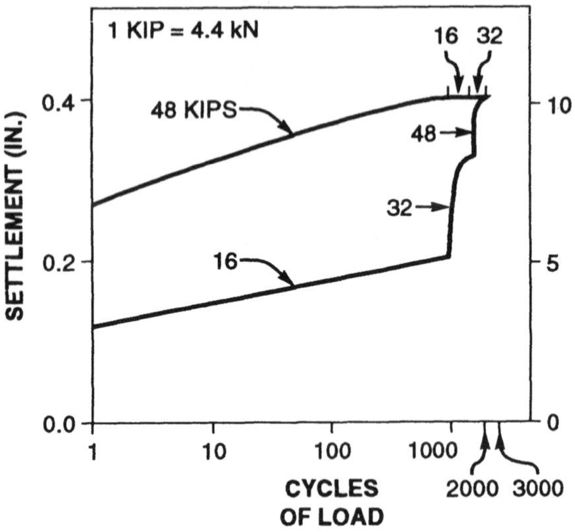

Fig. 6.5 Effect of mix of wheel loads on settlement

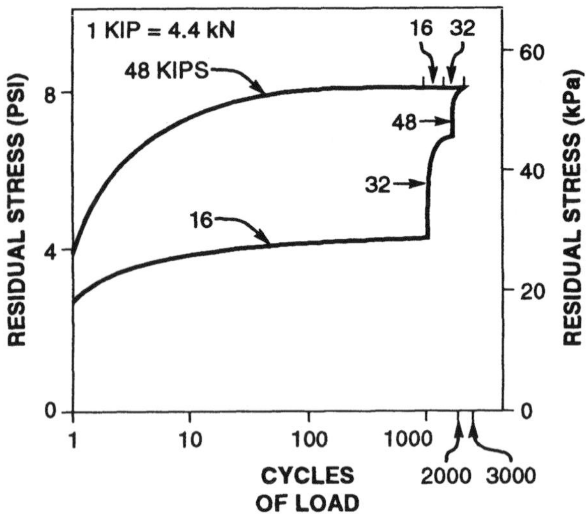

Fig . 6.6 Effect of mix of wheel loads on residual stress

unloaded state is termed the residual stress. The biggest change in horizontal residual stress occurs in the first cycle because the value prior to loading is less than 7 kPa (1 psi).

To demonstrate the effect of a mix of wheel loads, a thousand cycles each of 9, 18, and 27 kN loads (2000, 4000, and 6000 lb) were applied to the sleeper segment in the box test. These represent nominally 70, 140, and 1200 kN (16000, 32000, and 48000 lb) wheel loads on the rail. The test results showed that the order of load application did not affect the final settlement (Fig. 6.5). When the largest load was applied first, the smaller loads did not contribute additional settlement. When the loads were applied in increasing order, each load stage contributed significant additional settlement. The same trend was observed for the horizontal residual stresses as shown in Fig. 6.6. Staged loading triaxial tests conducted by British Rail (Ref. 6.3) showed similar effects of mixing load levels. The implication is that the largest load, which may be a result of wheel or rail defects, can be a major factor in track settlement even with a frequency of occurence as low as one in one thousand.

By running the simulation tests with five hundred thousand to two million load cycles, ballast degradation can be studied. Degradation is considered in two categories: 1) percent of ballast particles by weight in zones seven and eight (Fig. 6.2) that are broken into pieces of greater than 9.5 mm size (3/8 in.), and 2) percent of particles by weight that are less than 9.5 mm size (3/8 in.) size including fines (less than 0.075 mm size) . The first will be termed coarse breakage and the second fine breakage. In this test the ballast initially was screened to eliminate any particles less than 9.5 mm size (3/8 in.).

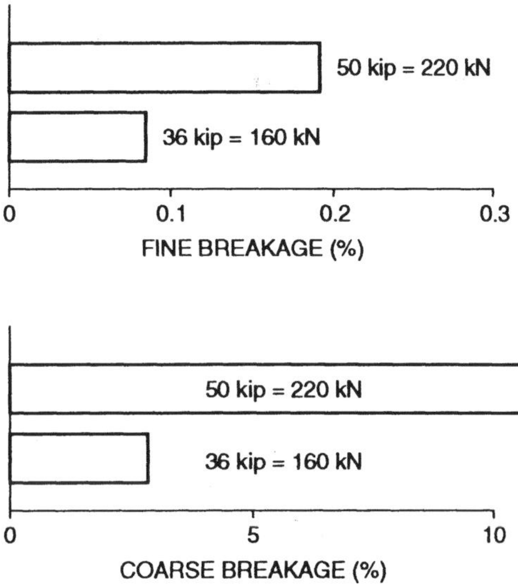

Fig. 6.7 Effect of wheel load magnitude on ballast breakage

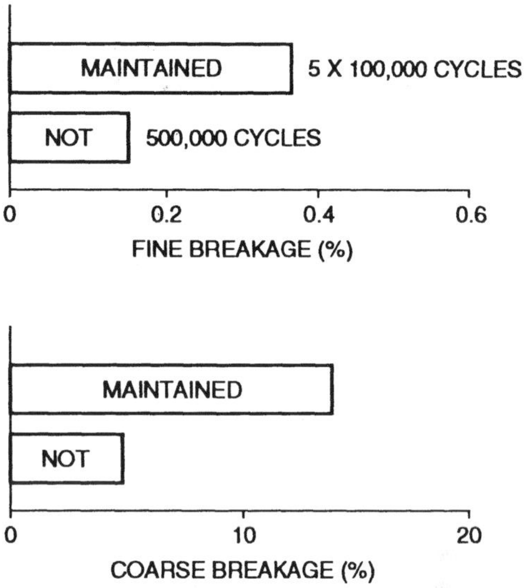

Fig. 6.8 Effect of maintenance on ballast breakage

In one example after one million cycles of wheel load, an increase in wheel load from an equivalent of 160 kN (36000 lb) to an equivalent of 223 kN (50000 lb), more than doubled the amount of fine particles generated and more than tripled the amount of coarse breakage (Fig. 6.7). In another test, the degradation after five hundred thousand cycles was compared with a test in which the ballast particles were rearranged every one hundred thousand cycles for a total of five hundred thousand cycles to simulate one effect of maintenance. For the ballast rearranged after each hundred thousand cycles the amount of fine material generated more than doubled and the amount of coarse material generated tripled (Fig. 6.8).

6.3. Layered System Demonstration

Railroad track structures incorporate a ballast layer over subgrade. The ballast for railroad track is a relatively uniform crushed stone with coarse gravel sized particles. The subgrade, on the other hand, is most likely to be a finer grained soil with the silt and clay components controlling the behavior. As a result the upper layer, when confined, has a higher stiffness than the lower layer. Although actual track conditions involve a more complex multi-layer system, this simplified description is sufficient for understanding the basic layer behavior concepts.

To design and evaluate the performance of track structures, analytical models are used for predicting recoverable (elastic or resilient) and permanent (plastic) deformations of the layers under repeated wheel loading. A problem generally found with the analytical solutions for the resilient behavior is that failure stress conditions are predicted in the bottom portion of the ballast layer under load. This results from a horizontal stress reduction (incremental tensile stress) when surface load is applied. This failure problem only occurs when the ballast layer is stiffer than the lower layer, which is normally the case for track systems. The contradiction is that such systems actually are able to carry many load cycles with no indication of failure.

Elastic layer theory clearly gives a negative (tensile) incremental horizontal stress at the bottom of a two-layered system in which the upper layer has a greater stiffness than the lower layer (Fig. 6.9). If so, failure will occur in this zone when the ambient (unloaded) horizontal compressive stress is too low to compensate. On the other hand, elastic half space theory also shows that the incremental horizontal stress at the same depth will be compressive if the entire deposit has the same stiffness or if stiffness increases with depth.

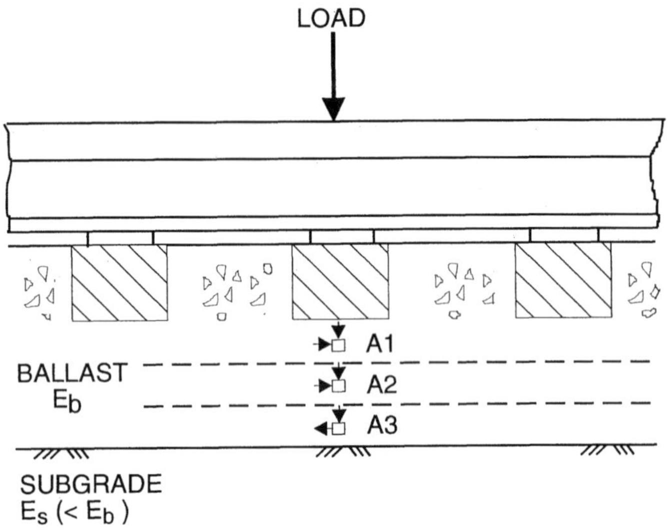

Fig. 6.9 Subdivided ballast showing incremental tensile stress at bottom of layer

The presence of substantial residual horizontal stresses in railroad ballast was observed in box tests as shown in Fig. 6.4. The values of K, the ratio of horizontal to vertical stress in the unloaded state were observed with the box to be as high as 11.

A negative incremental horizontal stress did not occur at any of the load measuring locations in the box, such as shown in Fig. 6.1, i.e., the horizontal loaded stress remained greater than the unloaded stress. However, the box conditions do not represent a stiff layer over a soft layer.

The stress and strain states in the tensil zone were investigated in detail in an instrumented model of a two-layered soil system (Ref. 6.2). The soil was placed in a cylindrical tank and loaded through a rigid circular plate at the center (Fig. 6.10). The upper layer was a coarse-medium, uniform, crushed quartz sand rained into the tank to a relative density exceeding 95 %. The lower layer was a kaolin clay compacted in layers with a hydraulic tamper at a moisture content about 3 % above optimum. Gauges were placed in the sand layer to measure vertical and horizontal stress and strain, primarily in the dashed region in Fig. 6.10. The vertical plate load and deformation were also measured.

A typical test consisted of 20 applications of load cycling (stage 1) between zero and 4.4 kN (1000 lb) followed by 30 applications of cycling (stage 2) between zero and 8.9 kN (2000 lb). The maximum load was selected to be well below the bearing capacity of the two-layered system.

The trends of residual horizontal strain (unloaded state after each cycle) are shown in Fig. 6.11. The strain increases at a diminishing rate with successive cycles

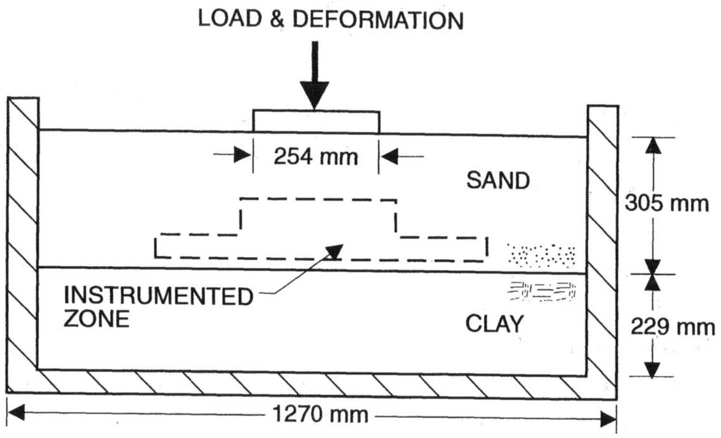

Fig. 6.10 Configuration of soil tank experiments

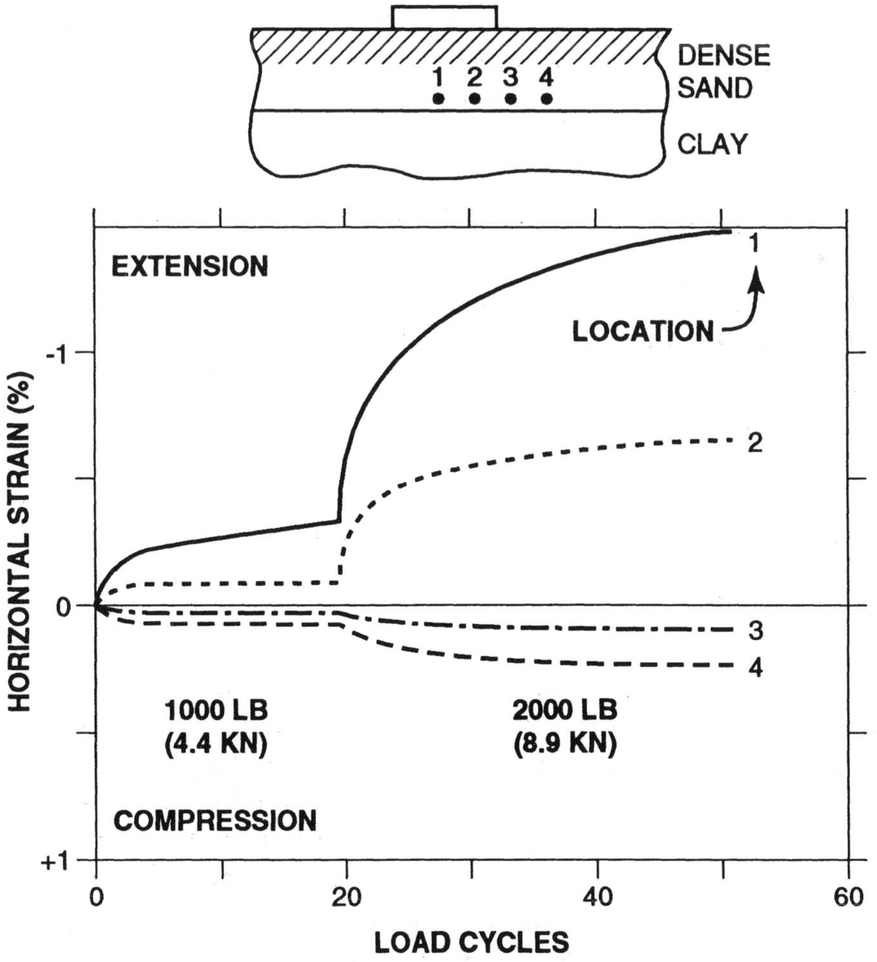

Fig. 6.11 Variation of residual horizontal strain along bottom of sand layer

within each load stage. The maximum extension strain occurs beneath the center of the loaded area. With horizontal distance out from the centerline the strain becomes decreasingly extension and then increasingly compression. Beyond the zone shown the residual compression strain will diminish to zero.

The horizontal strain variation with load at location 1 in Fig. 6.11 is shown in Fig. 6.12. The strain at zero load corresponds to the residual strain shown in Fig. 6.11. Large plastic strain occurs in the first cycle of each load stage, after which the soil response approaches an elastic state. This transition occurs more rapidly at the lower load level. Surface settlement and vertical strains follow the same trend as the horizontal strains.

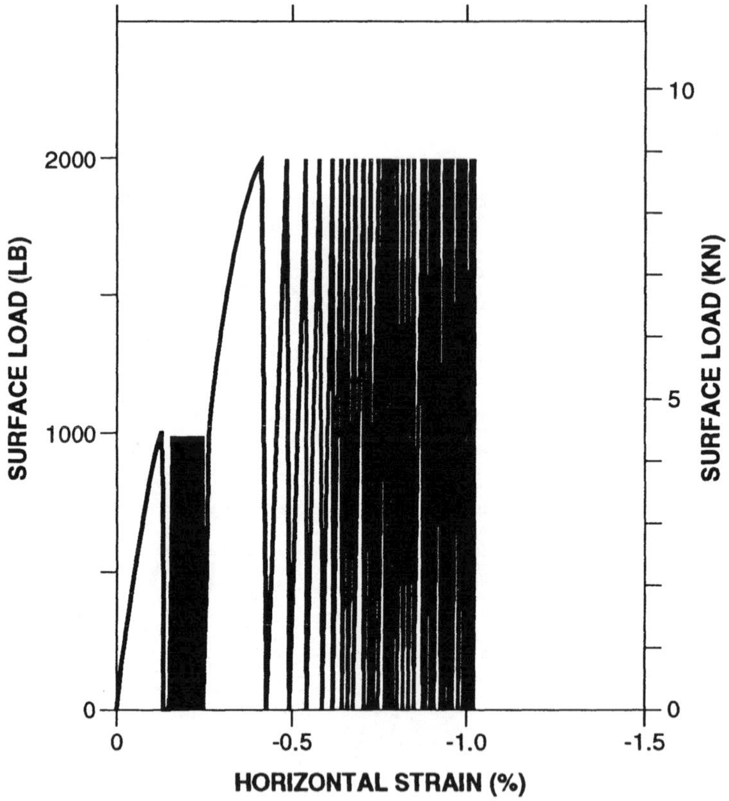

Fig. 6.12 Horizontal strain at location 1 at bottom of sand layer

An average of all values of horizontal stress in the bottom of the sand layer within 75 mm of the centerline was used to obtain each data point shown in Fig. 6.13. The horizontal stress increased substantially when the load was first applied in stage 1 and then, when the load was removed, decreased to a residual stress equal to about 50% of the loaded value. This residual stress was significantly greater than the initial horizontal stress existing after placement of the sand layer. During subsequent load cycles within the same stage the horizontal stress in the loaded state gradually diminished while the horizontal stress in the unloaded state (residual stress) rapidly increased. In about ten cycles the loaded and unloaded stresses became equal. Subsequently the unloaded stress was greater than the

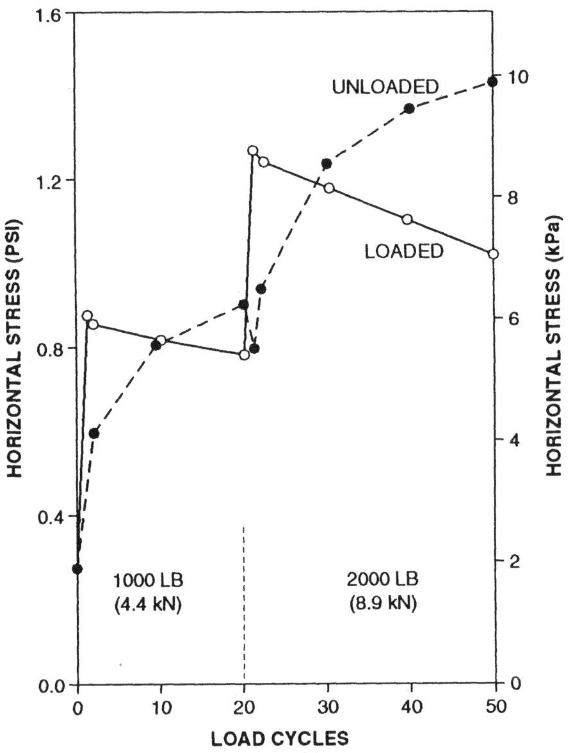

Fig. 6.13 Horizontal stress at location 1 at bottom of sand layer

loaded stress, indicating a negative or tensile incremental horizontal stress produced by the load. This process was repeated in the second stage of the loading except that when the load increase was first applied in the second stage the residual stress diminished.

The variation of loaded and unloaded horizontal stress with depth and distance from the centerline in the granular layer is shown in Fig. 6.14. The highest values of residual stress as well as the only negative incremental stress occurred at the bottom of the sand layer within 100 mm of the centerline. The incremental stress at about mid-depth (127 mm) was highly compressive, that is the loaded horizontal stress was always much greater than the unloaded horizontal stress.

The stress conditions in the tensile zone of a two-layered soil system are believed to be represented by Fig. 6.15. The letter V denotes vertical stress, and H denotes horizontal stress. In the subsequent discussion V and H will be given numerical subscripts corresponding to the numbers of the stress circles in Fig. 6.15. The results are representative of the case of dense sand over soft clay in which the upper sand layer is nominally more stiff than the clay. However, the sand modulus is highly dependent upon the stress state, and so under some conditions it can have a modulus as low as the clay.

The vertical stress in the unloaded state (V_1) is small. The horizontal stress prior to the first loading is a function of the layer placement methods. Assuming that a high degree

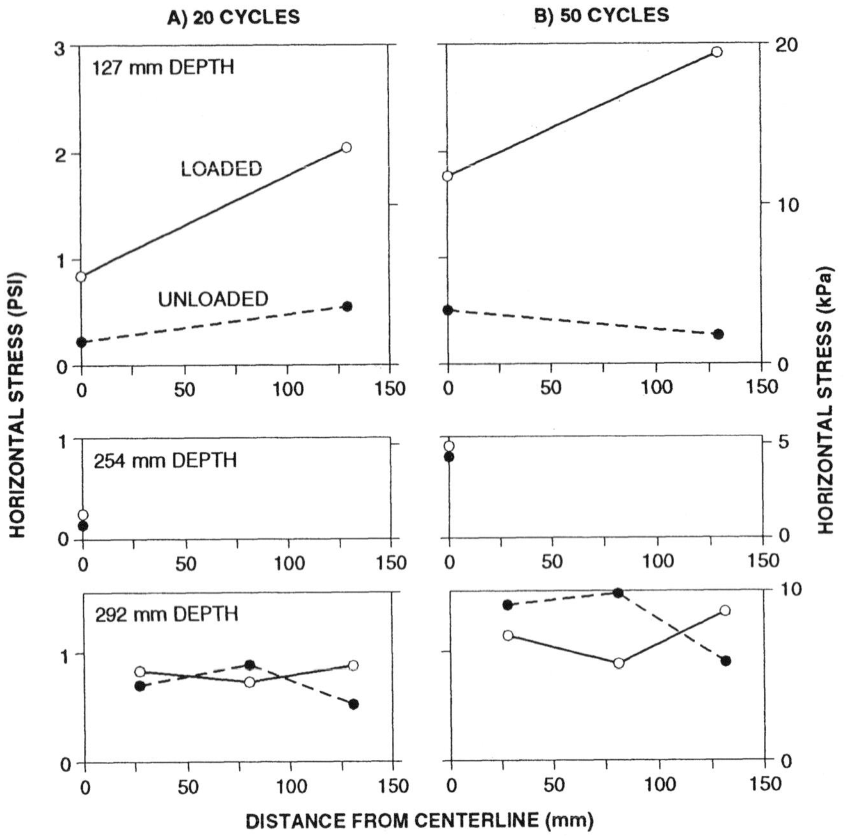

Fig. 6.14 Horizontal stress variations throughout sand layer

of mechanical compaction is not used, as in the case of sand raining, the initial horizontal stress in the unloaded state (H_1) is less than the vertical (V_1). However, the trend to be described will be valid even if H_1 equals or slightly exceeds V_1.

When load is first applied (Fig. 6.15), the sand develops a low stiffness because of the characteristic soil stress-strain relationship on the first loading and the tendency toward failure conditions at the bottom of the layer. Under these circumstances the horizontal stress increases to H_2 while the vertical stress increases to V_2. The incremental horizontal stress H thus is compressive. When the load is removed, the vertical stress reduces to V_3 which is approximately equal to V_1 , but the horizontal stress only partially rebounds to a residual H_3. The corresponding first cycle stresses thus follow path 1-2-3 in Fig. 6.15a.

Because the minimum principal stress is higher for the second cycle and also because the sand stiffness increases for reloading, the stiffness will be higher for the second cycle of loading. The increase in horizontal stress at the bottom of the layer (ΔH) will be smaller and the additional residual stress build up on unloading will be less.

After many cycles of the same applied load, approximately steady state conditions are achieved (Fig. 6.15b). The sand stiffness has increased substantially from the initial cycle,

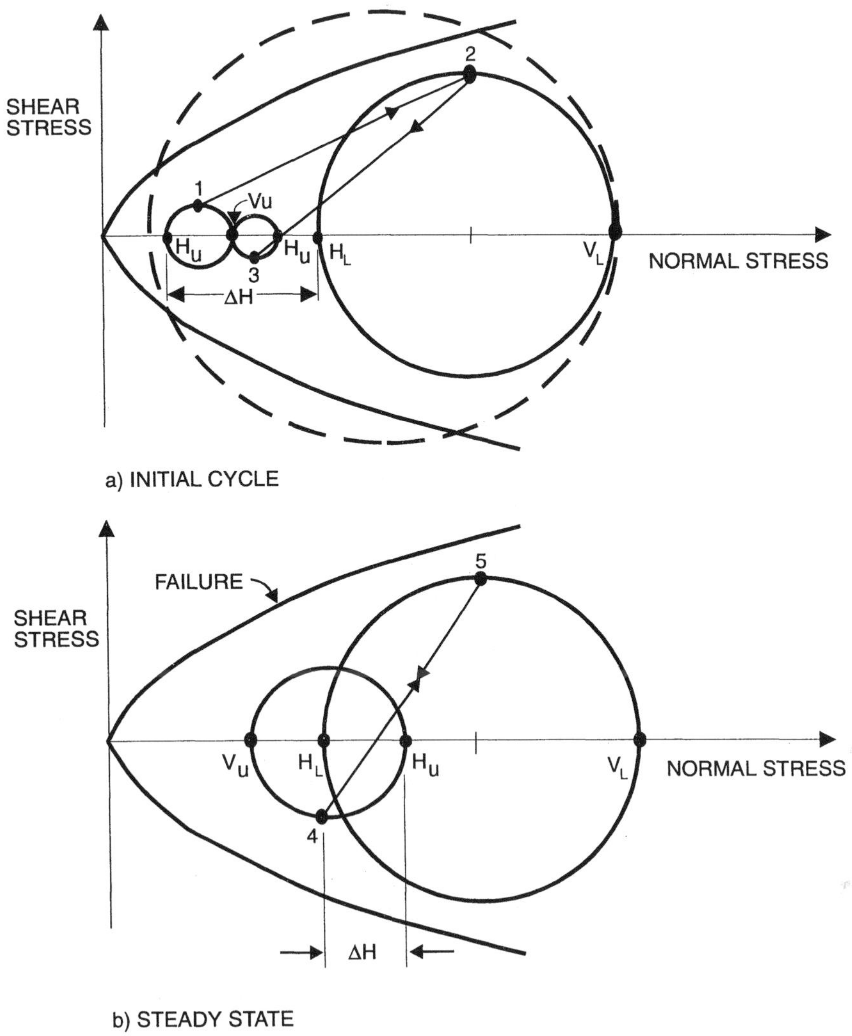

Fig. 6.15 Stress conditions in tensile zone

thus increasing the stiffness difference between the two layers. Permanent strain still accumulates slowly, but residual stress development is essentially complete so that the stresses alternate between the same loaded and unloaded states with subsequent cycles. The vertical stresses in the unloaded and loaded states (V_4 and V_5 , respectively) are shown about the same as initially, although the test data indicated an increase in both. When the load is applied horizontal stress decreases from H_4 to H_5 which means that the incremental horizontal stress ΔH is tensile.

Beginning with circle 3, shear stress reversal occurs with each load cycle. This means that on vertical unloading the internal shear stress resists the horizontal unloading and hence maintains the horizontal residual compressive stress. If the incremental horizontal stress was tensile on the first cycle of loading, then the loaded stress circle would be similar to the dashed circle shown in Fig. 6.15 and failure would have occurred. Although large strains do occur in the first cycle, the research shows that failure does not occur because ΔH is compressive.

Because ΔH is tensile in the bottom of the sand layer after steady state is reached, the residual horizontal stress, H_u , is the key factor limiting permanent deformation. This stress can be represented in the prediction model by KV_u where V_u is the vertical geostatic stress in the unloaded state, and K is the coefficient of earth pressure. The magnitude of K must clearly be much greater than unity and is limited only by passive failure conditions. The accuracy of permanent deformation prediction for layered systems thus will depend on the ability to predict K.

Analytical models for layered system behavior will not predict the development of residual stress because their soil constitutive relationships do not represent the inelastic behavior associated with the progressive changes in the early stages of cycling. Such models are more appropriate for predicting resilient layer response after many load cycles in which situation it is clear that failure conditions do not exist anywhere in the layer. Thus to avoid predicting failure at the bottom of the upper stiff layer and to properly model the layer response, residual stresses must be incorporated by superposition.

REFERENCES

6.1 Norman, G. M. and Selig, E . T. (1983). **Ballast performance evaluation with box tests.** *American Railway Engineering Association, Bulletin* 692, Proceedings Vol. 84, May, pp. 207-239.

6.2 Selig, E. T., Lin, H., Dwyer, L. J., Duann, S. W. and Tzeng, H. (1986). *Layered system performance evaluation.* Phase II Final Report, University of Massachusetts Report No. OUR85-329F, Department of Civil Engineering, University of Massachusetts at Amherst, Prepared for Office of University Research, U.S. DOT, April.

6.3 Wright, S. E. (1983). *Damage caused to ballast by mechanical maintenance techniques.* British Rail Research Technical Memorandum TM TD 15, May.

7. Properties of Ballast Material

The functions of ballast were listed in Chapter 2. Typical ballast material is gravel-size (nominal 20 to 50 mm diameter, with most particles between 6 and 64 mm diameter) crushed rock with durable particles. The ability of ballast to perform its functions is controlled by the particle characteristics together with the physical state of the assembly (structure and void ratio of assembly). No single characteristic controls ballast behavior. Instead the behavior is the net effect of combined characteristics.

Many relevant particle characteristics have been identified. These will be described in this chapter along with examples of their influence on individual properties, including strength, resilient modulus, and plastic strain. Then desired particle characteristics will be discussed, and ballast specifications which are used by various railroads will be given.

7.1. Tests for Particle Characteristics

Ballast material quality is defined by its particle characteristics. Many tests have been used to define these characteristics. These are listed in Table 7.1, grouped according to the general category of test. Each of these tests will be described and standard procedures referenced in this section.

7.1.1. Durability Tests

Currently, three abrasion tests are used by various railroads. These are the Los Angeles abrasion (LAA) test, the mill abrasion (MA) test, and the Deval, or British Standard, attrition test.

The LAA test has long been used as the primary abrasion test for ballast in North America. Even so, some railway engineers believe that the LAA test is not sufficient, and others believe that it is not useful. The Deval test is the primary test now used by British Rail. The MA test has been proposed as an abrasion test for North America which complements the LAA test. It is similar to the Deval test.

7.1.1.1 Los Angeles Abrasion

The LAA test (Ref. 7.1) is a dry test which is believed to measure a material's toughness or tendency for coarse breakage. The version of the test most suitable for ballast is ASTM C535. It consists of revolving 10 kg (22 lb) of dry material using one of three gradings with 12 steel balls weighing a total of 5 kg (11 lb) in a large steel drum for 1000 revolutions. Impact of the steel balls on the ballast particles causes crushing. The material is removed and the sample is washed on a 4.25 mm (No. 12) sieve. The LAA value is the amount of material less than the 4.25 mm (No. 12) sieve generated by the test as a percentage of the original sample weight.

Table 7.1 Tests for ballast particle characteristics

DURABILITY	• Los Angeles abrasion • Mill abrasion • Deval abrasion (wet and dry) • Clay lumps and friable particles • Crushing value • mpact
SHAPE AND SURFACE CHARACTERISTICS	• Flatness • Elongation • hericity • Angularity or roundness • Fractured particles • Surface texture
GRADATION	• Size • Size distribution • Fine particles content
UNIT WEIGHT	• Specific gravity • Absorption • Rodded unit weight
ENVIRONMENTAL	• Freeze-thaw breakdown • Sulfate soundness
CEMENTING CHARACTERISTICS	
IDENTIFICATION AND COMPOSITION	• Petrographic analysis • Chemical analysis • X- ray diffraction

7.1.1.2 Deval Attrition

In the Deval attrition test (Ref. 7.2) a specimen mass of 5 kg (11 lb) is rotated in a cylinder mounted on a shaft with the axis inclined 30 deg to the axis of rotation of the shaft. The cylinder is rotated 10000 times at a rate of 30-33 rpm. The inclined axis may provide some particle impact along with the rolling action found in the MA test. The attrition value is the mass of particles generated that are finer than the 2.36 mm size sieve (BS No. 7 sieve) expressed as a percent of the initial specimen mass. The test may be conducted with dry material or by adding an equal mass of clean water to the dry specimen.

The Deval test had originally been used in England for evaluating highway materials. This test was standardized in 1951 and adopted for use by the railroad industry. It is considered by British Rail (BR) to be their most important indicator of ballast performance in service because it can be used to define the minimum durability required to achieve the desired ballast life in the British climate with concrete sleepers. Since its adoption as a standard test, BR conducted an investigation of the effect of particle gradation, particle condition and presence of slurry on the results of the Deval test. The results are given in Ref. 7.3 and summarized in Ref. 7.4.

One of the results of the BR research was that the coarser the grading the greater the loss, i.e. more abrasion. An empirical relation was developed between the wet attrition value of smaller or larger size stone and the value which would be expected if the standard size had been tested. This relationship would be used when ballast material must be tested at a nonstandard gradation, e.g. when testing material taken directly from the track which cannot meet the Deval grading specifications.

Ballast which has been in service for a period of time will be worn and should have different abrasion characteristics than fresh material. To investigate the change in a material's abrasion characteristics due to wear, BR conducted two series of repeated wet attrition tests. In the first series, the slurry was removed from the cylinders after each 10000 cycles and the particles washed and returned to the cylinder. In the second series, the fine particles were retained in the cylinders after each 10000 cycles.

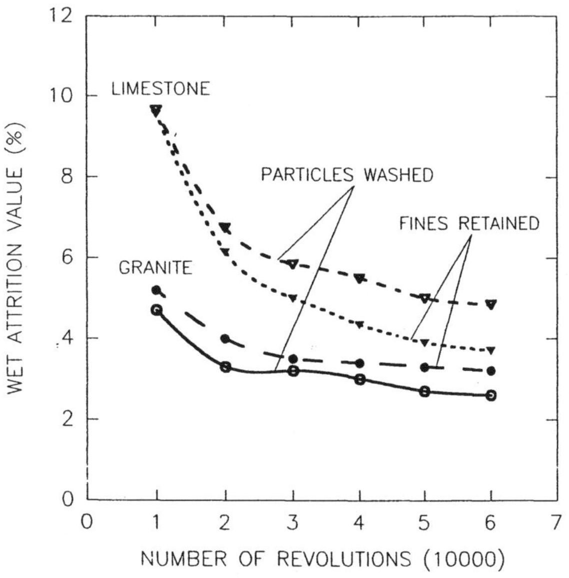

Fig. 7.1 Results of repeated wet attrition tests

In both types of tests (Fig. 7.1) the wet attrition value decreased significantly with each successive cycle, i.e. the resistance to attrition increased, as the particles became worn through testing. However, the rate of decrease was less for the limestone ballast when fines were retained than when the fines were washed out between tests. The reverse occurred for the granite ballast. This difference in material behavior has not been explained.

7.1.1.3 Mill Abrasion

The MA test is a wet abrasion test which involves revolving 3 kg (6.6 lb) of material of a specified gradation together with water about the longitudinal axis of a 229 mm (9 in.) external diameter porcelain jar for 10000 revolutions at about 33 rpm. Rolling of the particles causes wear without crushing.

The MA value is the amount of material finer than the 0.075 mm (No. 200) sieve generated by the test as a percentage of the original sample weight. In addition to this parameter, two other parameters were defined (Ref. 7.4) to further evaluate breakage of the material during the MA test. One of these parameters is total breakage, B, which is the percentage of material passing the sieve with an opening equal to one half the minimum particle diameter of the initial gradation. For example, for a gradation from 38 mm to 25 mm (1.5 in. to 1 in.), B is defined as the percentage of material passing the 13 m (0.5 in.) sieve. The other parameter is P, the percentage of total breakage which is less than the 0.075 mm (No. 200) sieve, and equals MA/B x 100%.

Information on the MA test is given in Refs. 7.4, 7.5, and 7.6. A standard ASTM test procedure has not been developed. An AREA committee is currently evaluating this test.

An investigation was conducted at the University of Massachusetts (Ref. 7.7) to determine the effect of gradation on the results of the mill abrasion test. Two materials were used, a basalt and a dolomite. These materials were chosen as examples of very good and fair ballast materials, respectively. The study included comparative effects of 1) coarse and fine uniform gradations, and broader gradations, 2) wet and dry test conditions, and 3) slurry and clean distilled water as the liquid phase. The results are discussed in Ref. 7.4 and summarized in Fig. 7.2. The larger the size particles the greater the MA abrasion value. This trend was reflected, too, in the comparison between the uniform and the broader gradations. The MA values for the broader gradations fell between the values for the uniform gradations for the sizes spanning the range of the broader gradations.

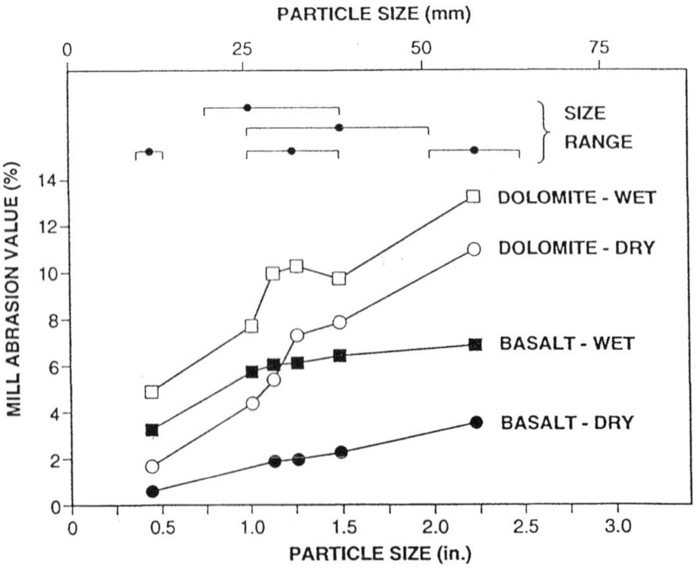

Fig. 7.2 Effect of particle size, rock type and water on mill abrasion value

The dry abrasion test conditions produced significantly less breakage and abrasion than the wet tests. This trend for the MA value can be explained in two ways: 1) water and fines will form a slurry which may increase the amount of abrasion, and 2) the particles in the dry tests as well as the inside of the jar become coated with fines; thus less abrasion could result because the particles are cushioned and not in a direct contact with either the jar or each other.

A study was made of the effect of slurry on the generation of fine material in the Mill Abrasion test. The gradation chosen was 38 mm to 25 mm (1.5 in. to 1 in.). Three conditions were examined: 1) fresh material with clean distilled water (standard test conditions), 2) fresh material with slurry from condition 1 tests, and 3) fresh material with slurry from condition 2 (i.e. greater concentration) slurry. The results show the same trend as for the BR Deval study. For the basalt material, the abrasion values increased with addition of slurry by nearly 10% compared to the "clean" test (condition 1), but the difference due to the amount of slurry concentration was minimal. However, for the dolomite material, both the condition 2 and 3 tests had lower MA values than the condition 1 (clean water) tests, even though the condition 3 test (greatest concentration of slurry) had a higher abrasion value than the condition 2 test.

The vast majority of fine material generated in the MA test is smaller than the 0.075 mm sieve (No. 200) and typically exceeds 90%. This is in contrast to the LAA test in which the majority of fine material is greater than 0.075 mm in size. The MA and LAA tests thus potentially are measuring different physical characteristics of the rock. For example, Raymond (Ref. 7.6) and Klassen et al. (Ref. 7.8) suggest that the MA test measures the rock particle hardness and the LAA test measures the rock particle strength or toughness. The LAA and MA tests were thought to be complementary and, hence, the Canadian Pacific Railroad characterizes (Ref. 7.8) the ballast by a combined index termed abrasion number (AN) defined as

$$AN = LAA + 5MA \quad . \tag{7.1}$$

7.1.1.4 Clay lumps and friable particles

The method of test for the approximate determination of clay lumps in the routine examination of aggregates is given by ASTM C142 (Ref. 7.9). This test describes clay lumps and friable particles as particles that can be broken with the fingers after the aggregate has been soaked in water for 24 ± 4 hours.

A related method of test for determining the quantity of soft particles in coarse aggregates on the basis of scratch hardness is given by ASTM C851 (Ref. 7.10). This test is used to identify materials which are soft, including those formed of a soft material and those which are so poorly bonded that the separate particles in the piece are easily detached from the mass. The test is not intended to identify other types of deleterious materials in aggregate.

7.1.1.5 Crushing test

The crushing test (Ref. 7.11) gives a relative measure of the resistance of an aggregate to crushing under a gradually applied load. A specimen is prepared in a steel mold 154 mm dia x 134 mm deep. A load is applied to the top of this specimen and is gradually increased to 390 kN (40 tonnes) in 10 minutes. The ratio of the weight of fines produced to the total sample weight, expressed as a percentage, is the crushing value.

7.1.1.6 Impact test

The impact value gives a relative measure of resistance to sudden shock loading or impact. For this test the prepared specimen is placed in a steel mold and is subjected to 15 blows or a 14 kg weight falling freely from a height of 380 mm. The ratio of the weight of fines produced to the total sample weight, expressed as a percentage, is the impact value.

7.1.2. Shape Tests

A description of shape tests and procedures is given in Ref. 7.12. A summary will be given in this section.

7.1.2.1 Flakiness or flatness

Flakiness and flatness are different names for the same shape characteristic. The British Standard (Ref. 7.13) defines a flaky particle as one in which the ratio of thickness to width is less than 0.6. The flakiness index is the percent by weight of flaky particles in a sample. The U.S. Army Corps of Engineers (Ref. 7.14) defines a flat particle as one with a thickness to width ratio of less than 1/3. Flatness is the percent by weight of flat particles in a sample. A more recent ASTM standard (Ref. 7.93), based on the Corps of Engineers method, provides a choice of three ratios: 1/2, 1/3 and 1/5.

7.1.2.2 Elongation

The British Standard (Ref. 7.13) defines an elongated particle as one with a length to width ratio of more than 1.8. The U.S. Army Corps of Engineers (Ref. 7.14) defines an elongated particle as one with a length to width ratio of greater than 3. In both cases the elongation index is the percentage by weight of elongated particles in a sample. A more recent ASTM standard (Ref. 7.93), based on the Corps of Engineers method, provides a choice of three ratios: 2, 3 and 5.

7.1.2.3 Sphericity

Sphericity is a measure of how much the shape of a particle deviates from a sphere. A perfect sphere has a sphericity of one. Sphericity, S_p, is defined as (Refs. 7.15 and 7.16):

$$S_p = \frac{\left(\frac{6V}{\pi}\right)^{\frac{1}{3}}}{L}, \qquad (7.2)$$

where V = volume of particle, and

L = diameter of the smallest sphere that would circumscribe the particle.

The measurement of particle volume may be made by water displacement.

Sphericity of a granular material is the average of the sphericities of the particles in a representative sample.

Marsal (Ref. 7.17) defined a sphericity shape factor, r_v, as

$$r_v = \frac{6V_a}{\pi\, d_r^{\,3}}, \tag{7.3}$$

where d_r = passing sieve size, and

V_a = average particle volume.

Average particle volume is defined by

$$V_a = \frac{W_a - W_w}{n\, \gamma_w}, \tag{7.4}$$

where W_a = weight of particles in air,

W_w = weight of particles in water,

n = number of particles, and

γ_w = unit weight of water

7.1.2.4 Angularity or roundness

Angularity, or its inverse, roundness, is a measure of the sharpness of the edges and corners of an individual particle. Roundness, ρ, is defined as (Ref. 7.15):

$$\rho = \frac{1}{N} \sum_{i=1}^{N} \left(\frac{r_i}{R} \right), \tag{7.5}$$

where r_i = individual corner radius,

R = radius of circle inscribed about the particle, and

N = number of corners on particle.

A projected image of the particle is used to obtain the measurements. Particles in a sample are grouped according to their angularity, with group categories ranging from angular to well-rounded. The average roundness for the sample, ρ_a, is a weighted average of the group roundnesses defined by

$$\rho_a = \frac{1}{N_t} \sum_{i=1}^{N_t} n_i \, m_i \,, \tag{7.6}$$

where n_i = number of particles in group i,

m_i = mid-point roundness of group i, and

N_t = total number of particles.

Visual estimates of particle roundness and sphericity are normally used in lieu of the more laborious method of image measurement. Charts to aid in this estimating are available, for example in Refs. 7.15, 3.12 and 3.13.

7.1.2.5 Shape factor

A shape factor for ballast was developed by Raymond in conjunction with research for the Canadian railroads (Ref. 7.6). His shape factor, S_f , is defined using all particles in a representative sample as

$$S_f = \frac{\Sigma L_i}{\Sigma T_i} \,, \tag{7.7}$$

where L_i = longest dimension of particle i, and

T_i = smallest dimension of particle i.

This shape factor is related to flatness and elongation characteristics.

7.1.2.6 Fractured particles

The Canadian Pacific Railroad has defined a test for percent fractured particles (Ref.7.8). The test divides a representative ballast sample into individual particle sizes. Each particle is then examined for fractured faces. A fractured face is a freshly exposed rock surface with a maximum dimension of at least 1/3 the maximum particle dimension. The included angle formed by the intersection of the average planes of adjoining fractured faces must be less than 135 deg for each of the faces to be considered as separate fractured faces. A particle is considered fractured if it has 3 or more fractured faces.

The percent of fractured particles in a sample, F_p, is defined by

$$F_p = \frac{W_p}{W_T} (100) \,, \tag{7.8}$$

where W_p = combined weight of fractured particles,

W_T = total weight of sample.

7.1.2.7 Surface texture

Surface texture relates to the frictional characteristics of the fractured surface. Surface texture is believed to have an important influence on ballast performance. However, a suitable direct measurement of this characteristic is not known. Instead this characteristic is considered to be incorporated in other tests such as the particle index test and the rodded weight test.

7.1.2.8 Particle index test

Although it is possible to separately measure the shape, the angularity, and the surface texture of individual particles, the effects of these on ballast stability are interrelated, and commonly a single number representing the shape, angularity, and surface texture of the aggregate is desired (Ref. 7.18). Huang (Refs. 7.19 and 7.20) developed a particle index test in which single-sized aggregate is rodded in a rhombohedral mold and the void ratio is compared with the void ratio of uniform spheres. Huang then compared the value of particle index with sphericity calculated according to the method developed by Wadell (Ref. 7.16) for gravel and crushed stone and found a good relationship between the two values for both materials. Subsequently the particle index test was modified to use a standard CBR (California Bearing Ratio) mold, and the test has been adopted as an ASTM Standard (Ref. 7.21).

With the CBR mold (152 mm diameter by 152 mm high) a maximum particle size of 19 mm (0.75 in.) is suggested. This is too small for ballast so a procedure modification would be required for use with ballast.

7.1.3. Gradation

The particle sizes in ballast (gradation) are determined by washing and mechanical sieving using procedures such as ASTM C117 (Ref. 7.22), C136 (Ref. 7.23) and D422 (Ref. 7.24). The gradation is normally presented on a cumulative frequency distribution plot with percent by weight passing a given size as the vertical axis and log of particle size as the horizontal axis. An example is given in Fig. 7.3. However, the gradation may also be shown as a frequency distribution plot with percent by weight of a given size (amount retained between 2 sieves) as a function of log of particle size. The example in Fig. 7.4 corresponds to the gradations in Fig. 7.3. The plot in Fig. 7.4 is most useful in showing the predominant sizes.

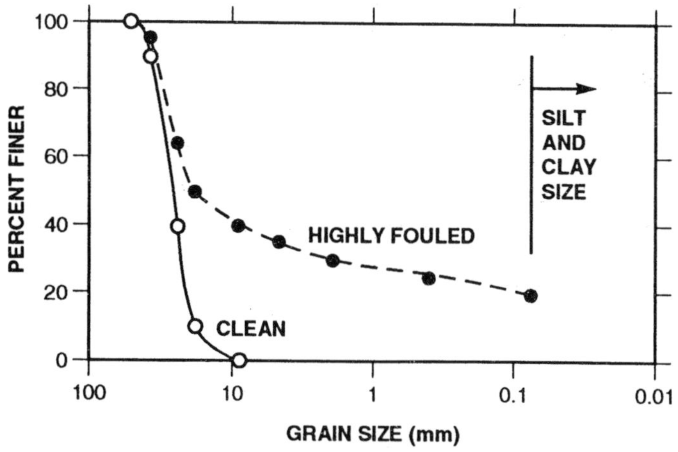

Fig. 7.3 Examples of ballast gradation curves

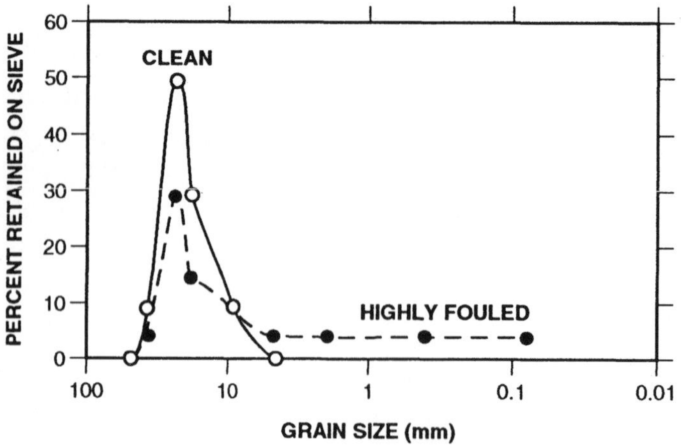

Fig. 7.4 Ballast gradation presented as a frequency distribution plot

A number of terms have been defined for qualitatively describing the shape of gradation curves. These are illustrated in Fig. 7.5. The term "uniformly-graded" means a relatively narrow range of particle sizes. New ballast may be considered to fit this category. The term "poorly-graded" has been used as an alternative to "uniformly-graded". However this alternative is not recommended because traditional ballast gradations are considered good for this application. The term "broadly-graded" means a wide range of particle sizes. The alternate term "well-graded" is not recommended because it, too, can be misleading. The desirability of a particular gradation depends on the application. A gap-graded material is one which has a relatively small amount of particles in a given size range within the total range of sizes for that sample. Fouled ballast is typically gap-graded.

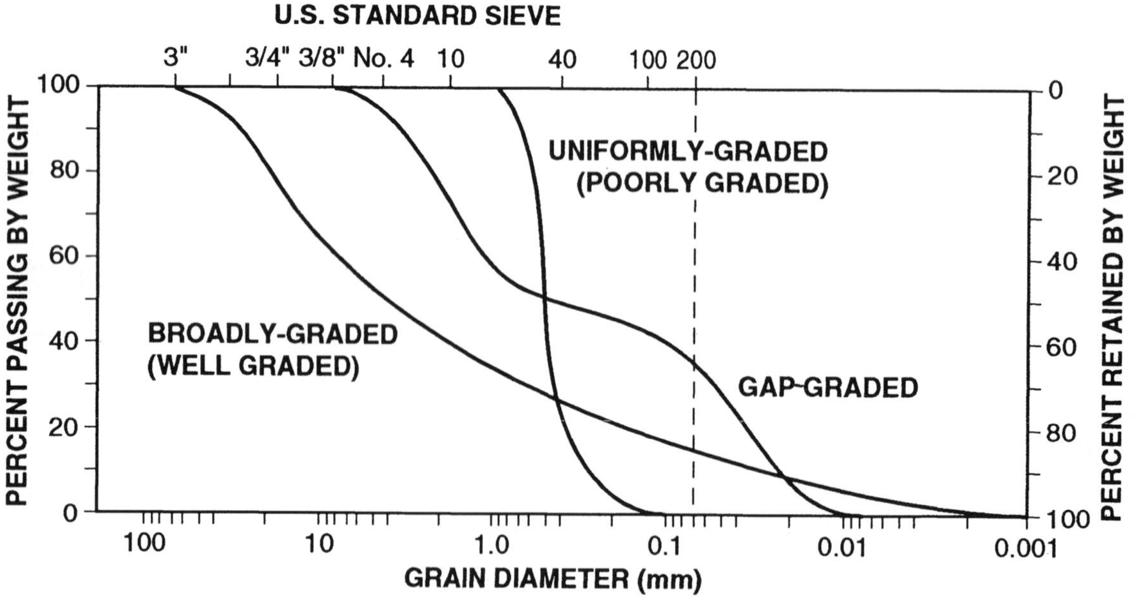

Fig. 7.5 Definitions of grain size distributions

Two gradation curve shape factors have been used in the Unified Soil Classification System (USCS) for providing a quantitative measure of the gradation. These are (Ref. 7.95) coefficient of uniformity, C_u , which really should be called coefficient of nonuniformity, defined as

$$C_u = \frac{D_{60}}{D_{10}}\ , \qquad (7.9)$$

and coefficient of curvature, C_c , defined as

$$C_c = \frac{(D_{30})^2}{D_{10}\ D_{60}}\ , \qquad (7.10)$$

where D_{10} = grain diameter corresponding to 10% passing by weight,

D_{30} = grain diameter corresponding to 30% passing by weight,

D_{60} = grain diameter corresponding to 60% passing by weight.

For gravel-size particles, in the USCS system a broadly-graded material is defined as one with $C_u > 4$ and C_c between 1 and 3. A $C_u < 4$ would be considered uniformly-graded if it is also not gap-graded.

Specifications for new ballast restrict the amount of fine particles. For ballast, fine particles may be considered any sizes on the small end of the gradation curve that are not desired, even though these small sizes may be permitted to a limited extent out of practical necessity. The term "fines" represents the subcategory of fine particles consisting of all particles smaller than the 0.075 mm sieve (No. 200). Thus "fines" means silt and clay sizes according to the USCS.

The degree of fouling (contamination) of ballast may be quantitatively represented from the gradation curve by the percent by weight of fine particles. This will always be equal to or greater than the percent fines.

Gradations were obtained for samples of ballast taken from a wide variety of track sites in North America (Refs. 7.25 and 7.26). Based on these data, representative gradations ranging from clean to highly fouled conditions were developed as shown in Fig. 7.6. Based on an examination of these data Selig proposed (see Tung, Ref. 7.26) a fouling index, F_I , defined as

$$F_I = P_4 + P_{200}\ , \qquad (7.11)$$

where P_4 = percent passing the 4.75 mm (No. 4) sieve, and

P_{200} = percent passing the 0.075 mm (No. 200) sieve.

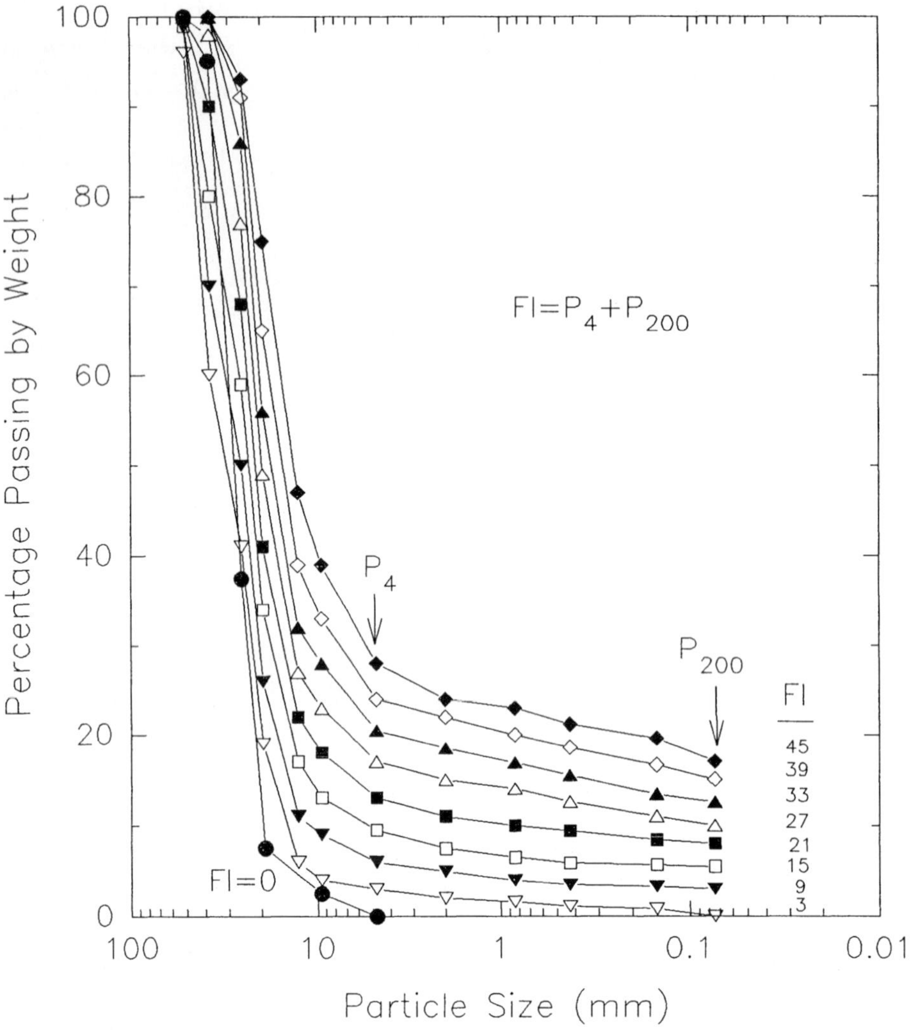

Fig. 7.6 Gradations representing ballast conditions from clean to highly fouled

Categories of fouling were then defined as given in Table 7.2.

Table 7.2 Fouling Index

Category	F_I
Clean	< 1
Moderately clean	1 to < 10
Moderately fouled	10 to < 20
Fouled	20 to < 40
Highly fouled	≥ 40

Percent fouling material for comparison, may be considered to be the percent passing the 9.5 mm (3/8 in.) sieve size.

7.1.4. Unit Weight

7.1.4.1 Specific gravity and absorption

An important factor influencing ballast unit weight is particle specific gravity. The ASTM test (Ref. 7.27) gives both bulk and apparent specific gravities. The amount of water that can be absorbed by the particles is also determined as part of this test. Water absorption is an indication of the rock porosity which relates to its strength as well as its tendency to break under freezing conditions.

Specific gravity is determined by a water displacement method. The following measurements are made:

A = weight of vessel containing specimen and water,
B = weight of vessel filled with water,
C = weight of saturated and surface dried specimen, and
D = weight of oven-dry specimen.

Water absorption, A_b (%), is defined as the weight of water absorbed by the particles divided by the dry particle weight, expressed as a percent. Thus

$$A_b = \frac{100\,(C - D)}{D} \quad . \tag{7.12}$$

Bulk specific gravity, G_b, is defined as the dry weight of particles divided by the weight of water in a volume equal to the total particle volume. Thus

$$G_b = \frac{D}{C - (A - B)} \quad . \tag{7.13}$$

The apparent specific gravity, G_a, is defined as the dry weight of particles divided by the weight of water in a volume equal to the total particle volume minus the volume of voids that will absorb water. Thus

$$G_a = \frac{D}{D - (A - B)} \quad . \tag{7.14}$$

7.1.4.2 Rodded unit weight

The dry unit weight of ballast may be determined by rodding the ballast to compact it in a rigid cylindrical container (Ref. 7.94). The container size depends upon the aggregate particle size. Typically for ballast the container volume will be 1 ft^3 (0.028 m^3), with the restriction that the height will be about the same as the diameter. The ballast is placed in three layers, each equal to about 1/3 of the container depth. Twenty five vigorous tamps with a steel rod are applied to each layer. The unit weight is defined as the weight of compacted aggregate filling the container divided by the container volume.

7.1.5. Environmental Tests

7.1.5.1 Freeze - thaw breakdown

Several versions of freeze-thaw tests have been used to evaluate aggregate resistance to disintegration by cycles of freeze-thaw (Ref. 7.18). In general, either partial or complete immersion can be used, and the fluid can be either water or a water-alcohol solution. Although the AASHTO procedure T103 (Ref. 7.28) does not recommend the number of cycles to be used, some authorities have recommended the use of as many as 50 freeze-thaw cycles. Raymond, et al. (Ref. 7.29) has used 58 cycles, with a 48 hour period per cycle. Resistance is measured by the change in gradation caused by the testing. Rapid freeze-thaw tests are open to criticism in that they do not simulate the actual rates of temperature change to which the ballast is subjected. However, a high number of cycles or a long cycle time is impractical and uneconomical in most approval tests because of the time and cost involved.

Particles with fractured surfaces may offer more opportunity for penetration of water and hence for freeze-thaw degradation. Also elongated, angular particles offer more surface area than do smooth, spherical particles.

7.1.5.2 Sulfate soundness

The sulfate soundness test for aggregates is used to determine their resistance to disintegration from freezing of absorbed water by subjecting the aggregates repeatedly to saturated solutions of sodium sulfate or magnesium sulfate. This method furnishes information helpful in judging the soundness of aggregates subject to freeze-thaw action, particularly when adequate information is not available from service records of the material exposed to actual weathering conditions. Test results by the use of the two salts differ considerably, so that care must be exercised in fixing proper limits in any specifications which require these tests.

The ASTM test specifications for soundness of aggregate by use of sodium sulfate or magnesium sulfate are given in Ref. 7.30. This test consists of five cycles of alternately soaking the aggregate in the saturated solution followed by drying. The weight of particles generated from breakdown during the test (defined as particles finer than the sieve size on which the original particles were retained) is measured. This weight expressed as a percentage of the initial specimen weight is the measure of soundness. Factors affecting the results of the soundness test include type of solution, number of cycles, void characteristics of particles, mineralogy of aggregate, and presence of fractured surfaces.

The soundness test relates to disruption of the rock by the growth of chemical crystals rather than by the growth of ice crystals during freezing and thawing and has been criticized because it does not simulate actual conditions (Ref. 7.18). Additionally the soundness test has been criticized because it is not reproducible in different laboratories, especially if sodium sulfate is used. West et al. (Ref. 7.31) concluded that the sodium sulfate test did not prove useful or reliable. Bloem (Ref. 7.32) found that magnesium sulfate produced more consistent results than did sodium sulfate, possibly because temperature variations affect the magnesium sulfate test less.

7.1.6. Cementing Characteristics

A study of ballast cementing tests by Feng (Ref. 7.33) provides the basis for this section. The term "cemented ballast" is used to describe a variety of conditions ranging from moist to dry, heavily fouled ballast to ballast conditions in which the coarse particles are bound together in a matrix formation. An example of the latter is the bonding of aggregates in concrete by portland cement. However, the railroad industry apparently generally means the former definition, i.e. moist to dry heavily fouled ballast, when referring to cemented ballast, and the source of fouling material is not specified.

The major constituents of portland cement are lime (CaO) and silica (SiO_2), both of which are minerals derived from rocks. Cement is produced by heating these in a kiln at a temperature exceeding 1500 deg C, and then grinding the resulting compounds into a powder. When mixed with water this powder hardens into a cement by hydration which is a chemical reaction combining the mineral compounds with H_2O. Such a cementing mechanism should not be expected to occur in ballast.

However, secondary or weak bonding can occur between fine particles in ballast from such mechanisms as van der Waal's forces, absorption of polar water molecules, electrostatic attraction, and mutual attraction by cations (Ref. 7.34). These bonding sources are only expected to be significant between the silt and clay size particles (fines). Another source of attraction is apparent cohesion from capillary water tension in partially saturated conditions. This mechanism can produce very high apparent cohesion in fines at low saturation conditions. These bonding mechanisms are the most likely to occur in ballast.

The determination of whether a chemical bonding type of cementing exists in ballast, or whether the cohesion is from capillary attraction or secondary bonding mechanisms, can be determined by soaking the ballast in water. Chemical bonding will remain strong whereas the other types of bonds will weaken and be lost. Both types require the ballast to be highly fouled with a significant portion of the fouling material being fines. Thus cementing cannot be a problem if the ballast is kept reasonably clean.

The original cementing value test for aggregate used in North America was established by Logan Walter Page of the U. S. Department of Agriculture in 1916 (Ref. 7.35). The actual cementing value was a blow count causing failure of a prepared briquette. In the 1930's, the American Railway Engineering Association revised this method, changing the evaluation criteria to the unconfined compressive strength of a prepared specimen (Ref. 7.36). Further modifications were made by Conrail (Ref. 7.37) and the National Crushed Stone Association in 1976 (Ref. 7.38), Amtrak in 1978 (Ref. 7.36), and Raymond in 1979 (Ref. 7.36). These procedures are described in Ref. 7.31.

In summary the test involves the following steps:

1) Generate fine particles, mostly fines, by crushing, grinding or abrading ballast particles.

2) Sieve the material to remove unwanted sizes.

3) Prepare a dough by mixing the fine particles with water.

4) Compress the dough under high pressure for a specified time in a forming mold.

5) Cure the specimens by drying.

6) Perform strength test.

No standard procedures for the ballast cementing test have yet been established. A study of the test by Raymond (Ref. 7.36) and by Feng (Ref. 7.33) showed that the results are highly dependent upon such factors as 1) how the fine particles are generated from ballast material to construct specimens, 2) the size of the fine particles, 3) the moisture content, compaction pressure, and curing conditions for specimen preparation, 4) specimen size, and 5) loading rate. These factors vary among the procedures used in the past, and hence the results by the various agencies are not consistent.

Correlation of the cementing test with field experience is lacking. Hence, an evaluation of the relevance of the cementing test is not available. Certainly the specimen preparation methods which have been used do not simulate field conditions.

7.1.7. Identification and Composition

7.1.7.1 Petrographic analysis

The value of petrographic analysis as a means of assessing and/or predicting behavior of an aggregate has been long recognized by the concrete industry. Techniques for evaluating aggregate for use in concrete and for examining hardened concrete have been established by ASTM in standards C295 (Ref. 7.39) for aggregate and C856 (Ref. 7.40) for hardened concrete. The purposes of this petrographic examination are 1) to determine the physical and chemical properties of the material which will have a bearing on the quality of the material for the intended purpose, 2) to describe and classify the constituents of the sample, and 3) to determine the relative amounts of the constituents of the sample, which is essential for the proper evaluation of the sample, especially where the properties of the constituents vary significantly. The value of the petrographic analysis depends to a large extent on the ability of the petrographer to correlate data provided on the source and proposed use of the material with the findings of the petrographic examination.

A method of numerical evaluation of the quality of aggregates used for concrete on the basis of petrographic analysis was developed by the Materials and Research Section of the Department of Highways (DOH) of Ontario (Ref. 7.41). This was based on extensive studies of field performance of aggregate used in concrete and bituminous pavements. With this method, a sample of aggregate is first examined for composition, i.e. rock type(s), and then assigned a number (factor) based on experience with that of similar rock type(s) and their observed field performance. Rock types considered to be excellent are given values close to 1 and those considered to be poor are given values close to 6. This number is then multiplied by the percentage of each rock type in the aggregate sample, giving a weighted value. The Petrographic Number, PN, for an aggregate sample is the sum of the weighted values for all components. This number is used as a quality rating for that sample. Limits have been set for acceptable PNs for various uses of aggregate.

After several years of application of this method to prediction and evaluation of aggregate performance, the DOH of Ontario drew several conclusions (Ref. 7.41).

1) Petrographic analysis is the most significant of all quality tests. The evaluation is based on actual performance and considers both the physical and chemical properties of the aggregate.

2) Petrographic analysis is the quickest test when time is an important consideration.

3) The results were found to be fairly reliable and to provide a better appraisal of aggregate performance than either abrasion or absorption alone.

4) The technique is adaptable to both the field and the laboratory.

5) The method can be used reliably to maintain the quality of an aggregate from a previously tested and known source.

Petrographic analysis is of major value in the selection of a suitable quarry for ballast and it can also be useful for prediction of the shape and character of the components of future ballast breakdown (i.e., the fines generated by breakage and abrasion of the ballast).

Several railroads use petrographic analysis as an aid in selecting ballast. For example, Canadian Pacific Rail (CP) has included petrographic analysis of potential ballast as a part of their standard ballast specification (Ref. 7.5). The CP requirements for the petrographic analysis of potential ballast include:

1) Description of rock types, including mineralogy, texture and structure,

2) Description of mechanical properties, including hardness, shape and type of fracture,

3) Description of chemical properties, including existing and potential chemical weathering,

4) Properties of the fine material, including shape and potential for weathering,

5) Estimation of index test results, and

6) Recommendation for special testing, if necessary.

Petrographic analysis involves the study of a rock in hand specimen as well as in thin section. Analysis of a material in hand specimen gives an indication of the approximate composition and texture of the rock. Thin section analysis allows a more precise assessment of the rock's constituents and texture. A thin section is a representative section of a rock that has been cut, mounted on a glass slide, and then ground to the standard thickness of 0.03 mm (0.0012 in.). At this thickness most minerals will transmit light. The thin section can then be examined using a transmitting light, polarizing microscope.

Most silt and clay size particles cannot be adequately identified using an optical microscope. A scanning electron microscope is useful in examining the shape, texture and composition of these particles using magnifications of up to 5000 times. Examples are shown in Figs. 8.30 and 8.31.

7.1.7.2 Chemical analysis

Chemical analysis to determine the type and proportion of constituent elements in a material may aid in identification and in estimating potential for chemical breakdown. Standard analytical chemistry techniques are available for this purpose. X-ray diffraction will also help in composition analysis. These techniques will be most useful for the fines in ballast since petrographic analysis will generally be adequate for visible rock particles.

7.2. Effects of Particle Characteristics on Behavior

The effects of some of the particle characteristics on mechanical behavior of granular materials, including ballast, have been examined by means of laboratory tests. Little information is available from field tests because of the large expense involved and the difficulty in controlling the variables.

7.2.1. Specific Gravity and Absorption

Although specific gravity is not necessarily related to aggregate behavior, Wadell (Ref. 7.16) indicates that three individual rock types having a low specific gravity (shale, sandstone, and chert) may display poor performance in concrete. Augenbaugh, et al. (Ref. 7.42) had a similar conclusion, finding that degradation of four different aggregate materials during compaction increased with decreasing bulk specific gravity. Raymond, et al. (Ref. 7.29) studies of the relationship between field breakdown and bulk specific gravity of ballast materials showed less breakdown for materials having a higher specific gravity. However, good correlation for field breakdown rating and absorption could not be established.

Brink (Ref. 7.43) reported that for a specific rock source, the absorption of the rock will be greater and the specific gravity will be lower for the more weathered material. Since the absorption of water by an aggregate has a relationship with the pore volume of the aggregate, Bloem (Ref. 7.32) believes it is more logical to evaluate the freeze-thaw resistance by using the simple absorption test rather than by the sulfate soundness test.

The absorption capacity of the aggregate is believed to be related to its weathering potential. However, based on the literature, there appears to be some controversy as to the significance of the test results (Ref. 7.18). Peckover (Ref. 7.44) feels that the absorption test offers little in addition to the soundness test and thinks a direct measure of the porosity of the rock is more important. Dalton (Ref. 7.45) found high testing variations (as much as 60 percent of the acceptable limit of 0.5 percent) for the absorption test. He also concluded that the absorption limit may allow approval of ballast which may break down too rapidly due to field weathering. Raymond et al. (Ref. 7.46) found good correlation between the freeze-thaw, soundness, and absorption test results.

7.2.2. Shape, Angularity and Surface Texture

Dunn and Bora (Ref. 7.47) tested a hard, crushed limestone aggregate in a special triaxial device. The particle size ranged from 4.8 mm to 38 mm (3/16 in. to 1.5 in.) and the percent flaky particles was varied from zero to 100% of the specimen. Any amount of flaky particles increased the shear strength, but the results suggest that the range of 25 to 75% flaky particles is better than 100%.

Gur, Shklarsky and Livneh (Ref. 7.48) used several different tests to evaluate the effect of flakiness on crushed material (type not indicated) ranging in size from 6.3 mm to 19 mm (1/4 in. to 3/4 in). The aggregate crushing value increased 2 1/2 times as the amount of flaky particles increased from zero to 100%. Over the same range the Los Angeles abrasion resistance value increased 4 times. Increasing the percent of flaky particles also increased the amount of breakage during compaction. Shear strength from triaxial tests was greater with flaky material than with nonflaky material. As a final test the aggregate was compacted in a box and then subjected on the surface to 9000 coverages with a rubber wheel. The rutting with the flaky material was roughly twice that with the nonflaky material. The increased rutting was attributed to particle alignment of the flaky material.

Siller (Ref. 7.49) found that the apparent cohesion intercept from triaxial tests on railroad ballast increased with increasing flakiness. He attributed this to increased particle interlock.

Eerola and Ylosjoki (Ref. 7.50) found that the triaxial shear strength of aggregate specimens increased in proportion to the ratio of particle length to thickness.

In a European study (Ref. 7.51) modulus of elasticity was calculated from loading tests on ballast specimens confined by steel rings. Ballast with flat particles generally gave a lower modulus of elasticity than with equidimensional particles.

Chen (Ref. 7.52) conducted triaxial tests on various sands and gravels with density ranging from loose to compact. Modulus was determined from the 25th cycle with loading to 30% of estimated strength. Strength was subsequently measured by loading to failure. The modulus decreased with increasing angularity, but the strength increased.

Holtz and Gibbs (Ref. 7.53) conducted triaxial tests on subangular to subrounded gravel and on sharp, angular crushed quartz rock, both with similar grading curves. As expected, the angular material had a higher strength.

Vallerga et al. (Ref. 7.54) conducted triaxial tests on both subrounded and angular sandstone aggregate at various void ratios. The strength of the angular aggregate was significantly higher than for the subrounded aggregate. At the same void ratios the difference in angle of internal friction was 4 to 8 deg. However at the same compactive effort the difference was reduced to about 1 deg. These authors also conducted triaxial tests on glass beads etched to provide a range of surface roughness. With increasing surface roughness the strength, measured by angle of internal friction, increased by about 8 deg.

Pike (Ref. 7.55) conducted shear box tests on 17 aggregates ranging in particle size from fine sand to coarse gravel. Increased angularity or surface roughness generally resulted in reduced dry density for the same compactive effort, but strength still increased.

Kalcheff (Ref. 7.56) showed that plastic fines lubricate the larger particles with a resulting shear strength reduction and increased strain at failure (Fig. 7.7a).

Holubec and D'Appolonia (Ref. 7.57) showed that increasing angularity of sands significantly increased the shear strength, but also increased the strain at failure and decreased the stiffness (Fig. 7.7b).

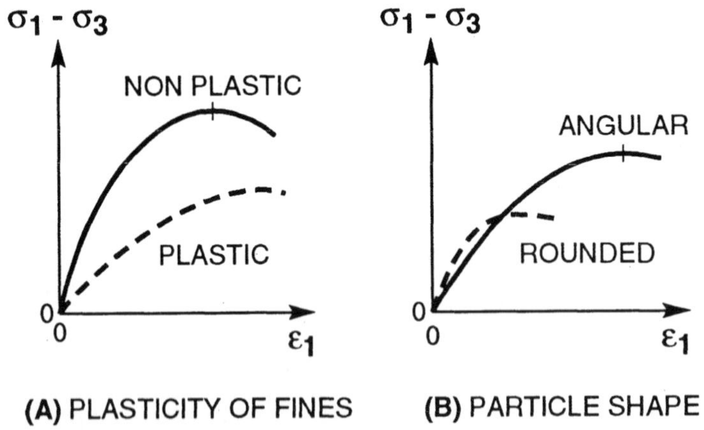

Fig. 7.7 Effect of plasticity of fines and particle shape on stress-strain characteristics

Edil and Luh (Ref. 7.58) showed that the shear modulus of sands, determined in the resonant column (vibration) test, increased with increasing roundness (decreasing angularity).

Norris (Ref. 7.59) showed that the shear strength of sand decreased significantly with increasing roundness.

Koerner (Ref. 7.60) conducted triaxial tests on sands of equal grading but different particle angularity and sphericity. He separated the measured effective angle of internal friction into frictional and dilational components. The dilational component was approximately independent of angularity, while the frictional component increased significantly with increasing angularity or decreasing sphericity for common relative densites. The same trends were previously shown by Kolbuszewski and Frederick (Ref. 7.61).

George and Shah (Ref. 7.62) compared triaxial test results on rounded gravel with the results on the same gravel after waxing the particles. The waxed particles produced a substantially lower strength. They also compared the triaxial results for platy crushed limestone with results for chunky crushed limestone. Although the same peak strength was obtained, the stiffness was less for the platy specimens. The authors did not indicate whether the above comparisons were made at common void ratios, however.

Thom and Brown (Refs. 7.63 and 7.64) showed an increase in resilient modulus with increasing particle surface friction angle for granular materials. They also showed that the resistance to plastic strain accumulation in repeated load tests increased with increasing particle visible surface roughness. Finally they show that increasing angularity and surface roughness increase shear strength.

7.2.3. Gradation and Size

Roenfeld (Ref. 7.66) conducted repeated load triaxial tests on limestone ballast. One set of specimens had a narrow range of particle size (coefficient of uniformity, CU = 1.14); the other set had a broader grading (CU = 4.1) but a slightly smaller mean size. The cumulative plastic strain for the uniform ballast was almost double that for the more broadly-graded ballast. Furthermore the particle degradation for the uniform ballast was four to five times greater than for the more broadly-graded ballast.

Klugar (Ref. 7.67) reported that replacing a small amount (<15%) of large particles in a ballast specimen by smaller particles increased the friction angle in shear box tests, while replacing a greater amount (>20%) caused a significant reduction in friction angle mobilized at a given displacement.

Rico, Orozco and Aztegui (Ref. 7.68) tested three different gradations of crushed basalt in a Texas triaxial cell. The comparisons were made at equal compactive effort. Thus void ratio and unit weight were not identical. A relatively uniform gravel size specimen gave a much higher strength than a relatively uniform sand size. For the same top particle size (38 mm or 1.5 in.) a more broadly-graded specimen gave a higher strength than a narrowly-graded specimen.

Marsal (Ref. 7.17) tested rockfill dam materials with a maximum particle size of about 150 mm (6 in.) using a large, high-pressure triaxial cell. He showed that shear strength increased as the gradation became broader for the same top size, even though the mean size decreased. Marachi, Chan and Seed (Ref. 7.69) came to the same conclusion for tests with common compactive efforts, although mineral composition and void ratio were not identical for all specimens.

Chen (Ref. 7.52) in tests previously described, found that increasing the coefficient of uniformity increased the shear strength, but did not affect the stiffness.

Kirkpatrick (Ref. 7.70) conducted triaxial tests on a sand with three different gradings in which the top and bottom sizes were kept constant while the mean size was varied. In general the shear strength increased as mean size decreased.

Leslie (Ref. 7.71) conducted triaxial tests on gravelly soils in which the minimum particle size was kept constant while the maximum size was increased, thus increasing mean size along with broadening the gradation. For comparable compactive effort, as mean size increased the shear strength decreased.

Triaxial tests on sands by Koerner (Ref. 7.60) showed little effect on shear strength of changes in coefficient of uniformity for common void ratios. In some cases increasing coefficient of uniformity at common relative densities increased the shear strength. However, strength increased as particle size decreased.

Triaxial tests by Marachi, Chan and Seed (Ref. 7.69) showed that strength increases as particle size decreases. However, Dunn and Bora (Ref. 7.47) in triaxial tests previously described, found that shear strength increased with increasing particle size. Finally triaxial tests by Vallerga et al. (Ref. 7.54) and by Holtz and Gibbs (Ref. 7.53) showed little effect of particle size on strength. Thus, the effect of particle size on strength is unclear. The same conclusion was indicated from the results of direct shear tests as summarized by Roner (Ref. 7.65).

7.2.4. Supplemental Triaxial Tests

7.2.4.1 Test description

To supplement results available in the literature, triaxial tests were conducted by Roner (Ref. 7.65) on nonflaky ballast specimens of several sizes and gradings and on flaky ballast specimens with both random and parallel particle orientations. The specimens were contained in a rubber membrane and a constant confining pressure applied through water. The ballast voids were also filled with water under a constant backpressure so that volume change could be determined from the amount of water flowing into or out of the specimen during the test. The effective confining pressure for all tests was 34 kPa (5 psi). Loading was provided by a compression machine at a constant rate of deformation.

The primary ballast used in the tests was a crushed quartzite. Its Los Angeles abrasion resistance was 20 and its mill abrasion value was 1.9. Because the quartzite did not contain sufficient flaky particles additional flaky particles were obtained from a sample of gneiss which had a Los Angeles abrasion resistance of 23 and a mill abrasion value of 2.0.

Four ballast gradings were prepared as given in Table 7.3.

Table 7.3 Ballast gradations for triaxial tests

Grading	Size Range (mm)	Size Range (in.)	Coefficient of Uniformity
1	38 to 13	1 1/2 to 1/2	1.47
2	38 to 29	1 1/2 to 1 1/8	1.03
3	29 to 19	1 1/8 to 3/4	1.03
4	19 to 13	3/4 to 1/2	1.03

Gradings 2, 3 and 4 are parallel and connected. Grading 1 consists of equal parts by weight of the other three gradings.

The regular (nonflaky) ballast specimens had 6% by weight of flaky particles according to the British Standard, but none according to the Corps of Engineers standard which was representative of the composition of the quartzite ballast as received. The first method defines flaky particles as having a width-to-thickness ratio of greater than 1.7, while the second method uses a width-to-thickness ratio of greater than 3. Thus the second is about twice as flat as the first.

The flaky ballast tested was composed of 96% by weight of flaky particles according to the British Standard, but only 4% by weight of flaky particles according to the Corps of Engineers Standard. The American Railway Engineering Association, prior to 1986, combined flat and elongated particles into a single definition which required the length-to-thickness ratio to be greater than 5. The flaky specimen had only about 3% by weight of flat and elongated particles by this definition. Only grading 1 was used for the flaky specimens because of the limited amount of flaky material available.

A range of void ratios was achieved by hand placing the particles combined with tamping, as needed, using a rubber-tipped falling weight device (Ref. 7.72). The flaky particles were placed either randomly or all parallel. The parallel plane was oriented at zero, 45 and 65 deg from horizontal, the latter representing the approximate orientation of the failure plane. Little variation in the void ratio could be obtained with the oriented specimens.

7.2.4.2 Test results

The maximum deviator stress (failure state) and corresponding angle of internal friction are shown as a function of initial void ratio in Figs. 7.8 and 7.9 respectively. No significant influence of particle size or gradation is evident for the nonflaky specimens within the range represented.

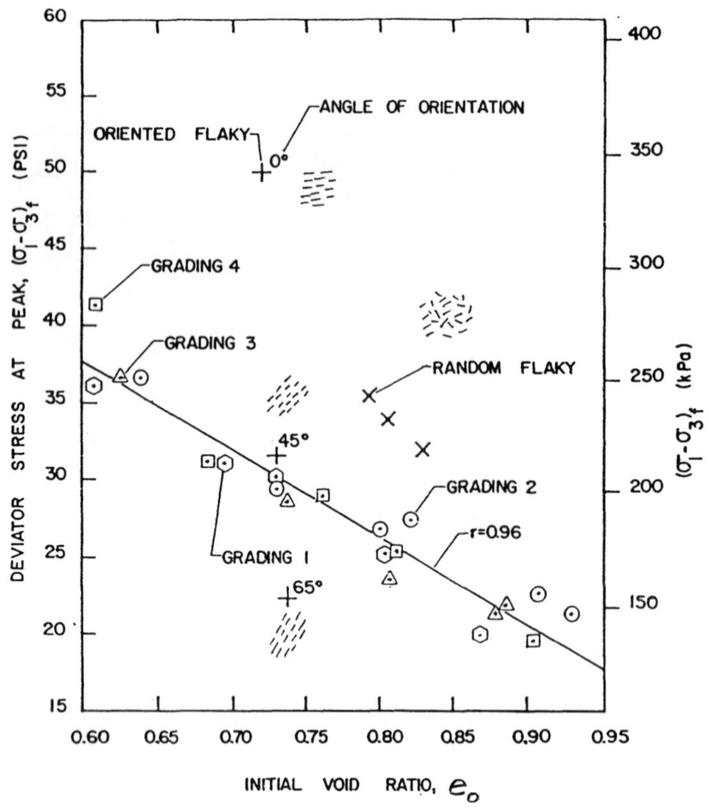

Fig. 7.8 Effect of initial void ratio on deviator stress at failure

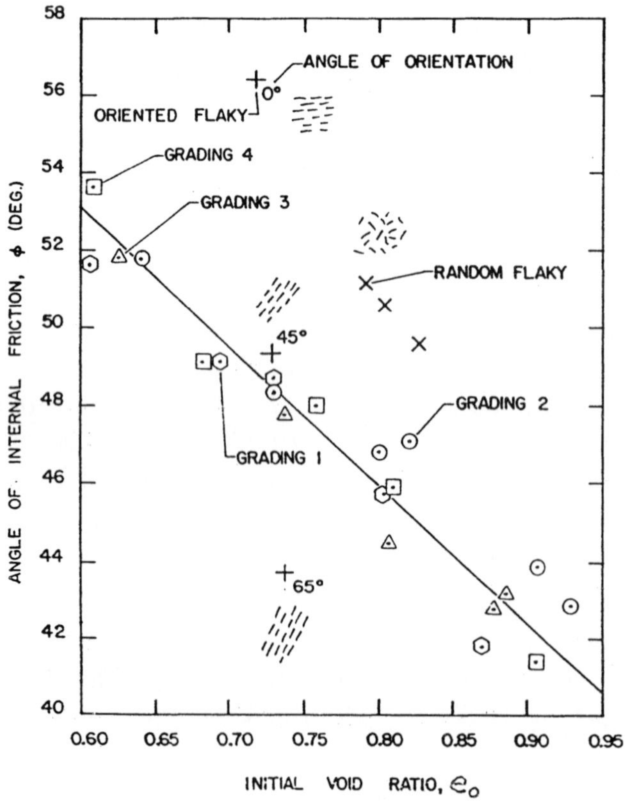

Fig. 7.9 Effect of initial void ratio on angle of internal friction

The volume change and axial strain at failure were used to calculate the secant Poisson's ratio at failure. The resulting values are shown as a function of initial void ratio in Fig. 7.10. Poisson's ratio decreases with increasing void ratio. All values exceed 0.5, signifying dilation even for the uncompacted specimens (highest void ratios).

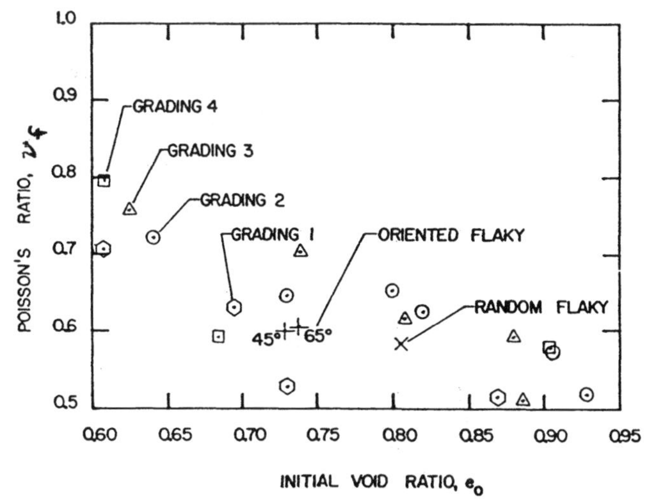

Fig. 7.10 Effect of initial void ratio on Poisson's ratio at failure

The maximum deviator stress and angle of internal friction for the flaky specimens are also shown in Figs. 7.8 and 7.9. The random flaky specimens had a strength that is significantly greater than the strength at the corresponding void ratios for the nonflaky specimens. The flaky specimen with horizontally oriented particles had by far the greatest strength of all specimens. Only by orienting the particles parallel to the failure plane did a substantially lower strength occur for the flaky specimens.

The Poisson's ratio values for all of the flaky specimens were consistent with those for the nonflaky specimens at the same void ratio (Fig. 7.10). Three flaky specimens are missing because of problems with volume measurement. It is presumed that these values would also have been consistent.

7.2.5. Conclusions About Several Particle Characteristics

The main problem in drawing conclusions about the individual effects of particle characteristics on aggregate specimen performance is that tests to determine these trends often involved changes in several significant variables simultaneously. For example, changes in size may also have included changes in particle composition, void ratio, gradation or angularity. In the following summary the effects of these independent factors are separated as well as possible by considering the conditions associated with all of the reviewed data as well as the supplemental tests conducted in this study. Some of the trends need further verification.

7.2.5.1 Shape, angularity and roughness

Any quantity of flaky particles, either randomly oriented or oriented other than generally parallel to the failure plane, increases the shear strength of the granular specimen. Orientation parallel to the failure plane, when a significant proportion of the particles are flaky, will cause a substantial strength reduction. The disadvantage of increased flakiness appears to be increased abrasion, increased breakage, increased permanent strain accumulation under repeated load, and decreased stiffness. A further study of these factors is needed.

Increased particle angularity, as is well known, increases the shear strength. However particle breakage increases and specimen stiffness decreases as well. At the same compactive effort a more angular material will tend to form a higher void ratio which will result in less strength increase than at a common void ratio.

Increasing particle surface roughness increases the shear strength of the assembly of particles. As with angularity this effect is greater at a common void ratio than at a common compactive effort.

No conclusions about the effects of elongation could be drawn from the literature.

7.2.5.2 Gradation

Strength was not affected by the gradation changes investigated by Roner (Ref. 7.65) for common void ratios. However, broadening the gradation by increasing the range of particle sizes with the same mean size should increase shear strength as a result of decreased void ratio that would exist with a broader grading.

Additional benefits of broadening gradation are decreased cumulative plastic strain under repeated loading and decreased particle degradation. However, the ballast stiffness will increase which could be a disadvantage if the resiliency becomes too low.

7.2.5.3 Size

The literature gives contradictory information about the effects of particle size on shear strength. Available results in some cases show an increase, in some cases a decrease, and in some cases no effect with increase in particle size. For the range tested by Roner (Ref. 7.65) mean particle size did not appear to affect strength when compared at common void ratios. However, increasing size will increase breakage.

7.2.5.4 Volume change

For all cases tested by Roner (Ref. 7.65) the ballast specimens expanded in volume as a result of shearing to failure. This means that the secant Poisson's ratio at failure exceeded 0.5 even for the uncompacted specimens. No distinctive affect on the relationship between Poisson's ratio and initial void ratio was detected as a result of changes in particle size, shape, or gradation.

7.2.5.5 Specific gravity and absorption

An increase in specific gravity may mean better durability under repeated load. Low absorption may indicate better weathering resistance.

7.2.5.6 Rodded unit weight

Rodded unit weight is a composite function of many other particle characteristics. Most important are size, size distribution, shape angularity, surface texture and specific gravity. an increase in rodded unit weight is desirable for increasing lateral stability of track.

7.2.5.7 Application to ballast

On the basis of strength considerations, broader ballast gradations are better than narrow gradations. However, other factors such as durability, fines storage volume, size segregation, and maintenance must also be considered in selecting gradation. More research on these factors is needed before an optimum grading can be established. Available information does not provide a basis for choosing an optimum ballast size. Factors other than stress-strain and strength characteristics may therefore govern the choice of size. Such factors include particle durability, and the movement and storage of fines within the voids.

Until 1986 the AREA specifications (Ref. 7.73) for ballast have limited the amount of flat or elongated particles (length to thickness ratio greater than 5) to 5% by weight. The flaky ballast tested by Roner (Ref. 7.65) is acceptable by this standard. However, only extremely distorted shapes would be eliminated by this specification. In 1986 the AREA adopted the Corps of Engineers definitions for defining flat (width to thickness ratio greater than 3) and elongated (length to width ratio greater than 3) particles. The percent of such particles allowed remains at 5% by weight. This requirement appears to be more restrictive, and hence better, than the previous AREA requirement, but the flaky ballast tested still is acceptable under this standard.

More research is needed to determine the important reasons for limiting the use of flaky ballast particles in track. Until such research is completed the best shape criteria and proper limits will not be known. However, if the effects of particle orientation on strength, and the effects of shape on stiffness that were indicated in this chapter for flaky ballast are to be avoided then an even more restrictive definition of shape would be required. The British standard should then be adopted, although perhaps a larger amount of such particles than 5% by weight could then be allowed.

7.2.6. Petrographic Analysis

Several investigators (Refs. 7.31, 7.74, 7.75 and 7.76) have attempted to correlate the results of petrographic analysis of the particles with field degradation. The results of petrographic analysis are greatly affected by the experience level of the petrographer and by the lack of quantitative analysis criteria. Because the nature of petrography is subjective, the results may have extreme variations.

An excellent discussion of the application of petrographic analysis to ballast evaluation is given by Watters et al. (Ref. 7.77). They note that the performance of rock ballast subject to the physical effects of traffic loading and to the chemical and physical effects of weathering, depends to a great extent on the characteristics of the rock which can be readily determined by petrographic analysis. Thus an experienced petrographer with knowledge of the requirements of ballast should be capable of at least a qualitative assessment of the performance potential of a particular rock choice. Furthermore an approximate estimate of the expected ranking in commonly used index tests (for example LAA, MA, sulfate soundness and absorption) also is possible. The big payoff in petrographic analysis occurs when it assists in avoiding the use of an unsatisfactory ballast with its associated high maintenance cost.

Watters et al. (Ref. 7.77) give examples of the relationships between petrographic results and rock particle properties, including characteristics of particles derived from breakdown. In addition they give examples of characteristics identifiable by petrographic analysis which, if present in abundance in a rock may represent "fatal flaws" and cause rejection of the rock for use as ballast. These examples are summarized in Table 7.4 taken from Ref. 7.77.

Table 7.4 Petrographic properties that may represent fatal flaws

Properties of Ballast Rock	Principal Deleterious Effect
MINERALOGICAL	
General high content of very soft minerals (e.g., clays, mica, chlorite)	Rapid physical degradation, clay and fine, mica-rich fines
Argillaceous sedimentary rocks (e.g., mudstone, shale)	Rapid physical degradation, clay-rich fines
Mica-rich metamorphic rocks (e.g., slate, phyllite, schist)	Rapid physical degradation, clay and fine, mica-rich fines
Igneous with deuterically altered feldspar	Rapid physical degradation, clay-rich fines
Sulfide-rich (> about 2 to 3 percent) (e.g., pyrite, pyrrhotite)	Oxidation of sulfide results in acidic conditions promoting chemical weathering of other mineral components
TEXTURAL	
Poor consolidation (in sedimentary and volcaniclastic rocks)	Rapid physical degradation by abrasion; susceptibility to freeze-thaw
High porosity (> about 5 percent) in sedimentary rocks	Degradation by freeze-thaw and abrasion if pores are large and abundant
Vesicularity (in volcanic rocks)	Degradation by freeze-thaw and abrasion
Friable texture in crystalline rocks	Rapid physical degradation by abrasion; susceptibility to freeze-thaw
STRUCTURAL	
Closely spaced joints, bedding partings, foliation	Rapid physical degradation by abrasion and freeze-thaw; generation of unsuitable particle shapes
PARTICLE SHAPE AND SURFACE CHARACTERISTICS	
Smooth particle surfaces (often due to rock texture)	Poor mechanical stability
Unsuitable particle shape	Poor mechanical stability; load fracture or elongated or tabular particles.

Boucher and Selig (Ref. 7.76) used petrographic analysis for evaluating the results of laboratory tests on a variety of ballast materials. The tests involved were Los Angeles abrasion, mill abrasion, cementing, and ballast box (Fig. 6.1). In general, petrographic analysis provided a reasonable explanation of the material performance in each of these tests. With further research petrographic analysis may play an important role in predicting field performance of ballast. But the success of this approach will most likely require an experienced petrographer who also has an understanding of the application.

7.3. Repeated Load

The behavior of a granular material under repeated loadingconditions is nonlinear, stress-state dependent, and very different from the behavior of the same material under monotonic loading condition (Chapter 3). During the first loading of a granular material such as ballast, the strain develops rapidly and is only partially recovered upon unloading. Each additional cycle contributes another increment of plastic or permanent strain (Fig. 7.11).

The magnitude of the plastic strain increment generally decreases with the number of cycles. The difference between the maximum strain under peak load and the permanent strain after unloading for each cycle is the resilient or recoverable strain. The resilient modulus of materials is defined as the repeated deviator stress divided by the strain. The amount of the resilient strain generally decreases with number of cycles.

7.3.1. Resilient Behavior

Under moderate levels of the same repeated load, the resilient strains become approximately constant after some number of cycles, and the material behaves elastically. Kalcheff and Hicks (Ref. 7.78) have indicated that only a few hundred cycles are necessary for this stabilization to occur. However, Brown (Ref. 7.79) and Morgan (Ref. 7.80) have shown that several thousand load repetitions are necessary to reach constant values of resilient modulus.

The magnitude of the resilient modulus is very much stress-state dependent. Several studies (Refs. 7.81, 7.82 and 7.83) have shown that the resilient response of unbound granular materials for tests of the type shown in Fig. 7.11 greatly increases as the confining pressure increases and is affected to a much smaller extent by the magnitude of repeated

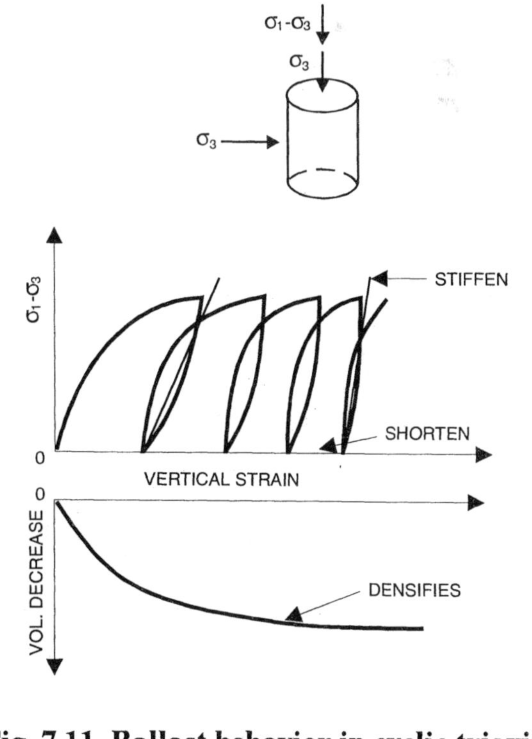

Fig. 7.11 Ballast behavior in cyclic triaxial test

deviator stress. This has led to the following commonly used relationship between resilient modulus and stress state:

$$E_r = K_1 (\theta')^{K_2} , \tag{7.15}$$

where E_r = resilient modulus,

K_1, K_2 = soil constants determined from laboratory test results, and

θ' = bulk effective stress in the loaded state.

The bulk effective stress equals the sum of the three principal effective stresses with the maximum load applied or

$$\theta' = \sigma_1' + \sigma_2' + \sigma_3' = 3\sigma_3' + (\sigma_1 - \sigma_3)_{max} . \tag{7.16}$$

Other relationships have also been considered. For example (Ref. 4.19) the following linear relationship was also found to be valid for repeated load tests on ballast:

$$E_r = K_3 + K_4 \theta' , \tag{7.17}$$

where K_3 and K_4 are soil constants.

Alva (Ref. 4.19) conducted repeated load triaxial tests on a crushed granite ballast for a range of confining pressures from 34 kPa (5 psi) to 145 kPa (21 psi) and a range of $(\sigma_1 - \sigma_3)$ from 69 kPa (10 psi) to 550 kPa (80 psi). Both compacted and uncompacted specimens were prepared. The results are given in Figs. 7.12 and 7.13 for comparison with Eqs. 7.15 and 7.17, respectively.

Although granular materials exhibit high values of resilient modulus even when the applied deviator stress level is near failure, as indicated in Ref. 7.84, the use of bulk stress cannot allow any distinction to be made between the effects of deviator stress and confining pressure. Furthermore, Eqs. 7.15 and 7.17 only represent a full unloading cyclic test (Fig. 7.11).

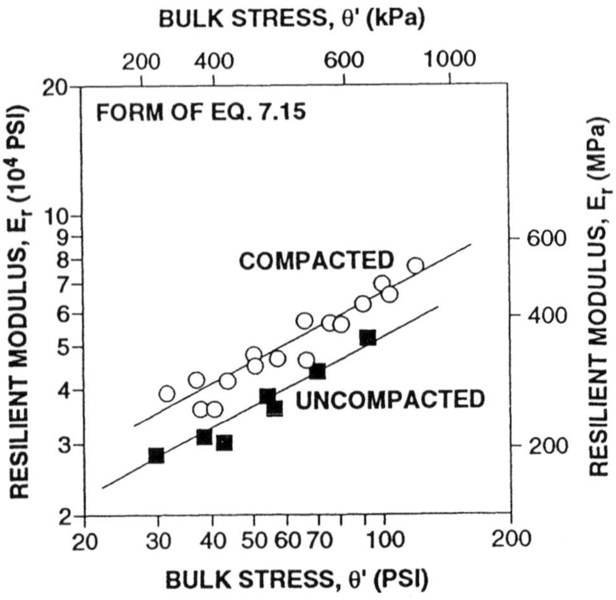

Fig. 7.12 Power relationship for resilient modulus

In order to obtain a more general relationship for resilient behavior of railroad ballast, cyclic tests were conducted (Ref. 7.85) that included partial unloading and shear stress reversal (Fig. 7.14). The cyclic shear stress reversal tests consisted of first applying an axial compressive stress followed by unloading to an extensional stress, and then reloading to isotropic conditions. Figure 7.14 shows that, for the same peak shear stress, as the cycling changes from partial unloading, to full unloading, to stress reversal, the resilient modulus decreases.

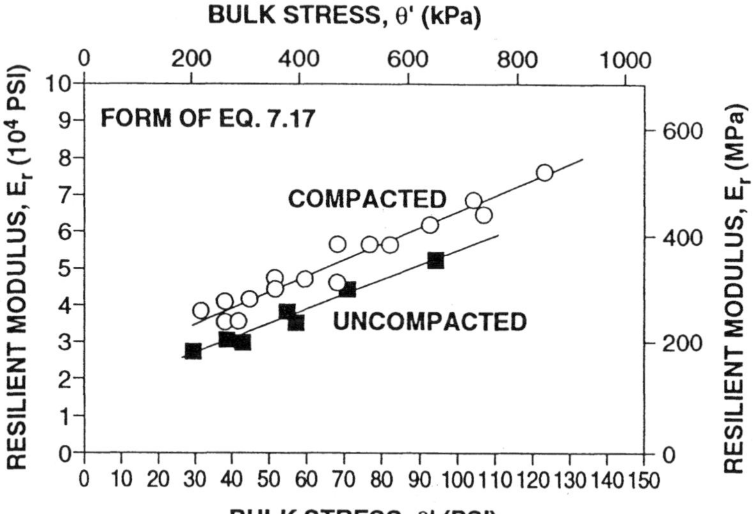

Fig. 7.13 Linear relationship for resilient modulus

Figure 7.15 shows the relationships between the applied stress levels and the vertical resilient strains for tests on a granite ballast. The resilient strains increase as the effective confining pressure decreases, for a given shear stress level. This trend is followed for both extension and compression. These data were used for characterization of the resilient behavior of the ballast materials.

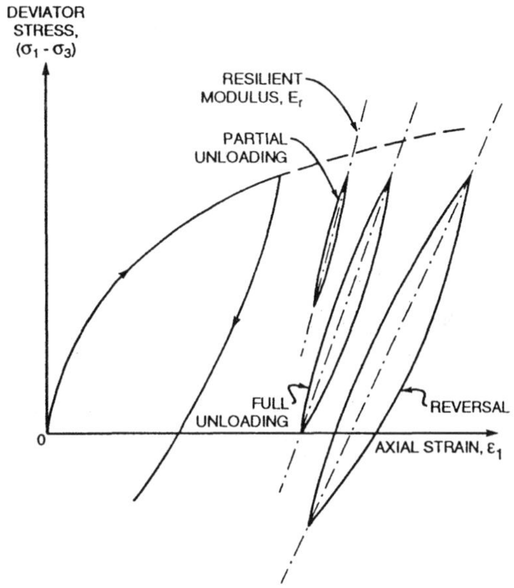

Fig 7.14 Effect of degree of unloading on resilient modulus

The relationship of Fig. 7.15 can be represented analytically for compression strains by

$$\varepsilon_{rc} = [0.00116\, q_c\, (\sigma_3')^{-1.15}]^{\eta} \,, \qquad (7.18)$$

where $\eta = 0.78\,(\sigma'_3)^{0.088}$,

ε_{rc} = resilient compression strain,

q_c = shear stress in compression (psi), and

σ_3' = constant confining pressure for cyclic test (psi).

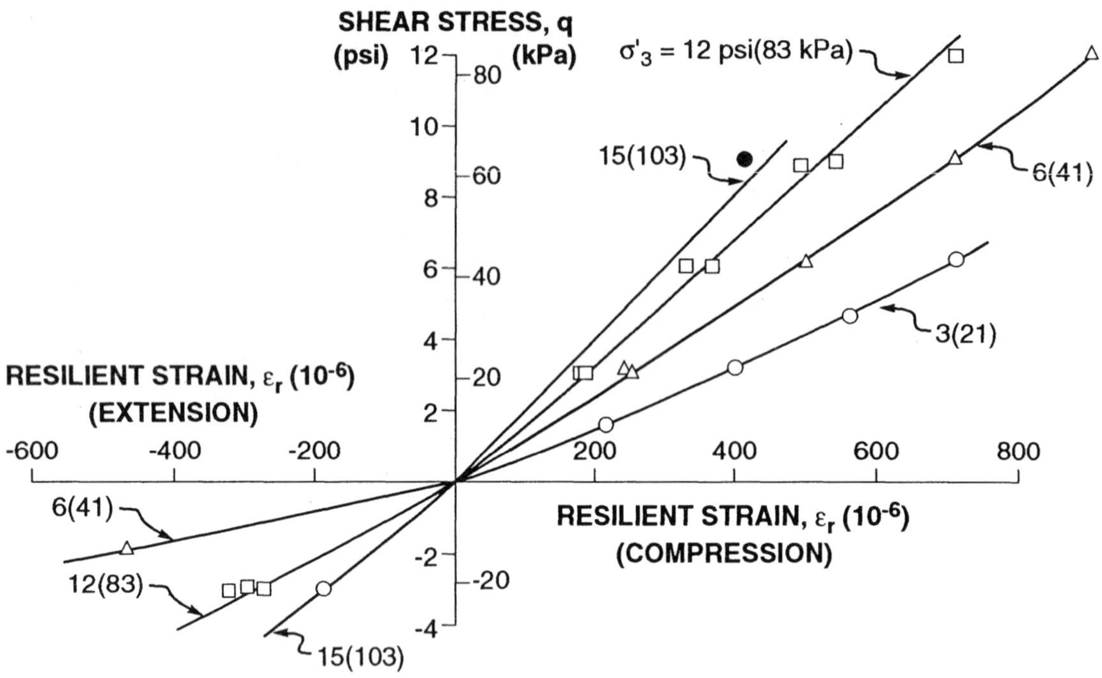

Fig. 7.15 Relationship of shear stress to resilient strain

For extension the relationship is

$$\varepsilon_{re} = 0.0039\, q_e \,(\sigma_3')^{-1.52}\,, \tag{7.19}$$

where ε_{re} = resilient extension strain,

q_e = shear stress in extension (psi),

σ'_3 = constant confining presure for cyclic test (psi).

The resilient modulus, E_r, is then calculated from

$$E_r = \frac{2(q_c - q_e)}{\varepsilon_{rc} - \varepsilon_{re}}\,. \tag{7.20}$$

Using lateral strain measuring instruments on the specimens, resilient Poisson's ratio can be also determined from cyclic triaxial tests. However, relationships between resilient Poisson's ratio and stress state are not as well defined as those for resilient modulus. Generally a constant value is assumed, even though this parameter can vary over a wide range with variation of cyclic stress or strain states for the same granular material, as shown in Fig. 7.16 (Ref. 7.86).

A more elaborate, but more accurate model for granular material resilient parameters is summarized by Brown and Selig (Ref. 3.16). This model uses resilient bulk modulus, B, and shear modulus, G, as a function of $(\sigma_1 - \sigma_3)$ and mean normal effective stress. These two resilient parameters were chosen because they relate to volumetric strain and shear strain, respectively, which are considered to be more fundamental than Young's modulus, E, and Poisson's ratio, v. However, E and v can be calculated from B and G as indicated in Chapter 3.

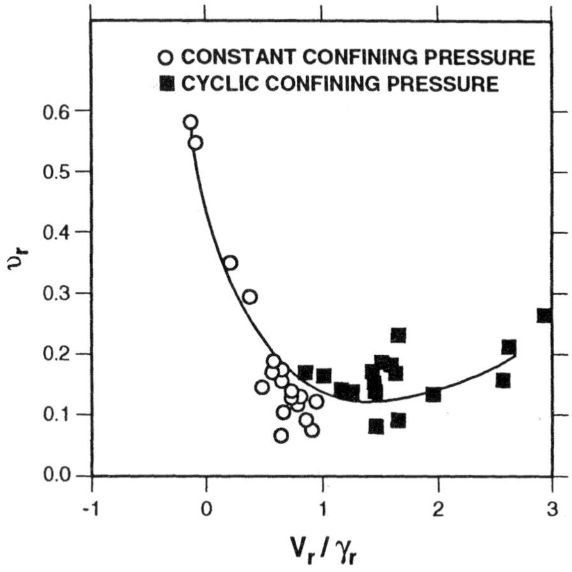

Fig. 7.16 Poisson's ratio as a function of volumetric to shear strain ratio

7.3.2. Plastic Behavior

The cumulative plastic deformation characteristics of ballast are a function of both the confining pressure and the cyclic deviator stress $(\sigma_1 - \sigma_3)$. An example is given in Fig. 7.17 (Ref. 7.87) for limestone ballast in a constant confining pressure test with full unloading. Increasing the stress ratio (ratio of deviator stress to confining pressure) increases the rate of strain accumulation.

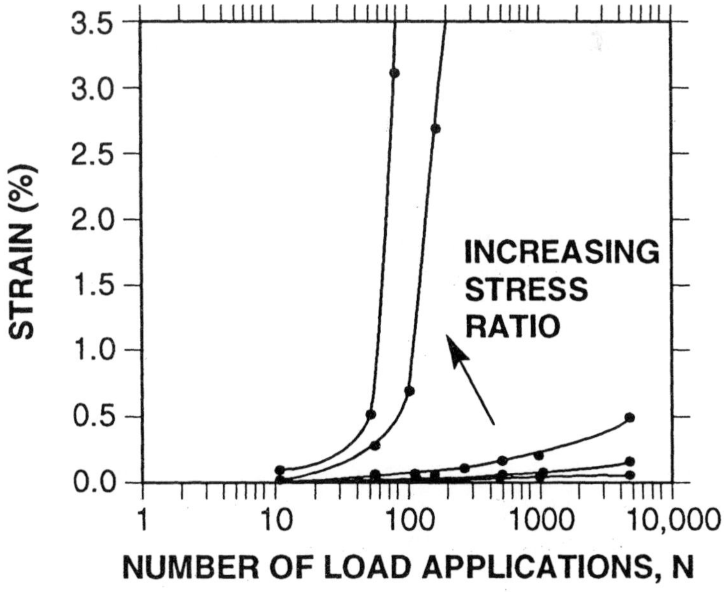

Fig. 7.17 Effect of stress level on permanent deformation response of ballast

The effect of initial density state on plastic strain accumulation is shown in Fig. 7.18 (Ref. 7.87) for constant confining pressure repeated load triaxial tests on limestone ballast with full unloading. Clearly specimen density has a significant effect, the lowest density producing the highest plastic strain in a given number of cycles, as expected.

Representative results of constant-amplitude repeated load tests with full unloading are shown in Fig. 7.19 (Ref. 7.88).

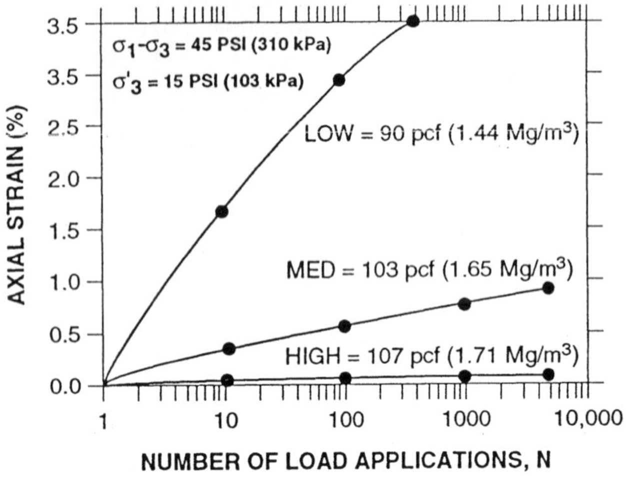

Fig. 7.18 Effect of density on permanent deformation response of ballast

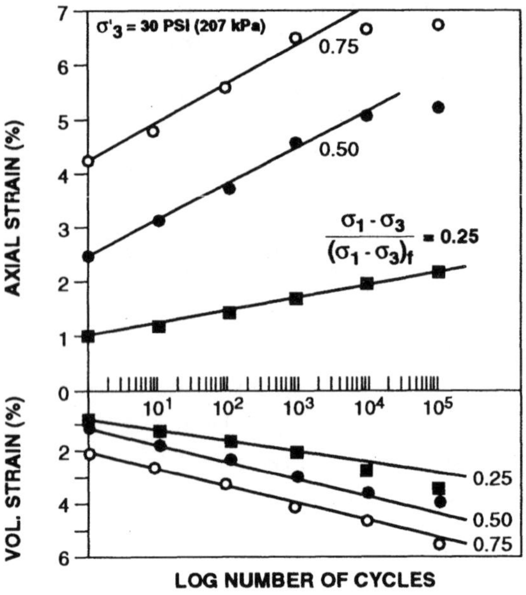

Fig. 7.19 Strain response of dolomite ballast as a function of number of load cycles

As the ratio of maximum applied deviator stress to failure deviator stress increased the plastic strain increased. The permanent strain after N cycles, ε_N, can be approximately related to the permanent strain after one cycle, ε_1, by the expression

$$\varepsilon_N = \varepsilon_1 (1 + C \log N) , \qquad (7.21)$$

where C is a material constant. Typical values are 0.2 and 0.4.

The Office for Research and Experiments, ORE, of the International Union of Railways (Ref. 7.89) undertook studies both in the laboratory and in the field to determine the behavior of the ballast layer in the track foundation under the action of repetitive loads. Laboratory repeated load triaxial tests were made on two ballasts. Different densities were tested. The results indicated that under controlled stress conditions, the ballast permanent strain can be predicted by the equation:

$$\varepsilon_N = 0.082 (100n - 38.2) (\sigma_1 - \sigma_3)^2 (1 + 0.2 \log N) , \qquad (7.22)$$

where ε_N = permanent strain after N loading cycles,

n = initial porosity of the sample,

$(\sigma_1 - \sigma_3)$ = deviator stress, and

N = number of repeated loading cycles.

The main observations from Eq. 7.22 were that:

1) The development of the permanent deformation of the ballast (deformation/load cycle) reduced considerably as the number of cycles increased.

2) The first load application produced a very large permanent deformation. The deformations produced by subsequent loading can be related to that produced at the first loading by Eq. 7.21 with C = 0.2.

3) The permanent deformation was very dependent on the initial compaction of the ballast, as represented by porosity.

4) The permanent deformation was proportional to the square of the applied deviator stress.

Shenton (Ref. 7.90) continued the laboratory testing undertaken by ORE to determine and quantify the effects of changes in the applied loading pattern. His trends were consistent with those of ORE. In addition he found:

1) The permanent strain is determined mainly by the largest load when two load levels are applied.

2) The permanent strain is reduced if full load removal is not allowed between load cycles.

Siller (Ref. 7.91) conducted 1000 cycle constant confining pressure repeated-load triaxial tests on a variety of ballasts. He also found that Eq. 7.21 provided a reasonable fit to the data, but values of C ranged from 0.25 to 0.4.

Constant confining pressure repeated-load triaxial tests were also performed on granite ballast in which the applied deviator stress, $(\sigma_1 - \sigma_3)$, was changed after each 1000 cycles. In each test a different sequence of load stages was used. Both full and partial unloading were involved.

Typical results for a staged test are shown in Fig. 7.20. Additional test results are given in Ref. 7.92). Some important trends identified by the staged tests are:

1) When the deviator stress was increased to a value greater than any past maximum values, the permanent strains continued to increase.

2) When the deviator stress was reduced to a value less than any past maximum values, negligible changes in permanent strain resulted with additional cycling.

3) Partial unloadings resulted in negligible increases in permanent strains with further cycles.

4) For any given effective confining pressure, the sequence of applied stresses did not affect the final values of permanent strain when the total number of cycles at each stress level was about the same.

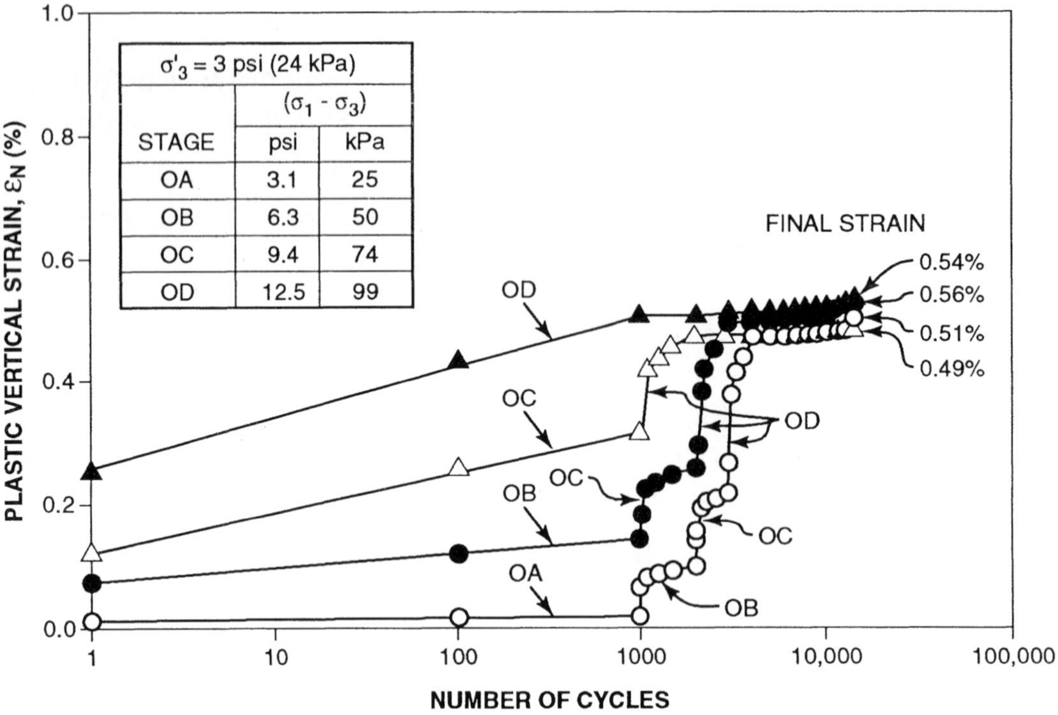

Fig. 7.20 Permanent strain accumulation for different loading sequences

Considering the accepted variability of sample preparation techniques and general ballast behavior, the strains measured at the end of all staged test series for similar effective confining stresses were in very close agreement. These ballast test results imply that the maximum wheel loads for a given maintenance life cycle will have a dominating influence on the amount of plastic strain accumulation.

Tests were also conducted with shear stress reversal (Ref. 7.92). The plastic strain at the end of the first cycle was always less than with no reversal as shown in Fig. 7.21. The plastic strain even can be negative (extension) if Δq is sufficiently larger than q_{max} . Regardless of the value of the first cycle plastic strain, the rate of strain accumulation was positive with further cycles, and the rate was always much greater with shear stress reversal than without (Fig. 7.22). The slopes of the lines in Fig. 7.22 are shown in Fig. 7.23 as a function of stress state to confirm this latter observation.

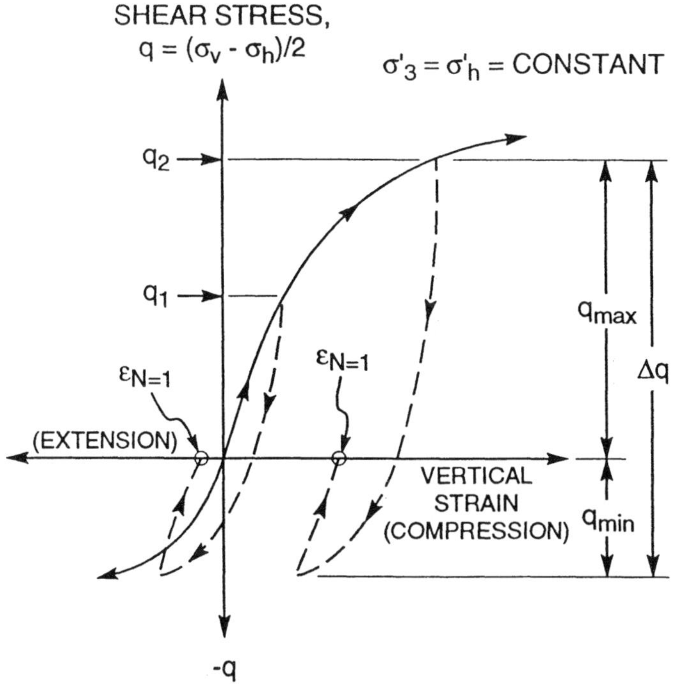

Fig. 7.21 Representation of stress-strain curves for shear stress reversal

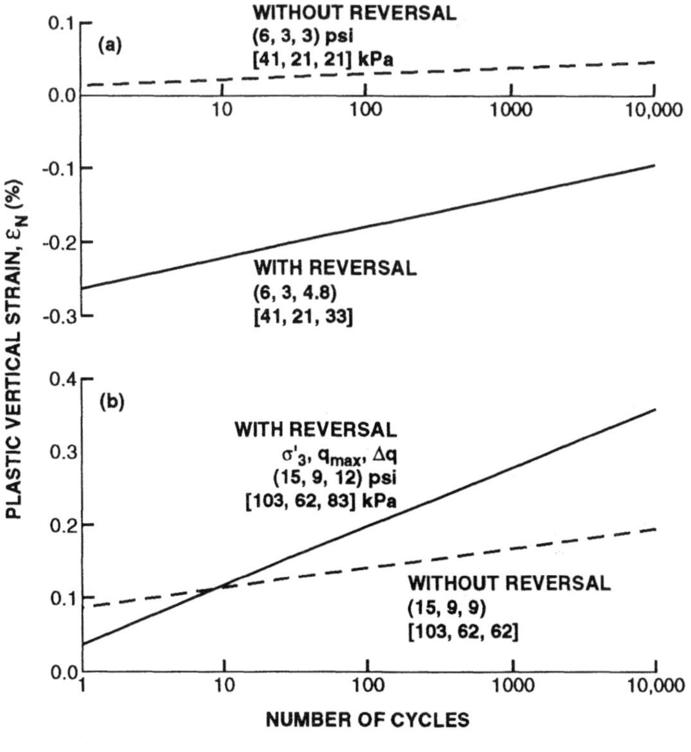

Fig. 7.22 Strain development rates for tests with and without shear stress reversal

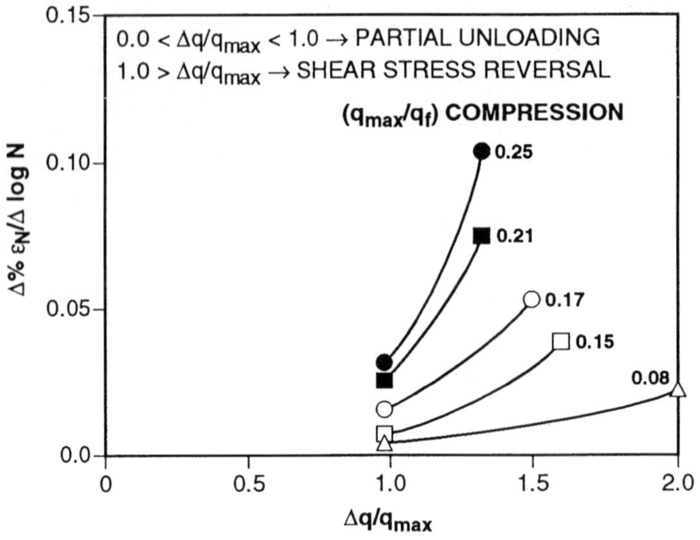

Fig. 7.23 Rate of strain accumulation as a function of stress parameters

REFERENCES

7.1 ASTM C535. **Resistance to degradation of large-size coarse aggregate by abrasion and impact in the Los Angeles machine.** ASTM Annual Book of Standards, Section 4, Construction,Vol. 04.03, *Road and Paving Materials.*

7.2 British Standard Institution (1951), *British Standard 812.*

7.3 Johnson, D. M. and Matharu, M .S. (1982). *Aspects of non-standard wet attrition testing of ballast.* **British Rail Research Technical Memorandum**, Report TM TD 7, November.

7.4 Selig, E. T. and Boucher, D. L. (1990). **Abrasion tests for railroad ballast.** *Geotechnical Testing Journal,* ASTM, Vol. 13, No. 4, December, pp. 301-311.

7.5 **CP Rail specification for ballast**. *Performance of aggregates in railroads and other track performance issues*, Transportation Research Record 1131, Washington, D.C., 1987, pp. 59-63.

7.6 Raymond, G. P. and Diyaljee, V. A. (1979). **Railroad ballast ranking classification.** *Journal of Geotechnical Engineering Division, ASCE,* Vol. 105, No. GT 10, October, pp. 133-153.

7.7 Boucher, D. L. (1987). *Petrographic analysis as an aid in the evaluation of ballast performance tests.* Report No. AAR86-336P, M.S. Degree, Department of Civil Engineering, University of Massachusetts, Amherst, MA 01003, September.

7.8 Klassen, M. J., Clifton, A. W. and Watters, B. R. (1987). **Track Evaluation and ballast performance specifications**. Transportation Research Record 1131, *Performance of aggregates in railroads and other track performance issues*, Washington, D.C., pp. 35-44.

7.9 ASTM C142. **Standard test method for clay lumps and friable particles in aggregates.** ASTM Annual Book of Standards, Section 4, Construction, Vol. 04.03, *Road and Paving Materials.*

7.10 ASTM C851. **Standard recommended practice for estimating scratch hardness of coarse aggregate particles.** ASTM Annual Book of Standards, Section 4, Construction, Vol. 04.02, *Concrete and Mineral Aggregates.*

7.11 British Standard Institution, **British crushing value test.** *British Standard 812.*

7.12 Roner, C. J., and Selig, E. T. (1992). *Definitions of particle shape factors and procedures for measurement.* Geotechnical report AAR84-316I, Department of Civil Engineering, University of Massachusetts, Amherst, June.

7.13 British Standard Institution (1983). **Methods for sampling and testing of mineral aggregates, sands and filters.** *British Standard 812.*

7.14 U.S. Army Corps of Engineers. **Method of test for flat and elongated particles in coarse aggregate.** *Test Standard CRD-C119-53.*

7.15 Pettijohn, F. J. (1957). *Sedimentary Rocks.* 2nd Ed., Chap. 2, Harper Bros., New York, 718 pp.

7.16 Wadell, H. (1933). **Sphericity and roundness of rock particles.** *Journal of Geology*, Vol. 41, pp. 310-331.

7.17 Marsal, R. J. (1967). **Large scale testing of rockfill materials.** *Journal of the Soil Mechanics and Foundation Engineering Divsion,* ASCE, Volume 93, No. SM2, Paper 5128, March, pp. 27-43.

7.18 Robnett, Q. L., Thompson, M. R., Hay, W. W., et al. (1975). *Technical data bases report, Ballast and Foundation Materials Research Program.* Report No. FRA/OR & D-76-138, prepared by University of Illinois for the U.S. DOT, Federal Railroad Administration, July.

7.19 Huang, E. Y. (1962). **A test for evaluating the geometric characteristics of coarse aggregate particles.** *ASTM Proceedings*, Vol. 62.

7.20 Huang, E. Y. (1965). *An improved particle index test for the evaluation of geometric characteristics of aggregates.* Michigan Highway Research Project No. 86546, Houghton, Michigan, July.

7.21 ASTM D3398. **Standard test method for index of aggregate particle shape and texture.** ASTM Annual Book of Standards, Section 4, Construction, Vol. 04.03, *Road and Paving Materials.*

7.22 ASTM C117. **Standard test method for materials finer than 75 micron sieve in mineral aggregates by washing.** ASTM Annual Book of Standards, Section 4, Construction, Vol. 4.03, *Road and Paving Materials.*

7.23 ASTM C136. **Standard test method for sieve analysis of fine and coarse aggregates.** ASTM Annual Book of Standards, Section 4, Construction, Vol. 4.03, *Road and Paving Materials.*

7.24 ASTM D422. **Standard method for particle-size analysis of soils.** ASTM Annual Book of Standards, Section 4, Construction, Vol. 04.08, *Soil and Rock.*

7.25 Collingwood, B. I. (1988). *An investigation of the causes of railroad ballast fouling.* M.S. degree project report, Report No. AAR88-350P, Department of Civil Engineering, University of Massachusetts, Amherst, May.

7.26 Tung, K. W. (1989). *An investigation of the causes of railroad ballast fouling.* M.S. degree project report, Report No. AAR89-359P, Department of Civil Engineering, University of Massachusetts, Amherst, May.

7.27 ASTM Standard C127. **Standard test method for specific gravity and absorption of coarse aggregate.** ASTM Annual Book of Standards, Section 4, Construction, Vol. 04.03, *Road and Paving Materials.*

7.28 AASHTO Freeze-thaw test.

7.29 Raymond, G. P., Gaskin, P. N., and Svec, O. (1978). **Selection and performance of railroad ballast.** *Railroad Track Mechanics and Technology*, A. D. Kerr, Ed., Proceedings of a Symposium at Princeton University in April 1975, Pergamon Press, pp. 369-386.

7.30 ASTM C88. **Standard test method for soundness of aggregates by use of sodium sulfate or magnesium sulfate.** ASTM Annual Book of Standards, Section 4, Construction, Vol. 04.03, *Road and Paving Materials.*

7.31 West, R. T., et al. (1970). *Tests for evaluating degradation characteristics of base course aggregates.* National Cooperative Highway Research Program Report 98.

7.32 Bloem, D. L. (1966). **Soundness of deleterious materials.** *Concrete and Concrete-Making Materials*, ASTM STP No. 196-A.

7.33 Feng, D. M. (1984). *Railroad ballast performance evaluation.* M.S. degree project report, Report No. AAR84-311P, Department of Civil Engineering, University of Massachusetts, Amherst, May.

7.34 Mitchell, J. K. (1976). *Fundamentals of Soil Behavior.* Wiley, N.Y.

7.35 U.S. Department of Agriculture (1916). *Bulletin* 347, pp. 15-23.

7.36 Raymond, G. P. (1979). **Examination of degraded aggregate, cement value test.** *Transportation Engineering Journal*, Proceedings of the American Society of Civil Engineers, Vol. 105, No. TE3, May.

7.37 Penn Central specifications M.W. 170, 4-69.

7.38 Kalcheff, I. V. (1976). *Memo to NCSA's* (National Crushed Stone Association) *Technical Research Subcommittee on Tests for evaluating degradation of railroad ballast materials*, December.

7.39 ASTM C295. **Standard practice for petrographic examination of aggregates for concrete**. ASTM Annual Book of Standards, Section 4, Construction, Vol. 04.02, *Concrete and Mineral Aggregates.*

7.40 ASTM C856. **Standard recommended practice for petrographic examination of hardened concrete.** ASTM Annual Book of Standards, Section 4, Construction, Vol. 04.02, *Concrete and Mineral Aggregates.*

7.41 Bayne, R. L. and Brownridge, F. C. *An engineering method of petrographic analysis for coarse aggregate quality*. Materials Laboratory, Department of Highways of Ontario.

7.42 Augenbaugh, N. B., Johnson, R. B., and Yoder, E. J. (1966). *Degradation of base course aggregates during compaction*. CRREL Technical Report, July.

7.43 Brink, R. H. and Timms, A. G. (1966). **Concrete aggregates: weight density, absorption and surface moisture.** *Concrete and Concrete-Making Materials*, ASTM STP 169-A, American Society for Testing and Materials, pp. 432-442.

7.44 Peckover, F. L. (1973). *Railway ballast material.* Office of Chief Engineer, Canadian National Railways, Montreal, Quebec, June 29.

7.45 Dalton, C. J. (1973). *Field durability tests on ballast samples as a guide to significance of the specification requirements.* Technical Research Centre, Canadian National Railways, Montreal, Quebec, January.

7.46 Raymond, G. P., Gaskin, P. N., Van Dalen, J., and Davies, J. R. (1974). **A study of stresses and deformations under dynamic and static load systems in track structure and support.** Canadian Institute of Guided Ground Transport, *Annual Report for 1973-74*, Project Report, Project 2.22, January, pp. 49-90.

7.47 Dunn, S., and Bora, P. K. (1972). **Shear strength of untreated road base aggregates measured by variable lateral pressure triaxial cell.** *Journal of Mechanics*, Vol. 7, No. 2, pp. 131-142.

7.48 Gur, Y., Shklarsky, E., and Livneh, M. **Effect of coarse-fraction flakiness on the strength of graded material.** *Proceedings of the Third Asian Regional Conference on Soil Mechanics and Foundation Engineering*, Haifa, Israel, Vol. 1, pp. 276-281.

7.49 Siller, T. J. (1980). *Properties of railroad ballast and subballast for track performance prediction.* Master's Project Report, Concrete Tie Correlation Study, Department of Civil Engineering, University of Massachusetts, Amherst, Massachusetts, December.

7.50 Eerola, M., and Ylosjoki, M. (1970). **The effect of particle shape on the friction angle of coarse-grained aggregates.** *Proceedings of the 1st International Congress of the International Association of Engineering Geology*, Paris, France, Vol. 1, pp. 445-456.

7.51 International Union of Railways, Office for Research and Experiments, Question D71 (1970). **Deformation properties of ballast (laboratory and track tests).** *Stresses in the rails, the ballast and in the formation resulting from traffic loads,* Report No. 10, Vols. 1 and 2, pp. 21-25.

7.52 Chen, L. S. (1948). **An investigation of stress-strain and strength characteristics of cohesionless soil.** *Proceedings of the Second International Conference on Soil Mechanics and Foundation Engineering*, Vol. 5, pp. 35-43.

7.53 Holtz, W. G. and Gibbs, H. J. (1956). **Triaxial shear tests on perviously gravelly soils.** *Journal of Soil Mechanics and Foundation Engineering Division, ASCE,* Vol. 82, No. SM1, Proceedings Paper 867, January, pp. 1-22.

7.54 Vallerga, B. A., Seed, H. B., Monismith, C. L., and Cooper, R. S. (1957). **Effect of shape, size and surface roughness of aggregate particles on the strength of granular materials.** ASTM STP 212, pp. 63-76.

7.55 Pike, D. C. (1973). *Shear-box tests on graded aggregates.* Transport and Road Research Laboratory, Department of the Environment, TRRL Report LR 584, Materials Division, Highways Department, Crowthorne, Berkshire, England.

7.56 Kalcheff, I. V. (1974). **Characteristics of graded aggregates as related to their behavior under varying loads and environments.** *Proceedings, Conference on Utilization of Graded Aggregate Base Materials in Flexible Pavements*, Illinois, March, 32 pp.

7.57 Holubec, I., and D'Appolonia, E. (1973). **Effect of particle shape on the engineering properties of granular soils.** *Evaluation of Relative Density and Its Role in Geotechnical Projects Involving Cohesionless Soils,* E. T. Selig and R. S. Ladd, Eds., ASTM STP 523, April, pp. 304-318.

7.58 Edil, T. B., and Luh, G. G. (1978). **Dynamic modulus and damping relationships for sands.** *Proceedings of the Special ASCE Conference on Earthquake Engineering and Soil Dynamics,* Vol. 1, June, pp. 394-409.

7.59 Norris, G. M. (1977). *The drained shear strength of uniform quartz sand as related to particle size and natural variation in particle shape and surface roughness.* University of California at Berkeley, PhD. Thesis.

7.60 Koerner, R. M. (1970). **Effects of particle characteristics on soil strength.** *Journal of Soil Mechanics and Foundation Engineering Division,* ASCE, Vol. 96, No. SM 4, Proceedings Paper 7393, July, pp. 1221-1234.

7.61 Kolbuszewski, J., and Frederick, M. R. (1963). **The significance of particle shape and size on the mechanical behavior of granular materials.** *Proceedings of the European Conference on Soil Mechanics and Foundation Engineering,* Vol. 1, Wiesbaden, Germany, pp. 253-265.

7.62 George, K. P., and Shah, N. S. (1974). **Dilatency of granular media in triaxial shear.** *Transportation Research Record* 497, pp. 88-95.

7.63 Thom, N. H. and Brown, S. F. (1988). **The effect of grading and density on the mechanical properties of a crushed dolomitic limestone.** *Proceedings of the Australian Road Research Board,* Vol. 14, No. 7, pp. 94-100.

7.64 Thom, N. H. and Brown, S. F. (1989). **The mechanical properties of unbound aggregates from various sources.** *Unbound Aggregates in Roads,* Jones and Dawson eds., Butterworth, London, pp. 130-142.

7.65 Roner, C. J. (1985). *Some effects of shape, gradation and size on the performance of railroad ballast.* M.S. degree project report, Report No. AAR85-324P, Department of Civil Engineering, University of Massachusetts, Amherst, June.

7.66 Roenfeld, M. A. (1980). *A study of mechanical degradation of a coarse aggregate subject to repeated loading.* Master's Thesis, University of Missouri, Rolla.

7.67 Klugar, K. (1978). **A contribution to ballast mechanics.** *Railroad Track Mechanics and Technology,* A. D. Kerr, Ed., Pergamon Press.

7.68 Rico, A., Orozco, J. M., and Aztegui, T. T. (1977). **Crushed stone behavior as related to grading.** *Proceedings of the Ninth International Conference on Soil Mechanics and Foundation Engineering*, Vol. 1, pp. 263-265.

7.69 Marachi, N. D., Chan, C. K., and Seed, H. B. (1972). **Evaluation of properties of rockfill materials.** *Journal of Soil Mechanics and Foundation Engineering Division,* ASCE, Vol. 98, No. SM 1, Proceedings Paper 8672, January, pp. 95-113.

7.70 Kirkpatrick, W. M. (1965). **Effects of grain size and grading on the shearing behavior of granular material.** *Proceedings of the Sixth International Conference on Soil Mechanics and Foundation Engineering*, Montreal, Vol. 1, pp. 273-276.

7.71 Leslie, D. D. (1963). **Large scale triaxial tests on gravelly soils.** *Proceedings of the Second Pan American Conference on Soil Mechanics and Foundation Engineering,* pp. 181-202.

7.72 Yoo, T. S., Chen, H. M., and Selig, E. T. (1978). **Railroad ballast density measurements.** *Geotechnical Testing Journal*, ASTM, Vol. 1, No. 1, pp. 41-54, March.

7.73 AREA (1990). *Manual for railway engineering.* American Railway Engineering Association, Washington, D.C.

7.74 Berard, J. (1973). *Petrographic study of ballasts.* Ecole Polytechnique, Montreal, Quebec, September.

7.75 Horn, H. M. and Deere, D. V. (1962). **Frictional characteristics of minerals.** *Geotechnique*, Vol. XII, No. 4, pp. 319-335.

7.76 Boucher, D. L. and Selig, E. T. (1987). **Application of petrographic analysis to ballast performance evaluation.** *Transportation Research Record* 1131, pp. 5-25.

7.77 Watters, B. R., Klassen, M. J., and Clifton, A.W. (1987). **Evaluation of ballast materials using petrographic criteria.** *Transportation Research Record* 1131, Performance of aggregates in railroads and other track performance issues, Washington, D. C., pp. 45-63.

7.78 Kalcheff, I.V. and Hicks, R. G. (1973). **A test procedure for determining the resilient properties of granular materials.** *Journal of Testing and Evaluation*, American Society for Testing and Materials, Vol. 1, No. 6, pp. 472-479, November.

7.79 Brown, S. F. (1974). **Repeated load testing of a granular material.** *Journal of the Geotechnical Engineering Division,* ASCE, Vol. 100, GT7, pp. 825-841, July.

7.80 Morgan, J. R. (1966). **The response of granular materials to repeated loading.** *Proceedings,* Third Conference of the Australian Road Research Board, pp. 1178-1192, Sydney, Australia.

7.81 Hicks, R. G. and Monismith, C. L. (1971). **Factors influencing the resilient response of granular materials.** *Highway Research Record* No. 345, pp.15-31.

7.82 Pell, P. S. and Brown, S. F. (1972). **The characteristics of materials for the design of flexible pavement structures.** *Proceedings, Third International Conference on Structural Design of Asphalt Pavements*, London, England, pp. 326-342.

7.83 Seed, H. B., Mitry, F. G., Monismith, C. L., and Chan, C. K. (1967). **Factors influencing the resilient deformations of untreated aggregate base in two-layer pavements subjected to repeated loadings**. *Highway Research Record* No. 190, pp. 19-57.

7.84 Raad, L. and Figueroa, J. L. (1980). **Load response of transportation support systems.** *Journal of the Transportation Engineering Division,* ASCE, Vol. 106, No. TE1, January, pp. 111-128.

7.85 Stewart, H. E. (1982). *The prediction of track performance under dynamic traffic loading.* PhD Dissertation, Department of Civil Engineering, University of Massachusetts, May.

7.86 Brown, S. F. and Hyde, A. F. L. (1975). **Significance of cyclic confining stress in repeated-load triaxial testing of granular material.** TRB, *Transportation Research Record* No. 537, pp. 49-57.

7.87 Knutson, R. M. (1976). *Factors influencing the repeated load behavior of railway ballast.* PhD Dissertation, University of Illinois at Urbana-Champaign.

7.88 Olowokere, D. O. (1975). *Strength and deformation of railway ballast subject to triaxial loading.* MS Thesis, Department of Civil Engineering, Queen's University, Kingston, Ontario, Canada.

7.89 Office for Research and Experiments, International Union of Railways (1970). *Stresses in the rails, the ballast and the formation resulting from traffic loads.* Question D71, Report No. 10, Volumes 1 and 2, Utrecht, Holland.

7.90 Shenton, M. J. (1974). *Deformation of railway ballast under repeated loading triaxial tests,* Soil Mechanics Section, British Railways Research Department, Derby, England.

7.91 Siller, T. J. (1980). *Properties of railroad ballast and subballast for track performance prediction.* M.S. Project Report, Department of Civil Engineering, University of Massachusetts at Amherst, Report No. FRA80-261P, December.

7.92 Stewart, H. E. and Selig, E. T. (1984). *Correlation of concrete tie track performance in revenue service and at the facility for accelerated service testing; Volume II, Predictions and evaluations of track settlement.* US DOT Federal Railroad Administration, Washington, D. C., Final Report, Report No. DOT/FRA/ORD-84/02.2, August.

7.93 ASTM D4791. **Standard test method for flat or elongated particles in coarse aggregate.** ASTM Annual Book of Standards, Section 4, Construction, Vol. 4.03, *Road and Paving Materials.*

7.94 ASTM C29. **Standard Test Method for Unit Weight and Voids in Aggregate.** ASTM Annual Book of Standards, Section 4, Construction, Vol. 04.02, *Concrete and Aggregates.*

7.95 ASTM D2487. **Standard test method for classification of soils for engineering purposes.** Annual Book of Standards, Section 4, Construction, Vol. 04.08, *Soil and Rock.*

7.96 Brown, S. F. and Selig, E. T. (1991). **The design of pavement and rail track foundations.** *Cyclic loading of soils: from theory to design*, O'Reilly and Brown Eds, Blackie, London, pp. 249-305.

8. Ballast Maintenance Cycle Characteristics

Under repeated loading from traffic, track progressively moves vertically and laterally causing deviations from the desired geometry. Because these deviations are generally irregular, ride quality decreases and the dynamic loads increase, causing increased geometry deterioration. In most present railroad maintenance practice, ballast tamping is used to correct track geometry effects that result from the repeated traffic loading. Tamping is the process of lifting and laterally adjusting track to the desired geometry while rearranging the upper portion of the ballast layer to fill resulting voids under the sleepers. This retains the sleepers in their raised position.

Tamping is the most effective way of correcting geometry faults. However, this desired objective is accompanied by some ballast damage from tamping, ballast bed loosening, and initially reduced resistance to rail lateral deplacement and buckling. The loosening results in further settlement with additional traffic, the degree of settlement increasing as the ballast deteriorates. Eventually tamping is again needed. Over a period of time fine particles derived from many sources accumulate in the ballast, a process known as "fouling". This impairs the functions of ballast including drainage and the ability to hold geometry after tamping. Eventually the ballast will need to be replaced or cleaned and returned to the track.

This reoccurring process can be termed the ballast maintenance cycle. In this chapter characteristics of the maintenance cycle will be presented and discussed. Chapter 14 describes the tamping machine and methods of application.

8.1. General Trends

To produce particle rearrangement tamping tools are inserted into the ballast between the sleepers and then are squeezed together beneath the sleepers (Fig. 8.1). Both steps are accompanied by vibration. This action causes some ballast particle breakage. The amount of breakage with each insertion may be small, but frequent tamping can add a significant amount to the breakdown caused by traffic. Some of the factors which affect the amount of breakdown are ballast type, tamping squeeze force,

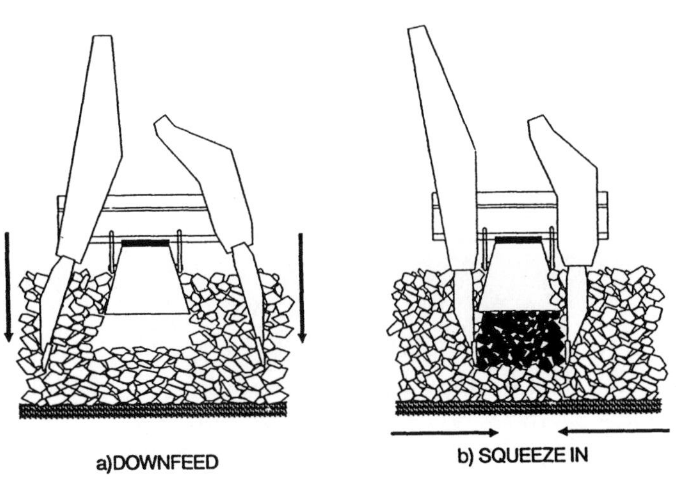

Fig. 8.1 Tamping action

tamping vibration characteristics, and number of tool insertions per machine pass. Thus the amount of damage can vary widely, but few data are available to quantify the effects of these factors.

The term "tamping" suggests compaction or densification. However, ample evidence exists that tamping loosens ballast. This loosening effect may require slowing train speed immediately after tamping because of temporarily reduced track lateral stability. The loosening effect is also the reason for a high rate of track settlement immediately after tamping. It is train traffic which causes the densification of ballast again. However some of this densification can be achieved during maintenance using compaction machines such as described in Chapter 14.

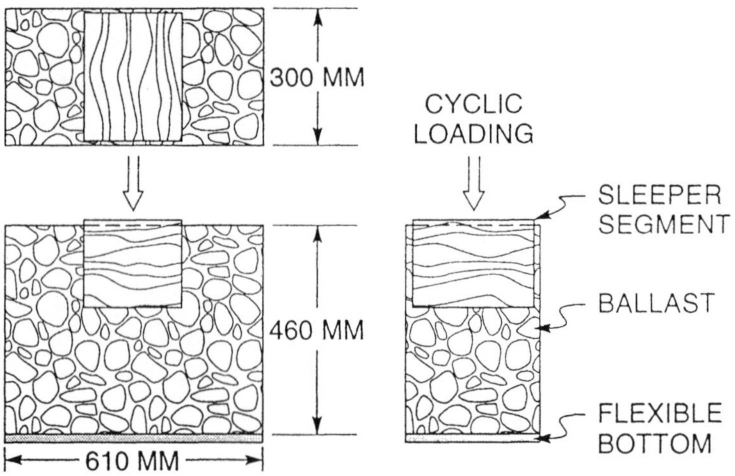

Fig. 8.2 Ballast box arrangement

In addition to loosening the ballast, particles are rearranged during tamping. This produces new particle contact points which causes increased breakage under additional traffic loads. This effect is demonstrated from ballast box tests designed to simulate field conditions. A schematic of the box set up is shown in Fig. 8.2. A wooden sleeper segment is repeatedly loaded with a force to produce the same average ballast contact stress as in the field. After a selected number of cycles the ballast is removed and the amount of breakage determined. Fine breakage represents particles less than 9.5 mm in size, while coarse breakage represents particles greater than 9.5 mm in size.

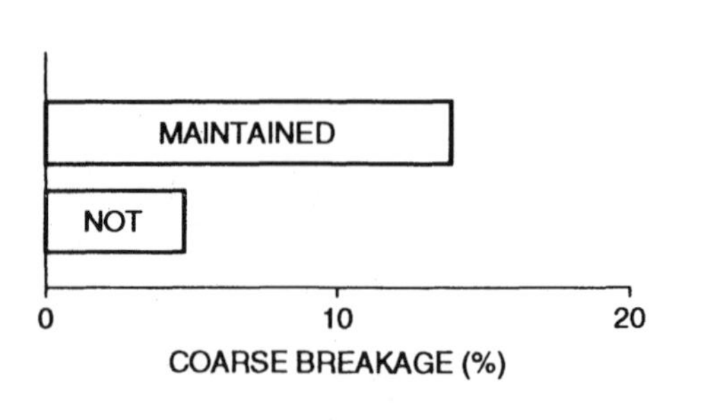

Fig 8.3 Effect of particle rearrangement on breakage

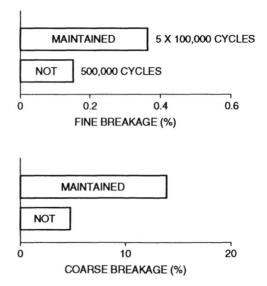

The results in Fig. 8.3 were obtained by applying 500,000 cycles of a vertical force representing a 160 kN (36000 lb) wheel load on the rail. In one test (designated "not") the particles were not disturbed during the 500,000

cycles. In the other test (designated "maintained") the particles were rearranged each 100,000 cycles to simulate the effect of tamping. The maintenance simulation more than doubled the breakage compared to no disturbance. Breakage from the tamping action itself is not included.

Repeated loading from traffic causes vertical plastic strains to accumulate in the ballast, subballast and subgrade layers. Track settlement results from these strains. In new construction significant contributions to settlement may come from the subballast and the subgrade because they have not previously been subjected to considerable traffic. The full depth of newly placed ballast will also produce a large share of the settlement. However, for track that has been in service for a long time, the layers that are not disturbed by tamping will usually contribute only a minor part to further settlement. This assumes a stable subgrade soil foundation. Hence the upper ballast layer which is disturbed periodically by tamping is generally the major source of settlement of track (Fig. 8.4).

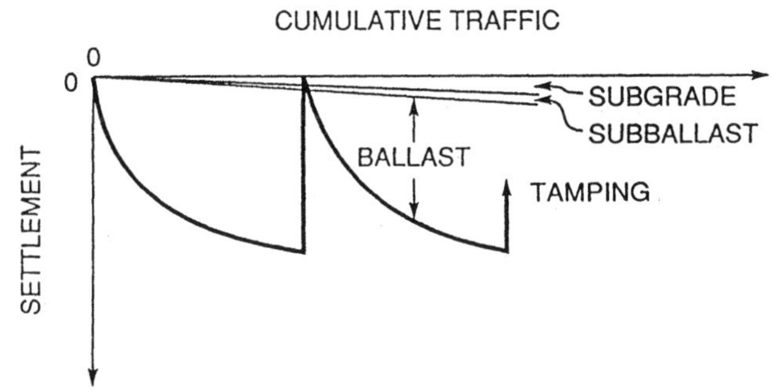

Fig. 8.4 Substructure contributions to settlement

The cumulative plastic strain is derived in part from a combination of material volume reduction (compaction) and lateral displacement under repeated load. In the ballast, particle breakage increases the volume reduction. Lateral displacement is restrained in part by horizontal residual stresses, especially in the granular layers, that develop from repeated loading. Tamping most likely reduces these residual stresses and hence indirectly increases the amount of settlement. These residual stresses have been demonstrated both in the box tests and in instrumented layered system studies (Chapter 6).

The trend for cumulative vertical strain in a layer with increasing load cycles, if tamping were not done, is shown in Fig. 8.5. This curve represents newly tamped ballast, newly placed subballast, or subgrade that has not previously been subjected to traffic. The solid portion of the curve assumes that the material properties remain about constant. However, as subgrade layers become saturated or as ballast becomes fouled (voids filled) the weakening of the material will cause large strains to again develop (dashed curve). This condition must be avoided because it will cause a rapid deterioration of track geometry.

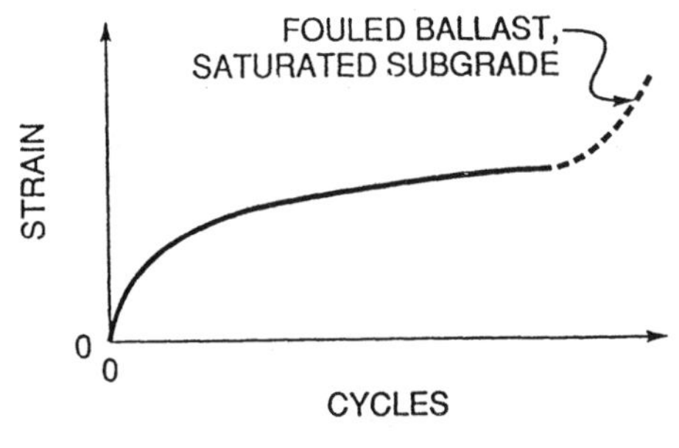

Fig. 8.5 Cumulative vertical strain in layer

One of the consequences of settlement is that a gap can develop in the unloaded track state between the base of the sleeper and the ballast. A gap can also occur from rail lift up ahead of the wheel. Impact can then occur as the gap closes rapidly with an approaching wheel load. If the gap fills with water and fine rock particles, a severe erosion process can add to the impact effects. This will increase the rate of settlement and the rate of ballast deterioration.

The importance of ballast drainage is clearly demonstrated by the above example. This means cleaning the ballast before it becomes too fouled to drain. This also means

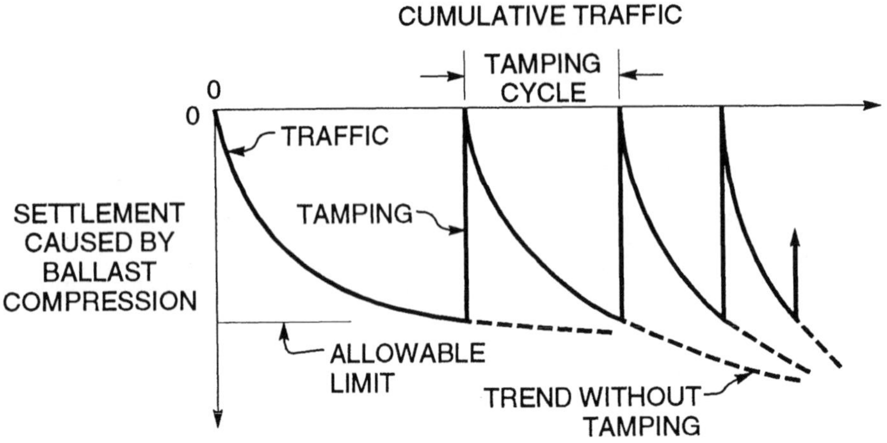

Fig. 8.6 Effect of progressive fouling on length of tamping cycle

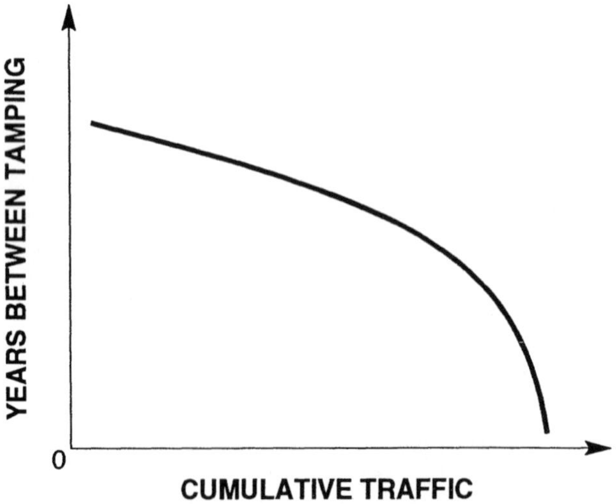

Fig. 8.7 Progressive decrease in time between tamping

minimizing gap development by tamping. If the ballast becomes highly fouled, however, the gap will reoccur rapidly. The consequences of the effects of progressive fouling of the ballast on the length of time between tamping (tamping cycle length) is illustrated in Figs. 8.6 and 8.7.

8.2. Field Measurement in Clean Ballast

Three methods will be used to show the physical state of ballast in the track. These methods are as follows:

1) Ballast density test (BDT),

2) Plate load test (PLT), and

3) Lateral tie (sleeper) push test (LTPT).

The ballast density test determines the in situ density or unit weight as a direct measure of compaction (Ref. 8.34). Conceptually, this test involves determining the volume of a membrane-lined hole excavated in the ballast by carefully measuring the amount of water required to fill the hole (Fig. 8.8). The density equals the weight of ballast removed divided by the volume of the hole.

Used in conjunction with this measurement is a reference density test, which provides the means to assess the relative amount of ballast compaction achieved in the field. This test involves compacting samples of ballast in a 305-mm (12-in.) diameter by 305-mm (12-in.) high steel container with a special rubber-tipped drop hammer. The reference density is essentially the maximum density achieved with this technique.

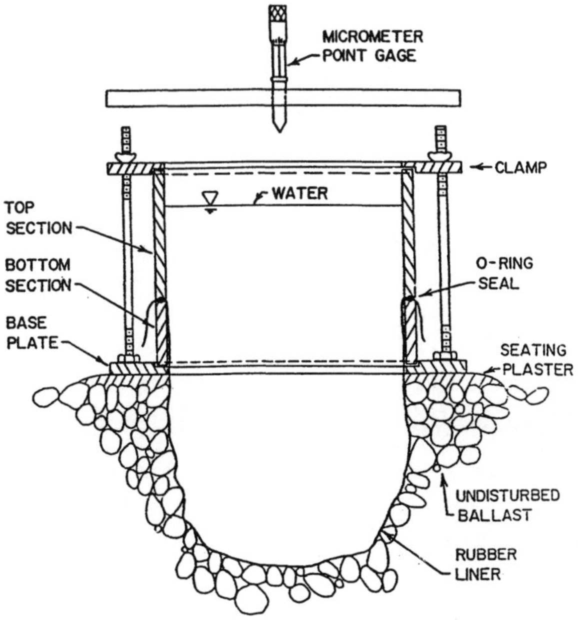

Fig. 8.8 Ballast density test

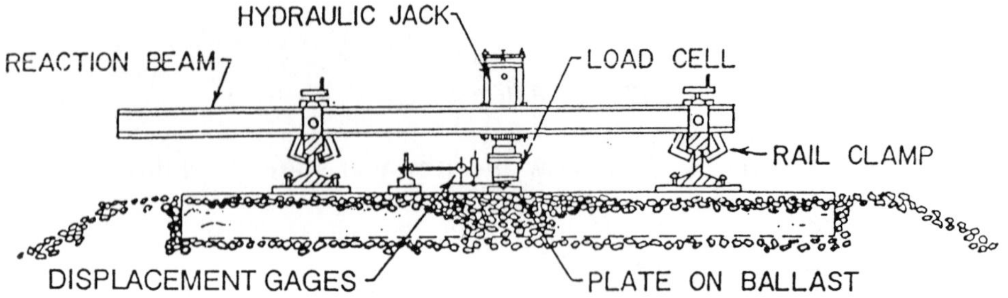

Fig. 8.9 Plate load test

The plate load test determines the vertical ballast stiffness as a measure of the effect of compaction on the ballast physical state (Ref. 8.1). A 127-mm (5-in.) diameter steel plate is seated on the ballast using gypsum plaster (Fig. 8.9). The plate is then loaded vertically and the measured contact pressure per unit plate settlement is taken as an index of ballast stiffness.

The lateral sleeper push test determines the resistance offered by the ballast to an individual sleeper displaced laterally as an indirect measure of the physical state and compaction (Ref. 8.2). In this test, the sleeper is displaced perpendicular to the rail after removing the fasteners so that the sleeper is disconnected from the rail and carries no vertical force other than its own weight (Fig. 8.10).

Results with all three of these methods will be used to illustrate the effects of maintenance and traffic on properties of relatively clean ballast. These results have been

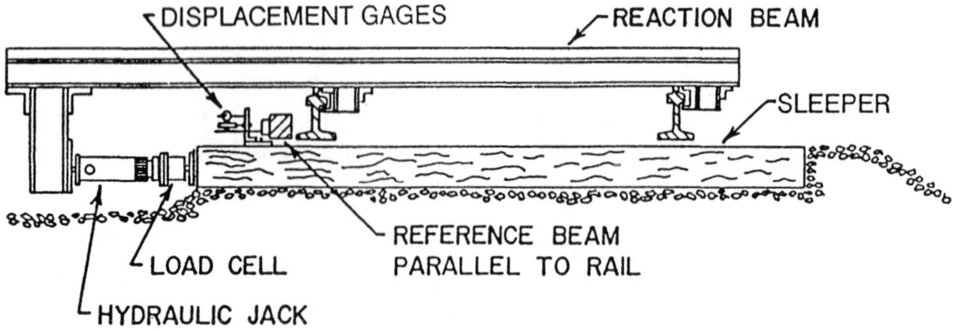

Fig. 8.10 Lateral sleeper push test

obtained in a number of field tests representing a variety of track and maintenance conditions, as well as in laboratory investigations.

8.2.1. Density

Density measurements were obtained in a variety of field situations. The observed trends have been summarized in Fig. 8.11. The four typical track conditions represented are:

1) after initial tamping during new construction or ballast replacement,

2) after compaction immediately following tamping,

3) after accumulation of traffic, and

4) after maintenance tamping.

Density is represented in Fig. 8.11 as percent compaction, which is the ratio of measured density to the laboratory reference density, expressed as a percent. The smooth boundary of the container used for the reference density tests gives a density value that is systematically lower than that given by the field measurement method, even though the actual density state in the two cases may be identical. Thus a percent compaction greater than 100 does not necessarily mean that the reference density test produces less compaction than occurs in the field.

The crib density measurements were obtained with essentially a full crib. Thus the crib density values represent ballast density between adjacent sleepers above the level of the sleeper bottom. The under-sleeper ballast density values were obtained in the sleeper-bearing area after emptying the cribs adjacent to a sleeper and then carefully removing the sleeper.

The initial tamping condition (Fig. 8.11a) represents a newly ballasted crib. The density along the crib was the lowest of any situation. Although the center density appeared to be slightly higher than the tamped zone, the density was relatively uniform along the crib. The density was greatest in the tamped zone under the sleeper. Apparently when the ballast is very loose, the vibratory tamping operation will densify the ballast. The density was

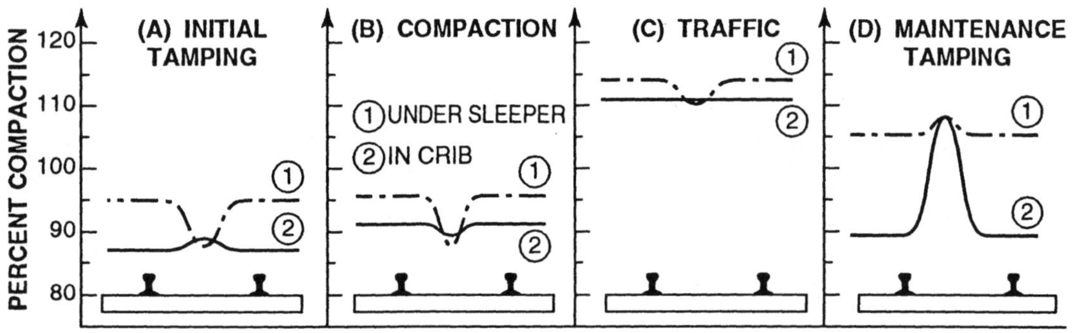

Fig. 8.11 Ballast density changes with track conditions

lowest in the center under the sleeper, where the ballast loosely fills in during the track raise.

After vibratory crib and shoulder compaction (Fig. 8.11b), the density is increased significantly in the crib near the rail where compaction is applied. A slight increase in density beneath the sleeper near the rails may also occur. However, no significant density change appears to occur in the center of the track.

In some cases, the uniformity in the measured ballast density from one sleeper to another appeared to be improved by compaction, compared to the tamped-only conditions. Thus the uniformity of ballast density distribution along the track could be one of the important benefits of using crib and shoulder compaction.

Traffic was observed to produce the greatest amount of compaction (Fig. 8.11c). Substantial increase in density occurred at all locations. The crib density was uniform and relatively high, apparently as a result of traffic vibration and perhaps cyclic loading from the sides of the sleepers. The density under the sleeper increased to a high level, particularly near the rail seat, where the ballast contact pressures were greatest. However, the center zone under the sleeper also was compacted by the traffic loading. Although the level of compaction did not reach the same percent as near the rails, the greatest increase in density occurred under the center of the sleeper where the density was initially the lowest before traffic.

The effects of maintenance tamping on the ballast density change can be seen in Fig. 8.11d. Compared to the measurements after traffic, ballast tamping in a track previously subjected to traffic consistently loosens the ballast layer regardless of the location of measurements. The density decrease was quite significant in the rail area, almost totally eliminating the compaction achieved by the traffic after initial tamping. Even in the center where no insertion of tamping feet was made, the ballast density is shown to have consistently been reduced, although by the smallest amount of any location.

The trends are summarized in Fig. 8.12a for the tamped zone near the rail, both under the sleeper and in the crib. The relative compaction for the dumped ballast state is shown for comparison. Tamping causes significant loosening compared to the compaction state produced by traffic both under the sleeper and in the crib. Loosening of the crib where the tamping tools enter is greater than under the sleeper, but the level of compaction remains above that of the dumped state.

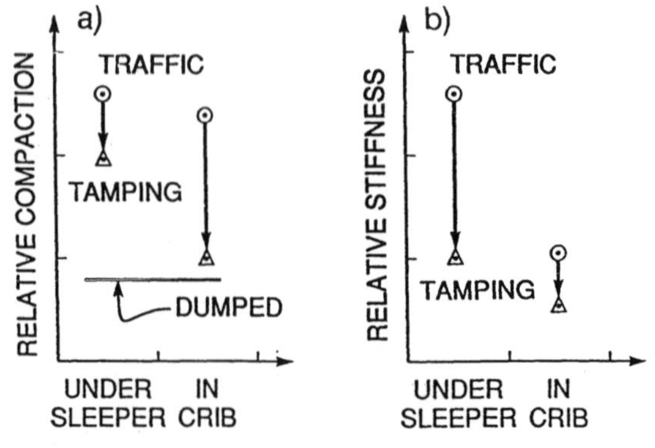

Fig. 8.12 Effect of tamping and traffic on ballast properties

The amount of density decrease from maintenance tamping appears to be dependent on various factors such as ballast type and condition, and also on the amount of raise during tamping. The greater the raise, the greater the density reduction.

8.2.2. Plate Stiffness

Ballast stiffness was measured in the field using the plate load test in track conditions similar to those for the density test. The results are summarized in Fig. 8.13. The stiffness was represented by the ballast bearing index, defined as the plate load per unit area required to displace the plate downward by a specified amount, which is usually 2.5, 5.1 or 7.6 mm (0.1, 0.2, or 0.3 in.) divided by this displacement.

The trends in Fig. 8.13 are similar to those for density. After initial tamping, the bearing index was low (Fig. 8.13a). The lowest values were in the center of the track. In the tamped zone under the sleeper, the ballast stiffness was much greater than in the crib, where the ballast was left in a loose state after withdrawal of the tamping tools.

Crib compaction approximately doubled the bearing stiffness in the crib where compaction was applied (Fig. 8.13b). A smaller increase occurred under the sleeper in the same area. This increase is believed to be partly caused by a ballast density increase under the sleeper and partly by increased lateral confinement from the compaction accomplished in the ballast directly below the crib. In the center of the track, which is away from the zone of compaction, little change in the ballast stiffness resulted from crib and shoulder compaction.

The addition of traffic greatly increased the ballast stiffness under the sleeper in the area near the rails, and also significantly increased the stiffness throughout the crib (Fig. 8.13c). A much less pronounced increase took place in the center beneath the sleeper.

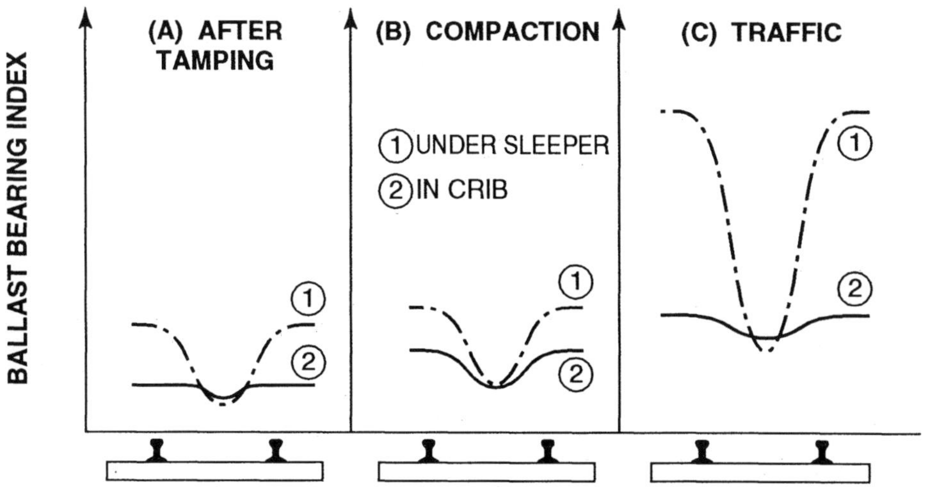

Fig. 8.13 Ballast bearing stiffness changes with track conditions

However, in time as center binding develops, the bearing stiffness will undoubtedly increase to a much higher level beneath the sleeper center.

Results are not shown for the condition following maintenance tamping. However, the observations for density indicate the expected trends. Tamping will decrease the bearing stiffness greatly in the zone of tamping penetration in the crib. The stiffness will also decrease in the tamped zone beneath the sleeper. The greater the raise, the greater will be the stiffness reduction beneath the sleeper.

The results are summarized in Fig. 8.12b. The stiffness was much higher in the bearing area beneath the sleeper than in the crib as a result of traffic. Also stiffness reduction due to tamping proportionately was much greater under the sleeper than in the crib, although the stiffness remained higher under the sleeper. But it is significant to note that traffic also increased the density and stiffness in the crib even though wheel loads are not applied to this area.

The ballast bearing stiffness is correlated in Fig. 8.14 with percent compaction. The ballast stiffness increases with density, but at an increasing rate. This trend indicates that small increases in density after some compaction has occurred that might be hard to detect with the density test, might still result in a significant stiffness increase.

Both the density and bearing stiffness values in the center of the track depend heavily on the track traffic and maintenance history. Results measured in the center of the track after any particular maintenance cycle do not necessarily represent the effects of that operation, because tamping and compaction are not performed in this location. A small raise will not result in much reduction in ballast stiffness and density in the center of the

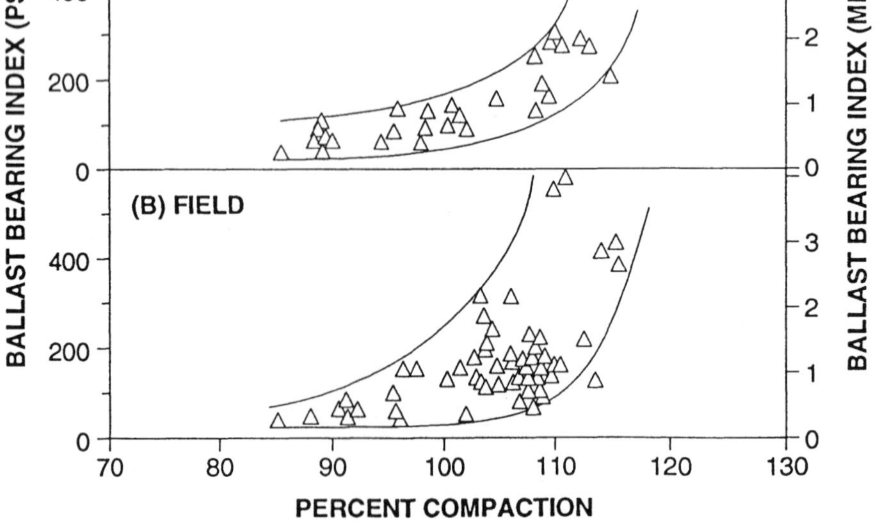

Fig. 8.14 Correlation of ballast bearing stiffness with percent compaction

track. However, a large raise will result in a big decrease because under the sleeper, the ballast is deposited loosely by rolling and sliding, and in the crib, the replacement ballast needed is also deposited in a loose state.

The density and plate test data were examined to estimate the amount of traffic required to produce the increase in ballast density and stiffness equivalent to that achieved during maintenance by crib and shoulder compaction. Unfortunately, insufficient data are available to clearly show the rate of change of ballast physical state with traffic, and field data variability is large. In addition, the results are a function of conditions prior to maintenance. However, the data suggest that roughly 0.9 to 1.8 GN (0.1 to 0.2 MGT) of traffic would produce the same change as the compaction.

8.2.3. Lateral Sleeper Resistance

The lateral tie (sleeper) push test (LTPT) has been widely used to evaluate ballast conditions and infer lateral track stability. The parameter used as an index is the force required to displace the sleeper a specified amount, typically 1 to 6.4 mm (0.04 to 0.25 in.).

Representative examples of lateral force-displacement curves for wood sleepers are compared in Fig. 8.15 for the conditions of tamping only, and crib and shoulder compaction following tamping. Resistance to displacement increases at a decreasing rate in both cases. However, the important observation is that compaction always significantly increases resistance above that present after tamping.

Proper interpretation of the LTPT requires an understanding of the factors contributing to the resistance. The three components of resistance provided by the ballast are the base, the crib, and the shoulder. An example of the relative magnitude of each

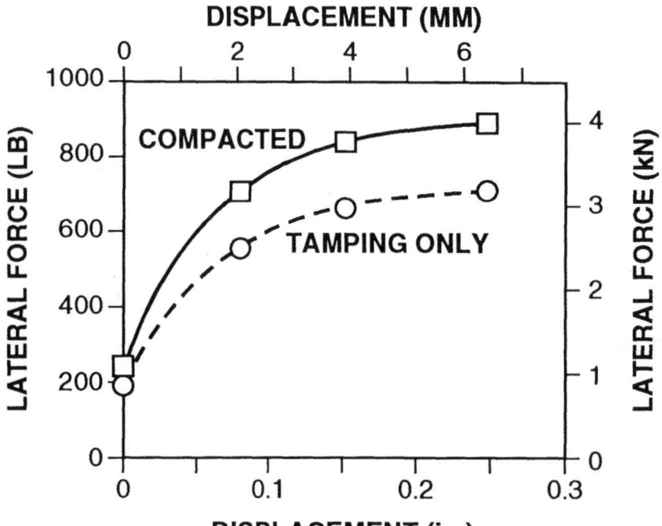

Fig. 8.15 Effect of compaction on lateral resistance

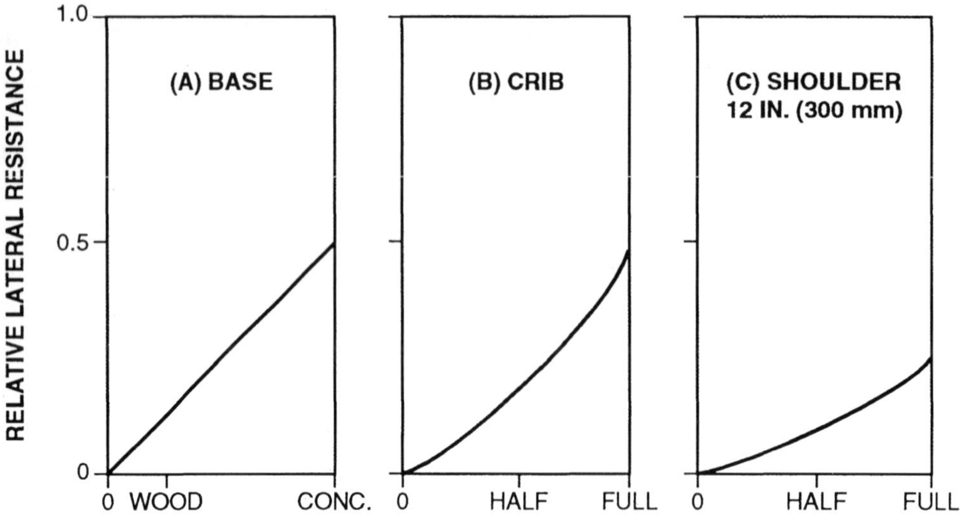

Fig. 8.16 Components of lateral resistance

component is shown Fig. 8.16 based on laboratory tests (Ref. 8.3). The lateral resistance is given as a ratio of an arbitrary value for comparison, because the actual values depend on the track condition and the specified lateral displacement.

The base resistance represents the force to overcome frictional resistance with a normal force equal to the sleeper weight. The friction coefficient will, of course, depend greatly on the roughness of the sleeper bottom. This component of resistance will increase with the sleeper weight (Fig. 8.16a).

The crib resistance is caused by frictional resistance on the sides of the sleeper under a normal force caused by lateral ballast pressure. The crib component increases directly with the depth of ballast in the crib (Fig. 8.16b). The illustration in Fig. 8.16b represents tamped-only ballast. The ballast pressure on the sides of the sleeper is greatly increased by compaction. Thus after compaction, the crib contribution will be even greater. However, vibration of the ballast or movement of the sleeper can relieve this pressure over a period of time, and thus diminish the effect of compaction observed immediately after maintenance.

The shoulder resistance is the net result of passive resistance pressure by the shoulder into which the sleeper is being pushed and the active pressure by the shoulder away from which the sleeper is being pushed. With only a relatively small lateral movement the active pressure should become negligible compared to the passive resistance. The shoulder resistance also increases directly with the height of the shoulder above the sleeper bottom and with the shoulder width (Fig. 8.16c). Compaction of the shoulder will increase this resistance further.

In the example given in Fig. 8.16, the base contribution to resistance, assuming a wooden sleeper, accounts for only about 14% of the total resistance. The crib accounts for

57%, i.e., more than half of the total, and the shoulder accounts for 29%. The lateral resistance under traffic will be much greater than for the unloaded sleeper because the base component will increase substantially without change in the crib and shoulder components. Thus the percent increase in lateral resistance from compaction should be much less under traffic load than shown in Fig. 8.16.

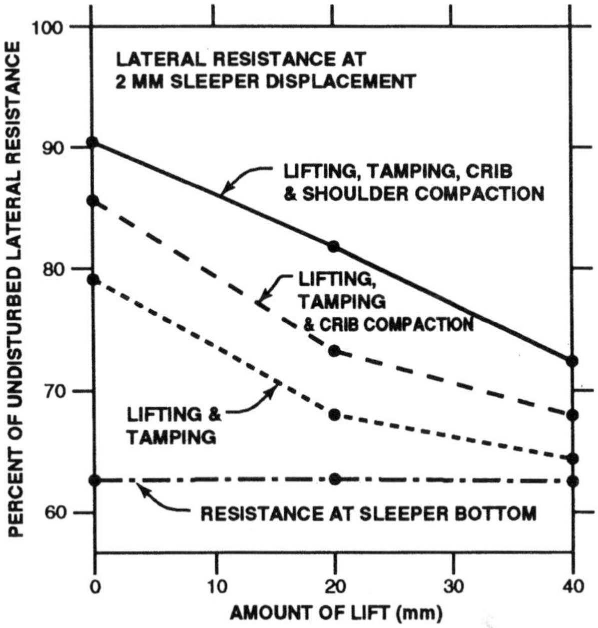

Fig. 8.17 Maintenance factors affecting lateral resistance

The lateral resistance after tamping also depends on the amount of raise during tamping. According to Fig. 8.17 (Ref. 8.4), as the lift increases from zero to 40 mm, the lateral resistance decreases from 80% to 65% of the undisturbed value before tamping (i.e. after traffic) as a result of tamping. Surface compaction of crib and shoulder restores some of this loss, as does dynamic track stabilization (see Chapter 14, Section 14.2). With the passage of traffic the lateral resistance gradually returns to the 100% value prior to tamping. Figure 8.17 also shows that crib compaction increases lateral resistance, and shoulder compaction further increases the resistance.

With the application of traffic, neglecting crib and shoulder compaction, all three components of resistance will increase (Fig. 8.18). The total lateral sleeper resistance will increase and the percent of total contribution from the crib will probably decrease. However, it is important to remember that these conclusions are for an unloaded track. For a loaded track, the percent contribution from the crib will be entirely different, and greatly reduced.

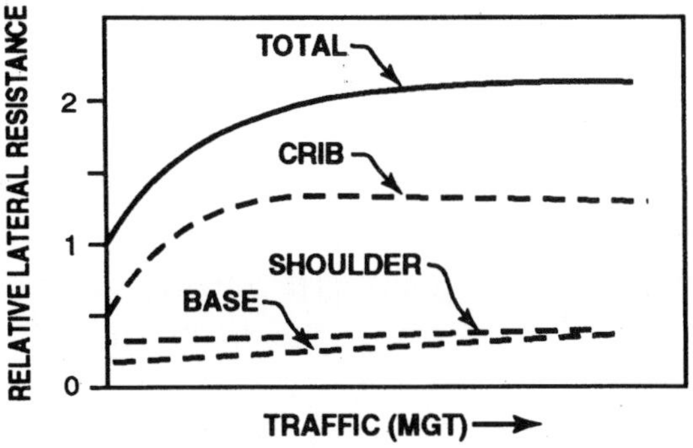

Fig. 8.18 Effect of traffic on lateral resistance

One of the accepted benefits of crib and shoulder compaction of ballast is to reduce any slow order time for traffic imposed because of the reduced lateral stability after tamping. All available relevant data were evaluated in order to estimate the traffic equivalent of crib and shoulder compaction on lateral sleeper resistance (Ref. 8.5). Variability of the data and the wide range of track conditions represented prevent precise conclusions from being drawn.

However, a reasonable interpretation is believed to be represented by Fig. 8.19. Crib and shoulder compaction increases lateral resistance by an amount that is equivalent to about 1.8 GN (0.2 MGT) traffic. Thus immediate operation of trains at the speed otherwise permitted after 1.8 GN (0.2 MGT) of slow orders could be initiated immediately after compaction. Low traffic density lines might benefit more than higher density lines, since the period of time for slow orders would be much longer on the low density lines. After 18 GN (2 MGT), the lateral resistance is about the same whether or not compaction was applied at the time of maintenance. After about 180 GN (20 MGT) traffic, the lateral resistance stabilizes.

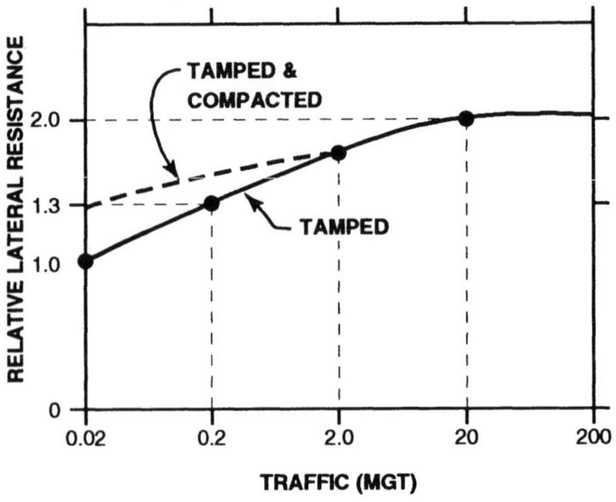

Fig. 8.19 Effect of traffic on compaction benefits

Slow orders should be more critical in maintenance operations involving undercutting and ballast replacement because the ballast bed will generally be much looser. Normally, the compaction process follows the final tamping-surfacing-lining operation with fully ballasted cribs. However, consideration should be given to compaction before the cribs are filled in order to provide a greater depth of penetration of the compaction effect into the ballast below the bottom of sleeper. The benefits of this approach might be particularly useful for the reballasted track. Test results have shown that the lateral sleeper resistance is greater when compaction is done on a fully ballasted crib than on a partially filled crib. To conclude from this that compaction is most effective using full cribs may be misleading, because in the LTPT measurements, a substantial part of the lateral resistance comes from the crib when the sleeper is unloaded. However, the bottom of the sleeper provides a much greater proportion of the resistance for loaded sleepers under train traffic. Possibly the benefits of compaction would be increased by a sequence which first provides compaction with low cribs, followed by crib filling and a second application of compaction. Further studies are needed to evaluate this possibility.

Another important application of crib and shoulder compaction is in conjunction with spot maintenance where loosening of the ballast has occurred around only some of the sleepers. The use of crib and shoulder compaction will reduce the physical state difference in the ballast between the disturbed cribs and the undisturbed cribs.

By far the most significant benefit of using crib and shoulder compactors that has been expressed by railroad users is associated with maintenance that must be done during hot weather. For a variety of reasons, track maintenance involving tamping may be impossible to defer until a sufficiently cool period. In such cases, immediate stabilization is a very important safeguard that even slow order traffic cannot provide.

Another approach for achieving compaction of ballast and restoring lateral sleeper resistance is to use the dynamic track stabilizer. This machine is discussed in Chapters 14 and 15.

Economically, the benefits of crib and shoulder compaction and dynamic track stabilization would be greatest if a substantial lengthening of the maintenance life of the track were to result. This would occur either if the rate of settlement were reduced by compaction or the uniformity of settlement were improved by compaction. In the case of the dynamic track stabilizer field information to confirm this benefit is beginning to accumulate.

8.3. Lateral Track Resistance

A review of lateral performance of track is given in Ref. 2.3, which cites a number of important references on the subject. The summary in this subsection is based in part on that review.

The previous section described the lateral resistance provided by the ballast for single sleepers detached from the rails. For two reasons, the lateral resistance of track to lateral forces is not equal to the sum of the lateral sleeper resistances. First the rail and fasteners provide some resistance (Fig. 8.20). The rail resistance comes from its bending stiffness in the horizontal plane. The fastener resistance comes from its torsional stiffness in the horizontal plane, although this is often very small. During buckling the rail contribution becomes zero because the rail produces the force causing the lateral displacement. Thus the ballast is the main source of lateral resistance.

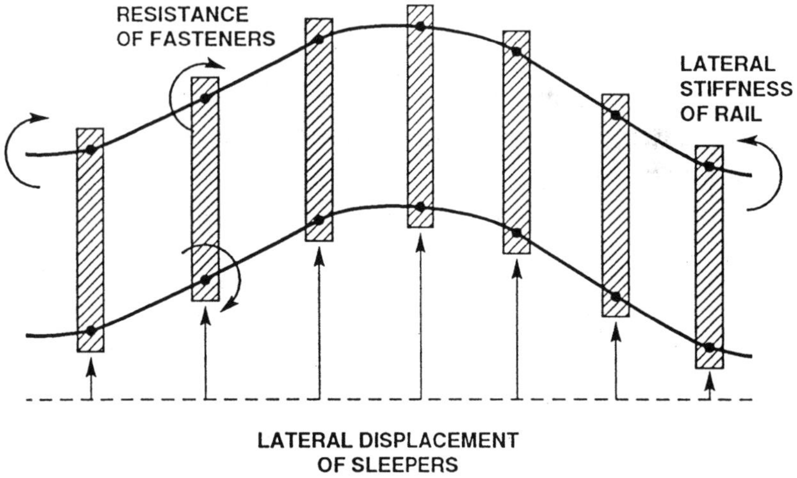

Fig. 8.20 Lateral stiffness contribution of superstructure

The second reason is that the interaction between sleepers will reduce the ballast resistance provided to each sleeper (Fig. 8.21). The amount of interaction depends on the sleeper spacing in relation to its dimensions. The passive pressure in the shoulder spreads out over a length of track considerably greater than the width of the sleeper. Therefore some portion of the shoulder resisting one sleeper in a single LTPT will be resisting two sleepers in a track panel displacement, thus reducing the resistance per sleeper. Also the crib contribution will be substantially reduced because the adjacent sleepers will not help restrain the crib ballast against shearing stresses applied to it from the sides of the intermediate sleeper. The base of the crib will also have shear stresses imposed from adjacent sleepers. Overlap of shearing resistance from the base of the sleepers will also reduce the base resistance per sleeper.

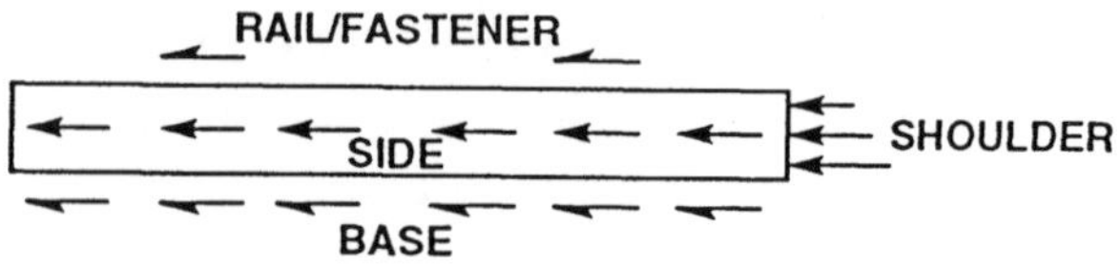

Fig. 8.21 Components of lateral resistance

The track lateral resistance under vertical train loading is also very different from that in the unloaded state, both in terms of the total resistance and the relative contributions of the crib, shoulder and sleeper bottom. In the presence of train traffic both lift up and downward force occur, which particularly influence the shearing resistance along the sleeper bottom.

Based on Refs. 8.4 and 8.6-8.13, DiPilato et al. (Ref. 8.40) derived the relative contributions to lateral resistance shown in Table 8.1, depending upon whether the track is verticallly loaded by a train or not vertically loaded . However, the track liftup ahead of the loaded axles will give results quite different from any of those in Table 8.1. The contribution from bottom resistance would be reduced or eliminated, so that the crib and shoulder would provide most of the lateral resistance. Klugar (Ref. 8.6) suggested that only passive end resistance and superstructure stiffness would contribute to lateral track resistance where sleepers were unloaded from the rail lifting wave. Obviously, as shown by Eisenmann's experiments (Ref. 8.7), the heavier the track the less the loss in lateral resistance from rail lift up.

Table 8.1 Contribution toward total lateral resistance

Lateral Resisting Force	Percent of Total Resistance	
	Unoccupied Track	Occupied Track
Sleeper Bottom/Ballast Bed	50 - 60	95 - 100
Sleeper Side/Ballast Bed	10 - 20	0 - 5
Passive End Resistance	30 - 40	0 - 5

A major factor influencing the passive end resistance is the effective end bearing area. This is defined by depth of ballast and effective width of sleeper. The effective width can be increased over that of continuous prismatic sleepers by several means (Ref. 8.40):

1) ears (Fig. 8.22),

2) dual blocks (Fig. 8.23), and

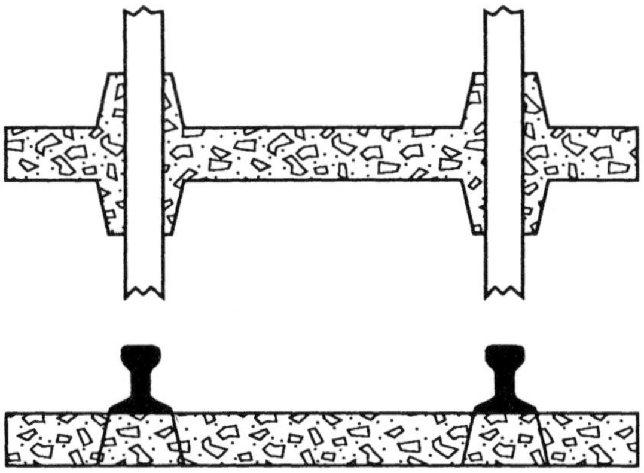

Fig. 8.22 Concrete ear

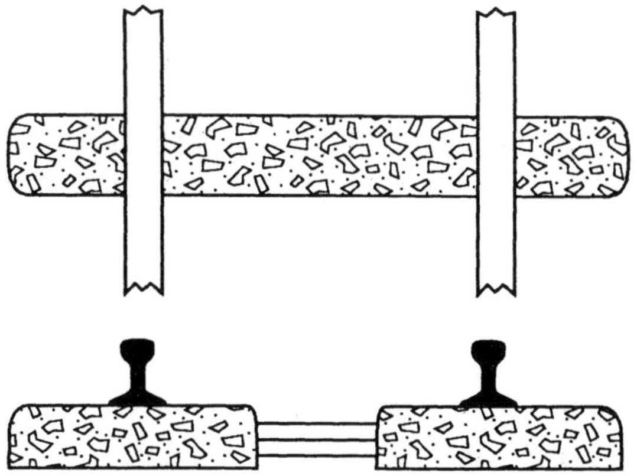

Fig. 8.23 Dual concrete block

3) safety caps (Fig. 8.24).

For the dual block sleeper the base and side contributions to lateral resistance will be reduced, hence offsetting at least some of the advantage of the increased end area. For the other two means, cost will be increased and track realignment will be difficult. Very likely the increased initial cost and maintenance cost will nullify the benefits of increased passive resistance through these means in most cases.

Shoulder width is a factor influencing passive resistance provided by the ballast. Studies reviewed in Ref. 8.2 indicate an increase in passive resistance of 10 to 40% with an increase in shoulder width from about 150 mm (6 in.) to about 300 mm (12 in.). Increasing

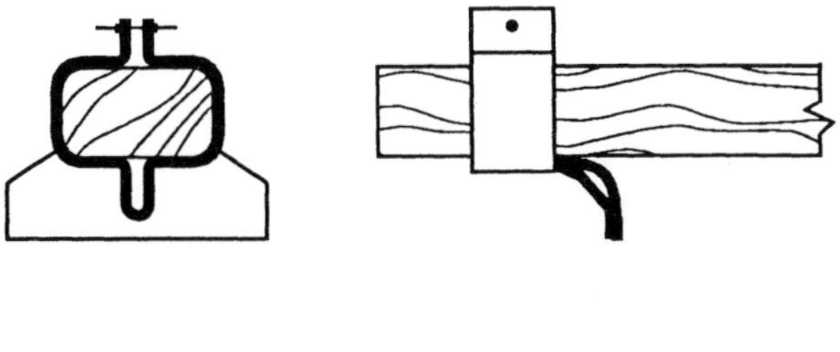

Fig. 8.24 Safety cap

shoulder width beyond about 300 mm (12 in.) to 400 mm (15 in.) did not seem to provide additional passive resistance. These relative effects must be adjusted for the proportional contribution of the shoulder to the total lateral resistance provided by the ballast to determine their influence on lateral resistance.

The lateral track restraining factors do not have equal reliability. When high temperatures prevail, there can be a tendency for the rail to buckle laterally. In such conditions, the contribution to lateral stability provided by the rail is clearly negative. Depending as it does upon rail to sleeper friction, the torsional resistance provided by the fastening can deteriorate with time, or can be zero by design. As a wheel load approaches a sleeper, the elasticity of the supporting ballast induces a procession wave which results in the sleeper being lifted by the rails away from the ballast surface. This results in a loss of the lateral frictional resistance generated between the ballast and the under side of the sleeper. Thus, with the sole exception of the lateral resistance provided by the shoulder ballast, the lateral resistance provided by all other sources is unreliable. It follows that great care should be taken to ensure that the ballast shoulder is maintained to its design profile. In this connection it is worth noting the mode of failure of the ballast shoulder when subjected to excessive lateral load from the sleeper shoulder. Failure is accompanied by the generation of a shear plane emanating from the bottom edge of the shoulder at an angle similar to that of the angle of internal friction of the ballast being used. It follows that the most effective shoulder ballast profile will be that which maximizes the length of the potential plane and the shear strength of the material along the potential failure plane.

8.4. Ballast Fouling

8.4.1. Sources

It is expected that new ballast will be placed in track with no more than 1 or 2% by weight of fouling components, that is particles of less than about 6 mm diameter. Over a period of time the amount of fouling material will increase, the rate of increase varying widely with circumstances.

A list of the many sources of fouling is given in Table 8.2. They are divided into 5 categories:

I) Ballast breakdown.

II) Infiltration from ballast surface.

III) Sleeper (Tie) wear.

IV) Infiltration from underlying granular layers.

V) Subgrade infiltration.

Table 8.2 Sources of ballast fouling
I. Ballast breakdown
a) Handling (related to IIA)
1) At quarry
2) During transporting
3) From dumping
b) Thermal stress from heating (desert)
c) Freezing of water in particles
d) Chemical weathering (including acid rain)
e) Tamping damage
f) Traffic damage
1) Repeated load
2) Vibration
3) Hydraulic action of slurry
g) From compaction machines
II. Infiltration from ballast surface
a) Delivered with ballast
b) Dropped from trains
c) Wind blown
d) Water borne
e) Splashing from adjacent wet spots
f) Meteoric dirt
III. Sleeper (tie) wear
IV. Infiltration from underlying granular layers
a) Old track bed breakdown
b) Subballast particle migration from inadequate gradation
V. Subgrade infiltration

These categories are also illustrated in Figs. 8.25, 8.26 and 8.27.

Fig. 8.25 shows an ideal substructure with a durable gravelly sand subballast on a stable subgrade and with good drainage. The subballast is assumed to be properly graded to prevent upward migration of the subgrade and subballast particles into the ballast voids. In this situation there are only three sources of fouling material. The first source, which is always present to some degree, is ballast breakdown. Of course part of this breakage will have occurred in shipping and handling the ballast, and hence would have been present when the ballast was delivered. The degree of the fouling from this source is not always greatest directly beneath the sleeper bearing area. This is partly because the breakdown is from mechanical weathering (freeze-thaw and temperature effects) and tamping as well as from traffic loading, and also because the fouling particles migrate as a result of tamping, traffic vibration and the movement of water. The second source is surface infiltration from

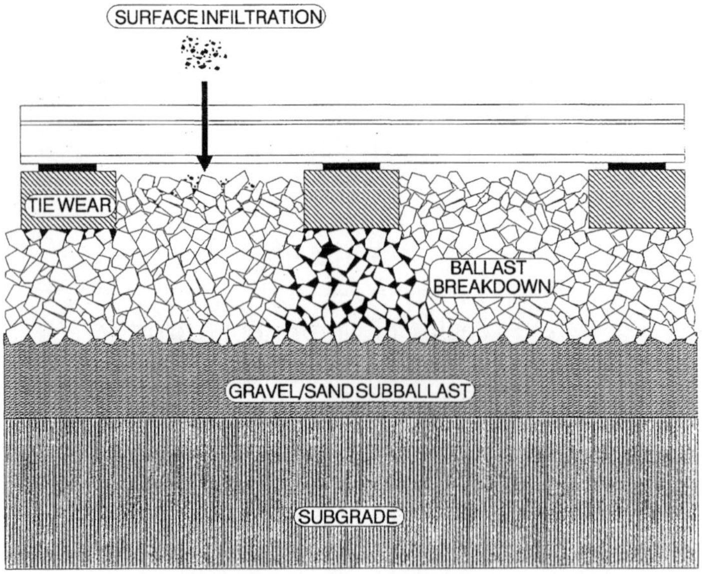

Fig. 8.25 Ballast breakdown, sleeper wear, and surface infiltration causing ballast fouling

car droppings and wind or water transported particles. The third source is sleeper wear, which can be appreciable for both wood and concrete sleepers.

Fig. 8.26 shows the layer beneath the ballast as the fourth source of fouling through particle migration into the ballast voids. This layer would normally be called the subballast because of its location and intended function, but it may be subballast by default, such as when it is created in track reconstruction as a remnant of the old road bed. Often it is the lower part of the old fouled ballast layer which has not been removed, or a layer of crushable cinders or slag. These materials can degrade under traffic to produce fine particles. Alternatively, the subballast may consist primarily of sand particles, which, even though durable, do not satisfy the gradation requirements to maintain separation from the ballast. In these three situations, migration is increased when water saturates the subballast because of the pumping action of traffic.

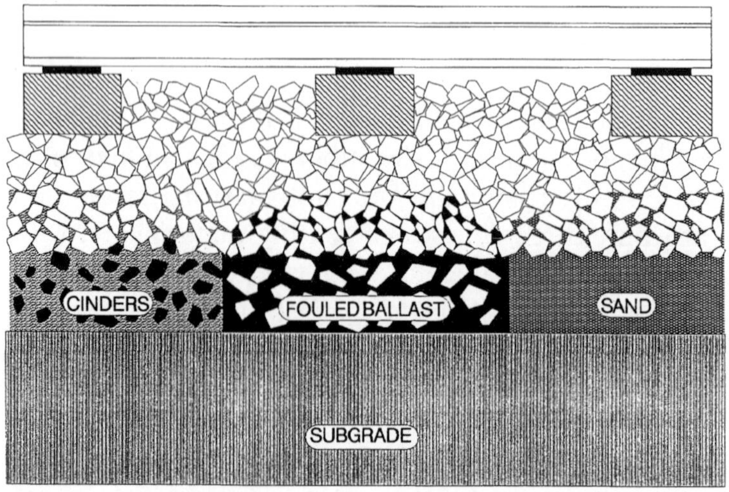

Fig. 8.26 Underlying granular layer infiltration causing ballast fouling

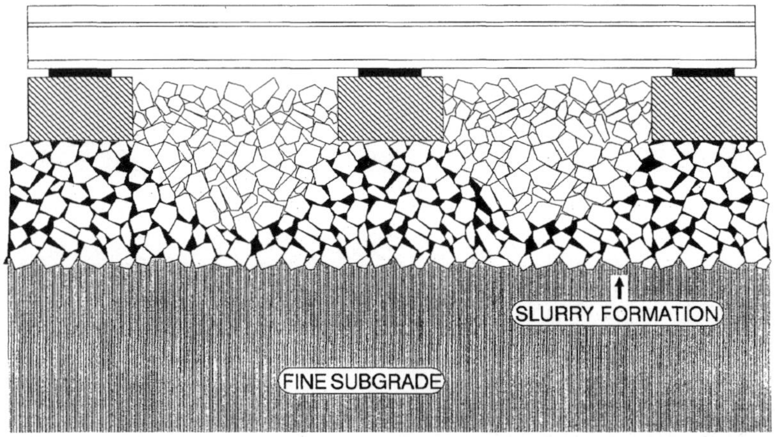

Fig. 8.27 Subgrade infiltration causing ballast fouling

The fifth source of fouling is the subgrade as shown in Fig. 8.27. Recent research at the University of Massachusetts (Ref. 8.15) as well as field observations indicate that this source is not a result of soft subgrade. In fact, the only location found in this study in which subgrade was the main source of fouling had a hard subgrade (mudstone). The track section was in a cut with water up to the ballast layer. The cause of the fouling at this location is believed to be the wearing away of the clay rich mudstone surface by repeated stresses from overlying coarse rock particles in the presence of water. Water not only helps weaken the hard mudstone surface, but also it mixes with the subgrade particles to form a clay slurry which pumps up through voids in the overlying ballast layer. For one section of track at this location geotextiles were placed below the ballast to prevent (or reduce) this source of fouling. However, the ballast in this section was as fouled as that at an adjacent section without geotextile. As a result it is apparent that the clay particles in the slurry were too fine to be blocked by the geotextile. The effective solution is a proper subballast (graded sand and gravel) combined with adequate external drainage. There is evidence (Ref. 8.15) that the primary role of the subballast in this case may be to prevent slurry formation by eliminating the subgrade attrition, rather than to prevent created slurry from migrating into the ballast voids. However, a further study of subgrade pumping is needed to clarify the mechanisms.

8.4.2. North America Studies

An extensive field and laboratory project was carried out by the Canadian Pacific Railroad to develop an improved ballast specification (Ref. 8.35). The researchers concluded that the main cause of ballast fouling on the CP system mainline tracks was ballast breakdown. This was attributed in part to the fact that a proper subballast layer was used to separate the ballast from the subgrade. A further study of the causes of ballast fouling was conducted by the University of Massachusetts to incorporate a wide variety of main line track conditions across North America (Refs. 8.36, 8.37 and 8.14). Many sites had mud showing at the ballast surface (see examples in Fig. 8.28), a condition generally thought

Fig. 8.28 Examples of mud pumping sites where the mud is not subgrade

by the railroad industry to be derived from fine subgrade soil underlying the ballast. This was found not to be true for the vast majority the muddy sites (Ref. 8.14).

8.4.2.1 Method of investigation

The following description of the methods of investigation represents the general approach; however, variations occurred in individual cases.

First, a site visit was made to observe conditions and to collect samples of ballast, subballast and subgrade. Whenever possible, a trench was dug with a backhoe to permit examination of the roadbed layers, and to facilitate sampling. However, in some cases only hand excavation was done, or cores were taken with Conrail's on-track boring machine (Chapter 14). The track characteristics, ground topography, and drainage conditions were documented, and information on traffic conditions and maintenance experience was obtained from the railroads.

In the laboratory, all samples were visually examined and then described according to ASTM D2488 (Ref. 8.38). The fouled ballast samples and at least the samples from one or two layers immediately below the ballast were wash-sieved and separated into as many as eight individual size fractions for compositional analysis. The wash water was carefully saved to permit recovery of the fines (silt and clay sizes).

For the fouled ballast samples, the particles smaller than 9.5 mm (3/8 in.) were assumed to represent the fouling components. The coarse fouling component consists of particles between 9.5 mm (3/8 in.) and 0.075 mm (No. 200 sieve) diameter. These are mainly sand sizes. The fine fouling component, termed fines, consists of those particles finer than 0.075 mm, which represent the silt and clay sizes.

The particles of each gradation were then subdvided by composition. This grouping is illustrated in Fig. 8.29. The composition of the ballast particles and the coarse-sand-size fouling material could be determined with a hand lens (10 - 20 x) while the identification of the medium- and fine-sand-size fouling material required a microscope (30 - 100 x). After the sorting was completed the percentage of each particle type in each size fraction was determined by weight, count or visual estimation.

Fig. 8.29 Particles of given size separated by composition

The silt- and clay-size particles could not be adequately identified using an optical microscope. Instead the scanning electron microscope (SEM) was employed to determine the shape characteristics of particles using magnifications of up to 5000 times. With this technique, particles of clay, rock and cinders could be distinguished (Figs. 8.30 and 8.31). X-ray diffraction analysis (XRD) also was performed to help determine the source of clay at a few sites. In cases involving carbonate ballasts, insoluble residue analysis facilitated identifying the fines composition. From the results of these tests the proportion of each fouling component in the total fines was estimated.

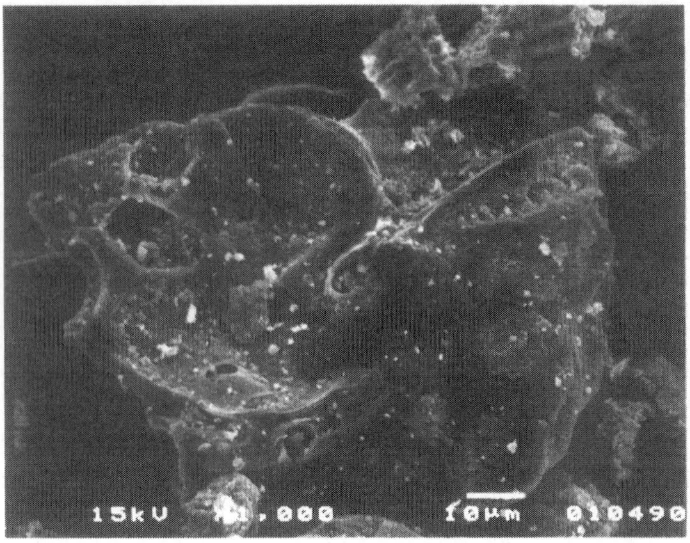

Fig. 8.30 SEM photograph of cinder particle magnified 1000X

The results of the laboratory investigation were compiled in terms of percent of each particle composition in each size fraction, and percent of each size fraction in the total fouled ballast sample. These values were plotted in the form of bar charts to help visualize the distribution of the various fouling components. Subsequently, the percentage of each particle composition in the fouling material (less than 9.5 mm or 3/8 in.) was calculated using the known sample gradation.

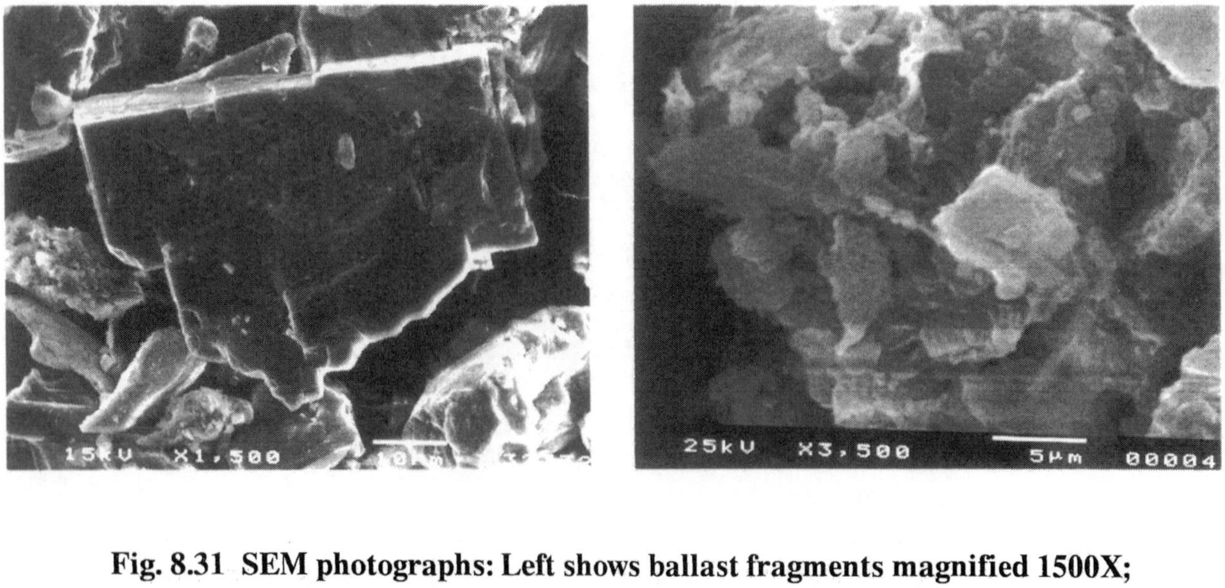

Fig. 8.31 SEM photographs: Left shows ballast fragments magnified 1500X; right is subgrade clay aggregate magnified 3500 X

An example is given in Fig. 8.32 which compares the composition of the fouled syenite ballast at one site with that of the subballast. The ballast sample was selected from a mud spot in the crib. The ballast was highly degraded with only 49% of the particles being of ballast size. The remaining 51% consists of 24% coarse fouling component and 27% fine fouling component (silt and clay sizes). Most of the fouling material (coarse and fine) was a result of ballast breakdown. However, an appreciable amount of subballast particles and wood from sleeper degradation were also present. The subballast is a natural, broadly-graded gravel with sand composed primarily of well-rounded quartz particles, but with a minor amount of coal and/or cinders. Hence, the subballast particles were easy to distinguish from the ballast particles. The fines components represent less than 1% of the total sample and so are not represented.

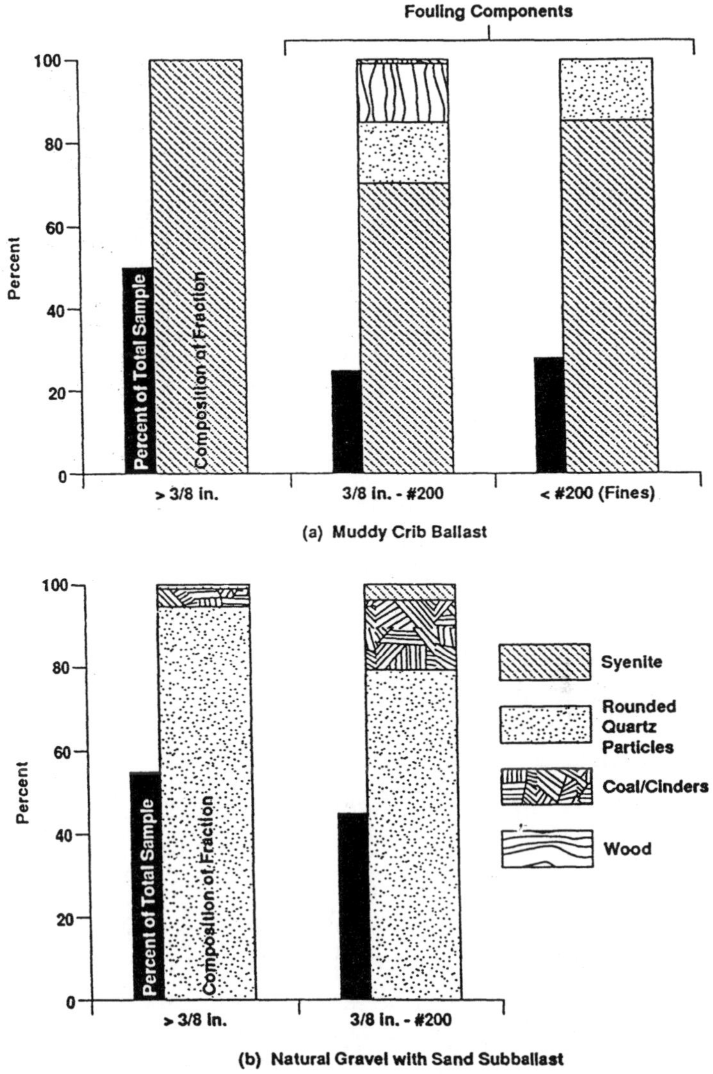

Fig. 8.32 Composition of ballast and subballast layers at a mud pumping site

The subgrade was a clay rich loess deposit described using the ASTM D2488 procedure (Ref. 8.38) as a sandy lean clay. About 40% of the soil was fine sand and 60% was silt and clay. The SEM and XRD analyses showed that there was little clay in the ballast fines and that the clay minerals in the ballast fines were compositionally distinct from those in the subgrade soil.

The final step in this research was to determine the source of the fouling components at each site. The field observations and laboratory data were considered together for this purpose. Ballast breakdown, infiltration from the layer immediately below the ballast, and surface infiltration source were evaluated first. If these sources did not correlate with the composition of the fouling material, then the search was extended to lower layers using the laboratory techniques described. In addition, the general site conditions were considered to insure that the conclusions about fouling were consistent with these conditions, and that alternative sources were not equally likely. As a result of this approach the interpretation of the fouling sources in most cases was straightforward.

8.4.2.2 Results

An average percent by weight of the fouling components for each of the 5 fouling categories in Table 8.2 was determined for each site. The result is shown in Fig. 8.33. By far the most important source of fouling was ballast breakdown. Next, but a much smaller contributor, is infiltration of particles from granular layers beneath the ballast. These were generally old road bed layers on which new ballast was placed. Surface infiltration sources were third. Subgrade infiltration was an uncommon source.

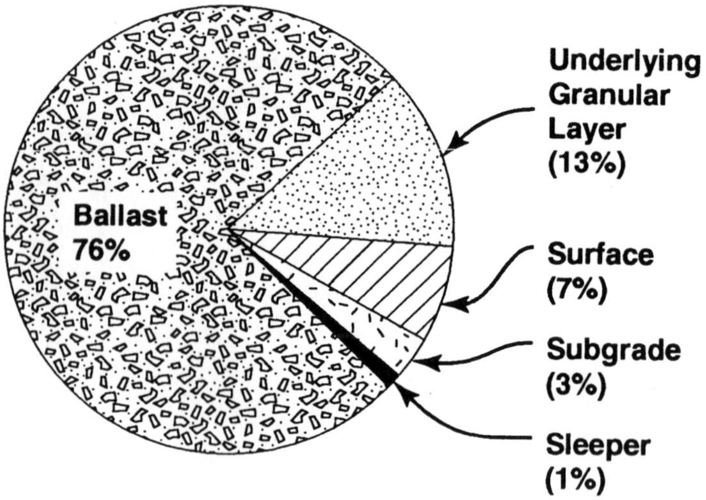

Fig. 8.33 Sources of ballast fouling for all sites combined

Although track subgrade was expected by many railroaders to be a major source of fouling, the results of this study indicate that it may be only a minor source. Clearly, the study shows that the presence of mud in the ballast does not indicate the existence of a subgrade fouling source. In fact, assuming that the mud is not from the subgrade is far more likely to be correct where a suitable subballast layer exists.. In general a combination of sources of fouling may be expected at individual locations, the mix depending on site specific conditions. The interpretation in any one case requires a proper subsurface investigation, since the clues cannot be obtained from surface inspection alone. However, a review of the fouling mechanisms described here will provide the basis for an accurate interpretation of the cause of fouling in most cases, without the need for the detailed laboratory investigation used in this research.

8.4.3. European Experience with Fouling

In Germany the main source of finer components is thought to be from the surface (category II) for two reasons (Ref. 8.16). First, a separation layer is required below the ballast to limit subgrade intrusion. Second, cleaned ballast removed from the track still has large particles with just worn corners.

Fouling from the subgrade (category V) results from overstressing at the ballast-subgrade contact points. The softening effect of water aggravates this situation and results in formation of a slurry. In the absence of a suitable separation layer, this slurry will pump up to the surface of the ballast to produce a 'pumping' failure. One of the difficult practical problems with drainage to prevent water accumulation is that the subgrade tends to settle most under the rails. This causes impermeable depressions which trap water that can produce subgrade softening, and lead to more settlement. This is a self-perpetuating situation which can only be cured by such means as regrading the subgrade surface, adding a suitable subballast and/or separation layer and ensuring drainage (Ref. 8.17).

When either the subgrade or the surface is the main source of ballast fouling, then the quality of the ballast material is less important than when ballast breakdown predominates. However, a slurry containing surface fines or larger particle movement caused by a soft subgrade can greatly increase ballast wear so that ballast quality may still be a consideration. It seems clear, however, that full advantage of a good ballast cannot be obtained without good supporting conditions.

British Railways engineers have examined sources of ballast degradation and from this have estimated the contributions of each to the amount of material of less than 14 mm size when the ballast has become fully fouled, i.e. when the voids are filled giving about 30% of the total sample by weight less than the 14 mm size. A representative estimate is shown in Table 8.3 (Ref. 8.18).

TABLE 8.3 British Railways sources of fouling

No.	Source	Degradation	
		kg/sleeper	% of total
1.	Delivered with ballast (2%)	29	7
2.	Tamping: 7 insertions during renewal and 1 tamp/yr for 15 years at 4 kg/tamp	88	20
3.	Attrition from various causes including traffic and conrete sleeper wear (Traffic loading: 0.2 kg/sleeper/million tons of traffic)	90	21
4.	External input at 15 kg/yr (Wagon spillage: 4.0 kg/ sq m/yr) (Airborne dirt: 0.8 kg/ sq m/yr)	225	52
	Total	432	100

Note that the tamping contribution is about the same as the other sources of attrition, and the largest source is external. British Railways has observed that the ballast particles remaining after removing the fouling material are still sharp and angular after 15 years of service which is consistent with their estimates that wearing of the particles is not the main source of fouling in their situation (Ref. 8.18).

8.4.4. Tamping Damage

British Railways has conducted several studies of ballast degradation from tamping. In one field investigation (Ref. 8.19), breakage of ballast was measured after 20 consecutive tine insertions with a tamping machine. The result was a 15 to 45% reduction in the

38-51 mm (1.5 to 2 in.) particle sizes, which size range initially provided 46 to 71% of the total specimen. Correspondingly the amount of particles of less than 13 mm (0.5 in.) size increased from about 1% before tamping to about 5% of the total specimen after tamping.

In another study (Ref. 8.20) conducted in the track laboratory at British Railways, approximately 2 to 4 kg of material less than 14 mm size was generated per tamp (single insertion) for a single sleeper. The ballast degradation is actually much greater because the breakage creating particles larger than 14 mm size are excluded from the measurement.

Figure 8.34 gives an example of measured results for two ballasts (Ref. 8.21) from Association of American Railroad field tests. The fine material generated is given as a weight percent of total ballast in the tamped zone.

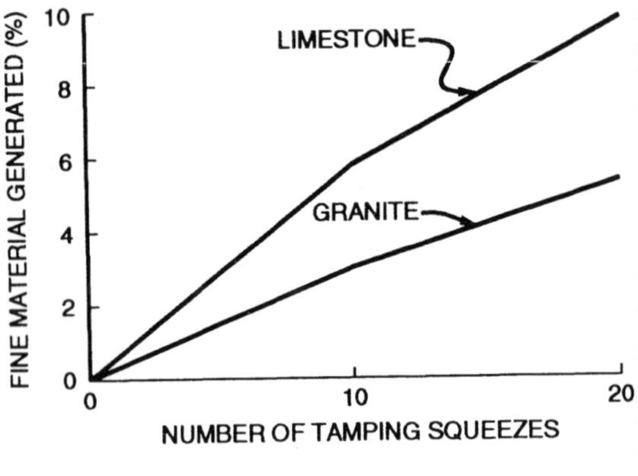

Fig. 8.34 Ballast breakdown from tamping

8.4.5. Hydraulic Erosion of Ballast and Sleepers

A particularly severe ballast and sleeper degradation problem has been documented and studied by British Railways (Refs. 8.22, 8.23 and 8.24). Severe failure of limestone ballast was observed mostly near welds (Ref. 8.24). The limestone particles were abraided to form a fine powder. This powder plus water formed a slurry which eroded the concrete sleepers and the ballast, causing voids under the sleepers. In such cases, termed washy spot failures, the fouling material has been found to contain particles from 10 mm down to clay size, with the fine sand size being predominant (Ref. 8.23). The sleeper attrition as well as most of the ballast attrition is believed to be associated with the high hydraulic gradients in the liquid slurry beneath the sleepers where voids (gaps) develop (Fig. 8.35). Speed of loading is believed to be more critical than axle ad in this situation, with increasing speed increasing the erosion (Ref. 8.22).

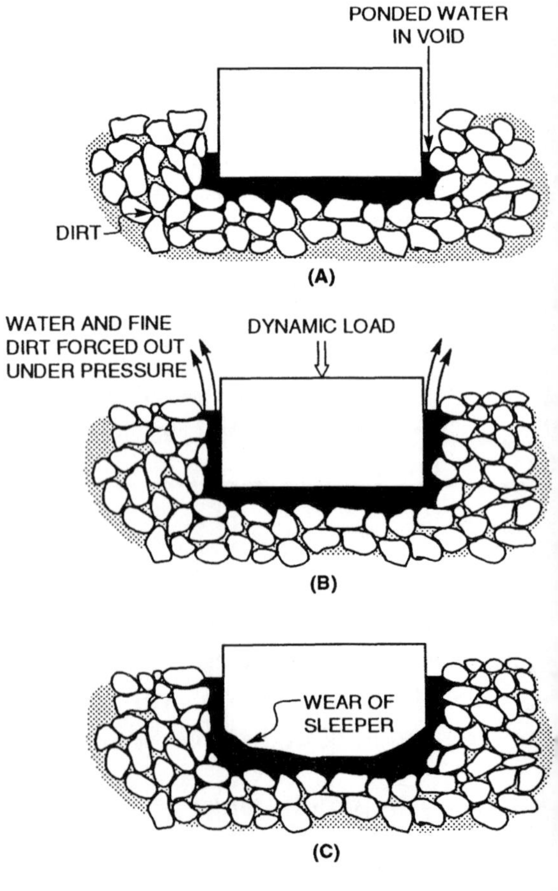

Fig. 8.35 Hydraulic action

Fig. 8.36 Jetting action displacing ballast

Under the influence of traffic loading, the sleeper is driven downwards giving rise to high fluid pressures within the dirty water beneath the sleeper. This excess fluid pressure dissipates itself by jetting sideways and upwards from beneath the sleeper. The higher the speed of the traffic the higher the loading rate and the higher the water pressure induced. For this reason, this type of failure is seldom associated with low speed lines.

This hydraulic erosion problem can also initiate from other fouling sources which also cause the ballast around the sleeper to become impermeable resulting in the ponding of dirty water.

The jetting action is sufficient to displace ballast particles from around the sleeper as shown in Fig. 8.36. This results in a reduction in the lateral resistance offered by the ballast to the sleeper.

The jetted material is highly abrasive and is able to erode concrete sleepers and in extreme cases, ultimately expose their prestressing wires. The products of this erosion add to the cause of the condition by further reducing permeability around the sleeper and adding to the abrasiveness of the jetted liquid. Wood sleepers will be abraded as well, although the wood particles do not add to the slurry abrasiveness.

Jetting action of the slurry can result in the mix of fine particles and water fouling the ballast surrounding adjacent sleepers (Fig. 8.37), thereby precipitating a further hydraulic erosion failure (see Table 8.2, Item IId). Such a failure mechanism can result in the failure condition spreading out from the joint sleepers (where the problem usually starts) to involve the whole track.

Fig. 8.37 Jetting action contaminating adjacent areas

The factors common to this type of failure in England are (Ref. 8.19):

1) Poor drainage.

2) Concrete sleeper to give high contact stress at particles.

3) Low wear resistant ballast material.

4) Void under sleeper resulting in impact and hydraulic action.

This problem seems to be most commonly associated with limestone ballast. There are at least two reasons given for this. First, limestone abraids more readily than other ballasts. Second, and perhaps more important, the limestone particles tend to adhere so that they remain in a zone around the sleeper where they trap water, restrict drainage and form an abrasive slurry which pumps with high velocity. Examples of such situations have been found where the ballast fouling is concentrated in the upper part of the layer with clean ballast below.

It would appear that this type of degradation accelerates as it develops because of the increase in amount and concentration of slurry as well as the magnitude of hydraulic velocity and impact force.

Ballast/sleeper attrition caused by ballast slurry can be prevented by using ballast material that is highly resistant to attrition. For this reason British Railways specify a maximum wet attrition value from the Deval test (Chapter 7) of 4%.

Ballast/sleeper attrition can be cured by ballast cleaning and/or ballast renewal using ballast that conforms to specification, particularly with respect to resistance to wet attrition. Good ballast drainage is also required.

8.4.6. Effect of Fouling

The effect of ballast fouling is to prevent the ballast from fulfilling its functions as described in Chapter 2. The specific effect depends on the amount and the character of the fouling material.

Sand- and fine-gravel-size fouling particles will increase the shear strength and stiffness of the ballast which adds to the stability and resistance to plastic strain as long as the coarse ballast particles still form the ballast skeleton. Frost protection will also be increased. However, void storage space and resiliency are reduced. Also surfacing and lining operations will become increasingly more difficult as the ballast voids become filled. Drainage will gradually decrease, but may remain adequate until the majority of the voids are filled. Segregation of particle sizes will occur during tamping, and when the voids become nearly filled, tamping will produce a looser ballast structure which will result in increased rate of settlement under further traffic. In general, a high degree of fouling from coarse sand and gravel particles will not cause significantly increased maintenance costs. Also the ballast can be readily cleaned.

The loss of performance mainly occurs when the fouling materials contain silt- and clay-size particles (fines). The quantity of these particles that will cause severe problems depends on the amount and size of the coarse-fouling components since these coarser particles reduce the void space, make the void sizes smaller, and combine with the fines to form an abrasive slurry. Clay particles alone will not form an abrasive slurry, but silt particles will. Both types of particles impede drainage and hence will increase the possibility of significant ballast deterioration, since water is a critical ingredient for severe ballast maintenance problems. The most important examples are: 1) hydraulic erosion, 2) subgrade attrition, and 3) loss of stability through particle lubrication. Ultimately when the degree of fouling with fines is high enough, the fines will control the ballast behavior and satisfactory geometry control will be impossible. As the degree of fouling with fines develops, tamping will become less effective in several ways: 1) when the fouling material becomes dry the ballast will be difficult to penetrate and rearrange, and the rearranged particles will be left in a looser state, and 2) when the ballast becomes wet, the particles' contacts will be coated with fines, lubricating them and the ballast will have a weakened structure after tamping. Maintenance will then be further increased because screening the ballast will be unsatisfactory, especially for clay fines, so that 100% replacement of the ballast may be necessary.

8.4.7. Minimizing Fouling

The following are examples of means to minimize fouling:

1) Care can be taken to ensure that no dirt is tipped with the new ballast when it is originally placed in position.

2) An attempt can be made to minimize wagon spillage.

3) Tamping can be kept to a minimum as a means of minimizing ballast damage caused by tamper tines.

4) A ballast having a high resistance to attrition can be used as a means of minimizing inter-particle attrition.

5) Poor ballast drainage, leading to increased inter-particle ballast damage and the production of fines resulting from wet attrition, can be avoided.

6) A blanket can be provided as a means of preventing local subgrade failure and slurry fouling resulting from pumping.

8.5. Track Settlement

Considerable opinion and evidence indicates that, in most track, ballast is the main source of both average and differential settlement between surfacing operations. This is known as short term settlement, as compared to long term settlement which is subgrade related. One reason is that, except for new track locations, subgrade has been subjected to traffic for decades and hence subgrade settlement from repeated loading accumulates very slowly. Adjustments for this are handled together with ballast settlements by surfacing operations. Another reason is that, where subgrade settlements are large, the ballast or subballast depth is increased to compensate which leads to reduced rate of subgrade settlement eventually.

Necessary conditions for ballast to be the main source of settlement are: 1) existence of filter/separation layer between the coarse ballast and fine subgrade, 2) a sufficiently strong subgrade or subballast/subgrade combination, and 3) good drainage of water entering from the surface. Otherwise track maintenace cycles could be dictated by subgrade problems.

8.5.1. Box Test Trends

Laboratory repeated load triaxial tests on ballast, as well as box tests on ballast, have shown that vertical plastic strain and corresponding specimen compression can be related to the number of cycles by a semilog relationship in the form of Eq. 7.21. However, an evaluation of settlement trends in track, as well as results from box tests carried out to large numbers of cycles suggests that the semilog relationship increasingly underestimates the cumulative plastic strain as the number of cycles increases. This has the effect of

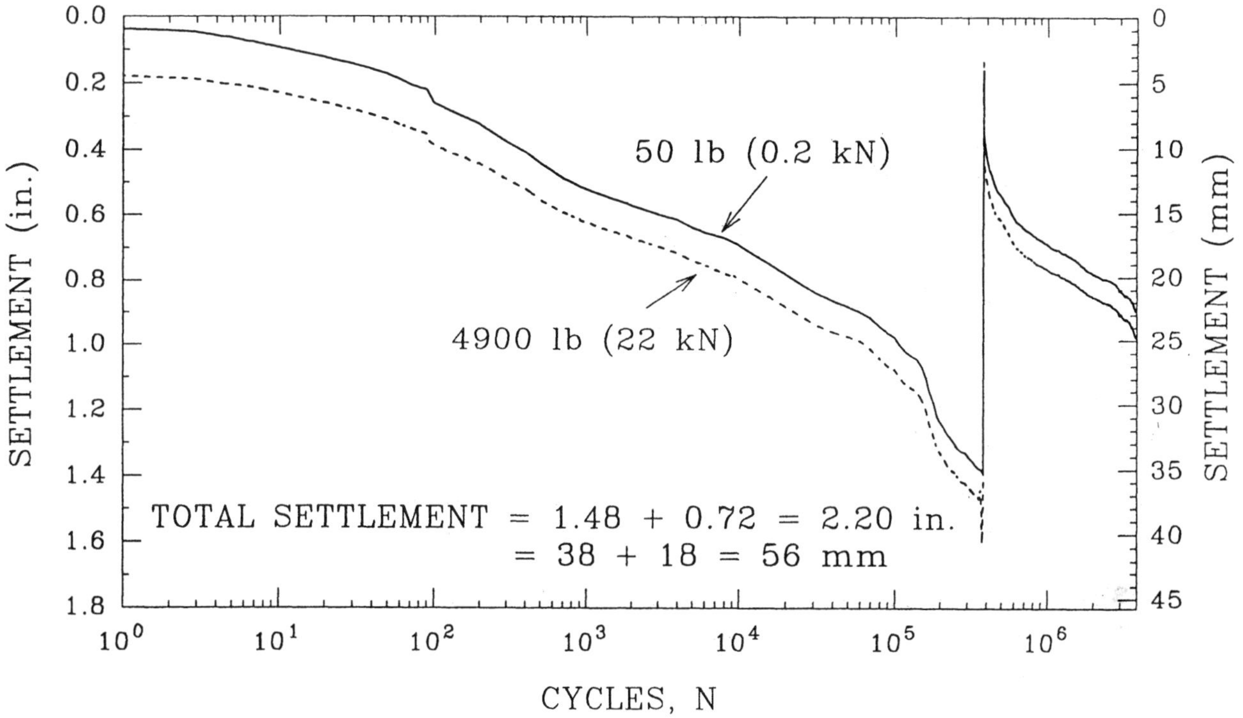

Fig. 8.38 Settlement as a function of log of cycles

overestimating the duration of the tamping cycle by underpredicting the rate of track settlement.

An example of box test settlement trends is given in Fig. 8.38 (Ref. 8.25). The ballast layer was 152 mm (6 in.) thick on a resilient base. A wood sleeper segment was used. The ballast was dry dolomite with an LAA of 29.7 and an MA of 8.6. The vertical repeated load alternated between a minimum seating load of 220 N (50 lb) and a maximum load of 21.8 kN (4900 lb). The maximum load was selected, based on GEOTRACK analysis, to give the same maximum sleeper-ballast contact pressure as that in a well-maintained mainline track subjected to an axle load of 347 kN (39 US ton) with no dynamic amplification. This corresponds to the 1110 kN (125 US ton) hopper cars in North America. The settlement in Fig. 8.38 increased continuously up to about 38 mm (1.5 in.) in about 400,000 cycles. At this point the sleeper segment was raised and ballast placed to fill the gap without disturbing the compacted ballast bed (unlike tamping). Then the loading was continued until a total of 3,840,000 cycles had been applied.

The total number of cycles in the box test example was selected to represent 2670 GN (300 MGT) of traffic. The relationship between MGT and number of cycles is given by

$$c_m = \frac{10^6}{A_t N_a} , \tag{8.1}$$

where C_m = number of load cycles/MGT,

A_t = axle load in tons, and

N_a = number of axles/load cycle.

The value of N_a was taken to be 2, giving 2 cycles per hopper car. Thus for 39 ton axle load C_m = 12,800, and 300 MGT requires 3,840,000 cycles.

The plot in Fig. 8.38 shows that the settlement trend with cycles up to the reset point is not semilog, but that settlement increases at a greater rate than a semilog relationship represents. The same data are plotted on linear axes in Fig. 8.39 to show that the settlement still does increase at a diminishing rate. The box test trends, such as shown in the example in Fig. 8.39, were compared to semilog, hyperbolic, parabolic and power mathematical relationships. The best trend overall was represented by a power relationship of the form

$$S_N = S_1 N^b \quad , \tag{8.2}$$

where S_N = settlement (layer compression after N load cycles),

S_1 = settlement from the first load cycle, and

b = exponent.

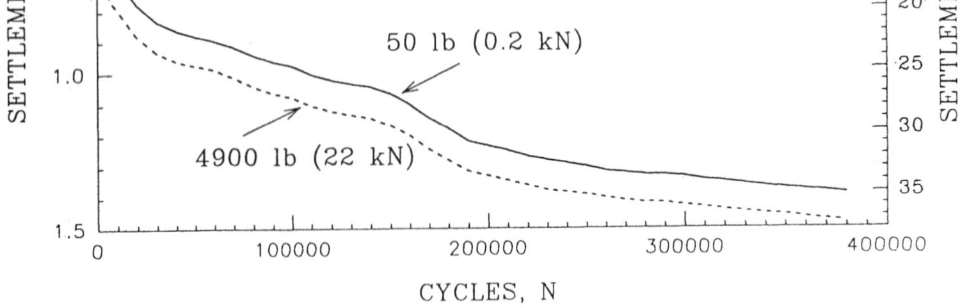

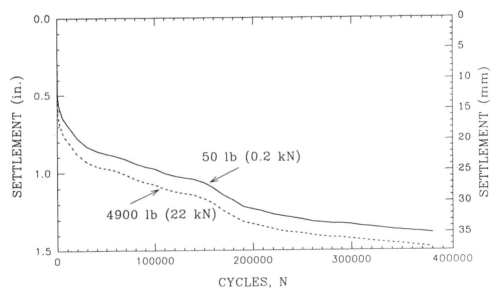

Fig. 8.39 Settlement increase with cycles

This trend is consistent with that observed by Shenton (Ref. 8.26). The linearized form of Eq. 8.2 is obtained on a log-log plot, which permits fitting the equation to the data. This is shown in Fig. 8.40 for the data in Fig. 8.39. The best fit power equation is

$$S_N\,(\text{in.}) = 0.17\,N^{0.17}\ . \tag{8.3}$$

Equation 8.2 may be rearranged to the form

$$R_S = \frac{S_N}{S_1} = N^b\ , \tag{8.4}$$

to observe the shape of the curve which is controlled by the exponent, b. The trends are shown in Fig. 8.41 for a range of b from 0.10 to 0.18.

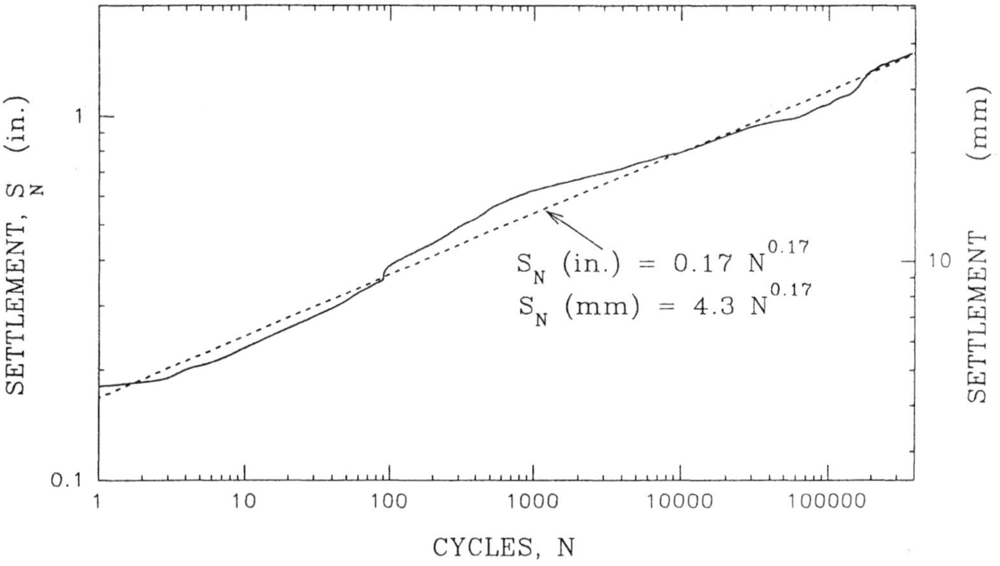

Fig. 8.40 Linearized settlement plot

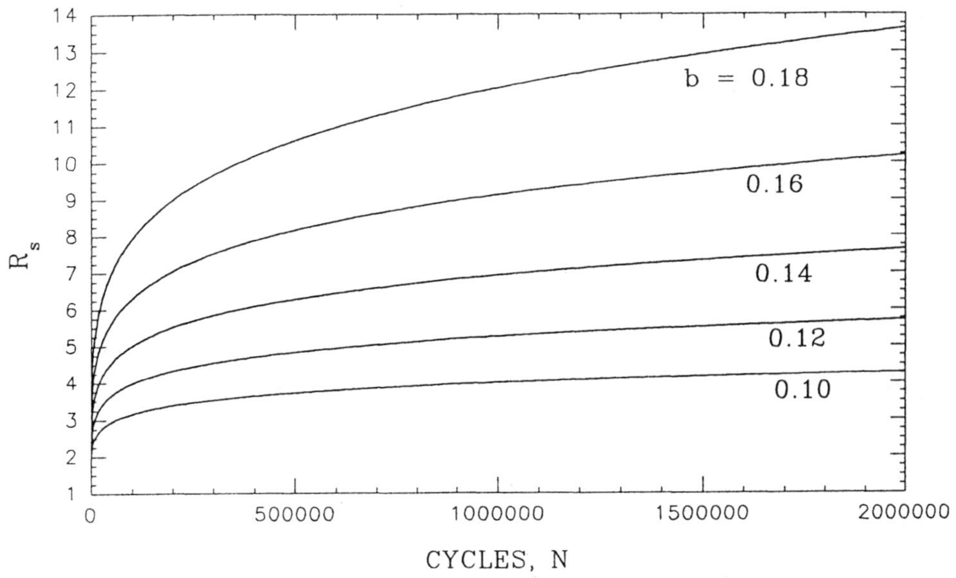

Fig. 8.41 Effect of exponent b

8.5.2. Field Measurements

Track settlement measurements were made at the Colorado test track (FAST) in which the contributions of the ballast, the subballast and the subgrade were distinguished (Refs. 8.27 and 8.28). The instrument arrangement is shown in Fig. 5.7. The test sections parameters are given in Tables 5.2 and 5.4. In section 18A (wood sleeper, tangent track, 21 in. or 533 mm of ballast depth) over 300 MGT (2670 GN) of traffic was applied without ballast tamping. The ballast strain accumulation with traffic is shown in Fig. 8.42 beginning immediately after construction of the track. The ballast as well as the subballast and the subgrade had only experienced construction traffic.

8.5.2.1 Ballast strain

Ballast strain was defined as ballast layer compression divided by the layer thickness. In the case of Fig. 8.42 the strain represents the top ballast only (the 12 in. or 305 mm directly beneath the sleeper). The settlement equation (Eq. 8.2) can be expressed in terms of strain by dividing both sides of the equation by the appropriate layer thickness. The result is

$$\varepsilon_N = \varepsilon_1 N^b \quad , \tag{8.5}$$

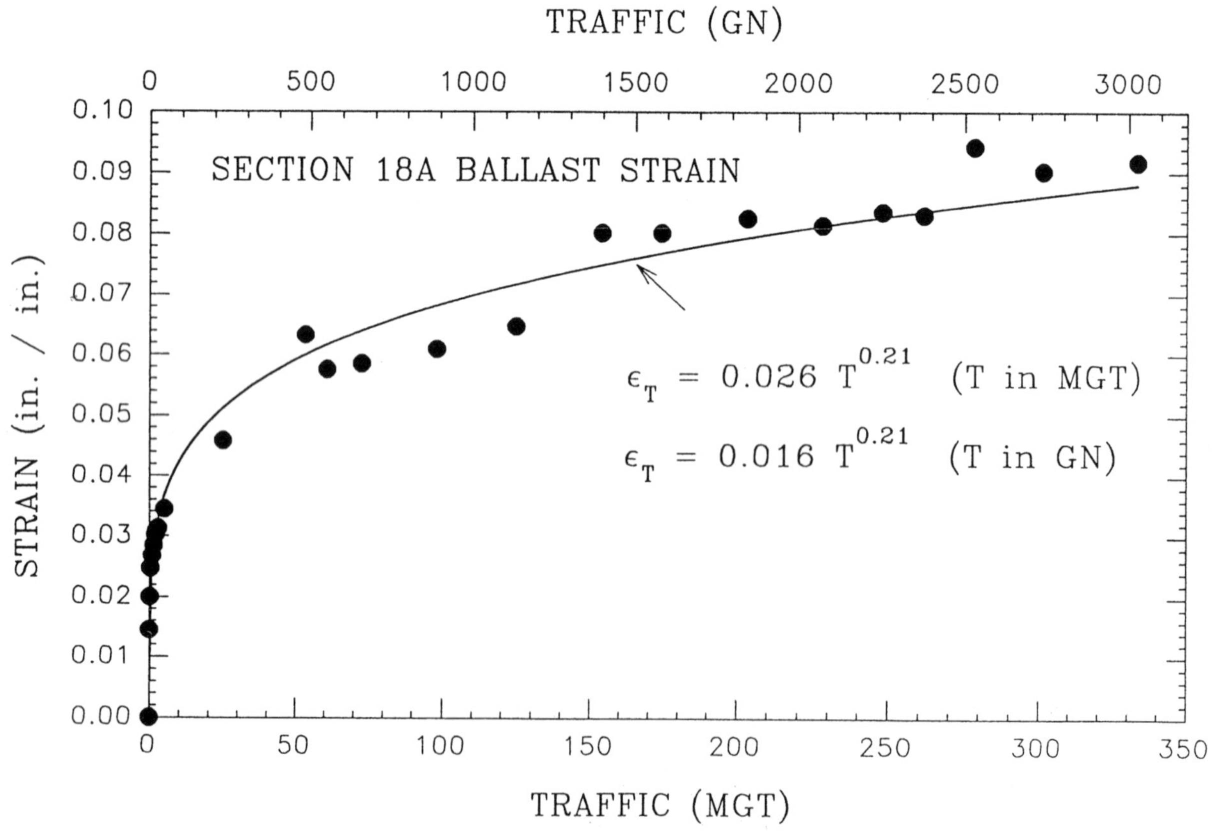

Fig. 8.42 Ballast strain at FAST track

where ε_N = ballast strain (layer compression divided by layer thickness) after N cycles,

ε_1 = ballast strain after the first cycle, and

b = exponent.

However, the data in Fig. 8.42 are in the form of strain as a function of MGT traffic rather than number of cycles. For Fig. 8.42 the form of the strain equation is

$$\varepsilon_T = \varepsilon_{T1} T^C , \quad (8.6)$$

where ε_T = ballast strain after T traffic in MGT,

ε_{T1} = ballast strain after 1 MGT of traffic, and

c = exponent.

But T and N are related by Eq. 8.1, i.e. $C_m = N/T$. Thus from Eq. 8.5

$$\varepsilon_N = \varepsilon_1 (C_m T)^b = (\varepsilon_1 C_m{}^b) T^b = \varepsilon_{T1} T^b . \quad (8.7)$$

If $N = C_m T$, then ε_N must equal ε_T, and so equating Eqs. 8.6 and 8.7 shows that b = c. In summary, converting from Eq. 8.5 to Eq. 8.6 is accomplished by setting

$$b = c , \quad (8.8a)$$

and

$$\varepsilon_{T1} = \varepsilon_1 C_m{}^b . \quad (8.8b)$$

For the ballast strain data in Fig. 8.42 the best fit power equation is:

$$\varepsilon_T = 0.026\, T^{0.21} , \quad (8.9a)$$

or

$$\varepsilon_N = 0.0035\, N^{0.21} , \quad (8.9b)$$

using C_m = 15000 cycles/ MGT based on a representative axle load of 33 US ton (294 kN) and 2 axles/load cycle (Eq. 8.1).

The average ballast strain measurements for the 5 tangent track test sections are plotted in Fig. 8.43 to show the observed variation. In all cases compression developed at a rapid rate through about the first 3 MGT (27 GN), and then at a much slower, but continuing rate thereafter, with further traffic.

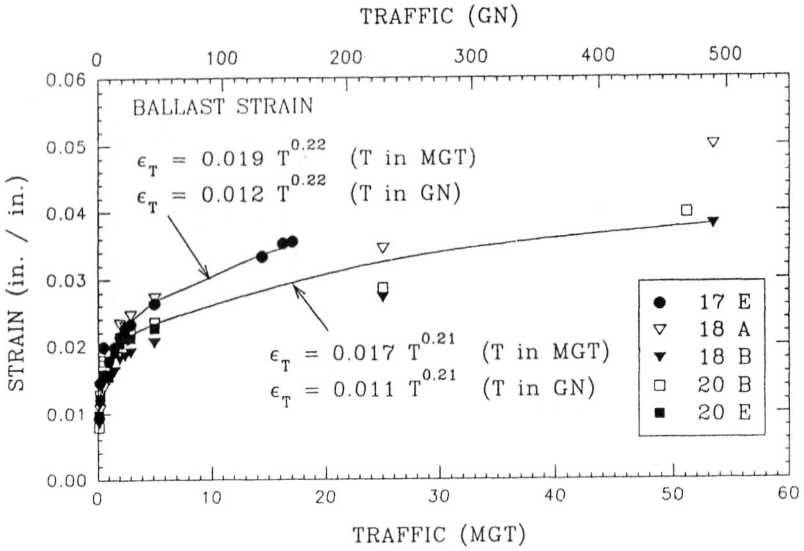

Fig. 8.43 Combined plot of FAST ballast strains

8.5.2.2 Subballast strain

A representative measurement of subballast strain (change in layer thickness divided by initial layer thickness) is shown in Fig. 8.44 for the first 50 MGT (445 GN) of traffic. The strain increases very rapidly with very little traffic, and then after about 3 MGT (27 GN) continues to increase at a gradual rate with further traffic. Log-log plots of the available subballast measurements for the 5 tangent track sections are shown in Fig. 8.45 for about 300 MGT (2670 GN) of traffic. The generally linear trend indicates that the power equation for ballast will reasonably represent subballast as well. The best fit power equation to all these data is also shown, and is given by

$$\varepsilon_T = 0.017\,T^{0.16} \,, \tag{8.10a}$$

or

$$\varepsilon_N = 0.0036\,N^{0.16} \,. \tag{8.10b}$$

The same data and equation are presented with linear axes in Fig. 8.46. After about 150 MGT (1340 GN) of traffic, the further strain increase appears to occur at a relatively slow and approximately linear rate. This can be represented by

$$\varepsilon_T = 2.6 \times 10^{-5}\,T \,, \tag{8.11a}$$

or

$$\varepsilon_N = 1.7 \times 10^{-9}\,N \,. \tag{8.11b}$$

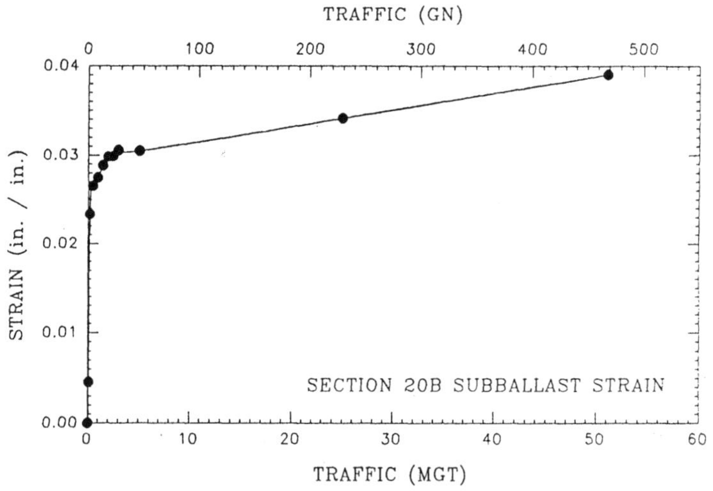

Fig. 8.44 Subballast strain at FAST track

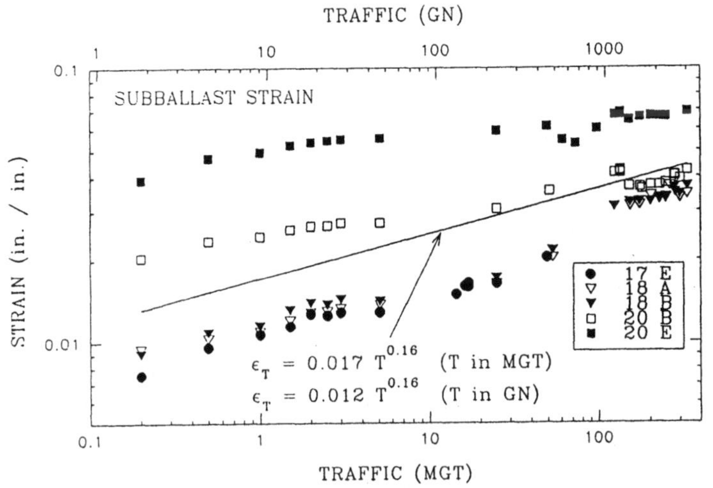

Fig. 8.45 Log-log plot of FAST subballast strain

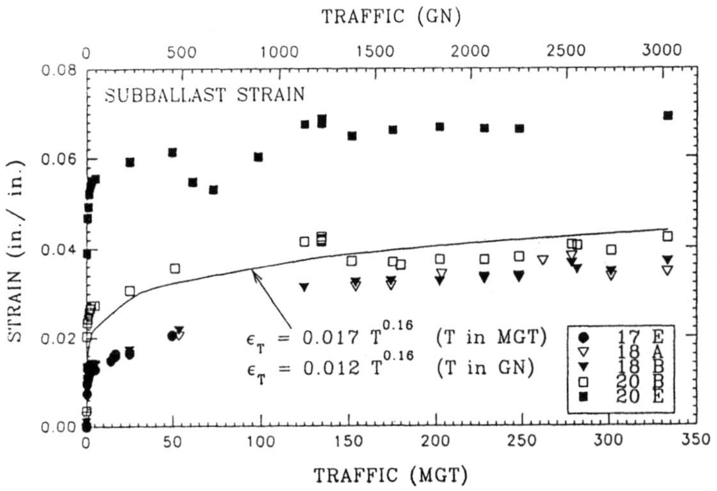

Fig. 8.46 Subballast strain plots with linear axes

8.5.2.3 Subgrade compression

Compression of the top 3 m (10 ft) of subgrade was measured. The subgrade soil at this site was a moist fine silty sand, lightly cemented in some locations. A representative curve for the first 175 MGT (1560 GN) is shown in Fig. 8.47. The available values for all sites up to 330 MGT (2940 GN) are shown in Fig. 8.48 to indicate the variability. Two best fit power equations are also shown, one for section 17E which showed the highest amount of settlement, and one for the remainder of the sections. These equations, obtained from Eqs. 8.5 and 8.6 after replacing strain by compression, are:

a) Section 17E--

$$S_T\,(\text{in.}) = 0.054\,T^{0.52}\,, \tag{8.12a}$$

or

$$S_N\,(\text{in.}) = 0.00052\,N^{0.52}\,. \tag{8.12b}$$

b) Other sections--

$$S_T\,(\text{in.}) = 0.039T^{0.37}\,, \tag{8.13a}$$

or

$$S_N\,(\text{in.}) = 0.0014\,N^{0.37}\,. \tag{8.13b}$$

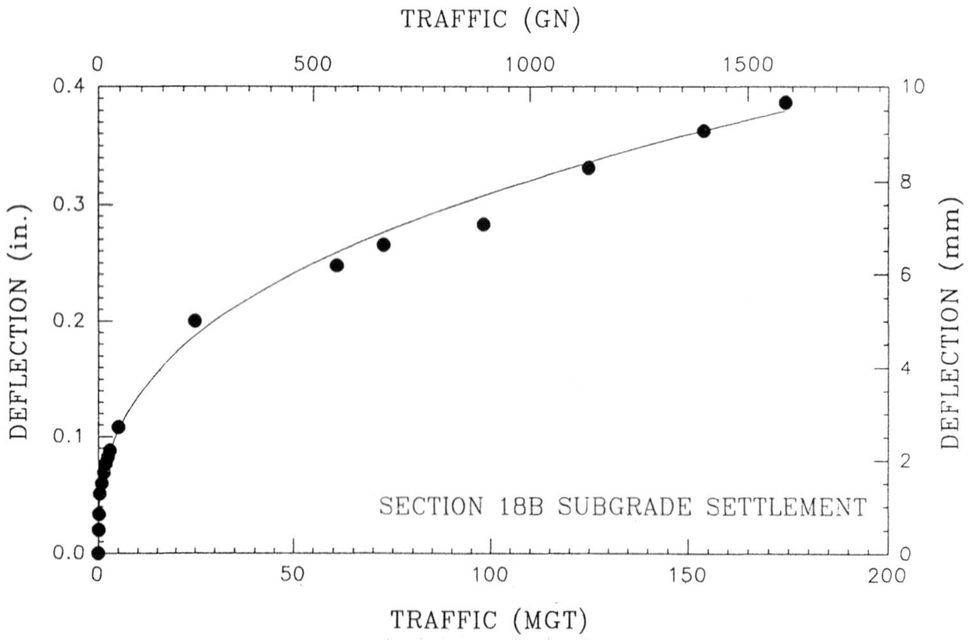

Fig. 8.47 Subgrade settlement at FAST track

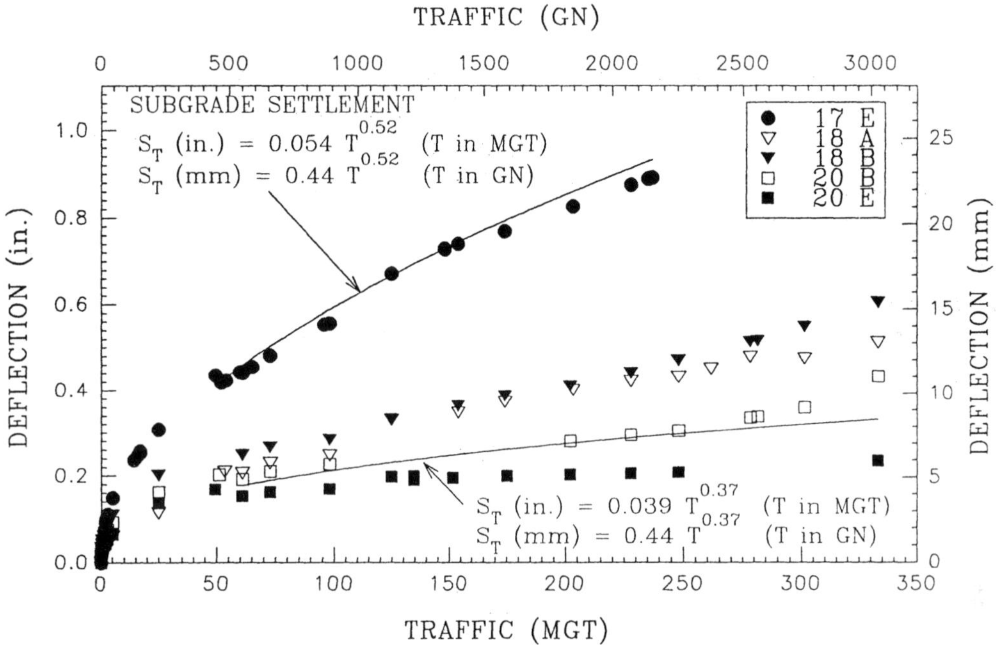

Fig. 8.48 Combined plot of FAST subgrade settlements

For the subgrade, 4 axles per cycle is assumed, so that C_m = 7600.

Assuming a linear trend beyond 50 MGT (445 GN), these equations become:

a) Section 17E--

$$S_T\,(\text{in.}) = 0.0026\,T\ , \tag{8.14a}$$

or

$$S_N\,(\text{in.}) = 3.4 \times 10^{-7}\,N\ . \tag{8.14b}$$

b) Other sections--

$$S_T\,(\text{in.}) = 0.001\,T\ , \tag{8.15a}$$

or

$$S_N\,(\text{in.}) = 1.4 \times 10^{-7}\,N\ . \tag{8.15b}$$

Equations 8.14 and 8.15 would be representative of the contribution of this subgrade to track settlement after the track had been in service for awhile.

8.5.2.4 Total settlement

The total track settlement caused by repeated loading, is equal to the sum of the compressions of the layers down to a depth of negligible traffic effect. Because total settlement varies along the track, differential settlement will accompany total settlement.

For the Colorado sites the substructure measurements extended only down to 3 m (10 ft) below top of subgrade. Although this was probably not deep enough to include all of the subgrade contribution, most of the settlement should be represented. Unfortunately top of rail survey data were not available for comparison with the calculated settlement from the substructure measurements.

Figure 8.49 shows the results for one of the sites (section 18B) up to 25 MGT (220 GN) of traffic. The ballast layer compression, S_B, is

$$S_B = \varepsilon_B H_B \ , \tag{8.16a}$$

where ε_B = plastic ballast strain, and

H_B = ballast layer thickness.

The subballast layer compression, S_S, is likewise

$$S_S = \varepsilon_S H_S \ , \tag{8.16b}$$

where ε_S = plastic subballast strain, and

H_S = subballast layer thickness.

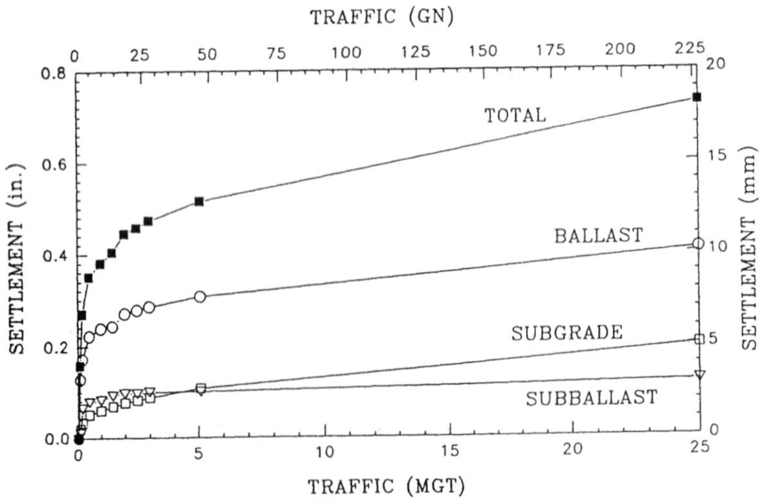

Fig. 8.49 Ballast, subballast and subgrade contributions to total settlements

Finally the subgrade layer (10 ft or 3 m) compression is S_G. Each of these three components is plotted in Fig. 8.49. The total settlement, S_L ,is then

$$S_L = S_B + S_S + S_G \tag{8.16c}$$

More than half of the total settlement is contributed by the ballast, but S_S and S_G are still significant. The reason is the track was new so that the subballast and subgrade had not previously been subjected to traffic. After tamping at a later time, the proportion of the ballast contribution to total settlement will be much greater because the undisturbed subballast and subgrade will continue their trends, while the top ballast will be loosened by tamping. The value of S_B after tamping at that site should be less than shown in Fig. 8.49 because the loosened layer thickness will be less than the full ballast depth, instead of the full depth as for the newly constructed track.

8.6. Effects of Fouling in Box Tests

8.6.1. Test Conditions

A metal ballast box was designed with the inside dimensions given in Fig. 8.2. The box was water tight to permit saturated ballast tests. The inside was lined with rubber pads for resiliency. The pads were covered with stainless steel sheets to protect the rubber from abrasion by ballast particles. Springs were attached to the sleeper segment to create a gap of 1 to 4 mm upon unloading which would simulate the effect of lift up and impact.

A series of tests was conducted with a hardwood sleeper segment to study ballast settlement and degradation with various conditions of water, fouling and impact (Ref. 8.29). The ballast used was a marble with the following properties:

- Bulk specific gravity = 2.75
- Absorption = 0.22%
- Los Angeles abrasion resistance = 39.3
- Mill abrasion loss = 8.1
- Magnesium sulfate soundness loss = 0.5%
- Gradation = AREA No. 4

Ballast was placed to a depth of 305 mm (12 in.) using lightly rodded layers. Specimens of fouled ballast were obtained by adding fine particles to the voids after each layer of coarse particles was placed. This was intended to represent the fouling process in the field. However the effect of mixing during tamping, in which the coarse particle contacts might initially be separated by fine particles, was not simulated. The fine particles, which were less than 1 mm in size, were obtained from crushing and abrasion of the coarse marble particles.

The equivalent wheel load simulated in the test was 160 kN (36000 lb). This required a maximum sleeper segment load in the box of 21 kN (4800 lb). The maximum number of cycles applied was one million which represents about 640 GN (72 MGT) of traffic. When large settlement developed, the sleeper segment was raised and the void filled with crib ballast particles.

8.6.2. Results

The effect of fouling index (Chapter 7, Section 7.1.3) on sleeper settlement with no gap (i.e., no loss of contact between sleeper and ballast) is shown in Fig. 8.50. Settlement increased as the amount of fouling material increased. Because the fouled ballast specimens for these tests were prepared in layers by first placing the coarse particles and then adding the fouling material to the voids, the loosening effect of tamping fouled ballast is not included in Fig. 8.50. If it were , the effect of an increasing amount of fouling material on settlement would be much greater. Also a much larger number of cycles would have amplified the difference in the effect of the amount of fouling material as well.

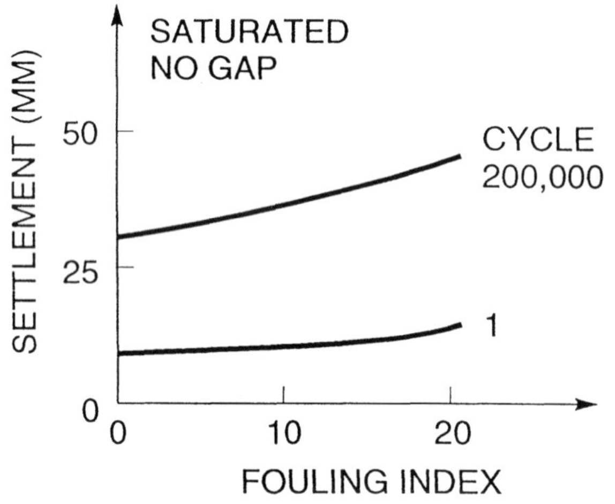

Fig. 8.50 Effect of fouling on ballast settlement

An order of magnitude increase in the amount of settlement occurred when a gap was created in the unloaded state (Fig. 8.51). This indicates the importance of preventing gap formation. If the ballast becomes highly fouled, settlement after tamping which produces the gap will develop rapidly, hence indicating need for cleaning the ballast.

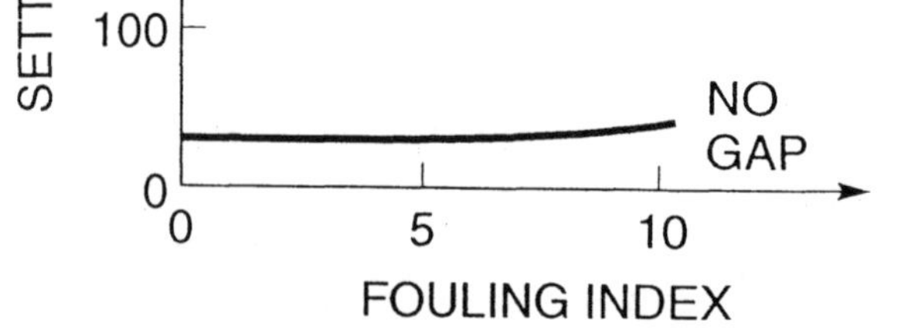

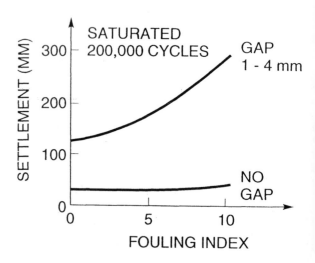

Fig. 8.51 Effect of gap on ballast settlement at 200,000 cycles

Figure 8.52 shows the effects of initial fouling index and gap on the further degradation of the ballast during the test. The results were obtained after one million cycles, except for the test with 9% initial fouling index with a gap. This test had to be terminated after 200,000 cycles because of the excessive cumulative settlement. If the sleeper segment had been repeatedly raised with crib ballast inserted to fill the gap as often as required to reach one million cycles, the degradation undoubtedly would have been much greater, as indicated by the arrows. The figure clearly shows that a greater degree of fouling generates increased fouling, and that a gap greatly increases the amount of fouling over that with no gap. The wood sleeper segment showed significantly increased wear, as well, with increased initial fouling and with a gap.

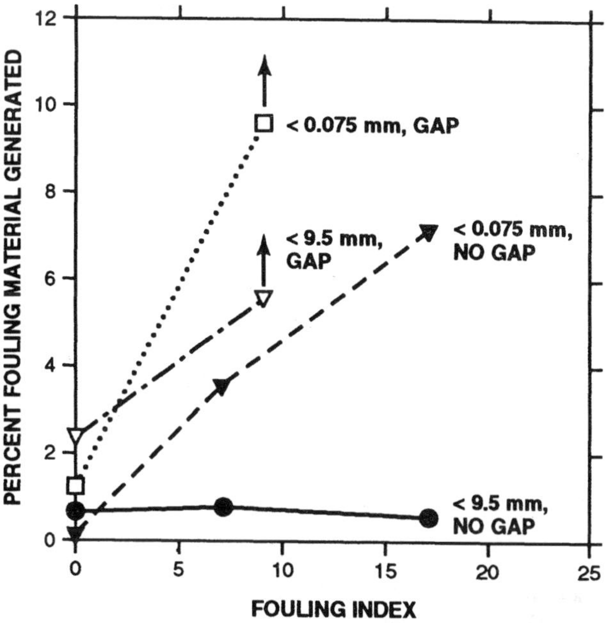

Fig. 8.52 Effects of fouling and gap on saturated ballast degradation

In the above comparisons the wheel load was constant. However, increasing settlement means increasing differential settlement (track roughness). Thus with increasing settlement dynamic wheel loads will increase for the same static wheel loads. Hence the rate of settlement will increase with the gap and with initial fouling even more than shown in the box tests.

8.7. Axle Load Effects

In some field tests conducted by British Railways (Ref. 8.20) no significant difference in track settlement was observed between 20 and 25 tonne (44-55 kip or 196-245 kN) axle loads. In tests in their track laboratory British Railways found (Ref. 8.20) that settlement increased approximately in proportion to load in the range 10 to 30 tonne (22-66 kip or 98-294 kN) axle load. For the same load, settlement also increased approximately in proportion to sleeper spacing. Finally, increasing the rail weight from 56 kg/m to 65 kg/m reduced settlement by 20%, presumably because of decreased ballast pressure.

The German railroad has studied the effect of ballast pressure on the maintenance cost (apparently related to time between surfacing operations). Eisenmann stated (Ref. 8.16) that maintenance cost increases with the 4th power of ballast pressure. As a result, when the German railroad decreased sleeper spacing to 600 mm, increased sleeper length to 2.6 m and increased rail weight to 60 kg/m, they found that the time between tamping increased from 1-2 yr to 4-5 yr.

British Rail concurs that the fourth power relationship between maintenance cost and ballast pressure is reasonable on the basis of the observed relationships between stress and cumulative strain in repeated load triaxial tests. This is illustrated by Fig. 8.53 which assumed that strain increases in proportion to stress and also that strain increases with the log of the number of cycles. If maintenance is required when a given cumulative strain level is reached, then the maintenance life or time between surfacing is proportional to, and the maintenance cost is inversely proportional to, the number of cycles to reach the designated strain, for a given traffic rate. As Fig. 8.53 clearly shows, the number of cycles increases much more rapidly than the stress decreases. Hence maintenance cost increases much more rapidly than ballast pressure. Using this method of analysis, data from repeated load triaxial tests and box tests on ballast actually predict that maintenance cost would increase with much more than the fourth power of ballast pressure.

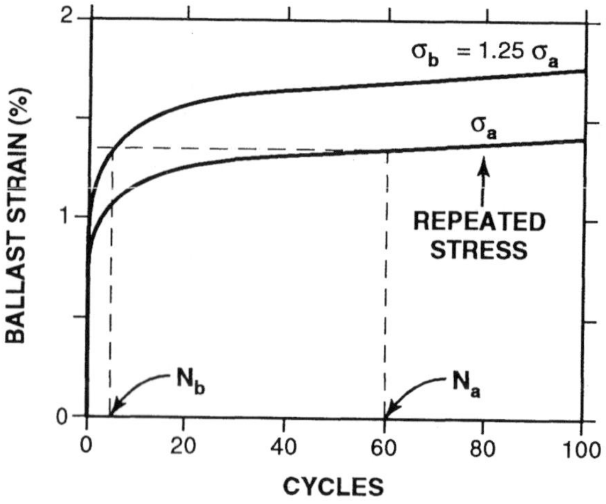

Fig. 8.53 Illustration of nonlinear relationship between ballast stress and load cycles for common cumulative strain

The AASHO (American Association of State Highway Officials) road tests (conducted in the United States) on flexible pavements also provided data relating axle load to number of cycles required to produce a given level of pavement damage (rutting and cracking). These data (Ref. 8.30) showed that the ratio of the number of cycles of a standard axle load (80 kN or 18000 lb) to the number of cycles of another axle load, W, producing the same damage was approximately proportional to the axle load ratio, W/80, to the fourth power. Thus this totally independent, but related, source showed trends consistent with those thought relevant to railway track performance.

Ballast box tests were conducted at the University of Massachusetts for both 36000 lb and 50000 lb (160 and 222 kN) equivalent wheel loads (Ref. 8.39). This represents a 39% increase in wheel load and would indicate a 370% increase in maintenance cost if the fourth power relationship were assumed. The test results gave an average increase in ballast degradation of about 650% using coarse breakage and 160% using fine breakage.

Although this analysis is a gross oversimplification of the factors defining surfacing life of track, it does indicate the highly nonlinear relationship and the potentially high cost of increasing axle load. For example assume the fourth power relationship and an increase from 100 ton cars to 125 ton cars. The axle load would increase by about 20%, but the maintenance cost would increase by 200%.

Some ballast box tests were conducted to further evaluate the effects of axle load on ballast settlement and degradation (Ref. 8.25). The ballast layer thickness was 152 mm

(6 in.). The sleeper segment was wood. The ballast was dolomite with an LAA of 29.7, an MA of 8.6, and an AREA 4A gradation with no particles of less than 9.5 mm (0.375 in.) size. The ballast was placed and rodded in layers and tested dry. If the settlement during a test reached 38 mm (1.5 in.), then the sleeper segment was removed, the crib ballast used to fill in the gap, and the sleeper reseated. The cycling was then resumed.

The axle loads and corresponding box test loads are listed in Table 8.4.

Table 8.4 Box test loads and cycles

Axle load		Box load		Cycles per 300 MGT (2670 GN)
Ton	kN	lb	kN	
23	210	2900	13	6510000
33	290	4200	19	4560000
39	350	4900	22	3840000
49	440	6200	28	3060000

A typical settlement trend from this test series is shown in Fig. 8.38. The total settlement is plotted in relation to the magnitude of the repeated load in Fig. 8.54. Settlement is given as a ratio to settlement at the 4200 lb (18.7 kN) load. Load is plotted as a ratio to 4200 lb (18.7 kN) load. The settlement increases at a diminishing rate with increasing load. The amount of fouling material generated is plotted as a function of repeated load in Fig. 8.55. Fouling is given as a ratio to fouling at the 4200 lb (18.7 kN) load. The fouling increases at an increasing rate with increasing load. The fouling material consists of all particles smaller than the 9.5 mm (0.375 in.) size. In these tests 30 to 50% of the fouling material was smaller than 0.075 mm, i.e. fines.

Figures 8.54 and 8.55 are plotted in the form of a damage equation given by

$$F_R = (P_R)^n \, , \tag{8.18}$$

where F_R = ratio of breakage or settlement at any load to that at the reference load,

P_R = ratio of any load to reference load, and

n = damage exponent.

With the reference load set at 290 kN (4200 lb), the damage exponent for settlement (Fig. 8.54) is 0.4 and that for breakage (Fig. 8.55) is 1.5. Further tests are being conducted to verify these trends.

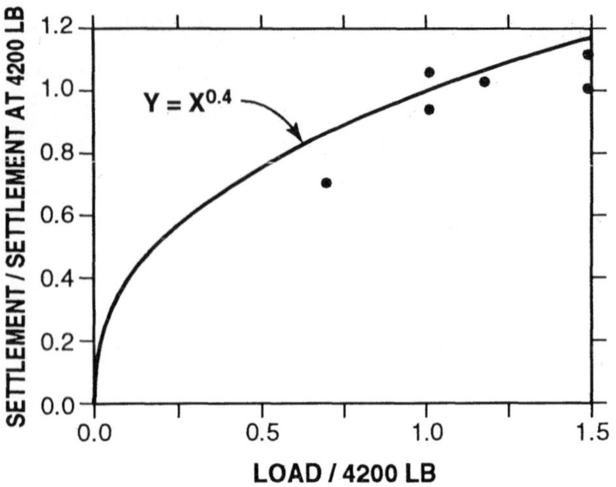

Fig. 8.54 Settlement ratio as a function of load ratio

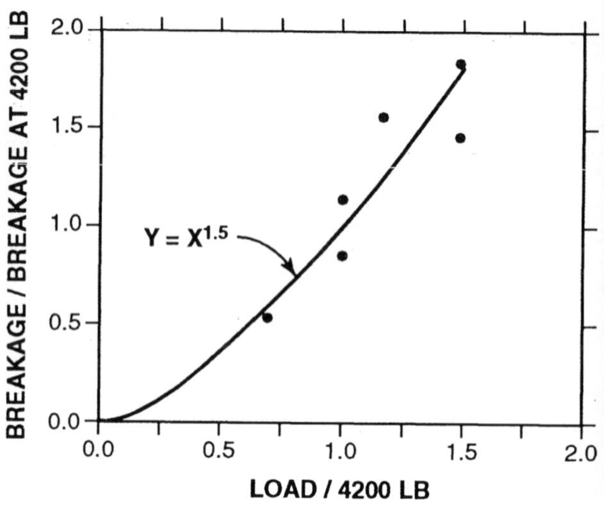

Fig. 8.55 Breakage ratio as a function of load ratio

8.8. Ballast Life

The Canadian Pacific Railroad (Ref. 8.35) has taken a unique approach to ballast selection. First the unacceptable rock is eliminated by petrographic analysis. Acceptable candidates are given a rating based on a combination of Los Angeles abrasion and mill abrasion test results. The resulting abrasion number is related to ballast life, represented by tons of traffic, as a function of gradation as shown in Figure 8.56. The gradations are given

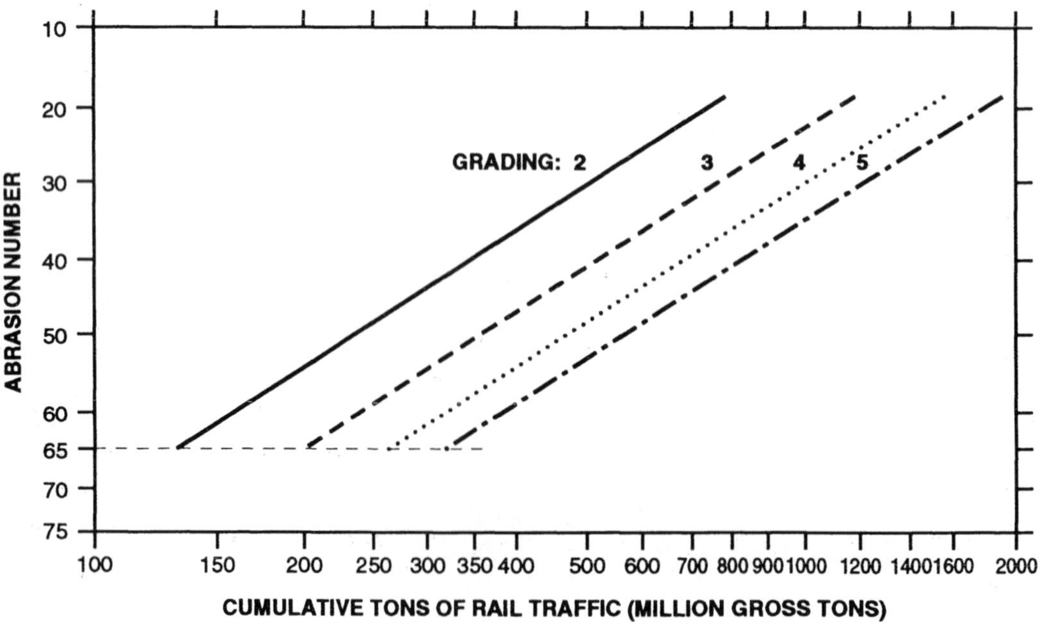

Fig. 8.56 Canadian Pacific Rail ballast life chart

in Table 8.5. The principal cause of ballast breakdown is assumed to be from traffic, based on fouled ballast studies on their system. Subgrade intrusion is minimized by using a proper subballast. The gradation is assumed to affect ballast life mainly because of 1) void storage capacity, i.e., a uniform ballast stores more finer particles than a broadly graded ballast, and 2) the observation that a coarse ballast gives less restriction to downward movement of products of degradation (Ref. 8.35). This means that a coarse, uniform ballast is better than a finer, uniform ballast, which in turn is better than a broadly-graded ballast.

Table 8.5 Gradations for CP Rail Ballast Life Chart

Ballast Grading	Maximum Size	Percent by weight smaller than specified sieve								
	mm(in.)	64mm (2.5 in.)	51mm (2 in.)	38mm (1.5 in.)	25mm (1 in.)	19 MM (3/4 in.)	13 mm (1/2 in.)	9.5mm (3/8 in.)	4.8mm (No. 4)	0.075mm (No.200)
2	2		100	90-100	70-90	50-70	25-45	10-25	0-3	0-2
3	2		100	90-100	70-90	30-50	0-20	0-5	0-3	0-2
4	2		100	90-100	20-55	0-5			0-3	0-2
5	2.5	100	90-100	35-70	0-5				0-3	0-2

Ballast gradings 2 and 3 shall be used for crushed gravel.
Ballast grading 4 shall be used for crushed gravel, crushed rock or slag.
Ballast grading 5 shall be used for crushed rock or slag.

Some data suggest that particle contact area rather than contact force controls ballast abrasion (Ref. 8.31). However other data show that breakage and abrasion are related to contact force (Ref. 8.39). Thus in cases where ballast breakdown from traffic is the primary source of fouling, the role of gradation needs to be clarified through further research.

A survey of North American railroad opinions (Ref. 8.32) suggested that freezing-induced ballast breakdown is not a major contribution to ballast fouling. The relative contributions of the other sources could not be determined from this survey. However a large percentage of these railroads found formation of slurry spots under high dynamic loads for most ballasts.

The same survey indicated that a broader ballast grading is generally believed to increase ballast life and reduce the rate of settlement caused by the ballast. The preferred way to broaden the grading is to add smaller particles, say down to 13 mm (1/2 in.) size, while limiting the largest size to 76 mm (3 in.). The broader grading is recognized to have a greater susceptibility to fouling because of smaller void spaces and a smaller void ratio. Whether or not it will foul faster depends on the source of fouling. This is one of the issues on which conflicting opinions exist and which cannot be sorted out with existing data. The resolution of this matter is important in order to define the optimum ballast grading.

A ballast durability study in track was conducted by the Association of American Railroads in cooperation with the Santa Fe Railroad (Ref. 8.33). Three ballasts were tested: a dolomite, a slag and a quartzite. Samples of these ballasts were taken each year for 7 years after ballast placement during which about 300 MGT (2670 GN) of traffic were applied. The degree of fouling was represented by the percent by weight of material smaller than 4.75 mm (No. 4 sieve).

At the end of seven years the amount of fouling material was 5 to 7% for the three materials. Thus none of the ballasts was close to being at the end of its life. Considering that about 1 to 3% of fouling material existed initially, the rate of fouling averaged 0.4 to 0.6% per year. Petrographic analysis of the samples after six years indicated that 75 to 90% of the fouling material was from ballast breakdown. The remainder came from external sources. Linear extrapolation of the rate of degradation from the first seven years to a condition of complete fouling gave unreasonably long ballast lives. This suggests that the degradation rate is not linear over the life of ballast. More reasonable ballast life estimates were obtained using the Canadian Pacific Rail ballast life chart (Fig. 8.56) after adjusting for the differences between the actual gradations and layer thicknesses and those used by Klassen in developing the chart. Also the average percent fouling for the three ballasts correlated reasonably well with their abrasion numbers (Eq. 7.1).

REFERENCES

8.1 Panuccio, C. M., Dorwart, B. C., and Selig, E. T. (1978). **Apparatus and procedures for railroad ballast plate index test.** *Geotechnical Testing Journal*, ASTM, September.

8.2 Selig, E. T., Yoo, T. S., and Panuccio, C. M. (1982). *Mechanics of ballast compaction-Volume 2: Field methods for ballast physical state measurement.* Final Report to U. S. DOT, Transportation Systems Center, Cambridge, MA, Report No. DOT-TSC-FRA-81-3, II, March.

8.3 Panuccio, C. M., McMahon, D. R., and Selig, E. T. (1982). *Mechanics of ballast compaction-Vol. 4: Laboratory investigation of the effects of field compaction mechanisms.* Final Report to the U.S. DOT, Transportation Systems Center, Cambridge, MA, Report No. DOT-TSC-FRA-81-3, IV, March.

8.4 Klugar, K. (1972). **Influence of track life on the lateral displacement of track.** Eisenbahntechnische Rundschau, Heft 11, November, pp. 446-449.

8.5 Panuccio, C. M., Yoo, T. S., and Selig, E. T. (1982). *Mechanics of ballast compaction-Vol. 3: Field test results f or ballast physical state measurement.* Final Report to the U. S. DOT, Transportation Systems Center, Cambridge, MA, Report No. DOT-TSC-FRA-81-3, III, March.

8.6 Klugar, K. (1974). **Track buckling experiments of the Austrian federal railroads with new types of ties.** ETR-Eisenbahntechnische Rundschau, Vol. 25, No. 3, pp. 70-75.

8.7 Eisenmann, J. (1976). **The significance of the rail lifting wave.** *Rail International,* No. 10, October, pp. 576-581.

8.8 Frederick, C. O. (1978). **The effect of lateral loads on track movement.** *Proceedings of Symposium on Railroad Track Mechanics and Technology,* A. D. Kerr, Ed., Princeton University, pp. 109-140.

8.9 Kerr, A. D. (1978). *Thermal buckling of straight tracks: fundamentals, analyses, and preventive measures.* Prepared for U.S. Department of Transportation, Federal Railroad Administration, Report No. FRA/ORD-78/49, September, 58 pp.

8.10 Kerr, A. D. (1978). **Lateral buckling of railroad tracks due to constrained thermal expansions —a critical survey.** *Proceedings of Symposium on Railroad Track Mechanics and Technology*, A. D. Kerr, Ed., Pergamon Press, pp. 141-169.

8.11 Klugar, K. (1976). **The static track lateral strength of loaded tracks.** *ETR-Eisenbahntechnische Rundschau*, Vol. 25, No. 4, pp. 211-216.

8.12 Prud'homme, A. (1967). **The resistance of the permanent way to the transversal stresses exerted by the rolling stock.** *Bulletin of the International Railway Congress Association,* Vol. 44, No. 11, November, pp. 731-766.

8.13 Prause, R. H. and Kennedy, J. C. (1977). *Parametric study of track response and performance.* Prepared for U.S. Department of Transportation, Federal Railroad Administration, Report No. FRA/ORD-77/75, December, 118 pp.

8.14 Selig, E. T. and DelloRusso, V. (1991). **Sources and causes of ballast fouling.** American Railway Engineering Association, *Bulletin* No. 731, Vol. 92, May, pp. 145-157.

8.15 Byrne, B. J. (1989). *Evaluation of the ability of geotextiles to prevent pumping of fines into ballast.* Master of Science Project, Department of Civil Engineering, University of Massachusetts, Amherst, MA, Geotechnical Report No. AAR89-367P, October.

8.16 Personal communications with Professor Eisenmann at Technischen Universitat Munchen and with Mr. Klaus Martinek and Dr. Oberweiler of Deutsche Budesbahn, April, 1986.

8.17 The Permanent Way Institution (1979). *British railway track-design, construction and maintenance,* Fifth Ed.

8.18 Personal communications with Dr. M. J. Shenton and Mr. D. M. Johnson at British Rail, Derby, Spring 1986.

8.19 Burks, M. E., Robson, J. D., and Shenton, M. J. (1975). **Comparison of Robel supermat and Plasser 07-16 track maintenance machines**. Tech. note *TN SM* 139, British Railways Board R & D Division, December.

8.20 Wright, S. E. (1983). *Damage caused to ballast by mechanical maintenance techniques.* British Rail Research Technical Memorandum TM TD 15, May.

8.21 Chrismer, S. M. (1989). *Track surfacing with conventional tamping and stone injection.* Association of American Railroads Research Report No. R-719, Chicago, Illinois, March.

8.22 Waters, J. M. (1974). *A review of the work of the soil mechanics section, December 1972-1973.* British Railways Board, R & D Division, Technical Memorandum SM7, January.

8.23 Johnson, D. M. (1982). *A reappraisal of the BR ballast specification.* British Rail Research Technical Memorandum, TM TD 1, October 1982.

8.24 Holmes, W. A. (1971). *Ballast failure-Duffield.* BR Report, July.

8.25 Selig, E. T. and Devulapally, R. V. (1991). *Heavy axle load box tests--series 2.* University of Massachusetts report to Association of American Railroads, November 26.

8.26 Shenton, M. J. (1985). **Ballast deformation and track deterioration**. *Track Technology for the Next Decade*, Thomas Telford, Ltd., London, pp. 253-265.

8.27 Selig, E. T., Yoo, T. S., Adegoke, C.W., and Stewart, H. E. (1981). *Status report-ballast experiments intermediate (175 mgt) substructure stress and strain data.* Report No. FAST/TTC/TM-81/03, by University of Massachusetts, for U. S. DOT Transportation Systems Center, Cambridge, MA, March 11.

8.28 Selig, E.T. (1981). *FAST substructure static strain and deflection measurements from 175 to 422 mgt.* Report No. TSC81 - 269I for U. S. DOT Transportation Systems Center, Cambridge, MA, April.

8.29 Chiang, C. C. (1989). *Effects of water and fines on ballast performance in box tests.* MS Degree Project Report No. AAR89-366P, Department of Civil Engineering, University of Massachusetts, Amherst, August.

8.30 Edwards, J. M. (1986). *Bituminous pavements: materials, design and evaluation.* University of Nottingham, Department of Civil Engineering, Lecture Notes, Lecture A, April.

8.31 Raymond, G. P. (1977). *Stresses and deformations in railway track.* CIGGT Report No. 77-115, November.

8.32 Johnson, D. M. (1985). *Investigation of practical ballast specification requirements.* Report No. WP-114, Association of American Railroads, February.

8.33 Chrismer, S. M., Selig, E. T., Laine, K., and DelloRusso, V. (1991). *Ballast durability test at Sibley, Missouri.* Association of American Railroads, Chicago, Illinois, Report No. R-801, December.

8.34 Yoo, T. S., Chen, H. M., and Selig, E. T. (1978). **Railroad ballast density measurements.** *Geotechnical Testing Journal,* ASTM, Vol. 1, No. 1, pp. 41-54, March.

8.35 Klassen, M. J., Clifton, A. W., and Watters, B. R. (1987). **Track evaluation and ballast performance specifications**. Transportation Research Record 1131, *Performance of aggregates in railroads and other track performance issues,* Washington, D. C., pp. 35-44.

8.36 Collingwood, B. I. (1988). *An investigation of the causes of railroad ballast fouling.* M. S. degree project report, Report No. AAR88-350P, Department of Civil Engineering, University of Massachusetts, Amherst, May.

8.37 Tung, K. W. (1989). *An investigation of the causes of railroad ballast fouling.* M. S. degree project report, Report No. AAR89-359P, Department of Civil Engineering, University of Massachusetts, Amherst, May.

8.38 ASTM D2488. **Standard practice for description and identification of soils (visual-manual procedures)**. Annual Book of Standards, Section 4, Construction, Vol. 04.08, *Soil and Rock.*

8.39 Feng, D. M. (1984). *Railroad ballast performance evaluation.* M. S. degree project report, Report No. AAR84-311P, Department of Civil Engineering, University of Massachusetts, Amherst, May.

8.40 DiPilato, M. A., Levergood, A.V., Steinberg, E. I., and Simon, R. M. (1983). *Railroad track substructure-design and performance evaluation practices.* Goldberg-Zoino and Associates, Inc., Newton Upper Falls, Massachusetts, Final Report for U. S. DOT Transporation Systems Center, Cambridge, Massachusetts, Report No. FRA/ORD-83/04.2, June.

9. Subballast Requirements

9.1. Introduction

The functions of subballast are listed in Chapter 2 (Section 2.1.5). Summarized, these are:

1) Reduce stress to subgrade.

2) Protect subgrade from freezing.

3) Keep subgrade and ballast separate.

4) Prevent upward subgrade fines migration.

5) Prevent subgrade attrition by ballast.

6) Shed water from above.

7) Drain water from below.

The requirements for satisfying these functions will be presented in this chapter.

9.2. Reduce Stress to Subgrade

The stress reduction function is an extension of this same ballast function. It is generally not economical to provide the entire stress reduction function by ballast alone because of the often large required depth between the sleeper and the subgrade surface. Because a subballast layer is required for other reasons and the nature of the subballast material makes it suitable for the structural function of subgrade stress reduction, the subballast layer thickness is considered as part of the total depth required between the bottom of the sleeper and the top of the subgrade. The basis for determining this total depth is explained in Chapter 10 which covers the subgrade requirements.

To serve as a structural material, the subballast must have a high enough resilient modulus, and a stable plastic strain accumulation characteristic under repeated wheel load. To achieve these properties the material must be permeable enough to avoid significant positive pore pressure build up under repeated load, must consist of durable particles, and must not be sensitive to changes in moisture content. The durability must be adequate to resist breakdown and abrasion from the cyclic stresses produced by the train loading. Such a material is represented by mixtures of sand and gravel particles composed of crushing and abrasion resistant minerals. These materials may be available natural deposits, or may be produced by crushing rock or durable slags.

The ASTM D1241 standard specification for aggregate layers for roads is used as a basis for subballast by North American railroads (Ref. 9.1). The coarse aggregate (> 2 mm) must not break up when subjected to freeze-thaw or wetting-drying cycles and must have a maximum loss in the LAA test using ASTM C131 (Ref. 9.2) of 50. Generally a higher LAA value is acceptable for subballast than for ballast because the traffic stresses are lower and the smaller size and broader gradation makes the particle contact stresses much lower than in the ballast. The fine aggregate must be free of vegetable matter and lumps or balls of clay. The fraction finer than 0.425 mm must have a liquid limit of 25 or less and a plasticity index of 6 or less. This fraction will generally have the characteristics of an ML soil (Fig. 3.4).

Resilient modulus and plastic strain characteristics of suitable subballast materials are similar to those presented in Chapter 3 (Section 3.6) and in Chapter 7 (Section 7.3). The discussion of analytical models for granular materials by Brown and Selig (Ref. 9.6) is particularly relevant for subballast materials.

Failure by rapid plastic strain accumulation under repeated load when the material becomes saturated and has insufficient permeability to prevent pore pressure build up is illustrated in Fig. 9.1 (Ref. 9.3). The permanent deformation represents rutting in wheel-tracking tests on four different granular materials. The highest quality material (G1) was highly compacted broadly-graded crushed rock with no plastic fines. The lowest quality material (G4) was a natural gravel presumably with sand and fines.

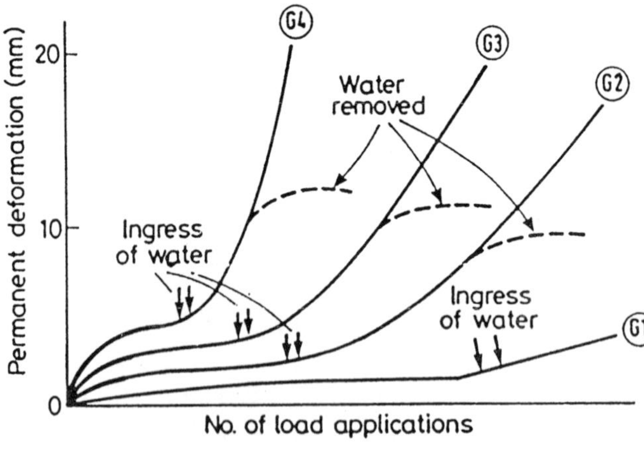

Fig. 9.1 Influence of water on permanent strain development in granular material

9.3. Frost Protection

When soil temperature can fall below the freezing point for water, then consideration must be given to potential problems associated with soil freezing. In climates where freezing conditions do not exist, then this section can be ignored.

When soil freezes the pore water converts to ice with the result that the resilient modulus greatly increases and the rate of plastic strain accumulation decreases. If the soil is not completely saturated, then the expansion of water upon freezing is accommodated by compressing the air voids, with the result that little soil volume change occurs. If no water migration occurs during the freezing process, then when the soil thaws the moisture content returns to its value prior to freezing, and the unfrozen soil properties will not have been altered significantly by the freeze-thaw cycle.

However when the pore water freezes a deficiency of adsorbed water is created around fine-grained soil particles. If the relationship between the rate of freezing and the rate of water movement is in a certain balance, then the ice crystals at the freezing front will grow into lenses of ice, pushing the soil particles apart. This results in local soil volume change by much more than the 10% maximum when water turns to ice. This process, called frost heave, can cause significant vertical differential displacement of pavement or track. Then when the ice melts, an excess of water remains which causes softening or loss of strength of the soil. During this period of thaw softening, severe plastic deformation can occur with resulting rapid loss of track geometry and accelerated damage to track components.

Three factors are needed simultaneously for the frost-heave-thaw-softening problem to develop: 1) freezing temperature in the soil, 2) source of water to feed the ice lens growth, and 3) a frost-susceptible soil. Good drainage helps limit the source of water. Insulation of the frost susceptible soil by a sufficiently thick covering layer of non-frost susceptible soil will prevent freezing temperature. The combined thickness of the ballast and the subballast will serve as insulation for the subgrade. The required thickness can be determined by local experience.

The subballast, however, must not be frost susceptible or it, too, must be protected by an insulating layer of ballast. Predicting the frost susceptibility of soil is difficult because of the complex thermodynamic and seepage processes involved. However general experience is that soils containing a high percentage of silt size particles are the most susceptible, with clays being next most susceptible. Sands and gravels without fines are not frost susceptible. Thus if freezing temperatures and a source of water can reach the subballast, then the subballast must have less than about 5% silt and clay size particles (fines) to prevent the indicated problems from occurring.

9.4. Particle Separation

The subballast must prevent intermixing of ballast and subgrade, and also prevent upward migration of subgrade particles into the ballast. Intermixing results from progressive penetration of the coarse ballast particles into the finer subgrade accompanied by the upward displacement of the subgrade particles into the ballast voids. This process can occur at any subgrade moisture condition. Intermixing has been observed with dry sand subgrade, for example. Upward migration of subgrade particles develops from at least three sources:

1) subgrade seepage carrying soil particles,

2) hydraulic pumping of slurry from subgrade attrition, and

3) pumping of slurry through opening and closing of subgrade cracks and fissures.

Prevention of intermixing and migration is achieved by using a proper subballast gradation. This is known as a separation function.

Research has been conducted to establish the gradation requirements for granular filters for drains associated with seepage of water from soil (Ref. 9.4). The problem of subgrade-ballast mixing is quite different because it is mainly produced by repeated traffic load rather than by seepage. However little research has been done to establish gradation criteria for repeated load situations. Thus the available criteria for drainage filters have been used, apparently with satisfactory results.

As described in Ref. 9.4 the filter criteria were first developed by Bertram in 1940 with advice from Terzaghi and Casagrande. Subsequent studies were made by the U.S. Army Corps of Engineers and the U.S. Bureau of Reclamation. The two resulting separation gradation criteria are:

$$D_{15}\,(\text{filter}) \leq 5\, D_{85}\,(\text{protected soil})\ , \tag{9.1}$$

and

$$D_{50}\,(\text{filter}) \leq 25\, D_{50}\,(\text{protected soil})\ , \tag{9.2}$$

where D_n is the sieve size (particle size) which passes n percent by weight of the total sample. For example D_{15} is the size for which 15% of the sample is finer.

The criterion in Eq. 9.1 causes the particles at the coarsest end of the protected soil (D_{85}) to be blocked by the particles at the finest end of the filter (D_{15}). This is illustrated in Fig. 9.2 which is adapted from Ref. 9.4. Assuming that no gaps exist in the grading of either the soil or the filter, the blocking action extends through the entire grading of both materials and a stable network of particles exists. The criterion in Eq. 9.2 helps to avoid gap-graded filters and create a filter gradation that is somewhat parallel to that of the protected soil.

For the medium to highly plastic clays without silt and sand the criteria in Eqs. 9.1 and 9.2 are relaxed for seepage applications to permit easier filter selection. In these cases the D_{15} size of the filters may be as large as 0.4 mm, and Eq. 9.2 may be ignored. However the filter must be broadly-graded, and to minimize the chance of filter particle segregation the coefficient of uniformity (D_{60} / D_{10}) must be 20 or less. These relaxed criteria may not be equally suitable for repeated load application.

Deviation from the above recommendations may be desired in some cases because of the difficulty in obtaining a suitable subballast gradation. In such cases laboratory tests can be conducted to test the filter capability of the subballast under repeated loading. Such a procedure is described in Chapter 12.

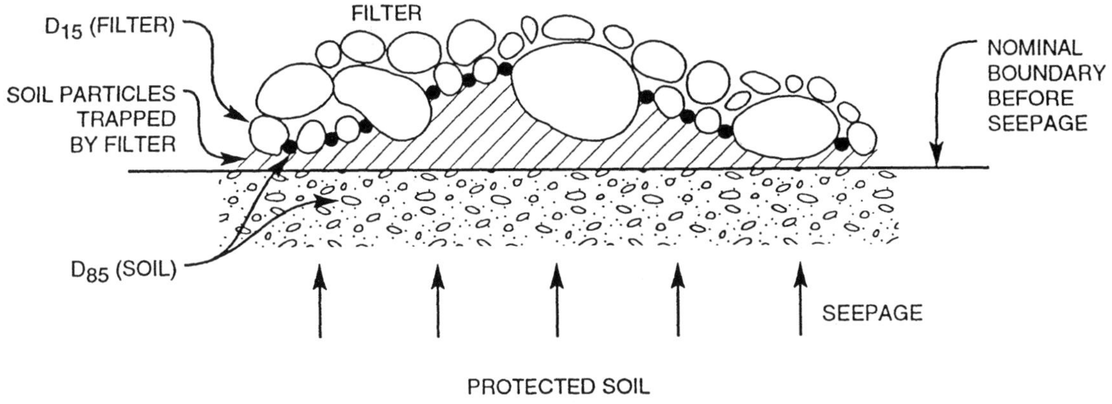

Fig. 9.2 Illustration of particle blocking action by granular filter

An example of the application of Eq. 9.1 to pavement filter design for drainage of upward seeping water is given in Fig. 9.3 from Cedergren (Ref. 9.4). The fine filter (curve 2) prevents the subgrade soil (curve 1) from pumping into the open-graded drainage aggregate (curve 3).

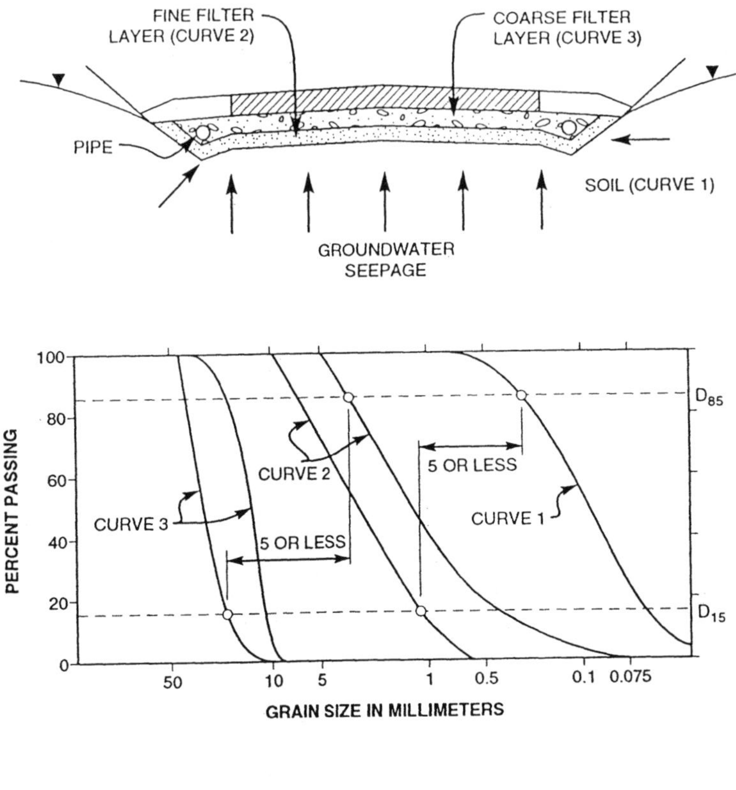

Fig. 9.3 Example of filter design for highway roadbed (reprinted from Ref. 9.4 with permission of the publisher)

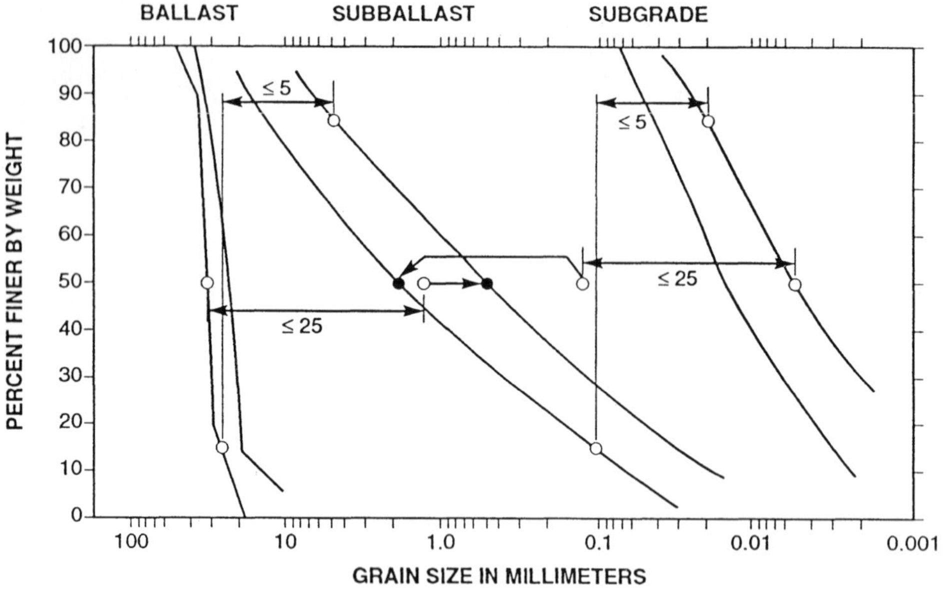

GRAVEL		SAND			SILT OR CLAY
COARSE	FINE	COARSE	MEDIUM	FINE	

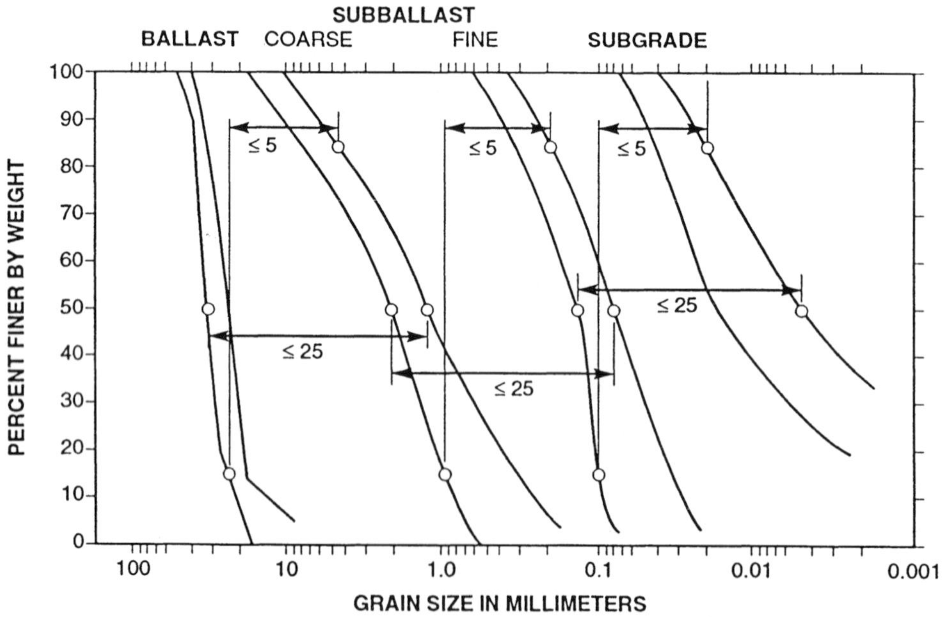

GRAVEL		SAND			SILT OR CLAY
COARSE	FINE	COARSE	MEDIUM	FINE	

Fig. 9.4 Examples of 1- and 2-layer subballast satisfying filter criteria

Not only must the subballast satisfy the criteria in Eqs. 9.1 and 9.2 in relation to the subgrade, but the criteria must also be satisfied in relation to the ballast. This is similar to the case in Fig. 9.3 if the subballast were considered to be represented by the fine filter (curve 2). This simultaneously places an upper and lower limit on the acceptable subballast gradation. In case a single subballast material cannot be found to fit this range of sizes for a particular subgrade and ballast, then a two-layer subballast may be used. The upper layer would be coarser to match with the ballast, while the lower layer would be finer to match with the subgrade. The relationship between these two subballast layers, of course, must also satisfy Eqs. 9.1 and 9.2.

An example of both a 1-layer and 2-layer subballast gradation in relation to a typical ballast gradation (AREA no. 4) and a fine-grained subgrade is given in Fig. 9.4. For the 1-layer subballast (Fig. 9.4a) a broad gradation ranging from fine gravel to silt size is required to satisfy Eq. 9.1. The criterion in Eq. 9.2 cannot be simultaneously satisfied. The 2-layer subballast (Fig. 9.4b) can easily satisfy both Eq. 9.1 and Eq. 9.2. Either the 1-layer or the 2-layer subballast shown will be satisfactory.

The subballast, especially the lower layer in the 2-layer system, is sometimes called a blanket layer. It usually consists primarily of sand size particles. A typical blanketing sand layer gradation used by British Rail is given in Table 9.1 (Ref. 9.5). It is similar to the 1-layer subballast in Fig. 9.4a. The recommended gradations in ASTM D1241 are also broadly-graded like that in Fig. 9.4a.

Table 9.1 Gradation of blanketing sand

Size (mm)	% by weight passing
14	100
2.36	80 (+/- 4) - 100
1.18	70 - 90 (+/- 4)
0.60	48 - 76
0.30	24 - 60
0.15	5 - 42
0.075	0 - 10

The aggregate particles must be durable so that the selected initial subballast layer gradations remain stable indefinitely.

9.5. Subgrade Attrition Prevention

When ballast is placed directly in contact with a fine-grained soil or soft rock subgrade, in the presence of water the ballast will wear away the subgrade surface and form a slurry. The slurry will then squeeze upward through the ballast voids. As will be demonstrated in Chapter 12, a subballast which satisfies the gradation criteria in Eqs. 9.1 and 9.2 will prevent this problem from occurring. One reason is that the high stresses at the ballast contact points on the subgrade surface are eliminated by the cushioning effect of the subballast. The subballast in contact with a fine-grained subgrade will either be primarily sand, if a 2-layer system, or will be broadly-graded down to a fine sand if a 1-layer system. In the latter case gravel particles present will not cause high contact stresses because they will be surrounded by sand particles.

9.6. Drainage

The subballast layer plays two roles in drainage of the track substructure. The first is to shed water entering from the surface. As will be shown in Chapter 13, proper ballast is much more permeable than subballast. Hence the primary outlet for water entering the ballast is the ballast layer itself. The subballast, then, serves as an underlying lower permeability boundary of the ballast to direct water away from the subgrade. To assist in this task the subballast permeability should be at least an order of magnitude smaller than that of the ballast, and have a surface sloped for lateral drainage.

The second role is to drain water seeping up from the subgrade, including that produced by excess pore pressure generated from cyclic stresses. Thus the subballast cannot be impermeable, but rather should have a permeability greater than that of the subgrade. The exceptions are when the subgrade is relatively permeable, such as a natural sand or sand-gravel layer, or when no upward seepage is expected such as on an embankment.

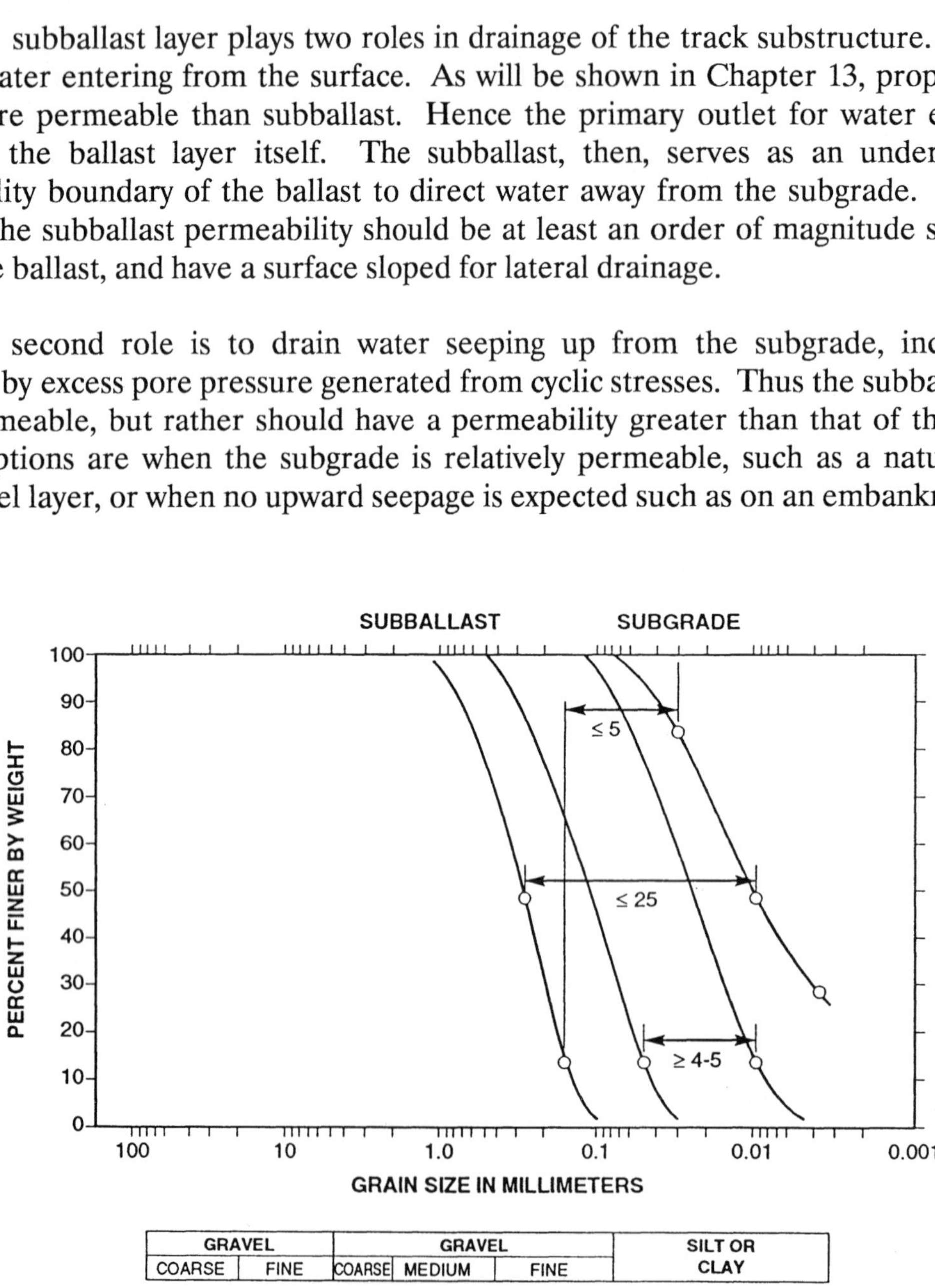

Fig. 9.5 Subballast grading requirements for separation and permeability

Thus to satisfy both roles, the subballast must generally have a permeability between that of the subgrade and that of the ballast. This requirement probably will be achieved just by satisfying the separation criteria of Eqs. 9.1 and 9.2. However an additional criterion, which was established along with those of Eqs. 9.1 and 9.2, is used to insure adequate permeability to drain an adjacent layer. This criterion is

$$D_{15}\,(\text{filter}) \;>\; 4 \text{ to } 5\; D_{15}\,(\text{soil being drained})\,. \tag{9.3}$$

An example of this criterion is shown in Fig. 9.5 together with the criteria in Eqs. 9.1 and 9.2. This criterion (Eq. 9.3) can be disregarded if drainage of water from the subgrade is not required.

A problem to consider is that a low subballast permeability cannot be achieved by allowing a high amount of fines in the material at locations where freezing is likely to occur, because this may make the subballast frost susceptible.

9.7. Placement of Subballast

The proper way to install the subballast is prior to placing the ballast and track. The subgrade surface would first be compacted and shaped for drainage. Then the subballast would be spread and roller compacted in layers at a suitable moisture content. Each layer should normally not be more than 100 to 150 mm (4 to 6 in.) thick prior to compaction. Placing and compacting a subballast layer with the track in place may be difficult to do. An example of an approach is described in Chapter 14, section 14.3.5. A high degree of compaction is desired but not essential, because traffic will finish the job. The main concern is to avoid disruption of the subballast when the ballast is placed, and to limit the penetration of the ballast into the subballast. Of course, the more the subballast is compacted when placed, the less the subsequent track settlement under traffic.

Each subballast layer of a different material or gradation should have a nominal compacted thickness of at least 150 mm (6 in.) to allow for construction variability and some subsequent compression under traffic. Theoretically a smaller thickness can be used to satisfy some of the subballast functions, but this runs the risk that at some locations the layer will be too thin.

REFERENCES

9.1 ASTM D1241. **Standard specification for soil-aggregate subbase, base, and surface courses.** *ASTM Annual Book of Standards*, Section 4, Construction, Vol. 04.08, Soil and Rock.

9.2 ASTM C131. **Standard test method for resistance to degradation of small-size coarse aggregate by abrasion and impact in the Los Angeles machine.** *Annual Book of ASTM Standards,* Section 4, Construction, Vol. 04.03, Road and Paving Materials.

9.3 Freeme, C. R. and Servas, V. (1985). **Advances in pavement design and rehabilitation.** *Accelerated testing of pavements*, CSIR, Pretoria, Republic of South Africa.

9.4 Cedergren, Harry R. (1977). *Seepage, drainage and flow nets.* John Wiley and Sons, New York.

9.5 Sperring, D. G. (1989). **Practical track design.** Railway Industries Association, Third Track Sector Course, Paper 3.4.

9.6 Brown, S. F. and Selig, E. T. (1991). **The design of pavement and rail track foundations.** *Cyclic loading of soils: from theory to design,* O'Reilly and Brown Eds, Blackie, London, pp. 249-305.

10. Subgrade Behavior

10.1. Introduction

As indicated in Chapter 2, subgrade has a major influence on track performance. The main subgrade function is to provide a stable platform for the ballasted track structure. To do this the following problems (failure modes) must be avoided. Summarized from Chapter 2, these are:

1) Excessive plastic deformation.

2) Consolidation and massive shear failure.

3) Progressive shear failure.

4) Excessive swelling and shrinking.

5) Frost heave and thaw softening.

6) Subgrade attrition.

This chapter will discuss the factors influencing each of these problems and means of solving them.

The depth of substructure influence on track stiffness is at least several times the length of the sleepers. Because the majority of this depth is subgrade, and because the subgrade resilient modulus varies by a large amount from point to point along the track, the subgrade probably has the most influence on track stiffness of any variable. This chapter will describe the resilient modulus of subgrade soils and discuss its effect on track performance. Correlations of track foundation modulus, u, with subgrade resilient modulus, E_r, will also be shown.

To analyze or predict track performance, to determine the cause of subgrade related performance, and to correct subgrade problems requires a knowledge of the subgrade conditions. The necessary extent of this knowledge depends on the specific problem. Relevant information includes: 1) the depths of the distinct soil or rock layers, 2) the identity of the materials, 3) the physical properties of the materials, and 4) ground water conditions. Unfortunately this information is not usually known. Hence a field investigation, often supplemented by laboratory tests, is required. This chapter will summarize means for obtaining this information.

10.2. Subsurface Investigation

Methods for identifying and classifying soils are described in Chapter 3. These should be used whenever an investigation of substructure layers is carried out in order to have an acceptable degree of precision in defining the materials encountered. This is the most important step in geotechnical engineering.

In preparation for a field investigation of subgrade at a particular site a study of available soil maps and geology maps with associated reports is useful in anticipating the materials likely to be encountered. Each railway engineering district would benefit from establishing a file of such information for their territory. This will save a great amount of time in preparing for individual field trips as they occur.

A limited subsurface investigation can be made by hand or by machine to excavate inspection holes at the side of the track (Fig. 10.1a). Generally one meter or so is about the practical limit for digging distance below the subgrade surface. Because conditions at the top part of the subgrade are often different beneath the track than at the side of the track, a better inspection can be made by removing the ballast shoulder (Fig. 10.1b). This permits

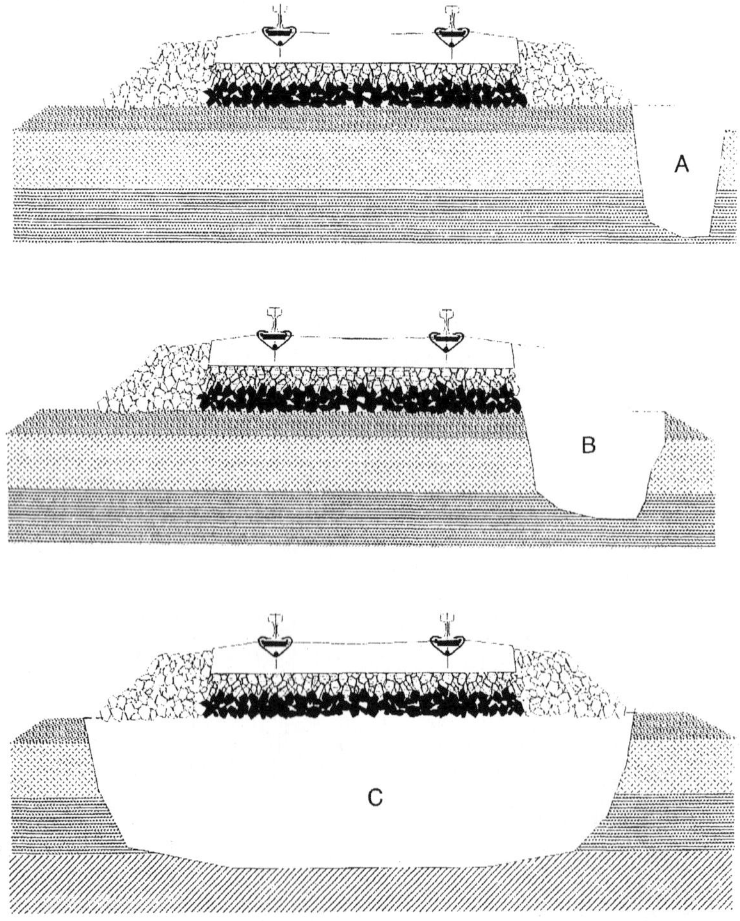

Fig. 10.1 Types of subgrade excavations

inspection of the ballast and subballast layers as well, which is usually necessary whenever subgrade information is required.

The next level of investigation involves excavating a cross trench from one side of the track to the other (Fig. 10.1c). This is important in fully evaluating the ballast, subballast and upper subgrade conditions because the conditions often vary with position across the track (such as beneath the rail compared with beneath the shoulder or beneath the center) and with depth. Subgrade depressions and ballast pockets, for example, will represent deviation from a plane boundary between layers. This is important information which will not be detected without a full (or at least half) width cross trench. Depths of excavation to at least 1200 mm (48 in.) below the top of the sleeper are often quite possible. Examples are given in Refs. 10.18 and 10.19.

A coring machine such as described in Chapter 14 can provide a more rapid sampling of layers at the center of the track and at locations outside the rails. The depth of penetration is usually limited to about 1200 mm (48 in.) below the top of the sleeper. The samples may be highly disturbed and so the layer boundary depths often are not accurate, but the general nature of the materials can be quickly observed.

Identification of subgrade layers and materials to depths below that possible with the coring machine or cross trench require use of standard drilling techniques. For soil layers, a hollow-stem auger or a standard penetration test sampling spoon (Chapter 4) inserted into either a cased or an uncased hole can be used. For rock layers, rotary core drilling is required. Specialists in this type of sampling may be consulted for advice and assistance.

Physical properties of the layers can be measured in the field using the in situ techniques described in Chapter 4. Direct measurement of such properties as resilient modulus, plastic strain and strength require the laboratory tests described in Chapter 3. For the subgrade materials undisturbed field samples will usually be required.

10.3. Resilient Modulus

The resilient modulus of coarse-grained soils (sands and gravels) has been discussed in chapters dealing with ballast and subballast. This same information applies to subgrade soils whose behavior is controlled by their coarse-grained components. This section will describe the resilient behavior of fine-grained soils, that is soils whose behavior is controlled by their silt and clay components. Then results of field studies involving resilient modulus of subgrade layers beneath track will be discussed.

10.3.1. Laboratory Test Results

An evaluation of resilient Young's modulus values for fine-grained soils is given in Ref. 10.1 based on published data from repeated-load triaxial tests. These data show that the resilient modulus of fine-grained soils is significantly dependent on:

1) Loading condition or stress state, which includes the magnitude of deviator stress and confining stress, and by the number of load cycles.

2) Soil type and initial structure, which for compacted soils depends on compaction method and compaction effort.

3) Soil physical state, which can be defined by moisture content and dry density.

Figure 10.2 shows a family of compaction curves relating dry density to moisture content at constant compactive effort. From the literature resilient modulus values were selected for optimum moisture content at the maximum dry density (peak point on the curves) and for the same dry density at other moisture contents. The results are shown in Fig. 10.3. A wide variety of soils were represented. All other parameters, including stress conditions and soil type, were constant for the pair of E_r and $E_{r(opt)}$ values representing each data point.

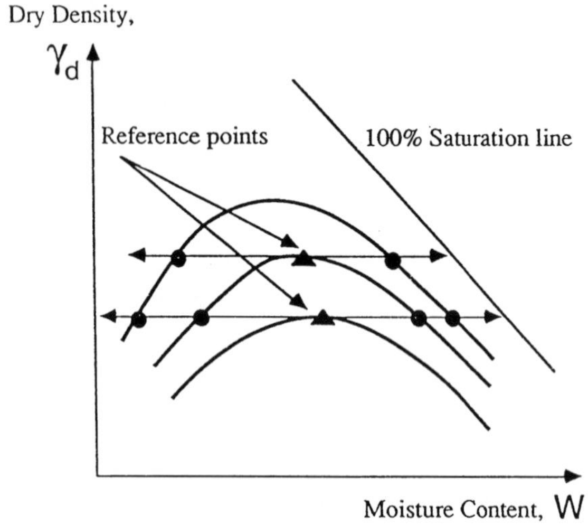

Fig. 10.2 Paths of moisture content variation with constant dry density

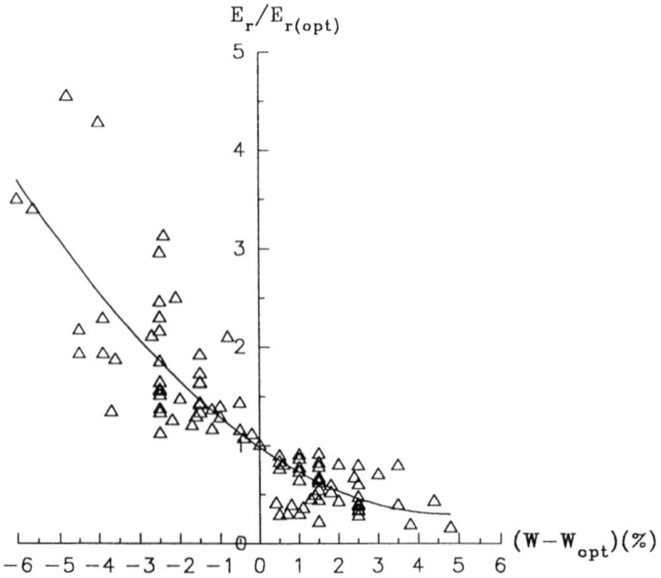

Fig. 10.3 Resilient modulus variation with moisture for paths of constant dry density

The best fit curve drawn in Fig. 10.3 is

$$\frac{E_r}{E_{r(opt)}} = 0.98 - 0.28\,(W - W_{opt}) + 0.029\,(W - W_{opt})^2 , \qquad (10.1)$$

where E_r = resilient modulus at any moisture content for the same dry density as $E_{r(opt)}$,

$E_{r(opt)}$ = resilient modulus at the peak point on the compaction curve,

W = moisture content (%), and

W_{opt} = optimum moisture content (%).

The resilient modulus decreases significantly with increasing moisture content.

Resilient modulus data were also obtained for soil conditions representing constant compactive effort as illustrated in Fig. 10.4. The results are shown in Fig. 10.5 along with the best fit curve given by

$$\frac{E_r}{E_{r(opt)}} = 0.96 - 0.18\,(W - W_{opt}) + 0.0067\,(W - W_{opt})^2 . \qquad (10.2)$$

The rate of decrease of resilient modulus with increasing moisture content on the dry side of optimum moisture content is less for these cases than for the cases in Fig. 10.3. This is a result of the influence of the increase in dry density with increasing moisture content in Fig. 10.5. Above optimum moisture content, the results in the two cases are similar.

Resilient modulus can be estimated for any dry densities and moisture contents using Eqs. 10.1 and 10.2 if a value at any state is known. The procedure, described in Ref. 10.1, involves dividing the difference between the state for which resilient modulus is known and the state for which it is desired into a combination of steps of either constant dry density or constant compactive effort (Fig. 10.6). Then Eqs. 10.1 and 10.2 are used to calculate the change in resilient modulus for each step.

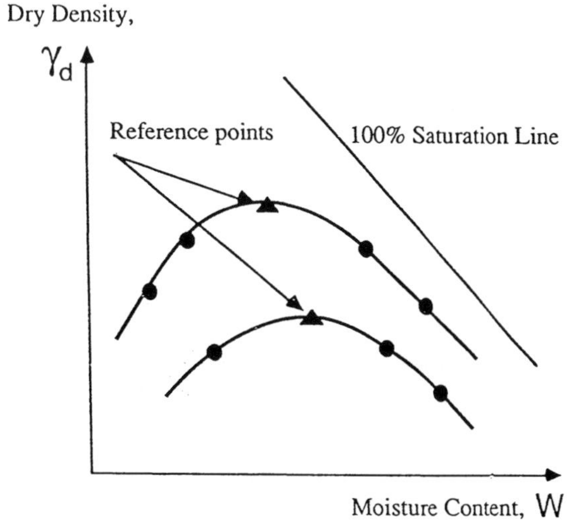

Fig. 10.4 Paths of moisture content variation with constant compactive effort

The cyclic stress state parameter which has been found to have the greatest effect on resilient modulus of fine-grained soils is deviator stress (maximum principal stress difference). Cyclic confining stress change has a

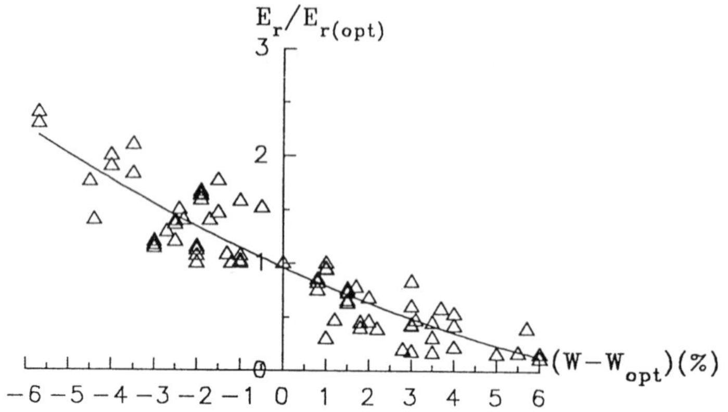

Fig. 10.5 Resilient modulus variation with moisture for paths of constant compactive effort

relatively small effect, and the resilient modulus tends to become constant with increasing numbers of cycles for soil conditions where plastic strain accumulation rate is small. Thus resilient modulus has been related to deviator stress by various mathematical expressions (Ref. 10.1). The equation which fit the examined data best was the bilinear model (Ref. 10.2). It is given by

$$E_r = K_1 + K_2 \sigma_d \quad \text{when } \sigma_d \leq \sigma_{di} , \tag{10.3a}$$

and

$$E_r = K_3 + K_4 \sigma_d \quad \text{when } \sigma_d \geq \sigma_{di} , \tag{10.3b}$$

where σ_d = deviator stress,

σ_{di} = break point deviator stress, and

K_1 to K_4 = model parameters which depend on the soil type and physical state (K_2 and K_4 are usually negative).

A model which uses only two parameters, but which has a somewhat decreased ability to fit test data, is given by (Refs. 10.3 and 10.4):

$$E_r = K \sigma_d{}^n , \tag{10.4}$$

where K, n are parameters which depend on soil type and physical state (n is usually negative). Brown et al. (Ref. 10.22) proposed incorporating effective confining stress, $\sigma_3{}'$, for saturated soils by

$$E_r = K (\sigma_d / \sigma_3{}')^n . \tag{10.5}$$

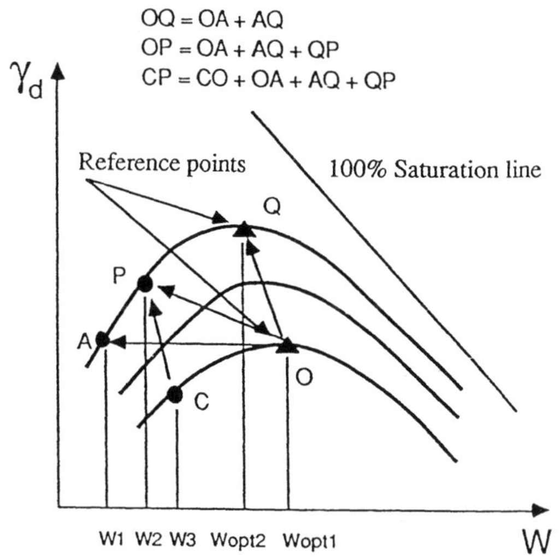

Fig. 10.6 Paths for determining resilient modulus for any condition

Examples of the fit of Eqs. 10.3 and 10.4 to test data are given in Figs. 10.7 and 10.8. Values of K and n in Eq. 10.4 at optimum moisture content and maximum dry density (i.e. peak point on constant compactive effort curve) were calculated for available data on fine-grained soils (Ref. 10.1). The parameter K ranged from 34,000 to 1,400,000 kPa (5000 to 200,000 psi) and n ranged from - 0.1 to - 0.9. The compaction test used to define the compaction curve is ASTM D698 (Ref. 10.6).

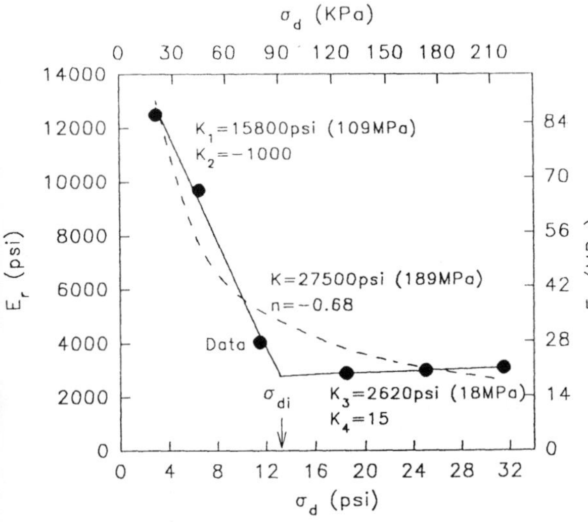

Fig. 10.7 Comparison of models with data from Ref. 10.5

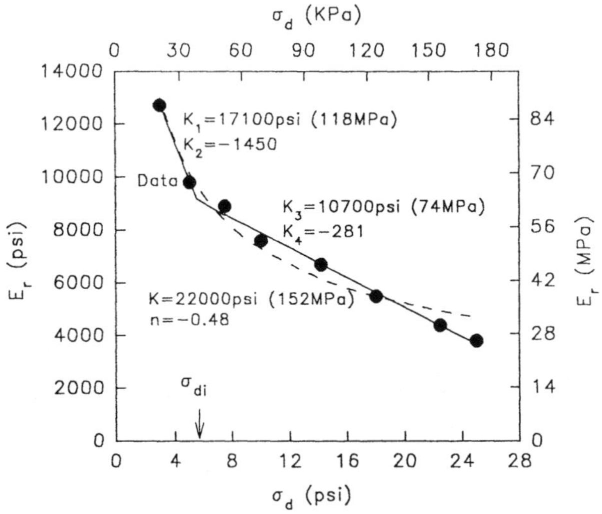

Fig. 10.8 Comparison of models with data from Ref. 10.2

Thompson and LaGrow (Ref. 10.7) suggest a relationship between resilient modulus at the break point stress, σ_{di} , and measures of the soil type as follows.

$$E_{r(opt)} = 4.5 + 0.01\, C_L + 0.12\, I_p \,, \tag{10.6}$$

where C_L = percent clay, which represents particles < 0.002 mm ,

I_p = plasticity index, and

$E_{r(opt)}$ = resilient modulus at optimum water content and 95% maximum dry density at the break point stress.

An estimate of the effect of other density states can be made using Eqs. 10.1 and 10.2 if the compaction curve is known.

10.3.2. Field Test Results

Subgrade layer resilient modulus values were estimated for eight sites in the United States by correlating top of rail vertical track stiffness measurements with cone penetration test profiles using GEOTRACK (Ref. 10.8). A wide variety of superstructure and substructure conditions were represented among these sites. Subgrade soils included silts, sands, and clays. The track stiffness measurements were conducted using both single point load tests and multiple axle vehicle load tests as described in Chapter 5. The track "foundation" modulus was calculated for each site using Eq. 5.13. The values of u ranged from 15 to 48 MPa (2200 to 7000 psi).

The steps involved in determining the layer resilient modulus values were as follows:

1) Divide the subgrade into layers based on the CPT profiles to a depth of at least 5 m below the bottom of the sleeper.

2) Select a representative average cone resistance value, q_c , for each layer.

3) Calculate the ratio, R_q , of the q_c for layer 1 to the q_c for every other layer.

4) Make a preliminary estimate of resilient modulus, E_r , for layer 1.

5) Calculate the preliminary E_r for the remaining layers assuming that for each layer E_r equals R_q for that layer times E_r for layer 1.

6) Use the preliminary subgrade E_r values, with other required parameters , to calculate vertical rail deflection for two load conditions: (a) seating, and (b) maximum.

7) Compare the calculated deflection change with the field measured deflection change.

8) Adjust the estimated E_r for layer 1 (and hence E_r for all layers) by trial and error until the calculated deflections agree with the measured deflections.

With this approach, the lower part of the load-deflection curve (see examples in Fig. 5.4) is assumed to represent slack in the system, while the upper part is assumed to represent actual layer resilient properties.

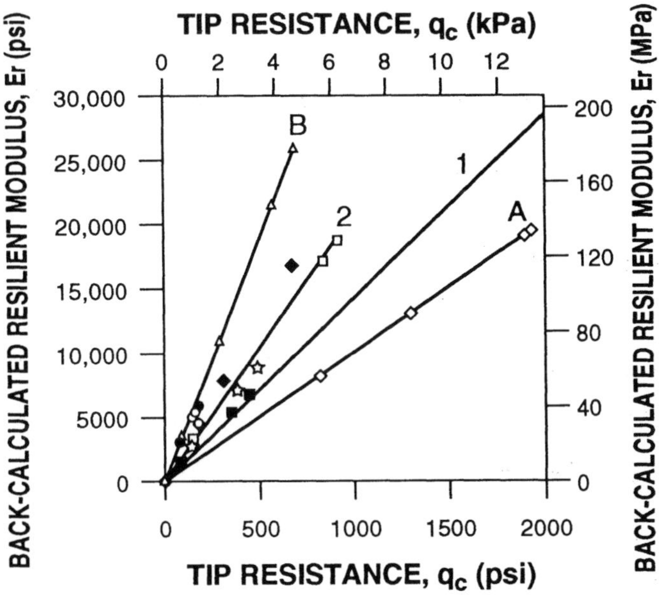

Fig. 10.9 Correlation between cone tip resistance and resilient modulus

The correlation of calculated resilient moduli, E_r, with cone tip resistance, q_c, is given in Fig. 10.9. Line 1 is the linear regression fit to all the data and is represented by

$$E_r = 14\, q_c \ . \tag{10.7}$$

Line 2 is a linear regression fit to the data after eliminating the extreme values indicated by lines A and B. Line 2 is represented by

$$E_r = 21\, q_c \ . \tag{10.8}$$

10.4. Subgrade Failure

Three categories of subgrade failure associated with train loading can be distinguished based on the mechanisms of failure:

1) Massive shear failure.

2) Progressive shear failure (also known as general subgrade failure).

3) Attrition (also known as local subgrade failure).

In addition for tracks placed on embankments, massive failure of the earth foundation under the embankment weight must be avoided. Normally this possibility only needs to be considered for newly constructed embankments. Methods of analyzing and treating this problem fall within the scope of normal geotechnical engineering practice and hence will not be covered here.

10.4.1. Massive Shear Failure

Massive shear failure is illustrated in Fig. 10.10. The driving forces are the weights from the train, the track superstructure and the unbalanced portion of the substructure. The resisting force is from the substructure layer shearing resistance. Because more of the failure zone is in the subgrade, then the subgrade strength properties have a big effect on the factor of safety against massive shear failure.

Progressive failure of the subgrade under repeated loading generally occurs at stress levels below that which will cause massive failure. Hence progressive failure should govern performance. Thus massive failure is likely to be a problem only when the subgrade strength diminishes because of increasing water content. This may occur for example at times of heavy rainfall and flooding. In such situations failure may occur even without train loading being present.

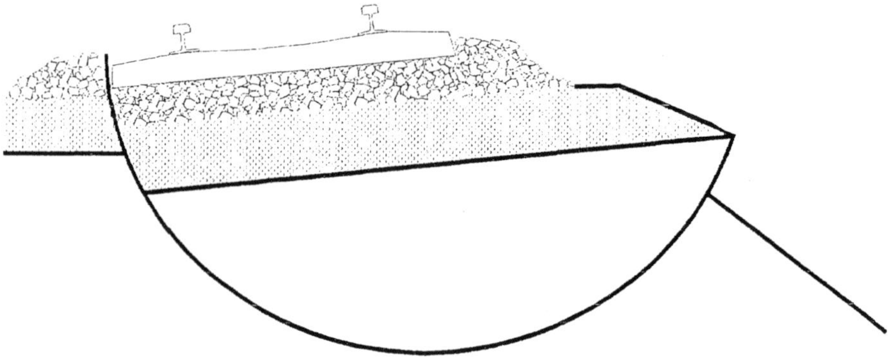

Fig. 10.10 Massive shear failure

Figure 10.11 shows cross-sections through a number of slopes that have sustained a slip failure in which the materials above and below the slip plane have moved with respect to each other resulting in shear. If the slip plane extends beneath the track, such movements will disturb the track geometry.

The prime requirement is the location of the slip plane. Once located, it is possible to confirm that the plane extends beneath the track, and thus, that the track movements being experienced are attributable to this cause. Simple, and widely used methods of directly locating a slip plane, and confirming that the slip is still active, are available.

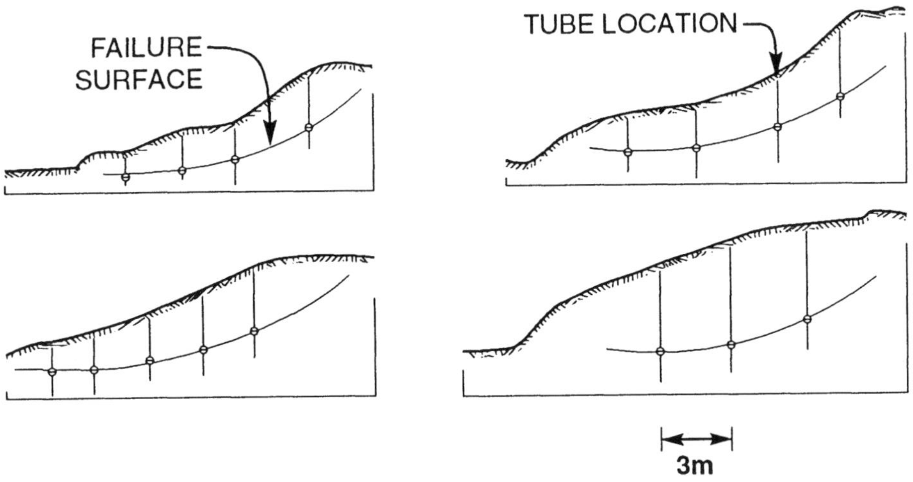

Fig. 10.11 Slip failure surfaces identified by tube method

For non-cohesive materials a driving dolly is used to drive a length of 19 mm (3/4 in.) diameter steel tubing, provided with an expendable point, vertically into the slope to a depth in excess of that at which it is considered the slip plane will be located. A length of plastic tubing (as used in domestic plumbing) is passed down the tube. The steel tubing is withdrawn, leaving the steel point in position.

For cohesive materials solid rodding, provided with a pointed end, is driven as described above. The rodding is withdrawn and the plastic tube is passed down the unlined hole formed by the rodding.

After a period of time, dependent upon the rate of movement of the slip, deformation of the plastic tubing will occur at the slip plane as a result of shear associated with relative movement of materials above and below the shear plane. This point is located by using a cord to lower a short steel mandrel down the center of the plastic tubing until the deformation point is reached-- the length of cord required being equal to the depth of the slip plane.

Figure 10.11 shows examples of circular and non-circular slip planes located in cut and embankment slopes by this technique, together with an indication of the points at which movement was detected.

Cut slope stability problems often arise because of ground water seepage into the excavated slopes. Other problems are weathering of freshly exposed soil and rock, and volume changes in expansive clays.

10.4.2. Progressive Shear Failure

Stresses imposed on the subgrade by the axle loads may be large enough to cause progressive shear failure (general subgrade failure). This condition will most likely develop in the top part of the subgrade where the traffic induced stresses are highest. Overstressed soil will be squeezed sideways from beneath the track and upwards to give the bearing capacity failure known in the United Kingdom as 'Cess Heave'.

Figure 10.12 shows diagrammatically how overstressed clay is progressively squeezed sideways and upwards. Figure 10.13 is a photograph of a section through a cess heave during its formation. Figure 10.14 is a photograph of a section through a fully developed cess heave.

Figure 10.15 shows how three initially straight columns of segmented tubing have been distorted by the further development of a general subgrade failure. The columns of segmented tubing were initially installed by threading them onto a straight steel rod and then driving them into position using equipment similar to that described in Section 10.4.1.

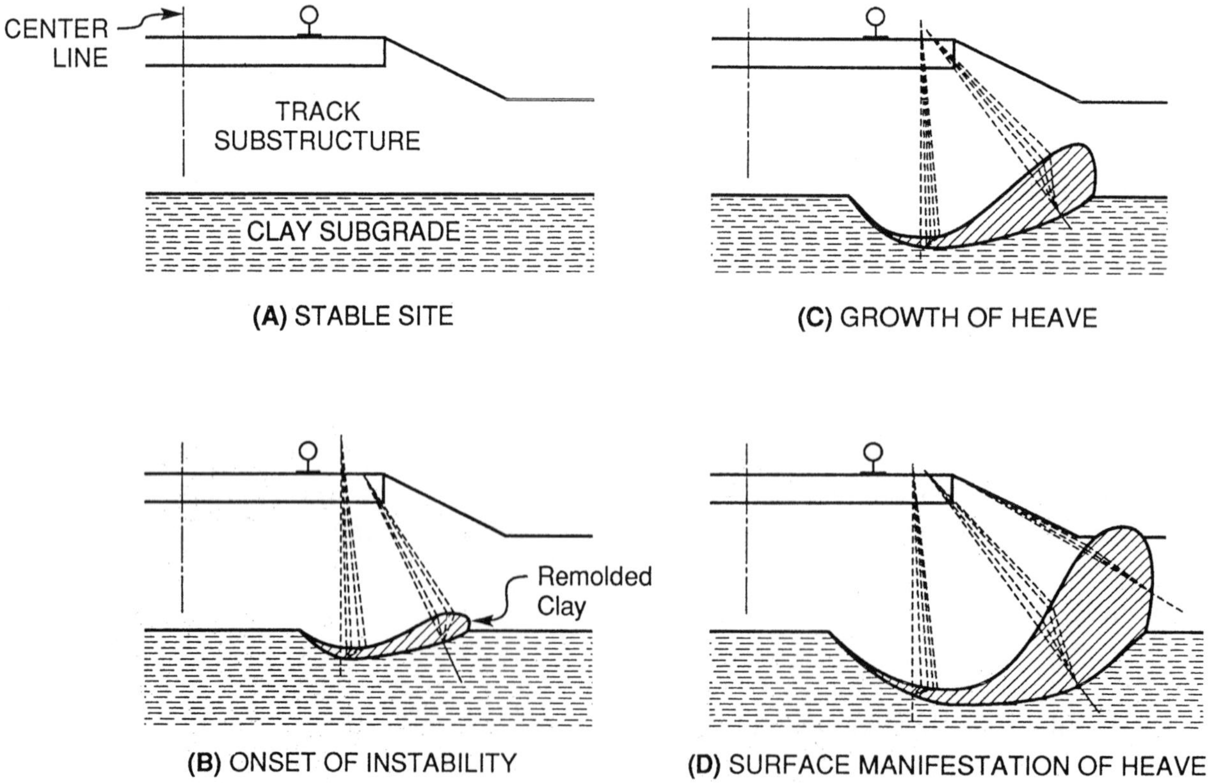

Fig. 10.12 Movement of overstressed clay

Fig. 10.13 Cess heave during formation

Fig. 10.14 Fully formed cess heave

Fig. 10.15 Distorted columns indicating cess heave failure

Once in position, the rod was withdrawn, leaving the segmented tubing in position. After a period of several weeks, the trench was excavated alongside the tubes to expose them as shown in the figure.

From a study of this, and a number of similar observations, it became clear that in this mode of failure soil movements are as indicated in Fig. 10.12.

The segmented tube technique described can also be used to establish whether or not a subgrade failure is taking place. To do this, columns of segmented plastic tubing are driven into the suspected failure site as described above. After a period of some weeks, the initially straight column of tube segements is probed with a straight rod. The location of an obstruction will indicate a distortion of the column of tube segments caused by an active 'cess heave'.

Figure 10.16 shows the disruption which is often caused to the track side drainage system as a result of cess heave. The heave can either physically destroy the track side drainage and/or, interpose an impermeable wall between the track and the drainage system, thus rendering it ineffective. Clearly, deprived of its drainage path, the water table within the track can rise resulting in an aggravation of the condition which originally gave rise to the cess heave.

Fig. 10.16 Disruption of track drainage from cess heave

The probability of a cess heave developing can be minimized by:

- Ensuring that an adequate depth of load distributing (granular) material exists between the underside of the sleeper and the surface of the subgrade.

- Ensuring that the drainage system maintains a low water table level.

As shown in Fig. 10.12, the heave of material at the lineside is matched by a corresponding depression beneath the track. This depression is reflected at the surface as a depression in the track, which is corrected by the addition of ballast beneath the track. This measure results in an increase in ballast depth and a corresponding reduction in soil stress at subgrade level which tends to improve stability. However the depression traps water which tends to reduce the potential improvement.

Progressive subgrade failure only occurs in fine grained materials exhibiting low values of internal friction. In coarse grained materials exhibiting high values of internal friction, the increase in shear strength associated with the applied normal stress exceeds the increase in associated shear stress.

If overhead clearances permit, the resistance of the subgrade to a progressive subgrade failure can be achieved by applying a general ballast lift to the track and thereby increasing the depth of ballast between the underside of the sleeper and the surface of the subgrade. This reduces the intensity of stress applied. Such a measure is of value if the stress levels imposed on the subgrade are to be increased; for example, as a result of increased axle loads.

A general ballast lift is not usually an acceptable method for restoring stability following a progressive subgrade failure, since it does not rectify the associated deformations in the surface of the subgrade. Such deformations could interfere with cross track drainage and result in the ponding of water on the surface of the subgrade.

The only solution to the problem could well be the removal of the ballast layer, excavation of the subgrade to a reduced elevation, and replacing ballast and subballast up to its original level. The resulting increase in granular depth will reduce the intensity of stress applied to the surface of the subgrade.

If these measures are utilized, it is usual to use the opportunity to replace or install a subballast layer to prevent the possibility of future local subgrade failure.

10.4.3. Attrition

As is indicated in Fig. 10.17 attrition (local subgrade failure) of the subgrade by the overlying ballast in the presence of water can result in the formation of slurry at the ballast/subgrade interface. Under certain conditions, cyclic loading associated with passing traffic can cause this slurry to be pumped up to the surface of the ballast. Such failures are normally associated with hard, fine grained materials such as clay, and soft rocks, such as chalk.

Figure 10.17A shows the sleeper resting in a bed of ballast which overlays a hard, fine grained subgrade material. The surface of the standing water is above the surface of the subgrade. Figure 10.17B shows how the slurry formed at the ballast/subgrade interface has been pumped to the surface of the ballast. Figure 10.17C shows a layer of blanketing material (subballast) interposed between the ballast and the subgrade as a means of preventing the occurrence of subgrade attrition failure.

As shown in Fig. 10.18, when the slurry reaches the sleeper/ballast interface, cyclic movements of the sleeper within the ballast bed result in the slurry being ejected from beneath the sleeper on to the surface of the ballast to give the condition known as "Pumping". The slurry, in this case from subgrade attrition, ultimately finds its way into the trackside drainage system which can result in premature blocking.

The surface manifestations of a pumping failure are very similar to those associated with ballast/sleeper erosion and/or the presence of dirty ballast around the sleeper. Its cause and its cure are however quite different. An examination of the fine grained materials present in the slurry can usually indicate its source and hence the type of failure being encountered.

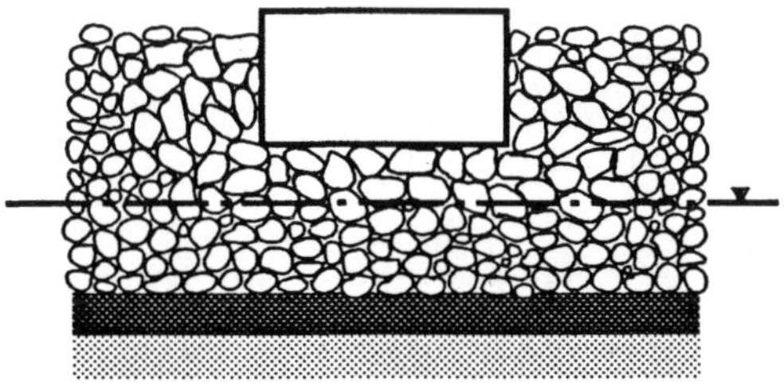

(A)

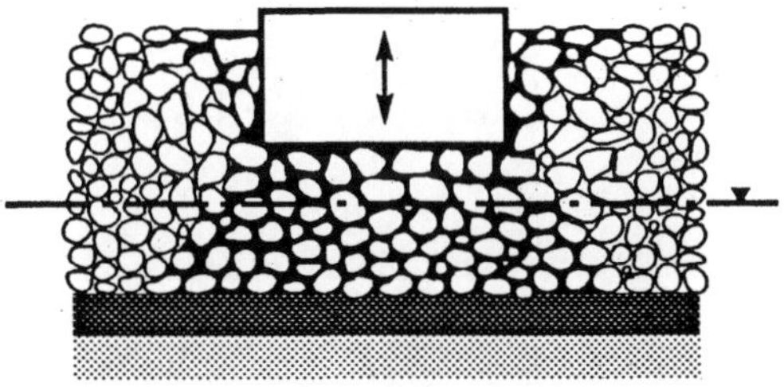

(B)

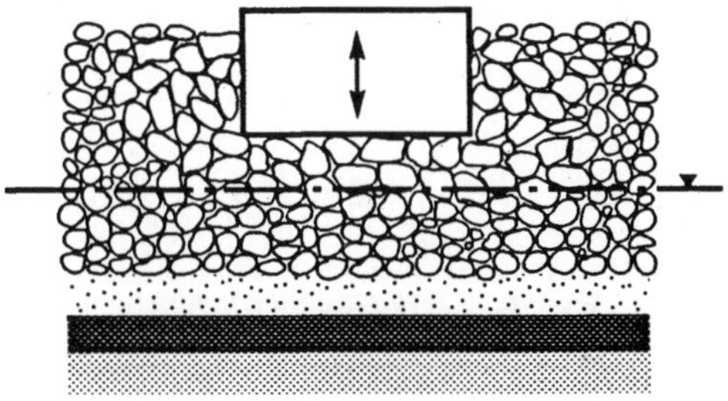

(C)

Fig. 10.17 Cause and prevention of subgrade attrition

Because of its association with high subgrade stresses and large sleeper movements, subgrade attrition failure is frequently found to originate at sleepers associated with rail joints.

Local subgrade failures accompanied by pumping can lead to a loss of lateral track restraint, which in turn can lead to a loss of correct horizontal and vertical track alignment, as a result of:

1) Displacement of ballast from around the sleeper by the jetting action of the slurry during ejection from beneath the sleeper.

Fig. 10.18 Pumping of slurry from subgrade attrition

2) Lubrication of the ballast/sleeper interface resulting in a reduction in sliding friction between the sleeper and the ballast.

3) Lubrication of the ballast particles resulting in a reduction in the shear strength of the supporting and shoulder ballast.

4) A local depression of the subgrade resulting from a loss of material associated with the erosion of the subgrade, and possible ponding of water within the depression.

Pumping failures can be prevented by placing a layer of blanketing material between the ballast and the subgrade as shown in Fig.10.17 to:

1) prevent the formation of slurry by mechanically protecting the fine grained subgrade from attrition and penetration by the overlying coarse grained ballast, and

2) prevent the upward migration of any slurry that forms at the subballast/subgrade interface, by virtue of the filtering properties of the subballast.

If possible, provide adequate drainage to ensure that the surface of the standing water is maintained below the level of the ballast/subgrade interface.

Temporary relief from the problems associated with subgrade attrition failure can be achieved by full ballast cleaning. However, ballast cleaning does not treat the root cause of the problem. A long term solution can only be achieved by placing a subballast layer between the ballast and the subgrade.

As a result of the investigative work undertaken on this subject, and the preventative and remedial works carried out, subgrade attrition failures on British Railways are now a rare occurrence.

10.5. Subgrade Settlement

Subgrade settlement can result from several causes:

1) Consolidation from weight of added earth or from ground water lowering (or conversely swell from removal of earth as in cut construction or ground water rise).

2) Shrinkage from moisture loss (and conversely swelling from moisture increase).

3) Underlying ground subsidence from sources such as mining and solution cavities.

4) Progressive deformation from repeated traffic induced stresses.

Consolidation settlement (or heave) can be analyzed and solutions provided by well-established methods of geotechnical engineering. Closely associated with consolidation is volume change from moisture content change. However consolidation involves expulsion of pore water from saturated soils, whereas shrinkage involves loss of water from unsaturated soils. Swelling and shrinking of soils is also a subject which is well known in geotechnical engineering practice.

Underlying ground subsidence problems also fall within normal geotechnical engineering practice.

Settlement by progressive deformation has been demonstrated in Chapter 8. Mechanisms of plastic deformation in the subgrade from repeated loading include: 1) volume reduction (compaction), 2) consolidation when repeated stresses cause pore pressure increase, and 3) shear deformation. Shear deformation associated with pore-pressure development is related to progressive shear failure described in Section 10.4.2 . A fourth mechanism is ballast void infiltration by subgrade particles due to a lack of a filter layer with resulting track settlement. This is a different problem than the problem of subgrade attrition with slurry formation. A proper subballast layer will solve this problem.

Settlement is usually nonuniform along and across the track. Hence excessive subgrade settlement can lead to unacceptable track geometry change. The wavelength of the geometry change depends on the source of the settlement and its pattern.

10.6. Frozen Subgrade

Problems associated with frozen subgrade have been discussed in Chapter 9. They include change in strength and stiffness, as well as frost heave and thaw softening.

10.7. Influence on Component Deterioration

The subgrade resilient modulus is likely to have a big effect on track component degradation. The subgrade has a big influence on track modulus. Track modulus affects the magnitude of rail and sleeper bending stresses, and hence affects their rate of fatigue. Track modulus also influences the stresses and deflections in the ballast. Increasing track modulus increases the ballast-sleeper contact stress and hence increases the rate of degradation. However decreasing the track modulus increases the ballast flexural deformations and hence increases ballast abrasion. Thus the ballast life is likely to be greatest at some intermediate track modulus. Little information to verify this trend is available.

10.8. Track Bearing Capacity

Methods of analyzing the potential for static bearing failure of track are reviewed in Ref. 10.9, which forms the basis for this section.

Determination of track load support capacity is essentially a multilayer bearing capacity problem. Several theoretical approaches are available to determine bearing capacity of a layered system. Analysis will show that a bearing capacity failure in railway track should not occur except when the subgrade strength is very low. Sections of railway track in coarse-grained subgrade are not likely to be susceptible to a bearing capacity failure. Thus attention can be focused on fine-grained subgrades. The bearing capacity equation developed by Meyerhof and Hanna (Ref. 10.10) represents a simplified two-layer approach to this problem. For more precise representation of the multilayer substructure and to incorporate surface geometry such as embankment slopes, a slope stability model can be used. Both approaches will be described, and results given.

10.8.1. Meyerhof and Hanna Method

Figure 10.19 illustrates the bearing capacity model for the Meyerhof and Hanna approach. The failure mechanism assumes that at the ultimate load the upper stronger granular layer, of approximately truncated pyramidal shape with vertical sides, is pushed into the lower weaker clay layer. In the case of general shear failure, the drained friction angle, φ', (assuming pore pressure is zero) of the upper granular layer and undrained cohesion, c, of the lower clay layer are mobilized in the combined failure zones. At the point of limiting equilibrium, the sum of the forces in the vertical direction yields an equation for the ultimate bearing capacity of a continuous (strip) foundation on a layered system. For a rectangular bearing area Meyerhof (Ref. 10.11) established correction factors

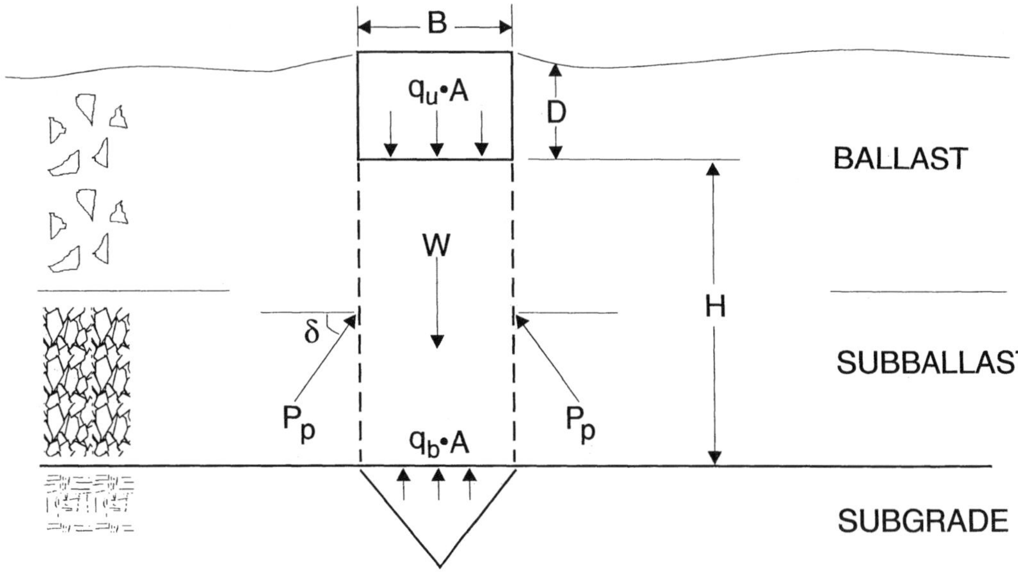

Fig. 10.19 Meyerhof and Hanna model for bearing capacity

for the strip solution based upon the dimensions of the bearing area. The modified Meyerhof and Hanna equation is:

$$q_u = C_1 c N_c + [C_2 \gamma H^2 (1 + 2D/H) K_s (\tan \varphi') / B] - \gamma H \le q_t , \qquad (10.9)$$

where q_u = average contact pressure at the base of the sleepers,

$C_1 = (1 + 0.2 B/L)$,

$C_2 = (1 + B/L)$,

B = bearing width,

L = bearing length,

c = undrained shear strength of the clay subgrade,

N_c = bearing capacity factor = 5.14,

γ = total unit weight of the overlying dense granular material,

D = depth of embedment of the bearing area,

H = thickness of the granular layer,

K_s = coefficient of punching shear resistance,

φ' = drained friction angle of the upper dense granular material, and

q_t = bearing capacity for infinite H, or essentially for $H >> B$.

The punching shear resistance coefficient K_s can be determined from the charts published by Meyerhof and Hanna (Ref. 10.10) or could be calculated using Eq. 2.2:

$$K_s \tan \varphi' = K_p \tan \delta \quad , \qquad (10.10)$$

where $K_p = \tan^2 (45 + \varphi'/2)$,

δ = angle of inclination of passive earth force P_p .

Meyerhof (Ref. 10.11) recommended using an average value of δ equal to $2\varphi'/3$.

Determination of bearing capacity of the track system requires the choice of a possible mode of failure so that dimensions of the bearing area can be established. Two alternative configurations are considered for bearing capacity analysis (Ref. 10.12):

Option 1: The geometry perpendicular to the track to consider local failure of an individual sleeper punching into the subgrade (Fig. 10.20).

Option 2: The geometry parallel to the track assuming that the track system is acting as a connected footing (Fig. 10.21).

These two cases were analysed using Eq. 10.9. Figure 10.21 shows the bearing capacity of a two-layer system , consisting of ballast on clayey subgrade, for a strip bearing area ($L = \infty$) for both options 1 and 2. The bearing capacity for the two-layer system increases as the thickness of the top ballast layer is increased, as long as the subgrade is weaker than the ballast and subballast. Also the bearing capacity is substantially less for option 2 (track system bearing), in general, compared to option 1 (local sleeper punching). The Meyerhof and Hanna approach actually underestimates the factor of safety for option 1 because the confinement provided by the adjacent loaded sleepers is not accounted for in the analysis. Thus the actual difference between options 1 and 2 is probably greater than indicated in Fig. 10.22.

Figure 10.23, representing option 2, shows the variation in bearing capacity as a function of length (L) of the bearing area and thickness (H) of the granular layer using Eq. 10.9, keeping the soil properties constant. The bearing capacity of the two-layer system increases with increase in the thickness of the granular layer. The bearing capacity also increases with decreasing length of the bearing area.

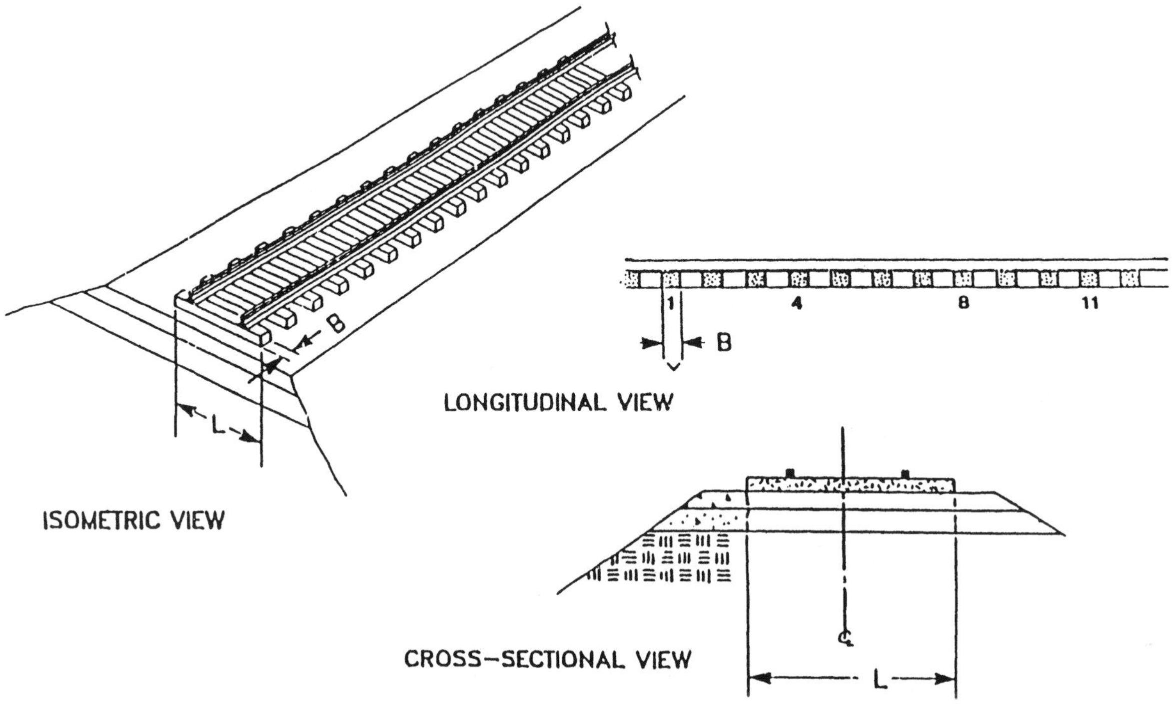

Fig. 10.20 Option 1: Geometry perpendicular to track

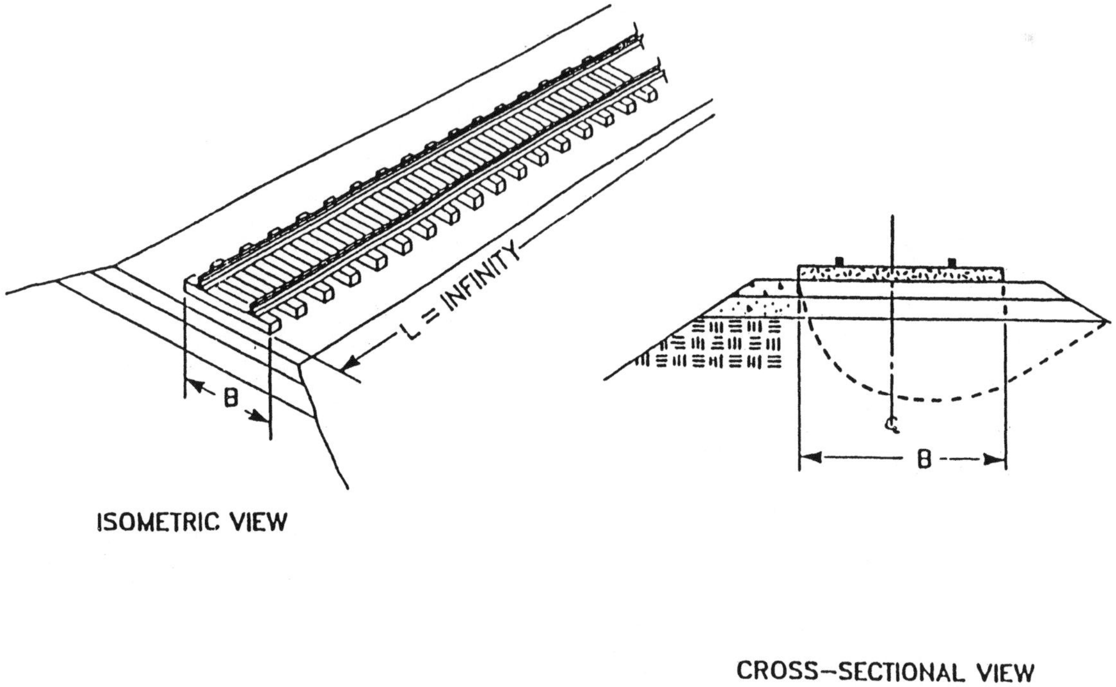

Fig. 10.21 Option 2: Geometry parallel to track

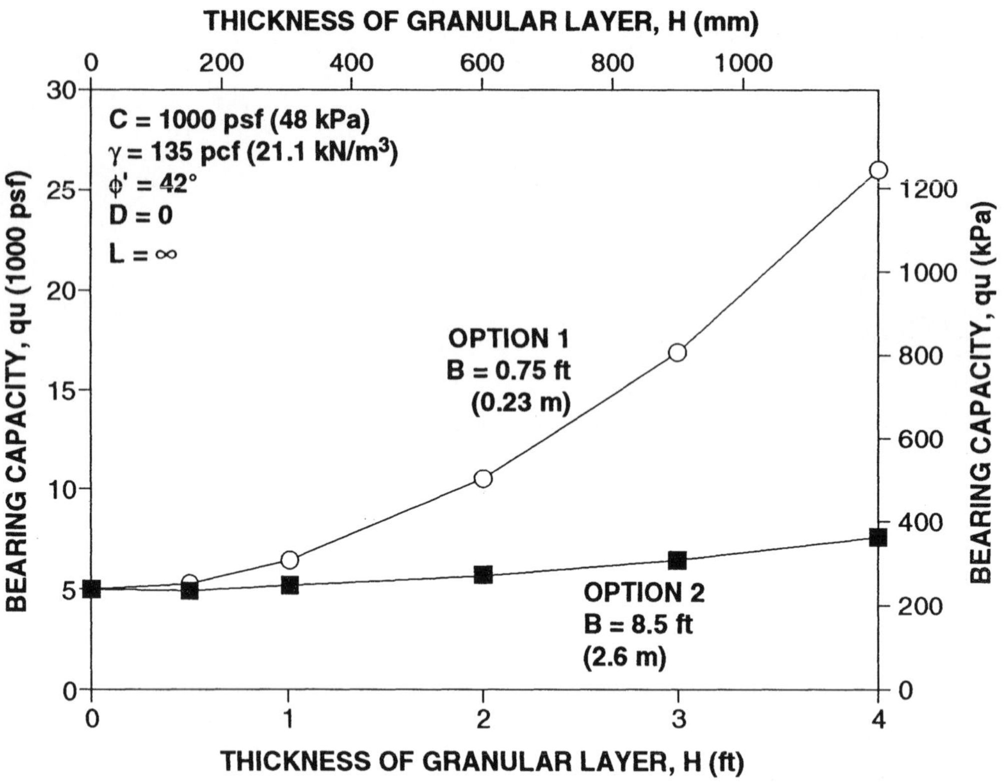

Fig. 10.22 Factor of safety for Meyerhof and Hanna options 1 and 2

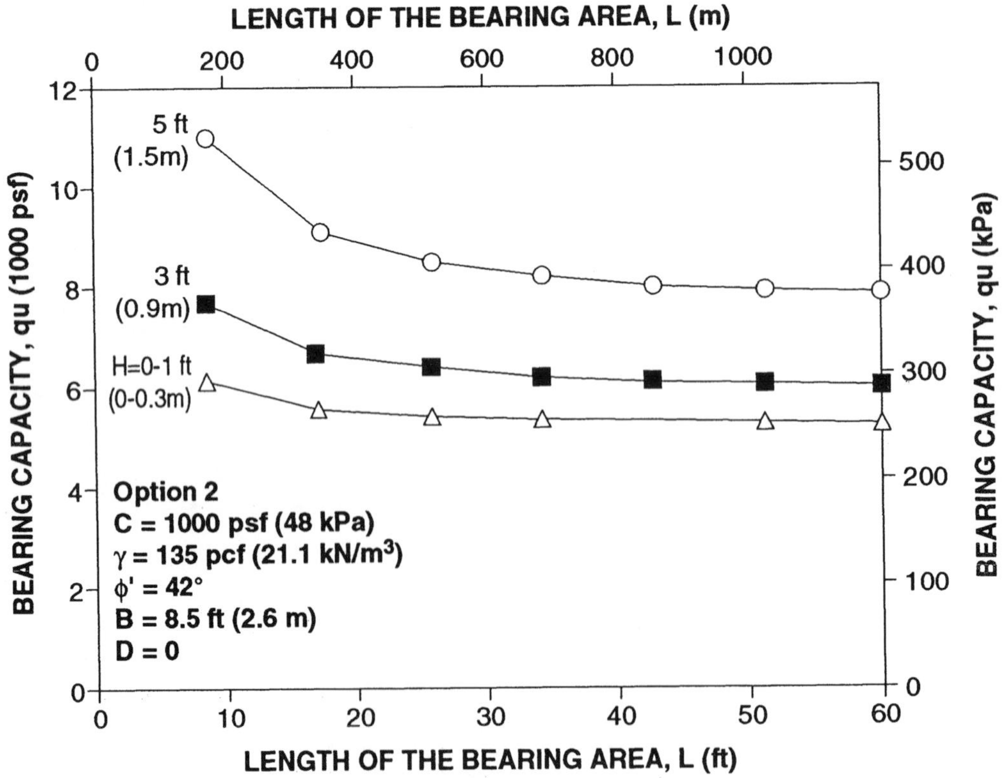

Fig. 10.23 Effect of bearing length and granular layer thickness on bearing capacity

10.8.2. Slope Stability Method

Bearing capacity approaches such as that of Meyerhof and Hanna are valid for the cases where the soil layers are horizontal and infinite in extent. To determine the effect of varying surface geometry, cut and fill slopes in particular, and for a more rigorous representation of a multilayer subgrade, slope stability analysis can be adopted for analyzing track bearing capacity.

The general solution of slope stability problems by a two-dimensional limiting equilibrium method determines the factor of safety against instability of a slope by the method of slices (Ref. 10.20). Computer programs are readily available for this purpose. The two-dimensional representation is a simplification which assumes that the loaded distance along the track is much greater than the length of the sleepers. This is a conservative assumption because it neglects end effects (shearing resistance in the transverse vertical plane) which will increase the bearing capacity.

An example of a failure surface for one analyzed track bearing capacity case (Ref. 10.9) is given in Fig. 10.24. The circle with the smallest factor of safety is shown (FS = 1.7). Each layer was assigned appropriate strength properties (total stress analysis was used) based on field and laboratory tests. The results were consistent with field load tests on the track.

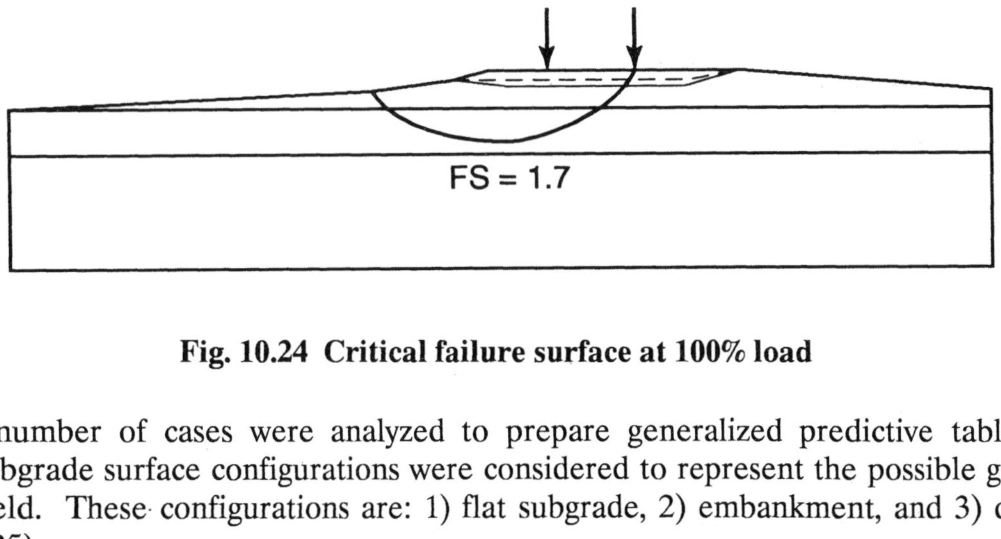

Fig. 10.24 Critical failure surface at 100% load

A number of cases were analyzed to prepare generalized predictive tables/plots. Three subgrade surface configurations were considered to represent the possible geometry in the field. These configurations are: 1) flat subgrade, 2) embankment, and 3) deep cut (Fig. 10.25).

Subgrade strength was varied to provide a range of factor of safety from less than one to a maximum of three for the designated load. A constant 600 mm (24 in.) thickness of the ballast and subballast layer was used in all the cases. Two variables were considered to evaluate the effect of changing the geometry of the subgrade surface. These variables are side slope, n, of the subgrade and the distance, W, of the starting point of the slope from the center of the loaded area (Fig. 10.26).

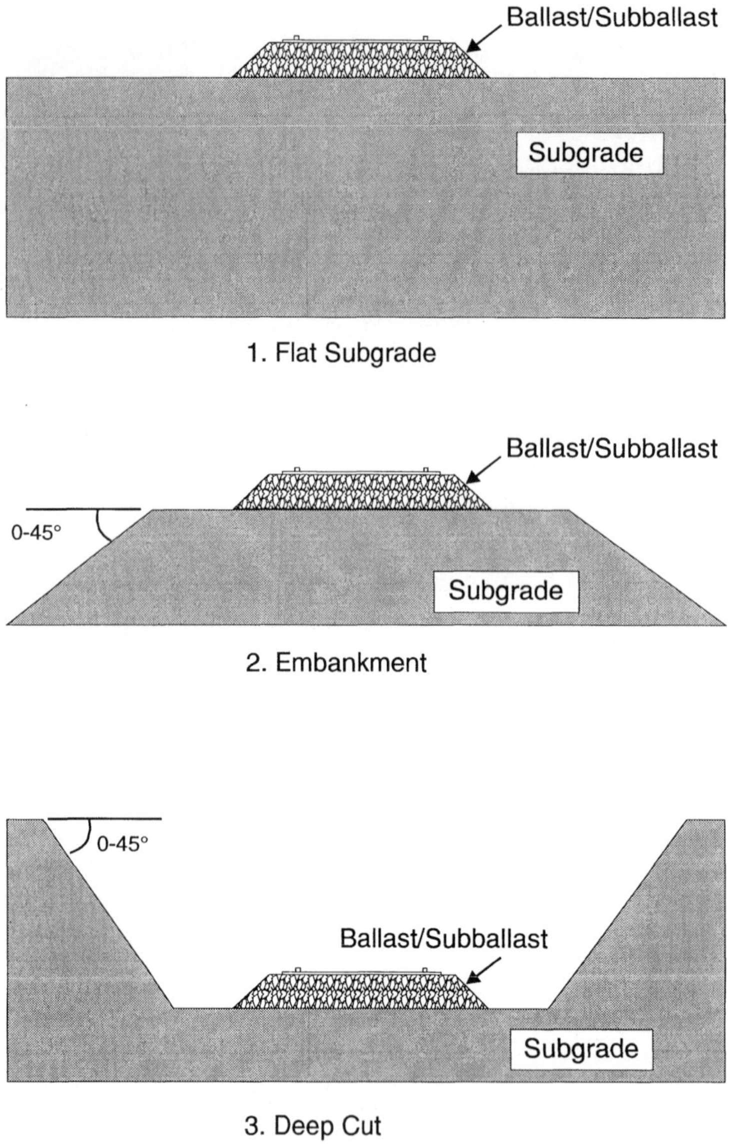

Fig. 10.25 Geometry options for slope stability analysis

A uniformly distributed load of 4880 lb/sq ft (234 kPa) was applied over the central 2.6 m (8.5 ft) length of the ballast layer which is the assumed length of the sleeper. The equivalent distributed static load produced by two adjacent bogies of 100 ton hopper cars used in North American railroads is 1400 lb/sq ft (67 kPa) which is about 30% of the analyzed load. The rated static weight of a loaded hopper car is 264,000 lb (1180 MN). Thus a factor of safety of 1.0 from the analysis provides a factor of safety of about 3.5 against bearing failure of the track under a loaded 100 ton car.

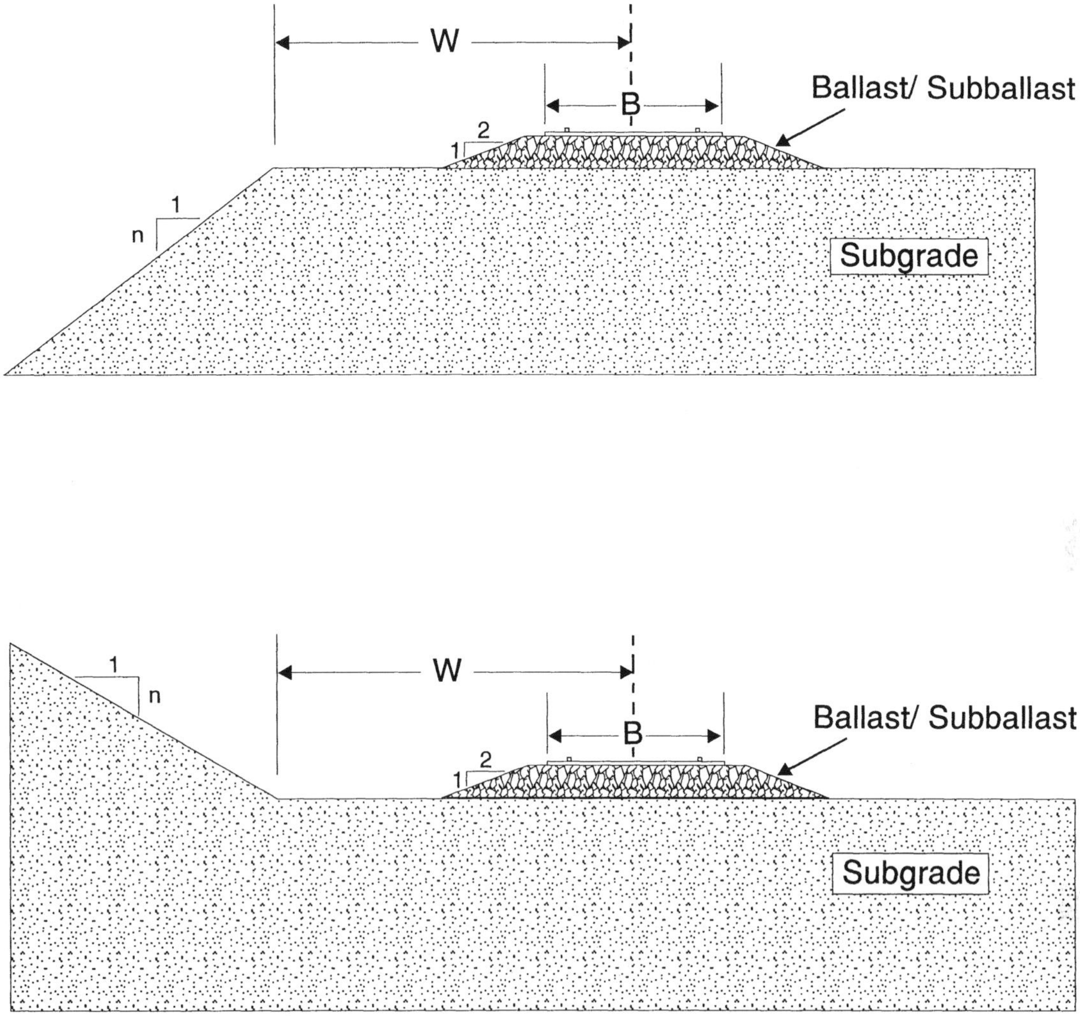

Fig. 10.26 Geometry variable parameters

All of these cases were analyzed and the minimum factor of safety was determined. The assumed input parameters are given in Table 10.1.

Table 10.1 Parameters for slope stability analyses

	Thickness, H		c	φ	γ	
	in.	mm		deg	lb/cu ft	Mg/m^3
Ballast/subballast	24	610	0	42	135	2.16
Subgrade	∞	∞	variable	variable	120	1.92

Table 10.2 lists the factor of safety for flat subgrade for different combinations of cohesion, c, and angle of friction, φ, for the subgrade (total stress parameters).

Table 10.2 Factor of safety by slope stability analysis for flat subgrade

c		φ (deg)				
kPa	lb/sq ft	0	5	15	25	30
0	0					1.1
5	100					1.3
10	200				1.3	1.4
24	500			1.2	1.7	1.9
34	700	0.8	1.0	1.4	1.9	2.1
40	900	1.0	1.2	1.6	2.1	2.3
48	1000	1.1	1.3	1.8	2.2	2.4
72	1500	1.6	1.8	2.2	2.7	2.9
96	2000	2.1	2.3	2.7	3.2	
120	2500	2.6	2.8			
144	3000	3.0				

To evaluate the effect of changing subgrade surface geometry, a few cases were analysed with the subgrade in an embankment and in a cut. To minimize the number of analyses only cohesive subgrades were considered. The most critical failure surfaces obtained for flat subgrade were used to determine the maximum value of W that needed to be considered for cases of a cut and an embankment. The width, W, was varied from this maximum value to the minimum possible, which was with the subgrade starting to slope from the toe of the ballast-subballast layer. Table 10.3 lists the factors of safety for the case of an embankment.

Table 10.3 Factor of safety by slope stability analysis for embankment

n	c = 1000 psf (49 kPa)			c = 1500 psf (72 kPa)		
	W, ft (m)			W, ft (m)		
	13.8 (4.2)	11.8 (3.6)	9.8 (3.0)	13.8 (4.2)	11.8 (3.6)	9.8 (3.0)
0	1.1	1.1	1.1	1.6	1.6	1.6
0.12	1.1	1.1	1.1	1.6	1.6	1.6
0.25	1.1	1.1	1.0	1.6	1.6	1.5
0.50	1.1	1.0	1.0	1.6	1.5	1.4
0.75	1.0	1.0	0.9	1.5	1.4	1.3
1.00	0.9	0.8	0.8	1.3	1.3	1.2

$\varphi = 0$ in all cases

As expected, a decrease in factor of safety occurs as the side slope of the embankment is steepened and as the offset distance is decreased. However for the cases considered, the reduction was at most 30% for slopes as steep as 45 deg. The factors of safety in the case of a cut were very close to those obtained for the flat subgrade. Hence the factor of safety for a cut can be obtained from Table 10.2.

Figure 10.27 illustrates the most critical failure surfaces obtained for each of the different surface geometries considered. In each case, the circle with the lowest factor of safety is shown along with the factor of safety. A comparison of these figures shows that the most critical failure surfaces in the case of flat subgrade and subgrade in cut extend at shallow depth as opposed to an embankment subgrade where the most critical failure surface is much deeper.

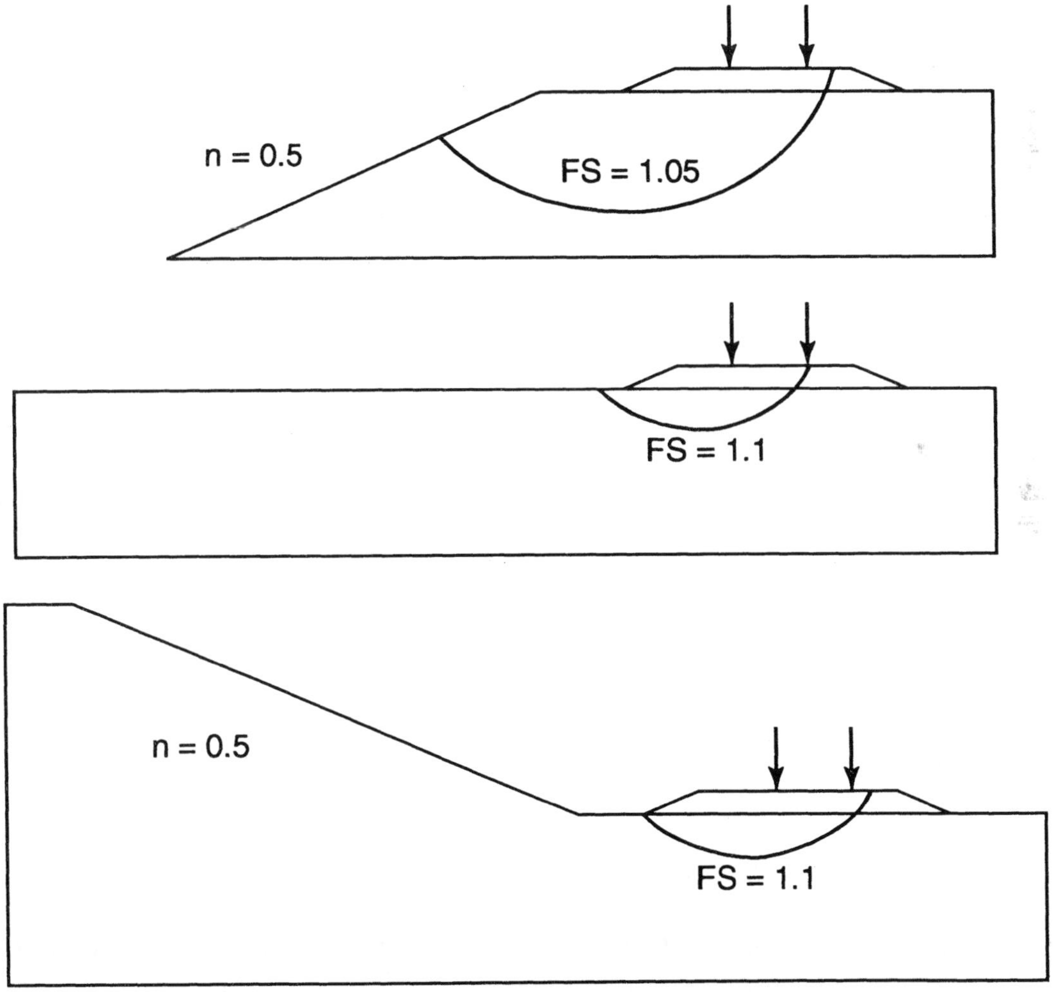

Fig. 10.27 Critical failure surfaces for three geometries

Figure 10.28 is the plot of factor of safety for different values of cohesion, c, and angle of friction, φ , for the flat subgrade taken from Table 10.2. As shown in Fig. 10.28, the factor of safety varies approximately linearly with increases in both cohesion and friction angle.

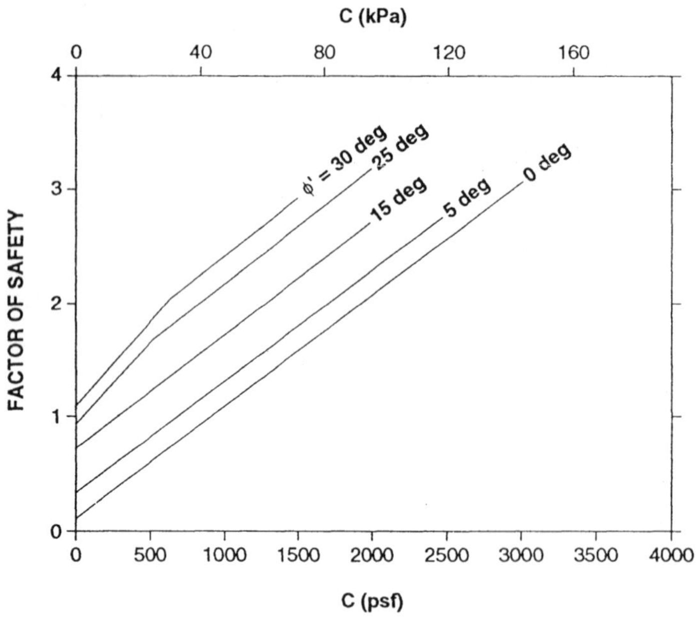

Fig. 10.28 Factor of safety as a function of subgrade soil properties for flat subgrade

The factor of safety is about the same for a soil with $\varphi = 0$ and c = 1000 psf (48 kPa) as for a soil with $\varphi =$ 30 deg and c = 0. The factor of safety is not zero when both cohesion and friction angle values are zero because of the small resisting force provided by the 610 mm (24 in.) ballast/subballast layer thickness.

A factor of safety of 0.3 in Fig. 10.28 means that under a loaded 100 ton car massive failure would be expected. This is because a multiple of 3.5 was used for the load compared to that for a 100 ton car. A factor of safety of 1.0 in Fig. 10.28 means that the 100 ton car is operating at a load level equal to 30% of the subgrade failure load, which is the highest it should be to prevent large plastic deformation under repeated loading.

The factor of safety obtained by Eq. 10.9 (Meyerhof and Hanna) is in agreement with that of the slope stability method when the subgrade has ($\varphi = 0$). For subgrades with $\varphi > 0$ the Meyerhof and Hanna approach is not suitable.

10.9. Required Ballast/Subballast Depth

A minimum ballast layer thickness is needed below the sleepers to provide for maintenance tamping and for void storage space. Guidelines for this will be given in other chapters. In addition a minimum subballast layer thickness is required for performing the functions of a separation/filter layer. This is discussed in Chapter 9. In addition to these considerations the combined ballast/subballast thickness must be sufficient to prevent progressive shear subgrade failure, and excessive rate of settlement through plastic strain accumulation in the subgrade. This section will describe methods for determining this thickness, which for simplicity will be referred to as granular layer thickness.

10.9.1. North America

By trial and error over a number of years minimum ballast and subballast layer thickness specifications have been adopted by North American railroads. For economic reasons these minimums are not likely to represent the worst subgrade condition on a given line, and so they cannot be taken as an adequate thickness in general. Because all of the major North American railroads have a wide range of subgrade conditions within their systems, a single design thickness for the granular layer is not appropriate.

According to Raymond (Ref. 10.13), for mainline track in North America the range of minimum ballast and subballast layer thicknesses among the major railroads is:

- Ballast: 6 to 12 in. (150 to 305 mm).
- Subballast: 4 to 12 in. (100 to 305 mm).

The AREA Engineering Manual specifies a 12 in. (305 mm) minimum ballast thickness and a 6 in. (150 mm) minimum subballast thickness. This results in a minimum granular layer thickness of 18 in. (460 mm).

To determine the pressure applied to the subgrade by the ballast, the AREA Manual gives four equations:

1) Talbot equation:

$$P_c = \frac{16.8\, P_m}{h^{1.25}} \quad . \qquad (10.10)$$

2) Japanese National Railways equation:

$$P_c = \frac{50\, P_m}{10 + h^{1.35}} \quad . \qquad (10.11)$$

3) Boussinesq equation:

$$P_c = \frac{6\, q_0}{2\pi\, h^2} \quad . \qquad (10.12)$$

4) Love's equation:

$$P_c = P_m \left[1 - \left(\frac{1}{1 + r^2/h^2} \right)^{3/2} \right] \quad . \qquad (10.13)$$

In these equations:

P_c = subgrade pressure (psi),

P_m = applied stress on ballast (psi),

h = ballast depth (inches, except JNR is in cm),

q_o = static rail seat load (pounds), and

r = radius of a circle whose area equals the sleeper bearing area, A_b (inches).

According to Ref. 10.21 the Talbot and the Japanese National Railways (JNR) equations are empirical. The JNR equation, while not so noted in the AREA Manual, was developed for narrow gage track. Talbot's formula was developed from a number of full-scale laboratory tests perfomed at the University of Illinois (Ref. 10.14). Several different types of ballast were tested, including sand, slag, crushed stone, and gravel, with pressures from applied static loads measured at various depths and locations under several sleepers. Wheel loads were not as large as those commonly encountered today.

The third and fourth equations are both based on the Boussinesq solution for stress in an elastic body due to an applied surface point load. Love's formula is an extension of the Boussinesq results, in which the load applied by the sleeper to the ballast is represented as a uniform pressure over a circular area equal to the sleeper bearing area, A_b.

The pressure, P_m, that the sleeper exerts on the ballast is defined by

$$P_m = \frac{L_R\,(FS)}{0.5\,A_b}, \qquad (10.14)$$

where L_R = rail seat load,

FS = factor of safety, and

A_b = effective sleeper bearing area.

For wood sleepers AREA gives an effective bearing area of about 2/3 the total sleeper area and recommends that P_m be a maximum of 65 psi (450 kPa). For concrete sleepers AREA uses the total sleeper area as the effective bearing area and recommends that P_m be a maximum of 85 psi (590 kPa), assuming that a high quality ballast is used.

As an alternative to the above approach, GEOTRACK (Chapter 5) can be used to estimate the subgrade surface pressure, P_c.

The last step for calculating the required granular layer thickness using Eqs. 10.10 to 10.13 is to determine the allowable subgrade bearing pressure. The AREA Manual recommends a universal 20 psi (140 kPa) limit for subgrade pressure for all soil conditions. Because subgrades actually vary widely in strength use of this universal value will lead to difficulty in soft soil conditions and be too conservative in strong soil conditions.

A more logical approach is to designate allowable subgrade bearing pressure based on the soil conditions. An example is given in Table 10.4 from Ref. 10.21 which was obtained by applying a factor of 0.6 to the allowable foundation bearing pressures given in Ref. 10.15. The factor of 0.6 was used to account for variations in sleeper support conditions.

Table 10.4 Allowable subgrade bearing pressures

Subgrade Description	In-Place Consistency	Allowable Pressure below Track	
		psi	kPa
Well graded mixture of fine and coarse grained soils: glacial till, hardpan, boulder clay (GW-GC, GC, SC)	Very compact	65 - 100	450 - 690
Gravel, gravel-sand mixtures, boulder-gravel mixtures (GW, GP, SW, SP)	Very compact	55 - 85	380 - 590
	Medium to Compact	40 - 60	280 - 410
	Loose	25 - 50	170 - 350
Coarse to medium sand, sand with little gravel (SW, SP)	Very compact	30 - 50	210 - 350
	Medium to Compact	25 - 30	170 - 210
	Loose	15 - 25	100 - 170
Fine to medium sand, silty or clayey medium to coarse sand (SW, SM, SC)	Very compact	25 - 40	170 - 280
	Medium to Compact	15 - 30	100 - 210
	Loose	8 - 15	60 - 100
Fine sand, silty or clayey medium to fine sand (SP, SM, SC)	Very compact	25 - 30	170 - 210
	Medium to Compact	15 - 25	100 - 170
	Loose	8 - 15	60 - 100
Homogeneous inorganic clay, sandy or silty clay (CL, CH)	Very stiff to hard	25 - 50	170 - 350
	Medium to stiff	8 - 25	60 - 170
	Soft	4 - 8	30 - 60
Inorganic silt, sandy or clayey silt, varved silt-clay-fine sand (ML, MH)	Very stiff to hard	15 - 30	100 - 210
	Medium to stiff	8 - 25	60 - 170
	Soft	4 - 8	30 - 60

The values in Table 10.4 are based on static loading conditions and are intended to provide an acceptable factor of safety against bearing failure and a satisfactory limit on settlement. However the bearing capacity for track has to consider not just a single load application, but also repeated loading. Allowable stress under repeated loading is much less than under static loading. Thus the recommended values in Table 10.4 may still be too high.

10.9.2. British Railways

10.9.2.1 Description of method

In the late 1960's British Railways cooperated with the Office for Research and Experiments of the International Union of Railways in a program of work concerned with the measurement of scatter of subgrade stresses and stress distribution under various combinations of ballast depth, sleeper spacing, sleeper type, and maintenance technique (Ref. 10.16).

The results of these tests confirmed the results of previous work. These are:

1) There is a linear relationship between sleeper loading and subgrade stress which is independent of speed and wheel arrangement.

2) The measured vertical stress distribution can be closely predicted by simple elastic theory and is not markedly different for timber or concrete sleepers.

3) Scatter in stress level between nominally identical positions in the track substructure is highly dependent on ballast condition.

Additional conclusions were:

4) The most important factor influencing vertical stress in the subgrade for a given ballast depth and rail/sleeper reaction, was the packing condition under the sleeper.

5) Packing condition was subject to a high degree of random scatter which tended to mask the difference between various packing methods.

6) The flexural rigidity of the sleeper was of secondary importance, and for practical purposes, similar vertical subgrade stress distributions were obtained for both timber and concrete sleepers.

7) In terms of vertical stress level at the surface of the subgrade per unit loading applied to the sleeper, sleeper spacing in the range 790 - 630 mm had a negligible influence.

In order to obtain suitable material parameters for use in a design method, it was necessary to adopt a relevant laboratory test. For this purpose, a form of the triaxial compression test was chosen in which the major (axial) principle stress was pulsed, and the minor (radial) principle stress was kept constant.

Figure 10.29 is an example of the results of repeated load triaxial compression tests in which the cumulative strain percent is plotted against the logarithm of the number of

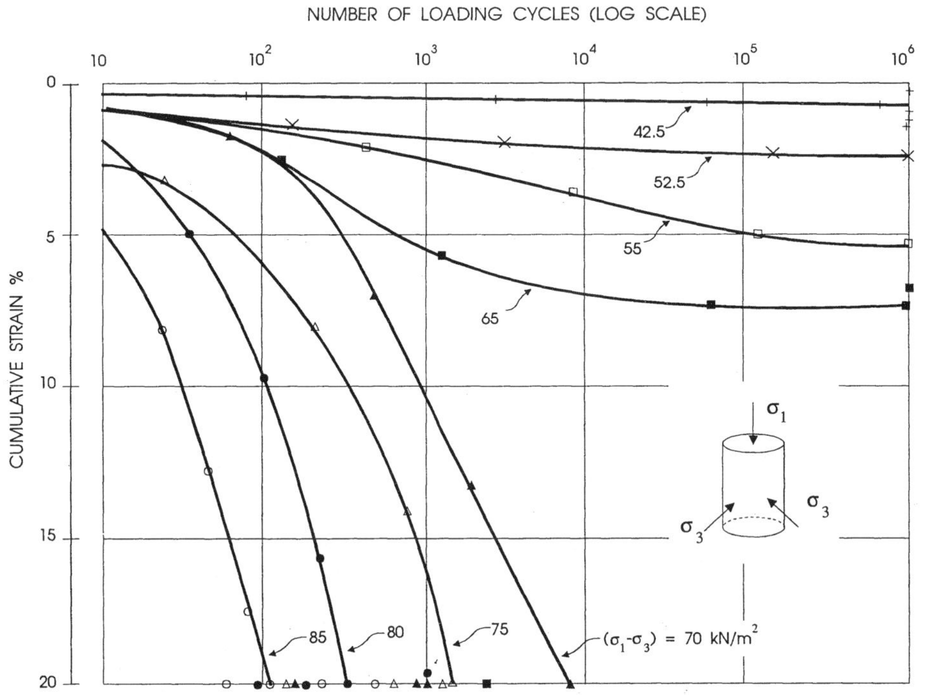

Fig. 10.29 Results of repeated load triaxial compression tests

sustained loading cycles. Each curve represents a different principle stress difference ($\sigma_1 - \sigma_3$).

It can be seen that the results fall into two distinct groups:

1) those in which the deformation is increasing at an <u>increasing</u> rate until complete failure of the specimen is reached, and

2) those in which the deformation is increasing at a <u>decreasing</u> rate and a stable condition is attained.

The principle stress difference ($\sigma_1 - \sigma_3$) separating these two groups has been designated the 'Threshold Stress'.

The results of tests carried out at three different mean effective ambient pressures (σ_3) indicated that the threshold stress was approximately a linear function of effective ambient pressure. The design procedure makes use of this relationship to allow for an increase of threshold stress, with depth of construction.

The track foundation design method pre-supposes that a balanced design is achieved when the stress induced in the subgrade by the heaviest commonly occurring axle load is equal to the threshold stress at that depth in the subgrade.

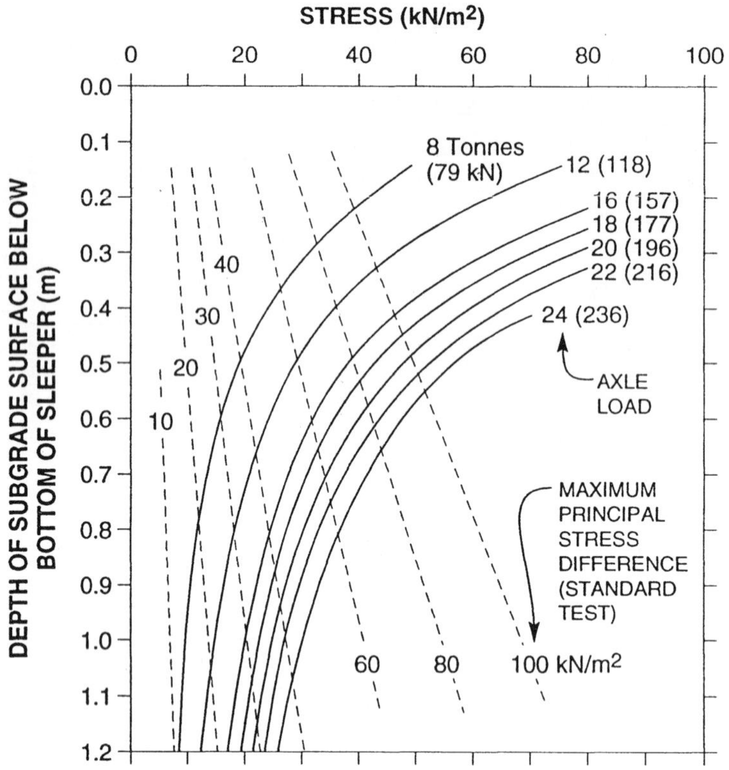

Fig. 10.30 Variation of subgrade vertical stress and threshold stress with depth

Figure 10.30 shows a family of curves (solid lines) relating stress to depth, for a range of axle loads. Also shown are a family of curves (dashed lines) relating threshold stress to depth, for a range of theshold stresses derived from the standard test which is carried out at a standard ambient pressure.

The points at which the two sets of curves intersect, represent the required design depth, i.e. the depth at which the induced stress is equal to the threshold strength. As an example, for an axle load of 16 tons, and a threshold stress of 60 kN/m^2, the required depth of construction is 600 mm.

Figure 10.31 shows a summary design chart, derived from Fig. 10.30 which relates required depth of construction to threshold stress, for a range of axle loads.

10.9.2.2 Field validation of design method

A series of correlated laboratory and field tests were carried out to check the validity of the design method. Test sites were chosen to cover geologically different types of clay. Sites were chosen in pairs to provide a site with subgrade failure and a stable site within a few hundred metres, and on the same line, thereby ensuring the same traffic loading and speed.

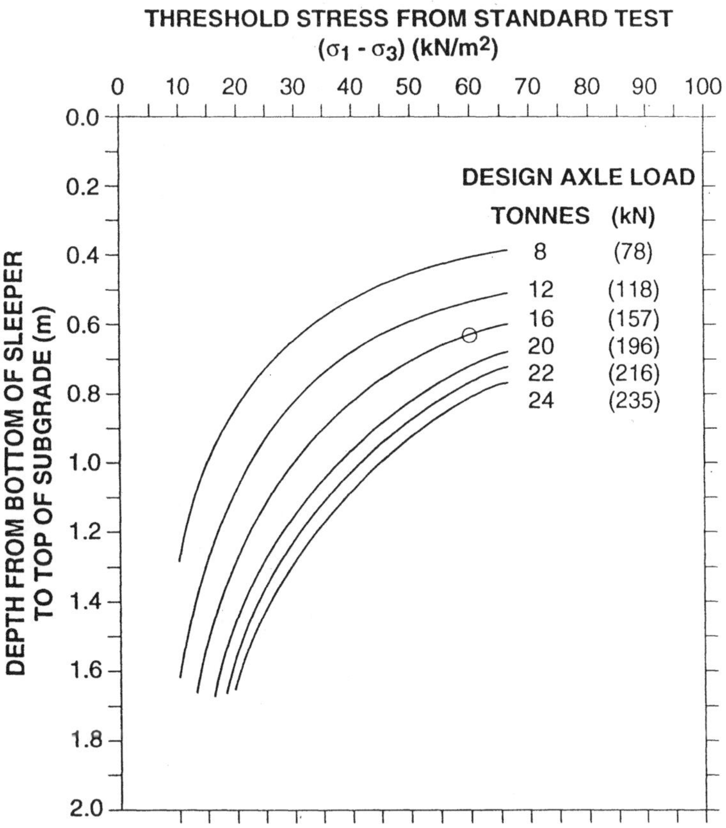

Fig. 10.31 Summary design chart

On one stable test site, the rail height was lowered such that from one end to the other, a range of ballast depths was obtained. The range was selected so as to give ballast depths varying from less than, to greater than the design depth. The site instrumentation used is shown in Fig. 10.32.

Typical results obtained from such instrumented sites are shown in Fig. 10.33. The family of curves shown at the top of the figure represents a stable site. The rate of settlement in terms of sustained axle loads is slow. The family of curves shown at the bottom of the figure, however, represents an unstable site. The rate of settlement is rapid.

For a number of sites, Fig. 10.34 shows the relationship between the rate of subgrade movement in mm/million axles, and difference between the actual depth to the subgrade and the required depth to the subgrade as indicated by the design method. The graph confirms that where the actual construction depth is greater than the design depth as indicated by the design method, the rate of subgrade movement is slow. Conversely, it shows that where the actual construction depth is less than the design depth, the rate of subgrade movement is rapid. Of particular interest is the site with two sets of points connected by arrows. This initially unstable site was stabilized by increasing the ballast depth to a figure in excess of that indicated by the design method. The result was a decrease in settlement rate from rapid to very slow.

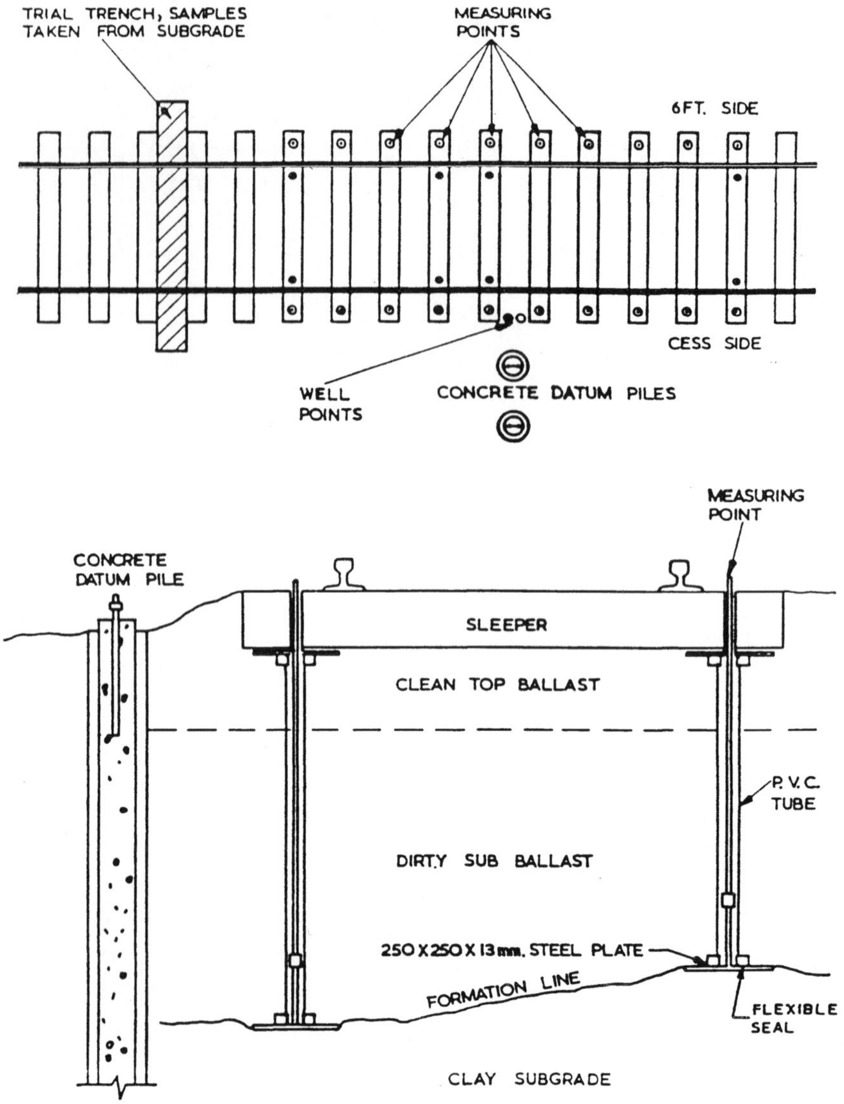

Fig. 10.32 Example of test site layout

It was considered by British Railways, that the field tests confirm the validity of the design method.

10.9.2.3 Application of method

In practice, British Railways adopts one or both of the following two procedures which take into account factors such as the probability of overload, body roll effects, etc., which were ignored in the earlier work (Ref. 10.17). As a consequence these procedures yield greater depths of construction.

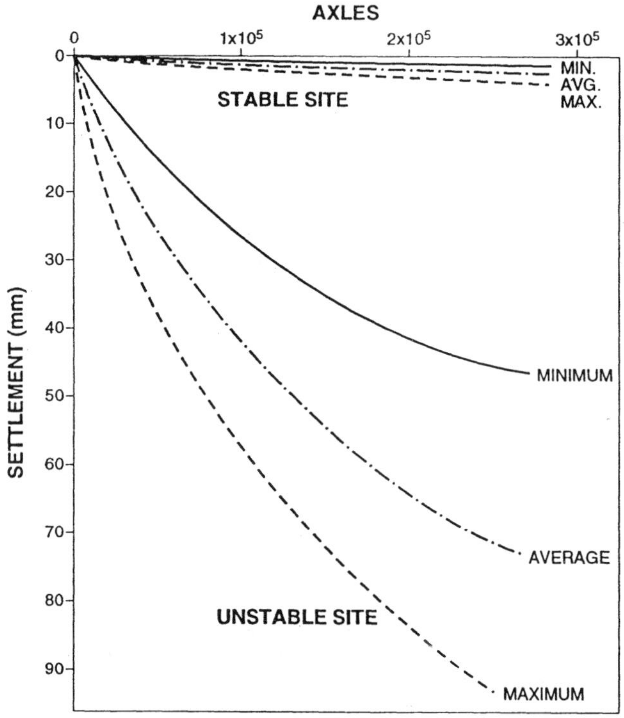

Fig. 10.33 Example of comparison of setttlement rates at stable and unstable sites

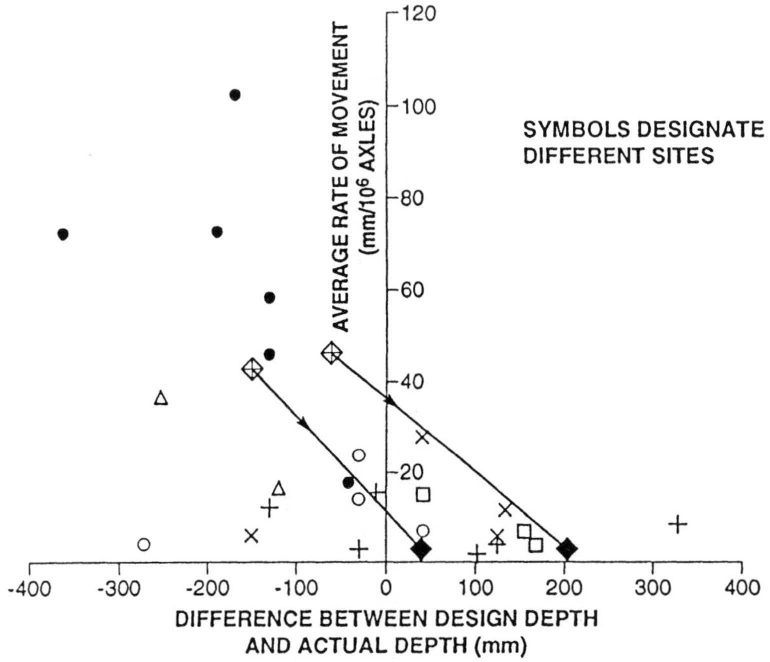

Fig. 10.34 Measured rate of settlement related to depth

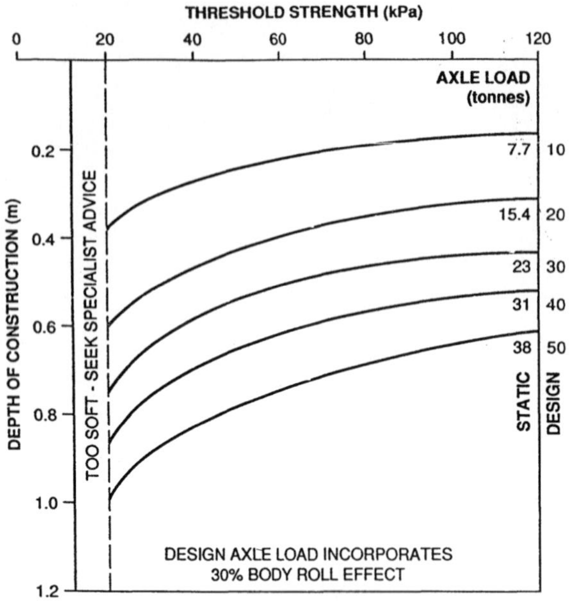

Fig. 10.35 Granular layer thickness based on subgrade strength

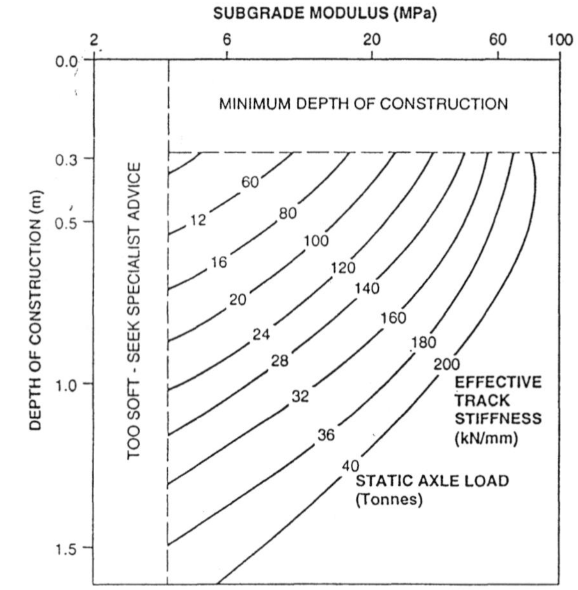

Fig. 10.36 Granular layer thickness based on subgrade modulus

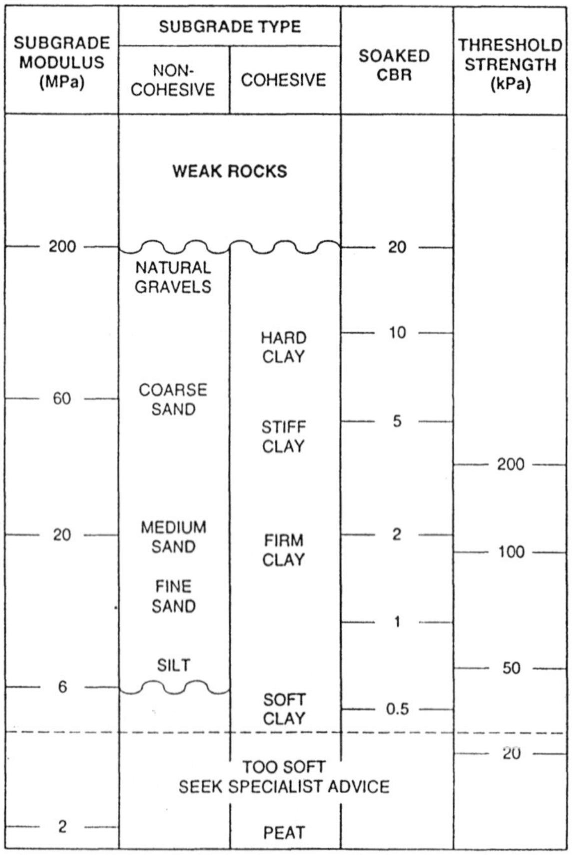

Fig. 10.37 Basis for estimating subgrade properties

1) The threshold strength is assumed to be equal to 50% of the ultimate compressive strength as determined from an unconfined or traditional triaxial compression test. The required depth of construction is then determined for the selected axle load using Fig. 10.35.

2) The subgrade resilient modulus, which is a measure of the elasticity of the soil, is derived from the slope of the stress-strain relationship as determined from the results of triaxial compression or plate bearing tests. The required depth of construction is then determined from Fig. 10.36.

The threshold strength or the resilient modulus of the subgrade material can also be estimated from the results of a CBR test, or from a physical description of the subgrade material by reference to Fig. 10.37. The required depth of construction can then be determined by reference to Fig. 10.35 or Fig. 10.36 as appropriate. For cohesive clays, the greater of the depths of construction derived from these two figures is adopted.

10.9.3. European Multi-layer Elastic Methods

According to a review of design methods in Ref. 10.21, a multi-layer elastic analysis is used by the Czechoslovak (CSD), Hungarian (MAV) and German (DB) federal railways. The approach is essentially as follows:

1) Thicknesses of the ballast and subballast are selected based on experience. A standard ballast thickness of 300 mm (12 in.) is used by the CSD and DB and 500 mm (20 in.) by the MAV.

2) An equivalent modulus for the entire substructure is determined based on the elastic moduli of the various layers.

3) The vertical stress with depth below the sleeper is determined using single-layer elastic theory empirically modified by the different railroads.

4) The vertical subgrade stress is compared with allowable stresses determined for the subgrade along the track. If the stress is greater than that allowable, the granular layer thickness is increased and the stress calculation is repeated.

REFERENCES

10.1 Li, D. and Selig, E. T. (1994). **Resilient modulus for fine-grained subgrade soils.** *Journal of Geotechnical Engineering*, ASCE, Vol. 120, No. 6, June, pp. 939-957.

10.2 Thompson, M. R. and Robnett, Q. L. (1979). **Resilient properties of subgrade soils.** *Transportation Engineering Journal*, ASCE, Vol. 105, No. TE1, January.

10.3 Moossazadeh, J. and Witczak, M. W. (1981). **Prediction of subgrade moduli for soil that exhibits nonlinear behavior.** *Research Record No. 810*, Transportation Research Board.

10.4 Pezo, R. F., Kim, D-S, Stokoe, K. H., and Hudson, W. R. (1991). **Aspects of a reliable resilient modulus testing system.** Transportation Research Board, 70th annual meeting, Washington, D. C., January.

10.5 Seed, H. B., Chan, C. K., and Lee, C. E. (1962). **Resilience characteristics of subgrade soils and their relation to fatigue failures in asphalt pavement.** *Proceedings*, First International Conference on the Structural Design of Asphalt Pavement, University of Michigan, Ann Arbor.

10.6 ASTM D698. **Standard test methods for moisture-density relations of soils and soil-aggregate mixtures using 5.5-lb (2.49-kg) rammer and 12-in. (305-mm) drop.** Annual Book of ASTM Standards, Section 4, Construction, Vol. 04.08, *Soil and Rock.*

10.7 Thompson, M. R. and LaGrow, T. G. (1988). *A proposed conventional flexible pavement thickness design procedure.* Illinois Cooperative Highway and Transportation at Urbana-Champaign, December.

10.8 El-Sharkawi, A. E. (1991). *Correlation of railroad subgrade resilient modulus with cone penetration test data.* MS degree report, Geotechnical Report No. AAR91-389P, Department of Civil Engineering, University of Massachusetts, Amherst, September.

10.9 Malviya, R. P. and Selig, E. T. (1989). *Prediction of track load capacity.* Project Report for Association of American Railroads, Report No. AAR 89-361F, Department of Civil Engineering, University of Massachusetts, Amherst, April.

10.10 Meyerhof, G. G., and Hanna, A. M. (1980). **Design charts for ultimate bearing capacity of foundations on sand overlying soft clay.** *Canadian Geotechnical Journal*, Vol. 17, pp. 300-303.

10.11 Meyerhof, G. G. (1974). **Ultimate bearing capacity of footings on sand overlying soft clay**. *Canadian Geotechnical Journal,* Vol. 11, pp. 223-229.

10.12 Sattler, P. J., and Fredlund, D. G. (1988). *The development of bearing capacity design charts for track system.* Final Report, University of Saskatchewan, Saskatoon, Canada, October.

10.13 Raymond, G. P. (1978). **Design for railroad ballast and subgrade support.** *Journal of the Geotechnical Engineering Division*, ASCE, Vol. 104, No. GT1, January.

10.14 AREA (1918). **First progress report, AREA-ASCE Special Committee on Stresses in Railroad Track,** *Bulletin*, AREA, Vol. 19, No. 205, March, 192 pp.

10.15 U.S. Navy (1962). *Soil mechanics, foundations and earth structures*. US NAVFAC Design Manual DM-7.

10.16 Heath, D. L., Shenton, M. J., Sparrow, R. W., and Waters, J. M. (1972). **Design of conventional rail track foundations**. *Proceedings*, Institution of Civil Engineers, Vol. 51, February, pp. 251-267.

10.17 Sperring, D. G. (1989). *Practical Track Design.* Railways Industries Association, Third Track Sector Course, paper 3.4.

10.18 Collingwood, B. I. (1988). *An investigation of the causes of railroad ballast fouling.* M.S. degree project report, Report No. AAR88-350P, Department of Civil Engineering, University of Massachusetts, Amherst, May.

10.19 Tung, K. W. (1989). *An investigation of the causes of railroad ballast fouling*. M.S. degree project report, Report No. AAR89-359P, Department of Civil Engineering, University of Massachusetts, Amherst, May.

10.20 Lambe, T. W. and Whitman, R. N. (1969). *Soil mechanics.* John Wiley and Sons, New York.

10.21 DiPilato, M. A., Levergood, A.V., Steinberg, E. I., and Simon, R. M. (1983). *Railroad track substructure-design and performance evaluation practices.* Goldberg-Zoino and Associates, Inc., Newton Upper Falls, Massachusetts, Final Report for USDOT Transportation Systems Center, Cambridge, Massachusetts, Report No. FRA/ORD-83/04.2, June.

10.22 Brown, S. F., Lashine, A. K. F., and Hyde, A. F. L. (1975). **Repeated load triaxial testing of a silty clay.** *Geotechnique* 25, No. 1, pp. 95-114.

11. Subgrade Improvement Alternatives

The importance of the subgrade to track performance was discussed in Chapter 10. In cases where the subgrade is too weak or has too low a stiffness, the resulting high cost of track maintenance may dictate the need to improve the subgrade conditions. The alternatives are listed in Table 11.1. There are four general groups. First is modification of the subgrade properties without removal or disturbance. Second is modification of properties by reconstruction and replacement. Third is strengthening with asphalt concrete. The fourth is slip stabilization. Each of these will be discussed. Prior experience, and often special equipment, is needed for successful application of these methods. Hence it is advisable to consult with specialists when subgrade improvement is needed.

Table 11.1 Subgrade improvement alternatives
I. Altering subgrade properties in place A. Grouting 1. Penetration 2. Compaction B. Lime slurry pressure injection C. Electrical treatment 1. Electro-osmosis 2. Electrochemical stabilization
II. Reconstruction and replacement A. Compaction B. Replacement C. Admixture stabilization 1. Cement 2. Lime 3. Bitumen 4. Fly ash
III. Asphalt concrete applications A. Full depth (replacing ballast and subballast) B. Underlayment (replacing subballast)
IV. Slip stabilization A. Drainage B. Retaining structures C. Slope change

Subgrade reinforcement techniques using geosynthetic products are also available. These will be discussed in Chapter 12.

There are numerous well-established methods for soft or loose soil improvement which are only suitable for new construction. These include heavy tamping, stone columns, preloading or surcharging with vertical drains, and displacement or replacement. These and other methods are described in Refs. 11.3 and 11.4. Such techniques are unlikely to be used with existing track and so will not be discussed in this chapter.

11.1. Altering Properties in Place

11.1.1. Grouting

Grouting involves injecting a material into the soil or rock to strengthen it, displace it, or to decrease its permeability. Two subcategories are penetration grouting and compaction

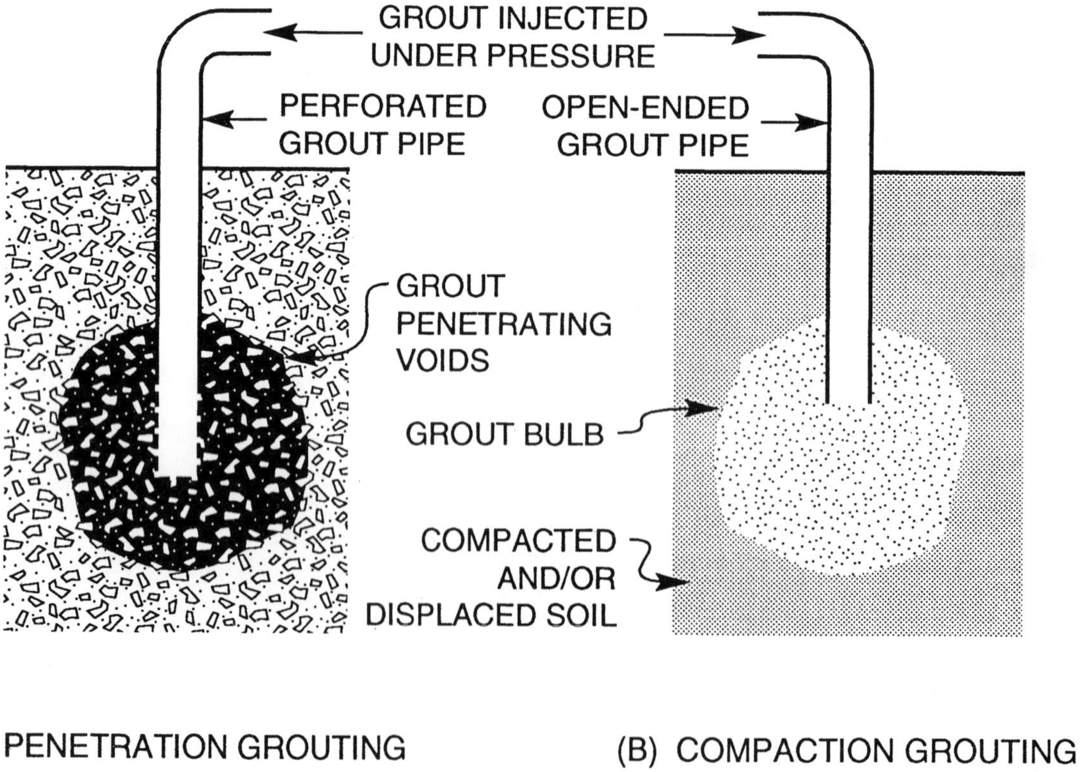

Fig. 11.1 Principles of penetration grouting and compaction grouting

grouting (Fig. 11.1). In penetration grouting the material is injected into voids or fissures in the soil or rock. Compaction grouting, used only for soil, displaces the soil to form a grout bulb. For unsaturated or high permeability (freely draining) soils compaction occurs from the grout pressure. In addition, the volume displaced by the grout bulb can be used to compensate for ground settlement. Hence compaction grouting has also been used as a soil jacking method.

11.1.1.1 Penetration grouting

Penetration grouting is intended to strengthen the soil by bonding the particles or to decrease the permeability by sealing voids, or both. For penetration grouting to work the soil or rock must be relatively permeable. Finely ground portland cement and water grout is used to strengthen and reduce the permeability of sands and gravels. Clay-water slurry grout is used to decrease the permeability of sands and gravels. Other ingredients may include sand and fly ash. Penetration grouting is not suitable for use with fine silts and clays. The cement, sand, ash and clay particles are too large to penetrate voids in silts and clays, or in sands and gravels with a silt and clay component of more than about 10%. Chemical grouts are solutions which can penetrate coarse silts, and are also suitable for medium to fine sands. Silicates are the most common chemicals used.

In situ and laboratory permeability tests as well as a grain size analysis will help judge the ability of grout to penetrate a soil. Strength tests are needed to estimate the potential strength improvement. Mitchell (Ref. 11.1) suggested that D_{15} of the soil being grouted should be at least 25 times the D_{85} of the particulate grout.

11.1.1.2 Compaction grouting

Compaction grout is a very stiff (low slump of 0 to 50 mm) mixture of cement, sand, fly ash, clay and water. The grout pipe must first be installed to the desired depth of the grout bulb. Then the grout is pumped at pressures ranging from generally 700 to 4000 kPa (100 to 600 psi).

This technique is not suitable for strengthening saturated cohesive soils because such soils are incompressible. At shallow depths, however, the ground can be displaced by the grout pressure. This may or may not be desirable.

11.1.2. Lime Slurry Pressure Injection

Lime is an effective additive for stabilizing clayey soils. Mixing lime with the clay reduces the clay's plasticity, reduces the clay's swelling potential in the presence of water, and increases strength. These changes result from chemical reactions involving cation exchange and cementing of the particles. The most effective way to stabilize clays with lime is in-place mechanical mixing which disperses the lime among the clay particles. This is one type of admixture stabilization which falls into category II methods. Because of the obvious limitations of this method for natural soil deposits beneath railroad tracks, pressure injection of the lime in the form of a slurry is attempted as an alternative.

The procedures for lime slurry injection are described in Ref. 11.2. The slurry is pumped into the soil through pushed in pipes at pressures between 340 and 1700 kPa (50 and 250 psi). A uniform mixing of lime and soil is not possible because the lime cannot penetrate the soil voids. Instead the lime penetrates through discontinuities such as fissures, cracks and permeable seams. The hope is that a network of lime seams will develop in the soil, from which the lime will diffuse into the zones around the seams. However, one of the problems with the method is that there is little control over the lime distribution.

The pressurized slurry may also fracture the soil and hence form new cracks to penetrate. This effect is illustrated in Fig. 11.2. Before application of the slurry pressure, the geostatic stress will be as shown in Fig. 3.17. For the soft clay soils likely to need stabilization K_o will be less than 1.0. This gives the initial total and effective stress Mohr circles as shown in Fig. 11.2a, assuming an initial pore water pressure of magnitude u_s. A slurry pressure of magnitude u_p is applied to the end of the pipe which causes a tensile stress in the soil around the circumference of the hole (Fig. 11.2b). When this tensile stress becomes equal to the minimum effective stress in the soil a crack will form on the plane perpendicular to the effective stress. When σ'_{ho} is the minimum principle stress, as in this example, the vertical plane will become a failure plane and a crack will be initiated upwards (Fig. 11.2c). Slurry pressure entering the crack can then quickly cause the crack to propagate to the surface and allow the slurry to escape into the ballast without penetrating the subgrade (Fig. 11.2) . Horizontal planes of weakness or K_o values greater than 1.0 would be required to cause the desired horizontal cracks.

A field soil exploration is needed to assess the effectiveness of slurry distribution. This would include determination of the presence of seams and planes of weakness that would lead to outward rather than upward distribution of the lime slurry. Laboratory tests are also needed to evaluate the potential soil property improvements. Some guidelines for these tests are given by Ledbetter in Ref. 11.2. However a full scale trial is the best way for evaluating the benefits.

The degree of soil improvement by lime slurry injection is difficult to predict. Significant improvements have been reported. However lack of effectivenss and even adverse effects have also been experienced. These adverse effects include the high water content and high pressure of the slurry which temporarily weakens the soil, and contamination of the surface and subsurface water systems by the escaping lime.

11.1.3. Electrical Treatment

11.1.3.1 Electro-osmosis

An effective way to improve soft clayey soils is to reduce the water content. Such soils are not free draining. Thus dewatering cannot be accomplished just by installing subsurface drains. Flow into the drains must be induced. Surcharging or loading the soil for a long period of time causes excess pore pressure which will dissipate by flow and result in soil strengthening by consolidation. This method is only suitable for new construction, however. An alternative, although expensive, method for existing track is dewatering through

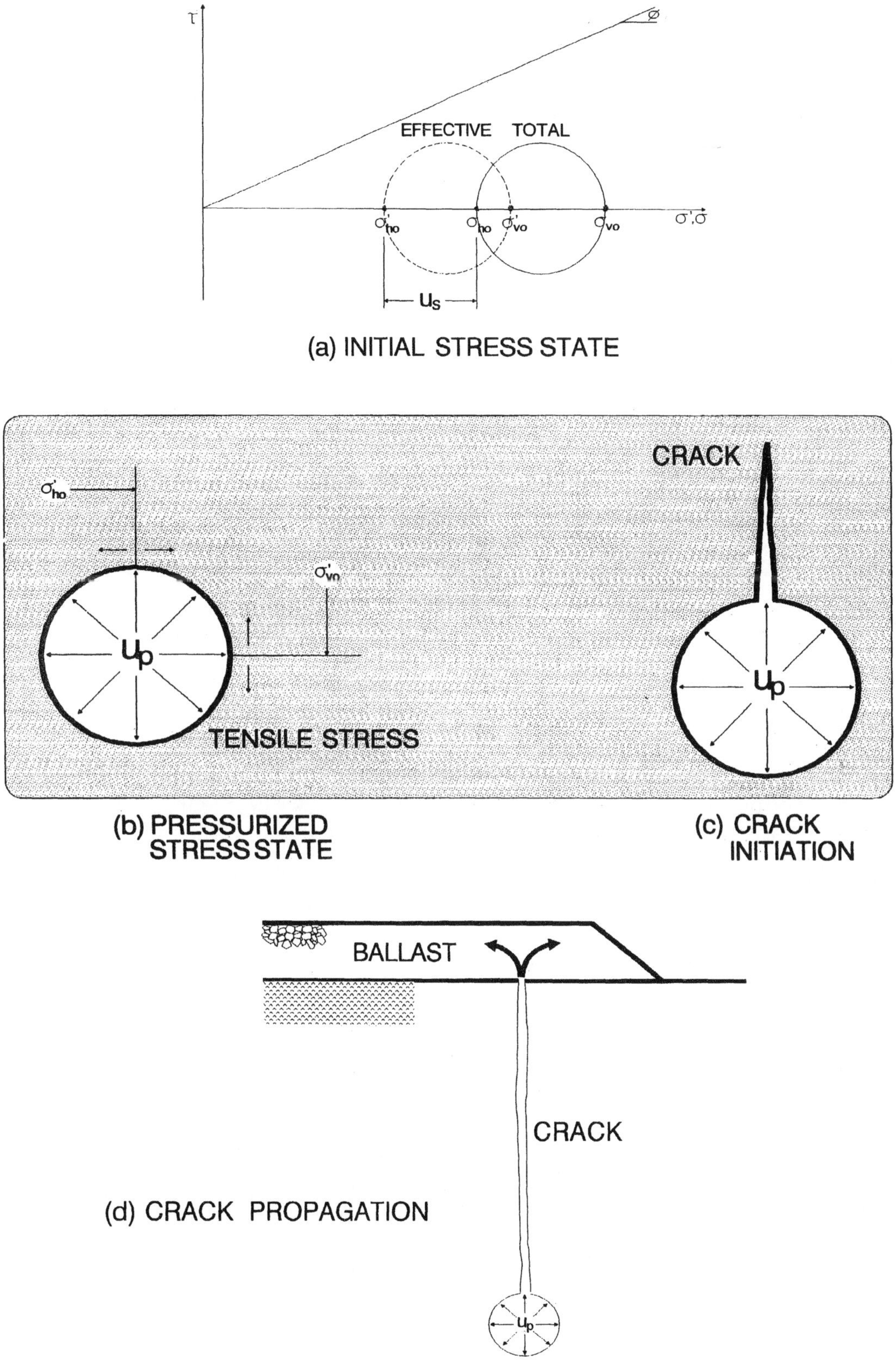

Fig. 11.2 Soil fracturing from slurry pressure

electro-osmosis (Ref. 11.3). With this method a direct electric current is passed through the soil between pairs of electrodes. Water molecules are caused to migrate from the anode into the drain surrounding the cathode, where they are removed. Electro-osmosis may be effective and economical for saturated and normally consolidated silts and clays with low pore water electrolyte concentration.

11.1.3.2 Electrochemical stabilization

Electrochemical stabilization uses the process of water movement by electro-osmosis to move dissolved salts through soil. In low permeability soils, such as clays and silts, it is possible to move solutions by electro-osmosis much more rapidly than by hydraulic pressure, as in permeation grouting. The cations in the infiltrated salts exchange with the cations existing in the clays to increase the clay shear strength and reduce its swelling potential (Ref. 11.4).

In a typical electrochemical stabilization system, anodes, consisting of perforated iron or aluminum pipe, are installed about 1 m to 5 m (3 ft to 15 ft) away from the cathodes. A solution, such as calcium, potassium, magnesium, or aluminum chloride, is fed into the anodes where it enters the soil through the perforated pipes. The solutions move toward the cathodes. Water collects at the cathodes where it is removed. In addition to being the injection location for the electrolyte solution, the anodes themselves will be electrically dissolved, forming iron or aluminum ions. Therefore, the anodes are gradually used up. The cathodes do not dissolve and can be recovered unaffected. High currents are required since current directly affects the rate of ion movement (Ref. 11.4).

Full scale field trials are usually desired to evaluate the effectiveness of the method. Selection of the anode type and electrolyte solution requires chemical analysis of the clay minerals. The method is very expensive and so is likely to be used only when other methods are not feasible. The main advantage of electrochemical stabilization, as for electro-osmosis, is that, because the pairs of electrodes would be installed on opposite sides of the track, the work can be done without disrupting the train service.

11.2. Reconstruction and Replacement

11.2.1. Compaction

Compaction is the process of soil densification by mechanical manipulation. Soil is a three-phase system consisting of a mixture of solid particles with the voids between them filled with water and air (Fig. 11.3). The densification is achieved by reduction in volume of the air voids. Therefore, it can be accomplished as quickly as the compactive effort is applied, since no expulsion of water is required. During compaction, the moisture content remains unchanged, neglecting wetting and drying caused by weather conditions, and the percent saturation increases.

If the soil is 100% saturated, then all voids are filled with water. Soils in this state cannot be compacted. They can only be densified by consolidation. Consolidation is the process of volume reduction in saturated soils that takes place gradually as pore water is expelled. In the unusual case that the loose soil is coarse sand or gravel, of course, consolidation time may be short.

Obtaining a greater soil unit weight is not a direct objective of compaction. Instead, the reason for compacting is to improve soil properties such as: increasing strength, increasing stiffness, decreasing compressibility, decreasing plastic strain under repeated load, decreasing permeability, and reducing swelling and shrinking. However, density is the most commonly used parameter for specifying the desired amount of compaction and for determining the state of compaction. This is primarily a consequence of historical tradition and convenience. An increase in density implies an improvement in the other parameters.

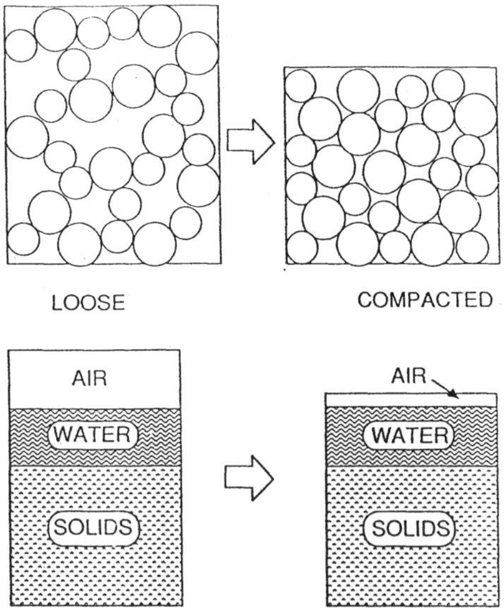

Fig. 11.3 Compaction as air void reduction

A given density, or a given compaction effort, does not produce the same magnitude of strength and stiffness properties for all soils. This is shown in Fig. 11.4 for three soil types. The percent compaction is the density expressed as a percent of the maximum dry density in the ASTM standard compaction test (Ref. 11.7). Effort is the work by the compactor per unit volume of soil expressed as a percent of the effort in the ASTM standard compaction test.

The nonlinear relationship between percent compaction and compactive effort is shown in Fig. 11.4a. The relative soil stiffness corresponding to the same percent compaction and effort are also shown. For the same effort, the percent compaction increases as the soil becomes more coarse-grained. Also to achieve the same percent compaction requires a much greater effort for clay, than for silt and in turn for sand. This demonstrates the concept of compactability or relative ease of compaction to a given percent compaction. The compactability of coarse-grained soils (sands and gravels) is much better than that of fine-grained soils (silts, and especially clays). Although not commonly appreciated, achieving the same percent compaction does not result in the same strength and stiffness for different soils. For the same percent compaction, the resulting soil stiffness increases from clay to silt to sand. The contrast is even greater for constant compaction effort (Fig. 11.4b). These are important considerations in determining the required compactive effort for a given application.

Compaction is most commonly done using surface rollers (static and vibratory). Because the roller contact area with the soil is small, the compaction stresses diminish rapidly with depth. This limits the effective depth of compaction with rollers to generally 300 to 600 mm (12 to 24 in.). To compact a greater depth with rollers requires removal and replacement in layers. The layers normally should be limited to 150 to 300 mm (6 to 12 in.) thick before compaction.

Deep compaction techniques are available (Refs. 11.3 and 11.4). One example is heavy tamping or impact compaction achieved by dropping a large heavy weight onto the soil surface. Another example is vibrocompaction in which coarse grained soils are densified by insertion of a vibrating probe accompanied by water jets. These methods would only be appropriate for localized deep deposits of soil. Both methods have been used with saturated soils because the methods induce excess pore water pressure which produces flow of water from the soil voids, with densification by consolidation.

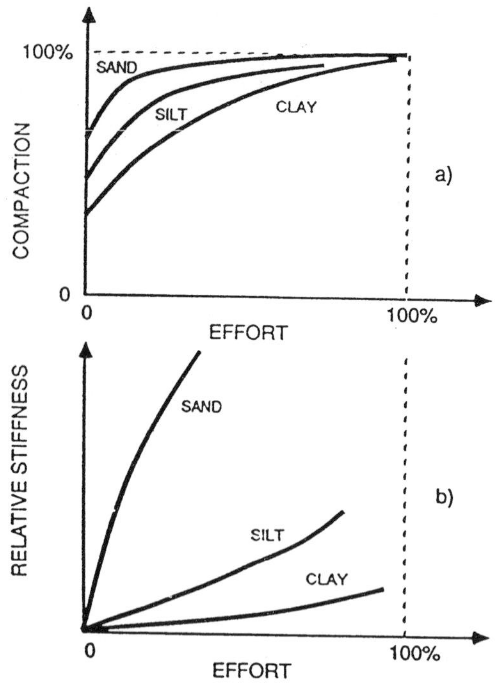

Fig. 11.4 Effect of soil type on variation of percent compaction and soil stiffness with compaction effort

11.2.2. Replacement

If saturated or nearly saturated, soils can be dried sufficiently, they can be removed and replaced by compaction in layers. This would more likely be feasible for coarse-grained soils or silty soils than for clayey soils, which are much more difficult to dry. Note, however, that coarse-grained or silty soils would probably not be saturated unless they were below the ground water table. In this case dewatering of the site would be required during the excavation and replacement process. If consolidation is not feasible, the saturated soils, particularly soft clays, would need to be removed and replaced with an acceptable soil which has a moisture content suitable for compaction.

Organic soils, that is soils whose properties are highly influenced by their organic content, are unacceptable as a track subgrade. These materials must be removed and replaced by an acceptable soil.

11.2.3. Admixture Stabilization

As a supplement to excavation and replacement with compaction, or to surface compaction, soil properties can be further improved by first mixing in other materials such as portland cement, lime, bitumen and fly ash.

Thorough mixing of the soil and additives is needed to get the benefit of the usually expensive material. Mixing is best accomplished by special mixing/pulverizing equipment that uses rotating tynes to break up the soil and mix it with the additive. Some of this special equipment excavates a layer of soil, transports it to a mixing drum, and deposits the modified soil out the rear end of the drum in a continuous moving operation. In some cases, particularly if bituminous stabilizers are used, the soil is excavated, transported to a central batch plant, and treated before being returned to the site. Conventional equipment is also frequently used; this includes disc or tyne harrows, bulldozers, ripping bars, motor graders, and scrapers. Although these types of equipment are less efficient, they generally cost less to mobilize than special mixers and are appropriate for small projects. Increasing or decreasing the exisiting moisture content may be required prior to adding the stabilizing material in order to achieve acceptable results.

The following summary of the installation techniques and benefits of each of the additives is based on Ref. 11.4.

11.2.3.1 Cement

Portland cement may be used to stabilize almost any soil except highly organic, high salt, or high sulphate materials. These components interfere with the portland cement reactions. Coarse gravels are not suited for cement stabilization because of their high cement content requirement, and heavy clays (CH) create mixing difficulties. Broadly-graded sand and gravel materials that possess a floating aggregate matrix, that is ones in which the fine fraction of the soil is inadequate to hold the coarse particles in place, are best suited for cement stabilization.

Generally cement stabilization should be limited to soils with less than 35 per cent passing the 0.075 mm (No. 200) sieve and a plasticity index of less than about 30 percent. Materials exceeding these percentages probably can be more efficiently stabilized with lime. In some cases, lime and cement can be combined to treat highly plastic material as described in the following section on lime.

Cement stabilization works by producing a very lean concrete called soil cement. The cementing action is produced by a hydration reaction within the portland cement components themselves. The hardened cement adds cohesion by bonding soil particles together. The principal difficulty in working with this method is that the cement begins to set in about two hours after wetting. Mixing and curing moistened soil cement must be completed before setting begins because disturbance after initial setting will decrease the final strength. If the in-place soil is dry, setting will be delayed until after water is applied to the soil-cement mixture.

Cement stabilization achieves the following subgrade performance improvements:

1) Increases strength including tensile strength.

2) Increases stiffness (modulus).

3) Increases resistance to wet-dry and freeze-thaw conditions.

4) Lowers permeability, except for clay, where the permeability is increased.

5) Decreases swelling in plastic clays.

The amount of cement required will vary with the soil type and required properties, but will generally fall within the range of 5 to 15% by dry weight of soil.

Cement stabilized soils placed over soils with a lower stiffness will develop horizontal incremental tensile stress at the bottom of the layer under vehicle loading (see Chapter 6). Unlike unbound granular material the cement stabilized soil cannot relieve this stress through development of compressive residual stresses. Thus if this tensile stress is high enough the stabilized layer will crack and lose some of its structural benefits.

11.2.3.2 Lime

Lime may be used to treat most soils that have at least some clay mineral fraction. Clay mineral content (finer than 2 microns) should be at least 7 percent by weight, and the plasticity index should exceed 10 per cent. Lime stabilization is used to decrease plasticity, improve workability, reduce swelling potential, and increase strength. Sometimes, lime is used to improve workability so that some other additive, such as cement or bitumen, can be mixed and compacted. Certain soils do not react with lime and therefore cannot be effectively treated with it. Laboratory testing is required to determine if a soil is lime reactive.

Soil stabilization with lime refers to the use of either of two chemicals, calcium oxide (CaO) called quicklime, or calcium hydroxide [$Ca(OH)_2$] called hydrated or slaked lime. These are produced from calcium carbonate ($CaCO_3$) or limestone by heating in a kiln.

Lime stabilization works by chemically combining the lime with the clay minerals and soil moisture to form calcium silicates, the same bonding compounds formed by portland cement. The clay minerals are a necessary component in the reaction. As they react, plasticity decreases immediately. This effect is dramatic for montmorillonite clays and minor in kaolins. Quicklime has another effect: the calcium oxide combines with water to form hydrated lime, releasing oxygen and causing excess water to be absorbed from the soil. However, quicklime is hygroscopic, that is it has strong affinity for water. This makes storage a problem because it tends to combine with moisture in the air. The hydrating reaction also releases substantial heat. This makes quicklime a hazardous substance to handle.

Lime also alters compaction characteristics. The maximum dry density of the treated compacted soil is less than, and the optimum water content is greater than, the untreated soil. The lower in-place density is more than compensated for by the cementing action, resulting in a higher shear strength. Because clayey soils often are wetter than the optimum moisture content in their natural state, increasing the optimum water content aids compaction. Unlike portland cement, the cementing reactions of lime take place slowly, so that working time prior to compaction is flexible. Pre-curing often results in a higher final strength.

Lime is sometimes used in combination with other agents. With portland cement or bitumen, lime is used to improve workability of the soil and promote mixing and compaction. Fly ash or other pozzolans are sometimes combined with it to provide the clay mineral component required for reactivity in clean granular soils.

Because lime reacts with the clay minerals in soils, the required quantity of lime is directly related to clay content. A range of 3 percent to 7 percent by weight of lime will treat most soils.

The principal advantage of using lime is that it produces immediate and dramatic improvements in the workability of clayey soils. Often, lime provides the only way to work with soft, wet clayey soil other than excavating the soil and replacing it. Also, the increase in strength from lime is gradual, so that timing between mixing and compaction is not as critical as with cement. However, if a rapid increase in strength is required, lime will not suffice. On the other hand, lime-treated soils are not as susceptible to cracking as soil cement.

11.2.3.3 Bitumen

Bitumen is used to cement soils that are mostly granular with limited plastic fines. In general, only soils classified under the USCS as GW, GP, SW, SP, GM, and SM are candidates. Wet soils usually are poor candidates, because adding the liquid bitumen stabilizer worsens characteristics of wet soils needed for compaction.

Soil-bitumen mixtures are made with asphalt cutback, emulsions, or foams. Cutback is asphalt dissolved in a volatile oil, such as kerosene or fuel oil, in about a 50-50 ratio. Emulsions are mixtures of about 55% to 60% asphalt and water with an emulsifier to promote suspension of the asphalt. Foams are produced by bubbling steam through hot asphalt. All these processes are used to promote mixing of the asphalt with the soil. The heaviest asphalt product that can be adequately mixed with the soil should be used. Bitumen stabilizes soil by cementing particles and soil aggregations together to form a cohesive mass. The solvent or water in asphalt products either must evaporate or be absorbed into the soil before the bitumen will cement.

Bitumen increases soil strength and decreases its permeability. However, if too much bitumen is added, strength will decrease as the thickness of the bitumen layers between soil particles increases. To achieve waterproofing, bitumen emulsion can be sprayed on the surface of a compacted subgrade and penetrate downward. The thin bitumen surface film limits infiltration of water. Used on "oiled" gravel roads for many years, this process may be effective for railroad track to treat soils that soften in the presence of water. A condition is that the groundwater level must be low enough so that the softening is not caused by water trapped beneath the bitumen.

11.2.3.4 Fly Ash

A lime-fly ash admixture has proved successful both for stabilizing special soils and for using waste products that are economically available. Fly ash is a finely ground silicous material (containing silica, SiO_2) that is most frequently produced from coal furnace flue dust or from blast furnaces. It also can be produced by grinding natural volcanic ash or sediments. When used in soil stabilization, fly ash is often combined with lime. In clean granular soils, the fly ash takes the place of naturally occurring clay with regard to its reaction with lime.

The closest competitor for lime-fly ash stabilization is portland cement, either alone or in combination with lime. Lime-fly ash may be beneficial in soils where the workability is improved by lime, but where there is insufficient reactivity to produce the necessary strength gain. Also it has a longer working time than cement.

11.3. Asphalt Concrete Applications

Asphalt concrete is a mixture of asphalt cement (bitumen) and crushed stone. The recommended mixture for track use is (Ref. 11.5) a dense-graded (broadly-graded) highway base course crushed stone with slightly more asphalt cement and mineral aggregate fines than in highway applications, and with 1 to 3% air voids in the compacted mat. Typically the maximum aggregate size is 25 to 37 mm (1 to 1.5 in.).

Two ways of using asphalt concrete (AC) in the track substructure are shown in Fig. 11.5 (Ref. 11.5). The AC replaces the top and bottom ballast and the subballast in Fig. 11.5a, termed full-depth asphalt, and it replaces only the subballast in Fig. 11.5b, termed underlayment asphalt. In both cases the crib and shoulder ballast remain.

Effectiveness of the AC design can be evaluated by considering the eleven ballast functions and the seven subballast functions in Chapter 2.

For the ballast:

1) Resistance to vertical forces by AC is probably greater than with ballast alone; resistance to lateral and longitudinal forces may be reduced because of less interlocking between the sleeper and the AC layer and between the crib and shoulder ballast and the AC layer than when ballast is present beneath the sleeper.

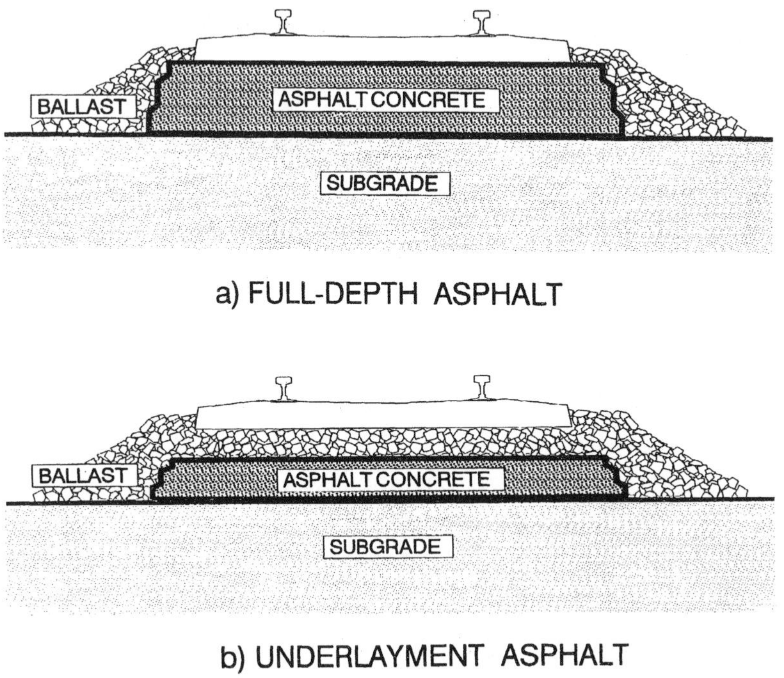

Fig. 11.5 Alternative hot mix asphalt trackbed cross-sections

2) Resiliency will be reduced by the AC because it has a higher modulus than ballast. Because most of the resiliency is in the subgrade in the example in Fig. 11.5a, the resiliency loss may not be a problem.

3) Void storage for fouling material is not provided by the AC, but it is not required.

4) Tamping is not possible with full depth AC. This is a major limitation of full-depth AC because if the vertical geometry changes as a result of subgrade movement, adjustment will be difficult.

5) The AC generally is impermeable and so will effectively shed water coming onto the track without first penetrating as in the case with ballast.

6) For a given layer thickness the AC will reduce stress to underlying layers more than will ballast as long as the AC remains structurally sound. Remaining sound means that cracking at the bottom of the AC layer must be prevented by proper design. This problem was mentioned for cement stabilized soils, and is discussed further in Ref. 3.16.

7) Insulation for frost protection will be provided by AC.

8) The AC will not promote vegetation growth.

9) Because the AC is covered by ballast, noise absorption will not be affected.

10) Electrical resistance will probably be comparable for AC with a ballast cover as for ballast alone.

11) Redesign/reconstruction of track will be much more difficult and expensive with the AC.

For the subballast:

1) The stress reduction function of the subballast can be handled by the AC, as well as that of the ballast, by proper AC design.

2) Frost protection could be less with the AC if the AC layer depth is less than the combined ballast/subballast depth because of the better load spreading ability of the AC.

3) Interpenetration of ballast and subgrade will be eliminated by the AC.

4) Upward migration of fines will be prevented by the AC.

5) Subgrade attrition by ballast will be eliminated by the AC.

6) Water shedding will be provided by the AC.

7) Drainage of water flowing upward from the subgrade will be inhibited by the AC. A granular drainage layer beneath the AC might be desirable rather than placing the AC directly on the subgrade. The granular layer also provides a platform on which to place and compact the asphalt.

In summary for the full-depth asphalt concrete design to be viable, two conditions must be met:

1) The reduction in track resiliency and the reduction in both lateral and longitudinal stability must be within acceptable limits.

2) The existence of the AC layer must essentially eliminate the need for geometry adjustment and reconstruction.

In addition a drainage layer beneath the AC should be provided if required.

Effectiveness of the underlayment AC design can be evaluated by considering the 7 subballast functions as well as the possible effect of the AC layer on lateral and longitudinal track stability. The subballast functions are satisfied essentially as discussed for the full-depth AC design. The big advantage of the underlayment design over the full-depth design is the ability to adjust the track geometry and tamp the ballast. Reconstruction would still be difficult, but if the AC design fulfills its function, this will be a distant future concern. Both designs in Fig. 11.5 have the AC covered by ballast. This serves to protect the AC from oxidization (hardening) and large temperature fluctuations which affect the AC properties such as stiffness.

The required thickness of the AC depends on the axle loads, the amount of traffic and the subgrade stiffness. Typical underlayment designs in North America have used 100 to 200 mm (4 to 8 in.) thick AC mat with 125 to 250 mm (5 to 10 in.) of ballast between the bottom of the sleeper and the top of the mat. A design guide is given in Ref. 11.6.

The majority of applications of AC in North American track have been for rehabilitating short sections of track such as crossing diamonds, turnouts, bridge approaches, tunnel floors and highway crossings (Ref. 11.5). These are often maintenance problem areas.

11.4. Slip Stabilization

This section will briefly discuss stabilization of potential slip failures that will affect track subgrade. The categories of failure location may be grouped as:

1) Cut slopes,

2) Embankment slopes,

3) Embankment foundations, and

4) Track subgrade.

Design of cut slopes, embankment slopes, and embankment foundations for new construction is beyond the scope of this book. Geotechnical practice is well established in this field. These designs should be done by a competent geotechnical engineer after a proper field investigation.

Design of track subgrade for new construction or reconstruction should be based on the information given in Chapter 10. Means of subgrade modification to improve its properties have been described in the preceding parts of this chapter.

The remaining topic of subgrade slip stabilization concerns cut slopes, embankment slopes and embankment foundations for existing track.

Cut slopes are slopes formed in the existing natural ground by excavation for the track right of way. Failure or movement of these slopes, which are usually close to the track, can cause movement of the track subgrade as well as interrupt traffic by depositing earth and rock on the track. The stability problem with existing cut slopes usually arises from water within the slope exerting pressure and decreasing the material's shearing resistance. The solutions include one or more of the following: 1) divert water coming into the slope from above to the extent possible, 2) install drains within the slope, 3) flatten the slope

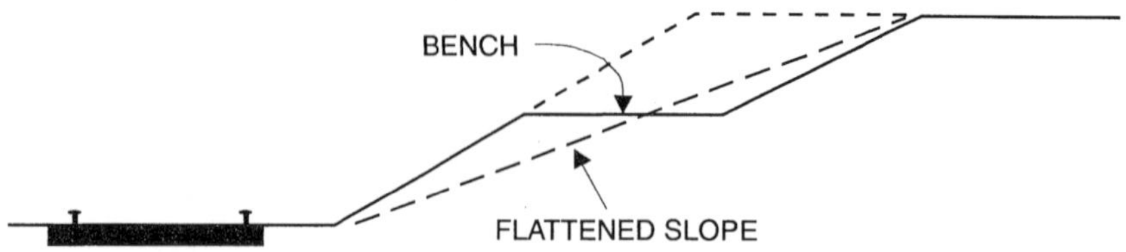

Fig. 11.6 Benching of cut slope

(Fig. 11.6), or 4) add tie back anchors or retaining structures. Cuts in expansive clays can give difficulty because of unloading from soil removal combined with exposure to wetting and drying cycles. Protection from moisture change or slope flattening may be required. Surface erosion of soil cuts from water flow or of rock cuts from weathering may also present problems, but these do not generally result in massive shear failure or subgrade disturbance. Surface protection is the cure in these situations.

Failure within existing embankments usually results from water infiltration which reduces the soil shear strength and increases its weight. The source of water may be heavy rainfall or flooding. Drainage of water off the embankment so that it does not infiltrate protects against the first source. In areas where flooding is expected the slopes may have to be suitably flattened or have berms (Fig. 11.7) to provide an adequate factor of safety against failure. Alternatively strengthening through the addition of retaining structures may be necessary.

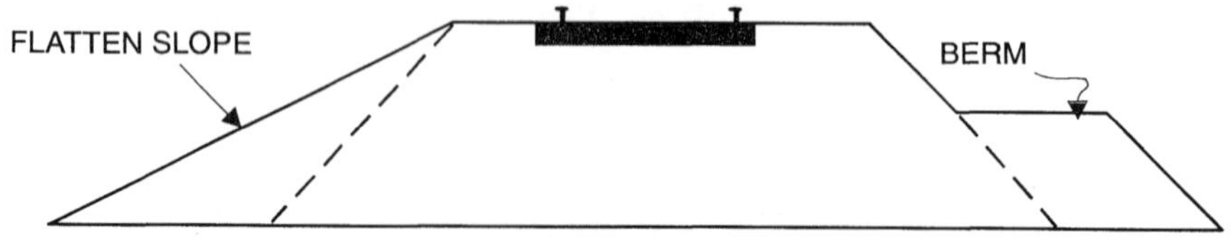

Fig. 11.7 Embankment widening

Erosion of existing embankment slopes will also decrease their factor of safety. Slope protection must be provided to prevent this problem.

Failure of the foundation soil upon which the embankment is placed usually is more likely with new construction since consolidation of the soil under the embankment weight will increase the foundation strength over time. However, situations such as increasing rainfall or flooding could weaken the foundation soil at a later time. Flattened slopes and berms on the embankment will reduce this potential force of failure as well.

Potentially the most critical situation exists when the embankment is placed on the side of a hill (Fig. 11.8). Embankment movement could occur by downhill sliding along surfaces through the embankment, or through the natural soil within the hill, assuming that benches were used to key the embankment into the hill. Movement may begin some time after original construction, probably as a result of changes in surface or subsurface drainage conditions. This problem must be corrected before failure occurs by providing proper drainage.

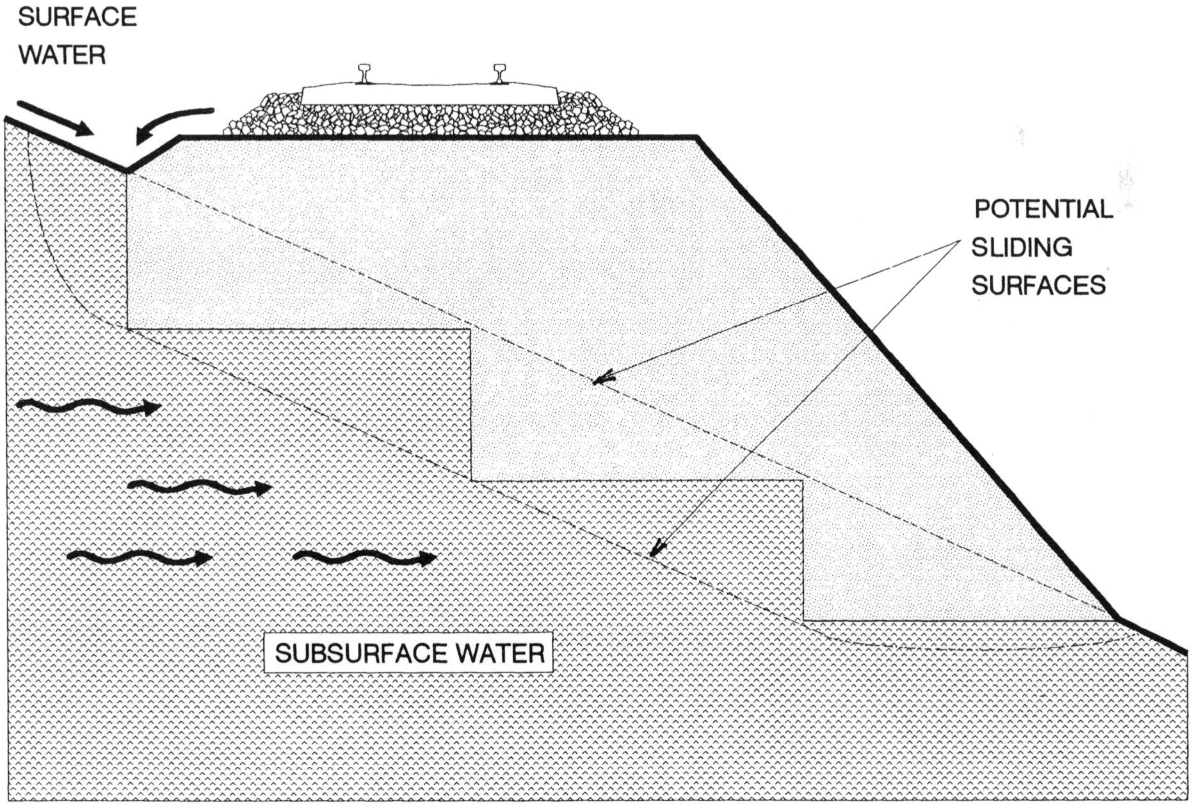

Fig. 11.8 Side hill fill problems

REFERENCES

11.1 Mitchell, J. K. (1970). **In place treatment of foundation soils.** *Journal of the Soil Mechanics and Foundations Division*, ASCE, Vol. 96, No. SM1, p. 86.

11.2 Blacklock, J. R. and Lawson, C. H. (original authors) as revised by Ledbetter, R. H. (1979). *Handbook for railroad track stabilization using lime slurry pressure injection*. U.S. Dept. of Transportation, Federal Railroad Administration, Washington, D.C., Report No. FRA/ORD - 77/30.

11.3 Mitchell, J. K. (1981). **Soil improvement--state-of-the art report.** *Proceedings,* 10th International Conference on Soil Mechanics and Foundation Engineering, Stockholm, Sweden, pp. 509 - 565.

11.4 Simon, R. M., Edgers, L., and Errico, J. V. (1983). *Railroad track substructure-materials evaluation and stabilization practices.* Final Report by Goldberg-Zoino and Associates for U.S. Department of Transportation, Federal Railroad Administration, Report No. FRA/ORD-83/04.1, June.

11.5 Rose, J. G. (1989). *Use of HMA in railroad trackbeds increasing.* Hot Mix Asphalt Technology, National Asphalt Pavement Association, Riverdale, Maryland, U.S.A., pp. 17-20, Winter.

11.6 NAPA. *Simplified design guide for hot mix asphalt (HMA) railroad trackbeds*. National Asphalt Pavement Association Report IS - 99.

11.7 ASTM D698. **Standard test methods for moisture-density relations of soils and soil-aggregate mixtures using 5.5-lb (2.49-kg) rammer and 12-in. (305-mm) drop.** Annual Book of ASTM Standards, Section 4, Construction, Vol. 04.08, *Soil and Rock*.

12. Geosynthetic Applications in Track

12.1. Types of Geosynthetics

Applications of geosynthetics represent a rapidly growing field of geotechnical engineering. Geosynthetics are products manufactured from plastics which are used in conjunction with soils and aggregates in constructed projects for purposes such as earth retention, drainage, and seepage control. Geosynthetics are desirable when they provide a more economical alternative for fulfilling the required function.

Consider, for example, the drainage of subsurface water. A granular filter is required which permits water to escape without loss of the soil from which the water is being drained. Meeting the gradation requirement illustrated in Fig. 9.5 with a natural material may be difficult, and with a crushed aggregate may be expensive. In addition construction labor costs may be high. In some cases a composite design incorporating a geosynthetic filter may be more economical.

This chapter will consider the geosynthetic applications in the track structure. Therefore, earth retaining structures and erosion control systems which use geosynthetics, although they are important, will not be included.

The types of geosynthetics and their functions are listed in Table 12.1. The five functions are illustrated in Figs. 12.1 through 12.5.

Table 12.1 Types of geosynthetics with functions

Type	Functions
GEOTEXTILES Woven Non Woven	Separation Filtration Transmission Reinforcement
GEOMEMBRANES	Isolation Separation Reinforcement
GEOGRIDS	Reinforcement
GEONETS	Transmission
GEOWEBS	Reinforcement
GEOCOMPOSITES	Combinations

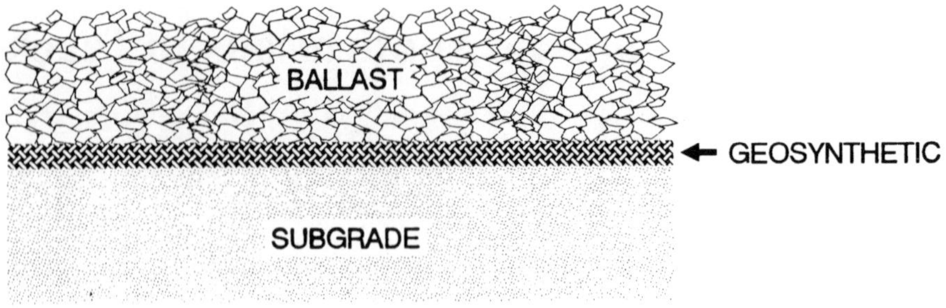

Fig. 12.1 Separation: prevents soil particles from intermixing

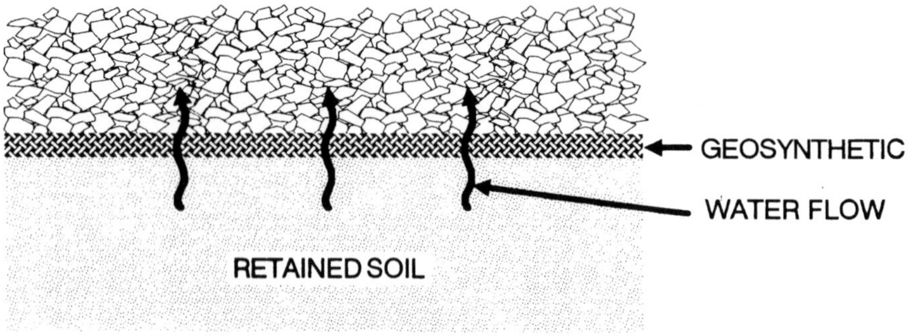

Fig. 12.2 Filtration: retains soil while passing water

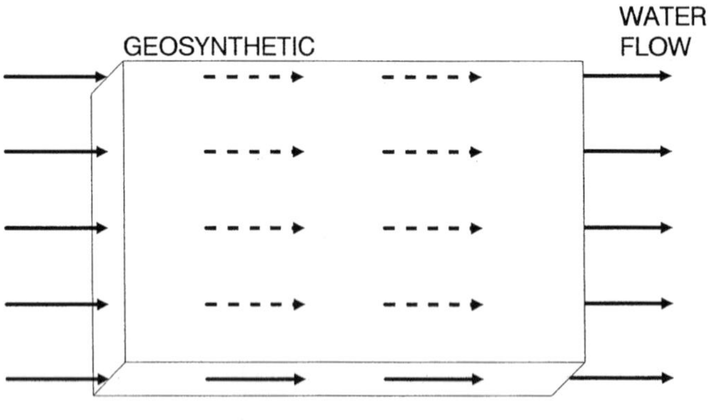

Fig.12.3 Transmission: transmits water in the plane of the geosynthetic

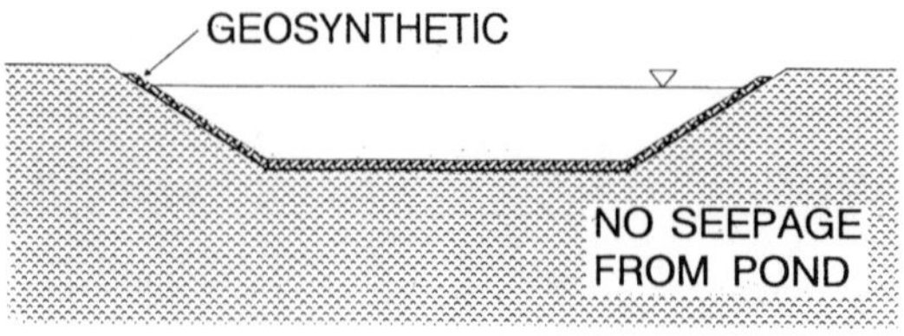

Fig. 12.4 Isolation: prevents fluid from passing from one side of geosynthetic to the other

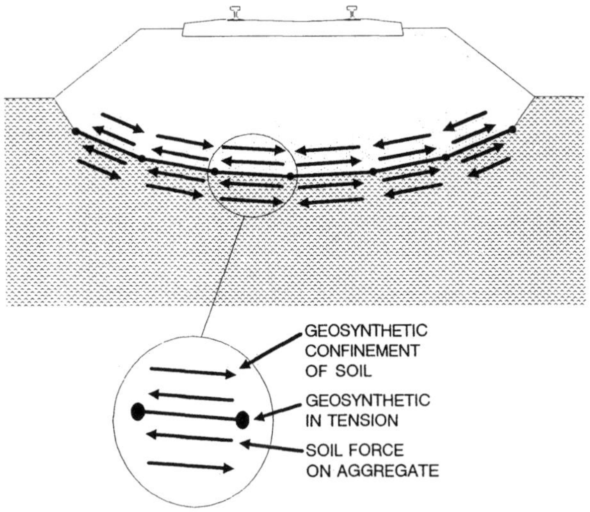

Fig. 12.5 Reinforcement: resists soil extension strain with confinement provided by tensile strength of geosynthetic

Geotextiles, also known as filter fabrics, are permeable textiles in either woven or non woven form. They have four potential functions: separation, filtration, transmission and reinforcement. When two adjacent layers of soil or aggregate have particle sizes which do not satisfy the separation criterion of Eq. 9.1, the soil with smaller particles will enter the voids of the soil with the larger particles. A porous geotextile with small enough opening sizes can keep these two soils separate. If the fabric has holes that are not too small, and if excess water pressure exists in the soil causing seepage, then the fabric will allow water to pass through while the soil particles will be retained. This is the filtration function. If the fabric is a thick enough non-woven type, then water can also flow or move by capillary action in the plane of the geotextile. This is the transmission function. Because the fabric has tensile strength, in situations where the soil strains are extensional in the plane of the fabric, then the fabric will provide some tensile reinforcement.

Geomembranes are plastic sheets without holes. They may have tensile reinforcement and also a rough surface texture for creating shearing resistance with the soil. In contrast to geotextiles, geomembranes are impermeable membranes which prevent fluids from passing from the soil on one side to the soil on the other. This is the isolation function, which is the primary function of geomembranes. Of course separation of the soil particles will also occur. Because the membrane has tensile strength, some reinforcement may also occur. However reinforcement is generally not a required function of geomembranes.

Geogrids are plastic sheets in the form of a grid with the strands stretched to align the long-chain polymer molecules for high strength and stiffness. The grids interlock with the soil to create tensile reinforcement when the soil strains are extensional in the plane of the grid.

Geonets are flat flow channels usually consisting of a hollow plastic core wrapped with a geotextile. Water will pass from the soil through the fabric, and then flow along the hollow core. Although the geotextile must serve a separation/filtration function, its purpose is to help the geonet perform its function as a transmission drain. The geogrid must be strong enough to maintain its flow channel under pressure from the soil.

Geowebs are plastic compartments with open top and bottom (Fig. 12.6). The geoweb is placed on the ground and filled with granular soil. Through their tensile strength the sides of the compartments provide lateral confinement to the soil, forming a load bearing reinforced soil layer.

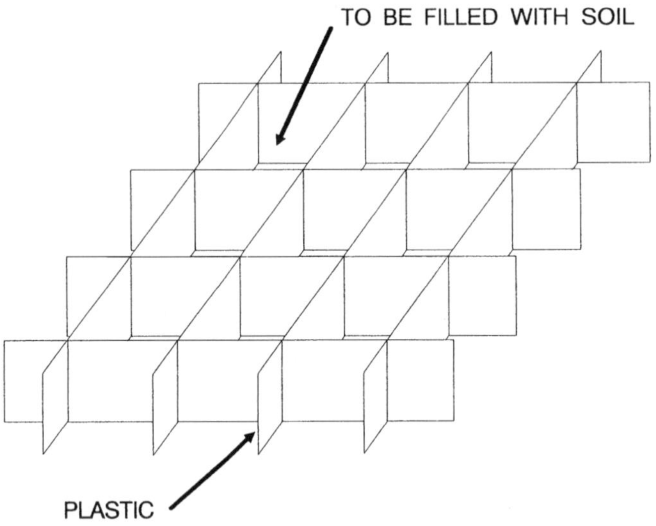

Fig. 12.6 Illustration of Geoweb

Geocomposites are combinations of other geosynthetics. An example is a geogrid bonded to a geotextile to provide better reinforcement capability than with the geotextile alone.

12.2. Geotextile Characteristics

12.2.1. Manufacturing Process

Geotextiles are manufactured from extruded fibers which generally are composed of one or more of the following plastics (Ref. 12.1): nylons (polyamides), olefins (polypropylene, polyethylene) and polyesters.

The actual fabrication techniques of geotextiles are based on traditional textile industry methods. The resulting geotextiles are generally classified into three categories: wovens, nonwovens, and knitted.

The manufacturing process usually includes two steps with an optional third. The first step consists of producing the fibers and yarns. The second step consists of combining these linear elements into fabrics. The optional third step involves surface treating of the fabrics in order to add chemical and structural stability (Ref. 12.2).

One way fibers are created is by extruding the melted polymer through dies. The fibers may be drawn to re-orient their molecules longitudinally for increased strength and stiffness. The fibers may also be combined into multiple fiber strands, or they may be crimped and cut. Another way to create the fibers is to extrude thin polymer film which is then slit into linear elements (Ref. 12.2).

The first synthetic fabrics were manufactured in the 1950's by weaving on a loom and then bonding the woven fibers to maintain their positions. Nonwoven fabrics are a more recent development, but these are now the most common for railway applications. Nonwoven fabrics are generally produced by randomly arranging the monofilament fibers as they are extruded from the dies to form a mat of the desired thickness. The mats may also be formed by overlaying cut monofilament fibers. Either way the fibers are then bonded at their contact points by heat, by adhesives, or by needle punching.

The characteristics of geotextiles are defined by the plastic polymer properties together with the manufactured fabric or mat properties.

12.2.2. Properties of Polymer Material

The plastic polymers consist of linked organic molecules which form long chains. The chains are combined into fibers for manufacturing the fabrics. The resulting material is not biodegradable and will not corrode like metals. However they are subject to forms of chemical attack and may be weakened by water absorption. They will also degrade in ultraviolet light. An ASTM test for ultraviolet effects is given in Ref. 12.3. This is a problem only when the materials are exposed in storage and handling prior to installation. Once buried the plastics are protected from ultraviolet light. Only exposed edges are subsequently vulnerable.

Plastics are viscoelastic which means that the value of modulus of elasticity is dependent upon the strain rate. Examples of this characteristic are creep under constant stress and stress relaxation under constant strain. If a high enough tensile stress is held for a long enough time then cracking and creep rupture will occur. This is an important consideration for localized tension situations such as can occur when fabric is in contact with ballast particles, and for general tension situations as when fabric stretch occurs from plastic deformation of the subgrade.

The chemical make up of the polymer molecules will have a large influence on the above plastic characteristics. Thus consideration must be given to the proper selection for each application.

12.2.3. Properties of Fabric

There are three categories of fabric properties which affect performance of geotextiles in track: 1) hydraulic, 2) separation, and 3) mechanical.

12.2.3.1 Hydraulic

Because geotextiles are porous they permit water to pass through from one side to the other. The ease with which water can flow defines the permeability property. An ASTM test for permeability is given in Ref. 12.4. Water can also flow in the plane of the fabric. This ability defines the transmissivity property. An ASTM test for transmissivity is given in Ref. 12.5.

Both of these hydraulic properties decrease significantly from those for a clean geotextile to those after the fabric has been in service while embedded in the track (Refs. 12.6, 12.7, 12.8). The reason for this loss is the clogging tendency from fine soil particles moving into the fabric. This is a most important effect to evaluate because adequate hydraulic properties are essential for most of the intended applications of geotextiles. Presently standard tests for clogging evaluation are still being developed.

Thick, nonwoven fabrics are far superior in their transmissivity characteristics to woven fabrics. However, in addition to clogging effects, squeezing or compressing the fabric by the soil pressure will decrease the fabric thickness, and hence the in-plane transmissivity.

12.2.3.2 Separation

Fabric openings are small in size and so can maintain separation between particles on the opposite sides. The ability to perform this function is defined by the fabric opening size in relation to the soil particle size, and also by the mechanisms attempting to cause soil particle movement. These mechanisms include: 1) water seepage forces, 2) water pressure from repeated load, and 3) soil skeleton pressure from repeated load. Many tests for separation ability of geotextiles have been for conditions of steady seepage. However some work has been done under pumping conditions which represent the latter two effects (Refs. 12.9 through 12.15). The results of this work will be reported later in this chapter.

The apparent opening size (AOS), also known as the equivalent opening size (EOS) of geotextiles is considered to be the property of most importance when defining the separation capability of a geotextile. The AOS of a geotextile indicates approximately the largest particle that would pass though the geotextile.

The ASTM test method (Ref. 12.16) to evaluate the AOS of a geotextile uses known sized glass beads of designated number and determines by sieving (using successively finer fractions) that size of beads for which 5% or less pass through the fabric in a given time.

The designated bead size is the size of the larger sieve of the sieve pair used to define the bead size. This means that beads designated No. 40 are beads which pass a No. 35 sieve (0.5 mm) but are retained on the No. 40 (0.425 mm) sieve. When the fraction passing is 5% or less of the sieved material, that bead size, which in turn corresponds to a U.S. standard sieve size, is designated to be the AOS of the geotextile.

12.2.3.3 Mechanical

The mechanical properties of the geotextiles which affect their ability to fulfill functions listed in Table 12.1 are:

1) tensile strength (Refs. 12.17, 12.18, 12.19),

2) tensile stiffness or modulus (Refs. 12.17, 12.18),

3) elongation at failure (Ref. 12.17),

4) puncture strength (Ref. 12.20),

5) tear strength (Ref. 12.21),

6) abrasion resistance, and

7) creep resistance.

The effect of temperature on these properties may be important as well, since the geotextiles can certainly be subjected to freezing temperatures in addition to much higher ground temperatures.

The main purpose of satisfactory mechanical requirements is to ensure that the geotextiles will withstand the rigours of handling and installation, and the forces encountered in service. The forces in service include puncturing by ballast particles, stretching to conform to the irregularities of the surface upon which the geotextile is placed, and stretching to resist soil movement. In cases where reinforcement is a required function of the geotextile, the tensile and creep properties are particularly important. Consideration must be given to the degradation in these properties over time from causes indicated in Section 12.2.2.

12.3. Geotextiles in the Role of Subballast

12.3.1. Effectiveness with Subballast Functions

One of the primary applications of geotextiles in track has been to fulfill some of the functions of subballast. In this capacity geotextiles have been used either in place of or to assist subballast. To evaluate this role it is necessary to re-examine the functions of subballast given in Chapter 9.

The first function listed is reducing stress to the subgrade. This requires a tensile reinforcement function of the geotextile. To do this the geotextile is placed at the bottom of the ballast to resist the incremental tensile stresses caused by the passing wheel loads (Fig. 6.9) by providing lateral confinement. In this capacity the geotextile would substitute for the residual stress development in the ballast. This role is particularly important on soft subgrades where stable residual stresses are hard to develop. The effect of this reinforcement would be to stiffen the ballast layer (reduce flexing) thereby spreading the vertical stress from the base of the sleepers more widely which reduces the vertical stress transmitted to the subgrade.

However the geosynthetic has limitations in fulfilling this role. First, it requires a lot of stretch to develop sufficient tensile reinforcement. Accompanying this stretch will certainly be significant track settlement and ballast depressions which are undesirable. Second, it has creep characteristics which will tend to relieve the reinforcement. Third, mobilization of tension in the fabric may be limited by slippage between the ballast and the fabric. Fourth, stretching the fabric may adversely affect its filter/separation function by increasing the opening size. Thus fulfilling the first subballast function should not be expected from the geotextile. A better approach would be to use a geogrid which is much more effective at tensile reinforcement.

The second subballast function is subgrade freezing protection. Geotextiles probably do not contribute to this task.

The third and fourth subballast function may be considered together. These are separation functions. This is the principal role expected of geotextiles. The geotextile opening size requirement is different from the gradation requirement of subballast. For the geotextile to work effectively as a separator the AOS must be less than the largest soil particles being restrained. With the largest particles blocked by the fabric, successively smaller particles will be blocked within the soil or aggregate by the larger particles and so a filter zone will be established adjacent to the geotextile.

The fifth subballast function is preventing subgrade attrition by ballast. Thick geotextiles can provide some cushioning. However the ballast will abrade the geotextile, particularly on hard subgrade, so any cushioning benefit will most likely be temporary. When the abrasion product is clay, and perhaps silt, the slurry will pass through the geotextile with at best a slower rate than with no geotextile.

The sixth subballast function is to shed water from above. The geotextile will not meet this requirement.

The seventh subballast function is to drain water from below including any excess pore pressure in the subgrade caused by the cyclic load. This is the permeability requirement. The geotextile can perform this function. Significant transmission capability will help here, both in terms of gravity flow and potential wicking (or capillary) action.

12.3.2. Separation/Permeability Requirements

The subballast functions which potentially can be satisfied by geotextiles are those involving separation and filtration. The geotextile must be fine enough to retain the underlying soil, and it must be permeable enough to prevent build up of pore pressure. While performing these functions the geotextile must not become clogged so that the permeability decreases with time to an unacceptable level.

Available specifications for meeting these requirements were reviewed by Byrne (Ref. 12.9). More detailed information is given in Refs. 12.1 and 12.22 to 12.26. None that he found was specifically for the repeated load situation, and none was adequate for all situations. Therefore in the following section examples of observed performance will be given to demonstrate the advantages and limitations of geotextiles for use in the track.

12.4. Observed Performance of Geotextiles

Geotextiles have been used frequently in railroad track substructure maintenance. Typically the applications are for localized mud problem areas such as: 1) wet cuts, 2) soft subgrades, 3) road grade crossings, 4) railroad track crossings , and 5) turnouts. An example is shown in Fig. 12.7. In these situations the cost of the geotextile is relatively small because of the small quantity involved. Thus very often little effort is devoted to assessing the need for the geotextile, and to a follow-up evaluation of its performance. Furthermore adequately documented cases of field performance with geotextiles from which to draw experience are practically nonexistent. However some examples will be given of field and lab investigations to illustrate the effects of geotextiles.

12.4.1. Laboratory Tests

12.4.1.1 Reinforcement

Cylindrical triaxial test specimens were constructed with a top half of ballast and a bottom half of subgrade. Comparisons with and without a geotextile between the ballast and subgrade showed greater strength and less plastic strain accumulation under repeated loading when the fabric was present (Refs. 12.27 and 12.28). However for reasons already discussed, field studies are needed to determine to what extent these benefits can be achieved in track.

12.4.1.2 Separation/filtration

A review of reported laboratory tests (Refs. 12.10 to 12.15) to evaluate the separation ability of geotextiles under repeated load was made by Byrne (Ref. 12.9). Typically these tests were conducted using an arrangement conceptually like that indicated in Fig. 12.8.

The observations may be summarized as follows:

Fig. 12.7 Geotextile being installed at fouled ballast site in a poorly drained cut

1) When the soil under the geotextile was clay, the repeated load caused the clay to pump through the geotextile.

2) Pumping of clay into the aggregate above occurred with all available geotextiles, which included fabrics manufactured and sold for this specific application; however, the rate of pumping varied with the geotextile characteristics.

3) The pumped slurry was formed at the contact points between the aggregate and the clay (through the geotextile).

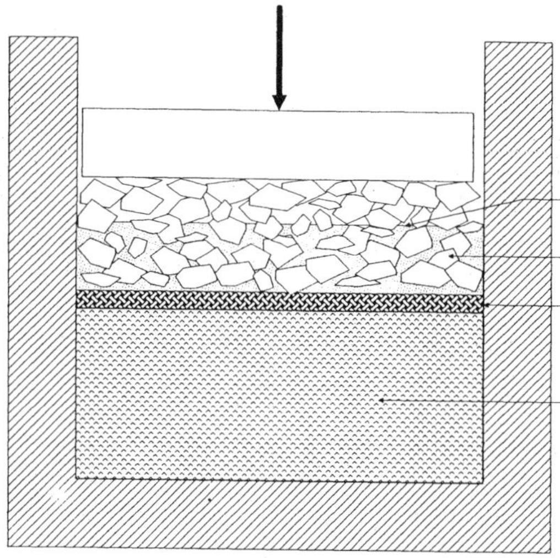

Fig. 12.8 Laboratory simulation of pumping

4) As the particle size of the aggregate decreased the amount of pumping decreased, probably because the contact stress was reduced by increasing the number of contact points per unit area.

5) The thickness of the nonwoven geotextile did not significantly affect the amount of pumping.

6) When the soil had a substantial sand component, the geotextiles acted as an effective separator.

7) A sand layer in place of the geotextile was effective in preventing clay migration into the ballast.

Byrne (Ref. 12.9) conducted some additional tests to evaluate and extend the above observations. His soil samples included: 1) a clay, 2) a silt with 10 to 15% fine sand, 3) a silt, and 4) the less than 13 mm (0.5 in.) size portion of a fouled ballast sample. In between the ballast and one of the four soil samples was either: 1) a heavy nonwoven needle punched geotextile with an AOS of about 0.075 mm, 2) a natural sand filter satisfying the gradation requirements for the clay alone, 3) a natural sand filter satisfying the gradation requirements for the clay and the ballast, 4) a natural sand filter covered by the geotextile, or 5) nothing. The gradations of the soils and the approximation for the geotextile are shown in Fig. 12.9.

The observations from these tests were:

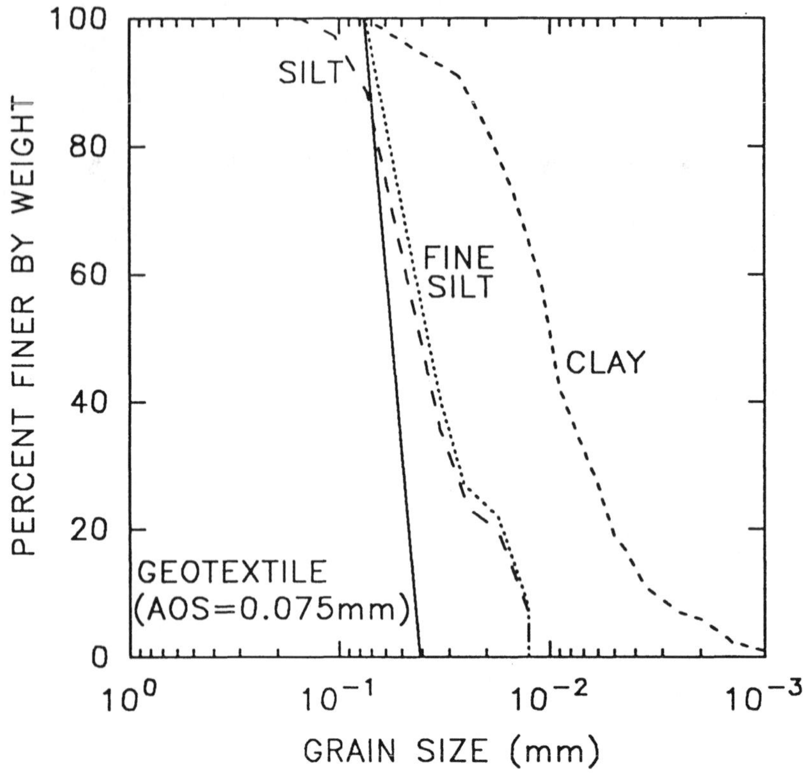

Fig. 12.9 Gradation of subgrade soils compared to geotextile

1) The geotextile alone did not prevent pumping of the clay into the ballast (Fig. 12.10 left), although it did slow the process compared with no geotextile.

2) The slurry formed at the surface of the clay; general softening of the clay layer did not occur.

3) The fine silt pumped through the geotextile into the ballast, but the amount was not as much as for the clay.

4) The silt with about 10-15% fine sand did not significantly pump into the ballast through the geotextile.

5) Decreasing contact stress at the soil surface decreased the amount of pumping of both the silt and the clay for the same ballast gradation.

6) The silt and clay particles near the surface of the fouling portion of the fouled ballast sample migrated through the geotextile into the ballast above. This created a

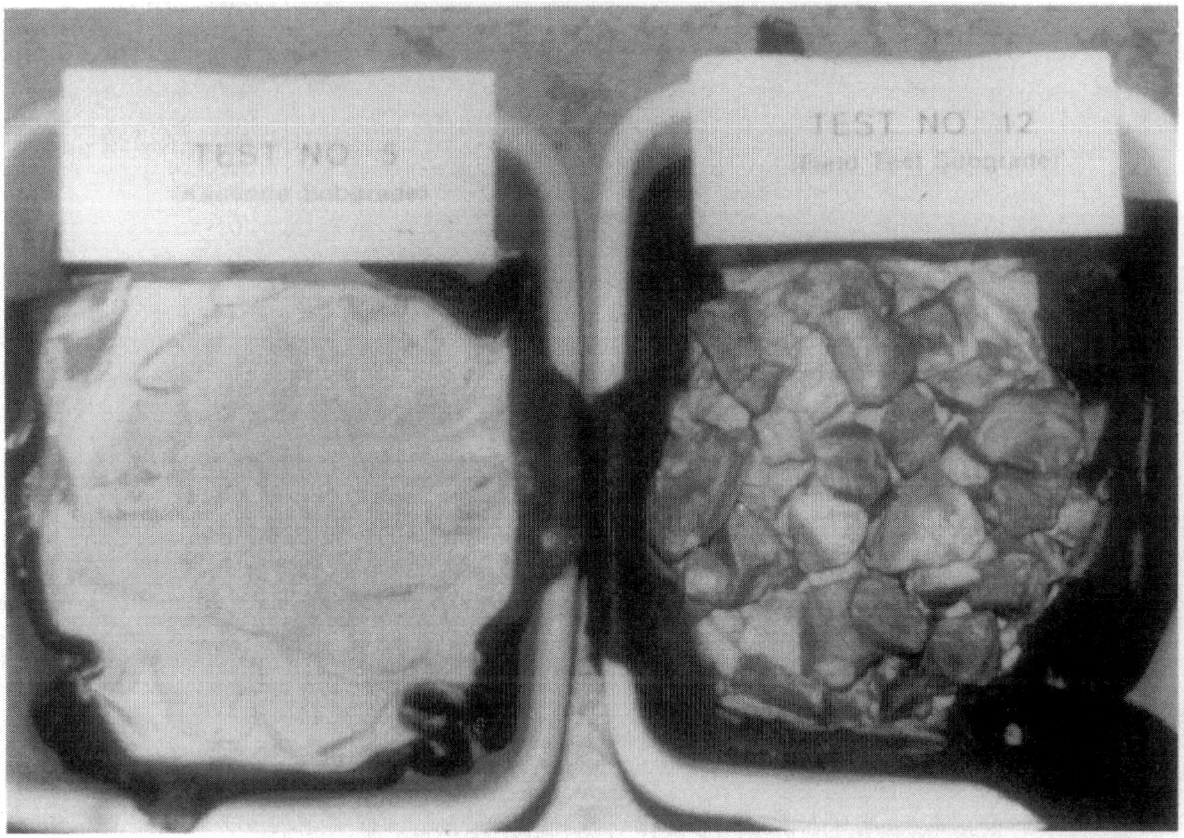

Fig. 12.10 Clay subgrade pumped through geotextile into ballast (left); ballast condition in test with geotextile over portion of fouled ballast (right)

granular zone (sand with gravel) just below the geotextile, forming a filter, after which little further pumping of the fines occurred (Fig. 12.10 right).

7) The sand filter which matched both the clay and the ballast prevented the pumping of clay into the sand and also prevented the sand filter from migrating into the ballast (Fig. 12.11).

8) The sand filter matching only the clay prevented the clay from pumping into the sand, but did not prevent the sand filter from migrating into the ballast.

9) The geotextile on top of the sand filter prevented the sand from migrating into the ballast.

10) The geotextile criterion for separation which appears to be consistent with available laboratory test data is that the AOS (mm) be less than D_{85} of the protected soil.

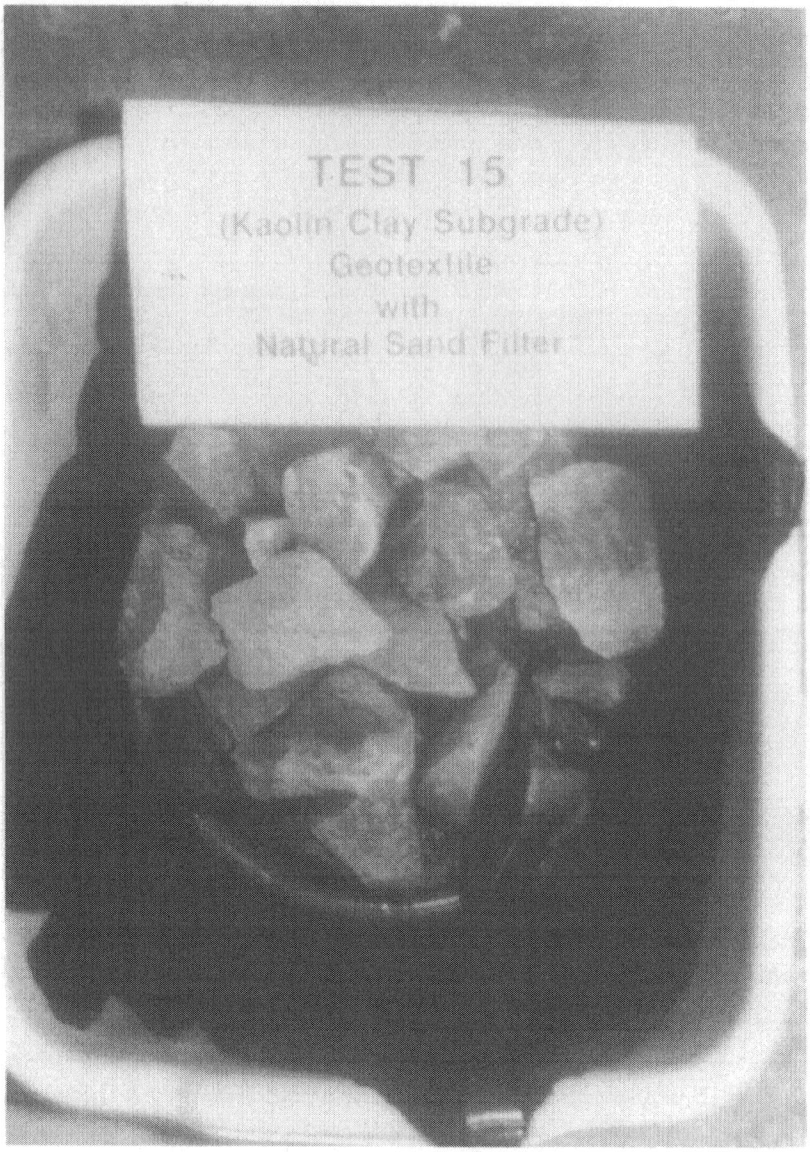

Fig. 12. 11 Condition of ballast after test with filter layer of sand and geotextile over clay subgrade

12.4.2. Field Studies

Raymond (Ref. 12.29) examined field sites in each of which a section of geotextile was placed adjacent to a control section with no geotextile. At one site a layer of sand was placed both above and below the geotextile. This site showed improved performance compared to the adjacent control section in which the ballast penetrated the subgrade. However the sand acted as the separator for the site with the geotextile, rather than the geotextile, which was in poor condition. At other sites the geotextile significantly reduced the quantity of subgrade fines pumped into the ballast. However these subgrades were primarily coarse-grained with 90% of the particles greater than the fabric AOS.

Haliburton (Ref. 12.22) warned of the possibility that clogging of the fabric might lead to loss in subgrade strength. This prediction was proved to be correct and was believed to be the cause of a train derailment in a track geotextile test section. In this test section a low permittivity geotextile was installed and the permittivity became lowered further by clogging over a period of 17 months in service. This caused an accumulation of water and excess pore pressure, which eventually led to the subgrade soil failure (Ref. 12.8).

Zimmerman (Ref. 12.7) tested samples of geotextile that had been in track for up to 48 months after installation. The results showed:

1) The permeability decreased by an average of 92% in 15 months, and then increased due to punctures.

2) The transmissivity decreased by an average of 85% in 15 months.

3) The tensile strength decreased by an average of 54% in 48 months.

4) The elongation at failure decreased 48% in 48 months.

5) The burst and puncture strengths decreased in 48 months.

Fig. 12.12 Mud generated from subgrade pumping at a site with geotextile

Williams et al. (Ref. 12.6) reported on a site in a poorly drained cut in which the fouled top ballast had been undercut and removed. Then the exposed fouled ballast surface was covered with various geotextiles followed by a layer of clean ballast. The geotextile properties were measured in the laboratory on samples of the new geotextile as well as after 1, 2, 6 and 12 months in track. Within three months of the installation a large decrease occurred in tensile strength, puncture resistance, transmissibility and permeability. Subsequent decreases were small.

Selig (Ref. 12.30) examined a site in which subgrade clay pumping was observed, even though a geotextile had been installed to prevent fouling of the ballast. The site was poorly drained but with water standing at the level of the sleeper bottom. The subgrade was a clay rich mudstone rock. The upper portion of fouled ballast was removed, the exposed surface covered with a geotextile, and clean ballast added. Slurry continued to develop and pump through the geotextile into the clean ballast. Within 2 years with about 15 MGT (133 GN) traffic per year, the slurry reached the surface and formed mud boils (Fig. 12.12). A core sample taken with the boring machine (Chapter 14) shows the geotextile between layers of fouled ballast (Fig. 12.13).

Fig. 12.13 Core sample showing geotextile

Fig. 12.14 Ballast which is separated from clay subgrade by sand layer

Fig. 12. 15 Removed samples of geogrid (left) and geotextile

Fig. 12.16 Slurry coming up through geomembrane and geotextile

Fig. 12.17 Removed sample of geotextile (left), geomembrane (right) and geogrid (bottom)

Selig (Ref. 12.31) examined a British Rail site in which serious subgrade pumping had been observed in the past. To evaluate alternative means to correct the problem, three years prior to the site visit the ballast had been removed by British Rail to a depth of 550 mm below the base of sleeper. The exposed clay subgrade was covered with various combinations of sand layers, geomembrane, geogrid and geotextile. Clean ballast was then added to bring the depth from sleeper to subgrade back to 550 mm.

At one location 50 mm of sand was deposited on the subgrade, on top of which was placed a geotextile and a geogrid. This was covered by 500 mm of ballast. After three years the ballast was still very clean (Fig. 12.14). Most of the 50 mm thickness of sand was in its original condition; only the bottom of the sand layer was darkened by clay. Samples of the geogrid and the geotextile cut from the inspection hole after three years are shown in Fig. 12.15. Both were in good condition.

At another location no sand was used. Instead a geotextile was placed, followed by an impermeable polyethylene sheet, on top of which a geogrid was placed. This composite was covered with 550 mm of clean ballast. After three years the ballast was filled with mud to

Fig. 12.18 Subgrade mud coming up through geotextile layers

above the sleeper bottom. A slurry of clay and water was present at the bottom of the ballast layer. The observation hole was dug down to the geosynthetic materials and left open. When trains passed, slurry squeezed up through holes in the geomembrane (Fig. 12.16). The slurry was most likely created by the action of the ballast on the clay surface in the presence of water trapped beneath the geomembrane. The trapped water was probably responsible for making this the worst case. Samples of the three geosynthetic materials removed from the hole are shown in Fig. 12.17. Note the holes in the polyethylene geomembrane.

At a third location a layer of woven fabric over a layer of nonwoven fabric was placed on the clay with no sand layer. These were covered with 550 mm of clean ballast. After three years the fabric appeared to be in good condition. However the ballast was still fouled with mud for most of its depth between the clay surface and the sleeper bottom (Fig. 12.18).

12.5. Geotextile In-Track Application Considerations

12.5.1. Functions

The two functions of geotextiles in track that are most likely to be of interest are:

1) Maintain separation of substructure soil and/or aggregate layers of such different particle sizes and gradations that they would otherwise intermix under repeated loading.

2) Permit seepage of water out of the track substructure layers while retaining the soil particles.

This second application requires that the geotextiles have adequate permeability and transmissibility to help drain away the excess water, while being fine enough to prevent loss of the soil particles. If the layer on the opposite side of the water source is permeable enough, then the water may only have to pass through the geotextile to escape. In this case geotextile transmissibility is not important. However if the opposite layer is not permeable enough, then lateral water flow is necessary. In this case transmissibility is important. While the transmissibility property has been demonstrated in the laboratory, field evidence of its potential still needs to be demonstrated.

The need for the filtration/transmissibility capability comes mainly from two sources: 1) excess pore pressure developed below the geotextile by repeated loading of the soil in a saturated state, and 2) upward seepage of ground water through the subgrade. The downward seepage of water into the layer above the geotextile should be shed or drained by the layer itself. However, it is conceivable that an underlying geotextile may help in some such cases through its transmissibility capability.

If reinforcement is required or desired, then geogrids would be the appropriate geosynthetic to consider.

Drainage of water away from the track is another function for geotextiles. This application will be considered in Chapter 13.

12.5.2. Alternatives

At least five different substructure interlayer alternatives involving geotextiles need to be evaluated. Each will be briefly described and assessed:

1) Between ballast and subgrade as a subballast alternative (Fig. 12.19a).

As a minimum for this alternative to work, the granular layer thickness requirement would have to be provided by the ballast alone, the geotextile AOS would have to be appropriate for the subgrade, and good drainage through the ballast must exist. This still leaves the problem of avoiding subgrade attrition by ballast which may require a blanket layer of sand beneath the geotextile. This is alternative number 3. This first alternative geotextile application is likely to be satisfactory only when the subgrade has a substantial sand component. With a gravel component as well, the subgrade may not need a filter/separation layer.

Protection of the geotextile from abrasion and puncturing by the ballast may require a subballast-type protection layer on top of the geotextile. This situation becomes alternative number 2.

2) Between subballast and subgrade (Fig. 12.19b).

This alternative would be meaningful if the subballast were too coarse to meet the gradation requirement for subgrade protection while the geotextile had a fine enough opening size (AOS). This alternative might also be useful if the subballast had a low permeability while the geotextile provided good lateral transmission of water seeping from the subgrade. However the geotextile would need to have high resistance to clogging over a long period of time.

3) Between the ballast and subballast (Fig. 12.19c).

For this alternative the subballast must satisfy the gradation requirement for protecting the subgrade. The use of a geotextile on top of the subballast would substitute for the gradation matching requirement between the subballast and the ballast. Thus a geotextile would normally be required if the subballast were only sand, such as the lower subballast (fine filter) layer in Fig. 9.4.

If protection of the geotextile from the ballast requires a covering layer of subballast, this situation becomes alternative number 5.

4) Between replacement ballast and the surface exposed by undercutting (Fig. 12.19d).

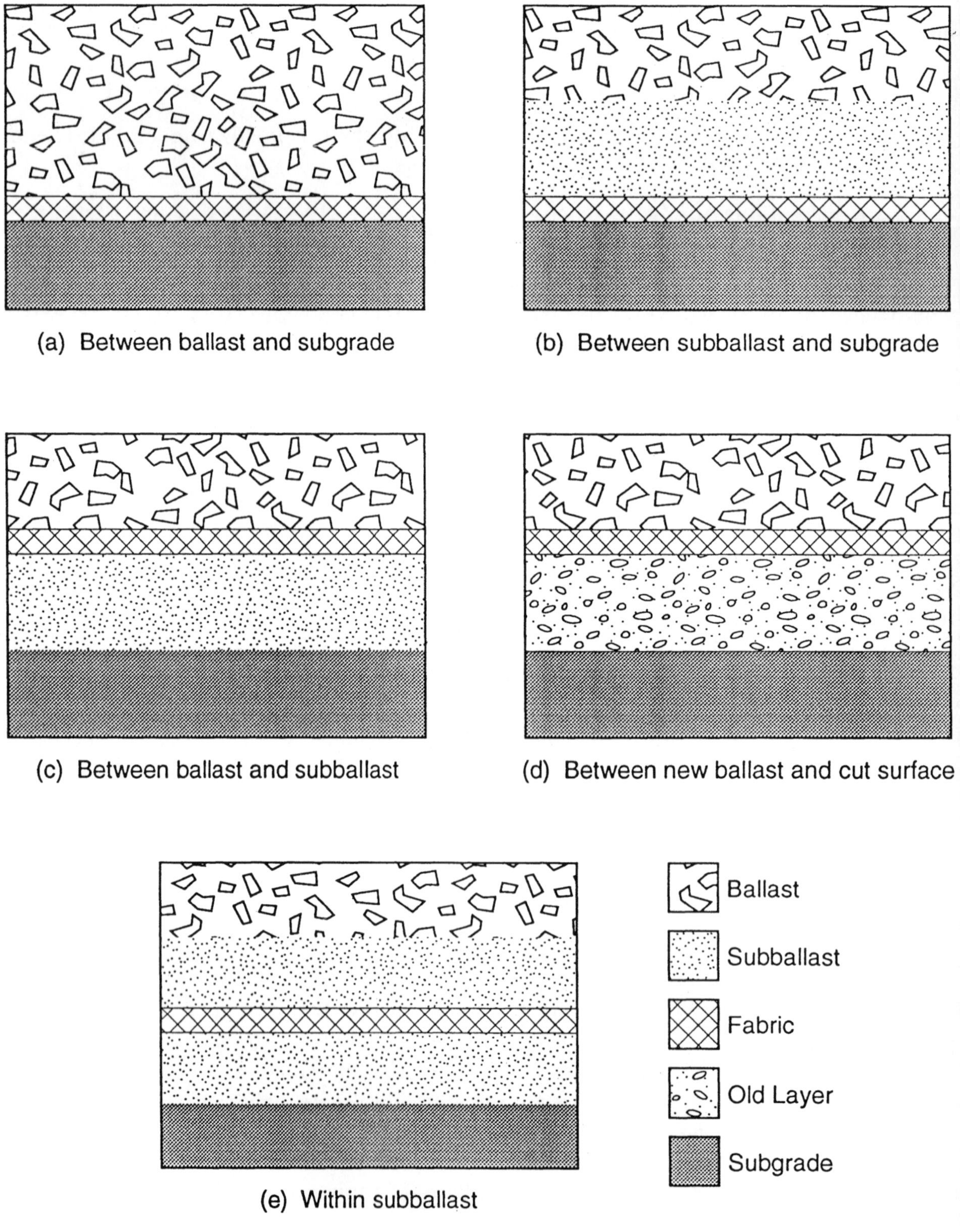

Fig. 12.19 Geotextile alternatives in track substructure

If the surface exposed by undercutting is either subballast or subgrade, then this alternative becomes the same as either number 3 or number 1, respectively. Often, however, the exposed surface is fouled ballast and this can be gap graded. A geotextile can help prevent the fouling material from moving up into the clean ballast. In this case the AOS of the geotextile should be selected to match the underlying fouling component gradation, ignoring the ballast-size particles in the fouled ballast.

Protection of the geotextile by a covering layer of subballast material would make the situation similar to that for alternative number 5 (Fig. 12.19e).

5) Creating a subballast layer by encapsulating a geotextile within the subballast (Fig. 12.19e).

The subballast in this case would have to satisfy the separation requirements of both the subgrade and the ballast and be at least 50 mm thick on each side of the geotextile. If this is done, then the geotextile seems to have no functions.

For all of the alternatives using a geotextile, the geotextile should be at least 300 mm (12 in.) below the bottom of the sleeper to provide reasonable protection against tamping damage and to avoid interference with future ballast undercutting/cleaning operations.

A common application for geotextiles has been a part of the cure for muddy ballast conditions. The muddy ballast is removed down to some level, and the surface covered with a geotextile before replacing the ballast. This is often representative of alternative number 4, but it could be number 1 or 3. The assumption made is that the mud is coming from the subgrade. This solution is often unsuccessful as a cure for one of two reasons: 1) the mud is not coming from the subgrade, or 2) if the mud is coming from the subgrade it is too fine grained to be stopped by the geotextile.

The lesson to be learned from past experience is that the decision to use geotextiles should be based on a proper site investigation, that involves the following steps:

1) Determine the cause of the problem by a field examination of substructure conditions together with necessary laboratory tests and analysis.

2) Evaluate possible roles of geotextiles as well as alternatives not using geotextiles.

3) Do an economic analysis of the alternatives.

4) Conduct a field trial of the best alternatives, if prior experience can not be used as a guide.

5) Monitor performance for a sufficient period of time to assess the results.

6) Document the results of the investigation for future reference.

12.5.3. Requirements

Both woven and non woven geotextiles are available. For the application alternatives indicated, non woven fabrics appear to be the only acceptable choice. These need to possess the following durability characteristics:

1) High tensile strength.

2) High puncture resistance.

3) Large elongation without rupture.

4) High abrasion resistance.

5) Resistance to chemical degradation.

6) Treatment to prevent ultraviolet degradation.

To provide the separation/filtration/drainage functions, the geotextiles require:

1) AOS less than the D_{85} of the soil being protected.

2) Thick mat for high transmissivity.

3) Evidence of acceptable clogging resistance.

Adjustment must be made for gap graded soils for determining the effective D_{85} to use to obtain the AOS. Consider also that non woven fabrics may not be available with a small enough AOS to handle silts and especially clays.

As an example, suggested values for some of the above properties are given in Ref. 12.25. For other properties tests are still being established. Much more experience is needed to determine the magnitude and proper combination of the property values.

Care must be taken in geotextile installation to avoid damage which will impair proper functioning. As a start a smooth surface should be prepared on which to lay the geotextile. This surface should be sloped for drainage. Ruts, holes, and sharp objects should be eliminated as much as practical. Exposure time to ultraviolet radiation should be as short as possible. Construction equipment should not drive over the geotextile, and care must be exercised in the tasks of replacing the track structure, reballasting and tamping to avoid damaging the newly placed geotextile.

Double layers of geotextile should be avoided because of the possibility that slip planes can develop between the layers.

Finally, an adequate drainage system must be established to carry water away from the track. Guidelines for this are given in Chapter 13.

The only way to ensure that the separation/filtration function of subballast is permanently achieved is to use a properly graded layer of durable granular material. Geotextiles, however, are often easier to select and install and so their use may result in a lower initial cost. On the other hand geotextiles have a limited duration of improvement. Thus, geotextiles may have to be replaced periodically, a difficult process which may offset the benefit of their ease of installation. Further, geotextiles cannot fulfill all of the functions of subballast. Careful thought is required to determine the most economical solution.

12.6. Geomembranes

12.6.1. Characteristics and Manufacturing

Geomembranes are impermeable, flexible plastic sheets which are manufactured from neoprene, polyvinyl chloride, chlorinated polyethylene, or other plastics. If high strength is required, the membrane typically is bonded to a nylon or polyester mesh. Membranes suitable for use below track are available in thicknesses ranging from about 0.15 mm (0.006 in.) to at least 3.0 mm (0.12 in.).

Geomembranes are produced through processes such as direct extrusion, blown film or spread coating. These methods produce a sheet of geomembrane varying up to widths of at least 10 m and lengths up to 500 m. Geomembranes often undergo further plant fabrication into wider or specially shaped panels prior to delivery to field sites. Recognizing that the function of geomembranes is isolation, the seaming process which joins geomembrane sheets, either at the site or during fabrication, is considered critical. Most geomembrane leaks occur in seams. Accordingly, quality assurance during placement and seaming is typically considered to be the most critical phase of the installation.

Some geomembranes are manufactured in situ by spraying asphalt, bitumen, or a molten polymer directly onto a carrier surface (typically a non-woven geotextile). This technique has the advantage of no seams; however, the quality control of thickness and temperature is sometimes difficult, especially on slopes.

12.6.2. Applications

One application of geomembranes is to prevent infiltration of water from the surface into the subgrade. The membrane is placed at the subgrade surface and is protected from puncturing by installing it between two sand blankets of at least 50 to 100 mm (2 to 4 in.) thickness each. The membrane must not come into contact with ballast particles. Provision must be made to drain water reaching the membrane surface from above.

A second application is to block moisture coming from the soil below. This may be desirable to prevent loss of water in high swelling soils which change volume significantly with change in moisture content. The first application given above prevents increase in water content.

A third application is to interrupt the capillary rise of ground water. This capillary rise is necessary to form ice lens (Chapter 9). In this case the geomembrane must be placed at a depth that is above the groundwater level yet is below the maximum depth of freezing. This application usually is practicable only when an embankment of frost-susceptible soil is constructed over an area of shallow groundwater. Protection of the membrane against puncturing is still required.

A fourth application is to intercept ground water flow or prevent lowering of the ground water table by drainage of track in a cut where the ballast and subballast extend below the water table. Such a need existed in the city of Boston (U.S.A.) where lowering of the water table would have damaged surrounding structures by causing soil consolidation beneath the buildings. The design (Ref. 12.32) employed is illustrated in Fig. 12.20 for a situation in which the ground water table is above the base of the sleepers. Excavation was made for the ballast and subballast, and for an internal drain. The excavation was lined with a geomembrane having a geotextile on both sides for protection of the membrane and for drainage. The internal geotextile assisted in conveying water to the drain through lateral

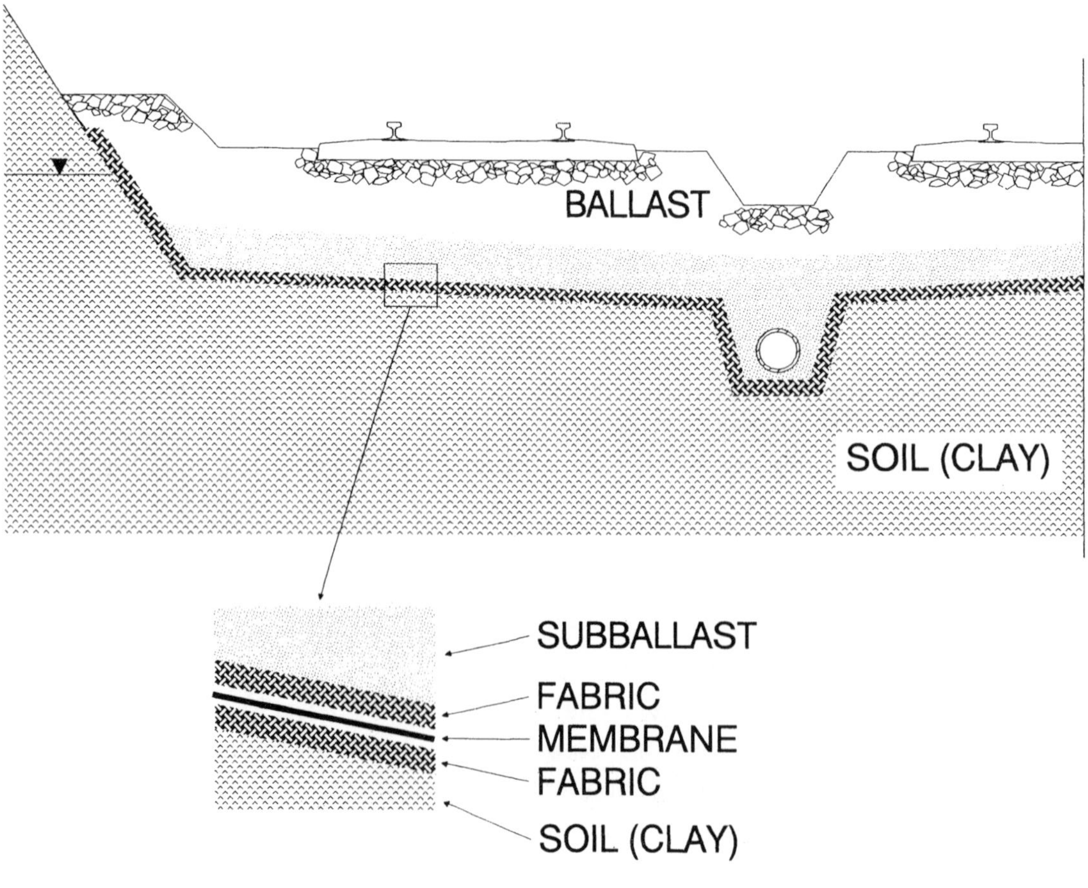

Fig. 12.20 Application of geomembrane with geotextile in track

transmission. The external geotextile was relied upon to dissipate any excess pore pressure (above the ground water pressure) which developed in the clay beneath the geomembrane.

One severe problem will occur if excess water is trapped beneath the membrane. This could result from repeated loading of saturated soils or upward flow of ground water. Unless a drainage layer is provided beneath the membrane, this water will create a slipping surface at soil-membrane interface, and cause local failure of the subgrade because of soil shear strength reduction. This problem was observed with clogged geosynthetics in Section 12.4.2.

Another problem can occur when leaking seams or holes develop in the membrane. If excess water reaches these locations, concentrated hydraulic action through the openings can create local soil erosion. This can be more severe than if there were no membrane. Such a situation was observed with the punctured membrane in Section 12.4.1. The effects of placing geomembranes may be hard to predict accurately. Therefore full-scale field tests should be considered to evaluate the benefits and determine any problems.

12.6.3. Requirements

The main requirements of geomembranes for the described applications are the following:

1) Resistance to chemical and biological degradation.

2) Ability to bend and stretch without damage.

3) Resistance to puncturing.

4) Protection against ultraviolet radiation.

5) Ability to make leak proof seams in the field.

6) Ability to proof test field seams.

7) Acceptable effect of temperature on properties.

8) Acceptable interface friction/adhesion characteristics.

Although the geomembranes have tensile strength and will therefore resist soil elongation in the plane of the membrane, this is not a function to be relied upon. The reason is the generally low shearing resistance (adhesion and friction) between the plastic membrane and the soil. This is especially true in the presence of water. This problem has been readily observed with geomembranes used on pond slopes. Here sliding of the membrane and soil on top has frequently occurred. Membranes are now available with roughened surfaces to increase the interface shearing resistance.

12.7. Bituminous Spray Alternative

An alternative method for producing a relatively impermeable water barrier between layers is application of a bituminous coating by spraying emulsified asphalt.

The spray will not penetrate fine-grained soils and stabilized soils except for fissures and holes. Therefore only a thin water barrier can be achieved. Ballast placed on top of the coating will penetrate and produce the possibility of leakage points. Thus a protective layer of sand placed over the bitumen is advisable. A more effective barrier may be to first place a layer of sand or subballast on the subgrade and then spray the surface with a bitumen having a consistency that will allow substantial penetration of the granular material pores. The spray may also be applied to the exposed surface behind the undercutter as part of ballast cleaning and renewal. When substantial penetration of the bitumen is achieved the barrier is provided by the bitumen filled voids rather than by the surface coating. As long as the layer strains do not disrupt the network of filled voids, the barrier will remain intact. To avoid developing waterpockets above the bitumen coating, the surface must be properly shaped for shedding water before the spray is applied. The adverse effects of trapping water beneath the coating, which were discussed for geomembranes, also apply for bituminous coatings.

The advantage of a bitumen spray over a geomembrane is the lower cost of the spray installation. The material is modest in cost and may be easily applied with readily available equipment. As with the other techniques, a proper investigation is required to determine the suitability of this bituminous coating method.

REFERENCES

12.1 Koerner, R. M. (1986). *Designing with geosynthetics.* Prentice-Hall, Englewood Cliffs, New Jersey.

12.2 Fluet, J. E. (1985). *Geosynthetic products and applications.* Presentation at conference on Geotextiles and Geomembranes in Civil Engineering, Boston, Mass., U.S.A., 18 May.

12.3 ASTM D4355 (1990). **Standard test method for deterioration of geotextiles from exposure to ultraviolet light and water (Senon-Arc type apparatus),** *1990 Annual Book of Standards,* Vol. 4.08, Philadelphia, PA.

12.4 ASTM D4491 (1990). **Standard test method for water permeability of geotextiles by permittivity.** *1990 Annual Book of Standards,* Vol. 4.08, Philadelphia, PA.

12.5 ASTM D4716 (1990). **Standard test method for constant head hydraulic transmissivity (in-plane flow) of geotextiles and geotextile related products.** *1990 Annual Book of Standards,* Vol. 4.08, Philadelphia, PA.

12.6 Williams, N. D., Grubert, P. A., Jang, D. J. and Gallup, R. A. (1988). *Performance of railroad track embankment using geotextiles.* Georgia Institute of Technology Report for Hoechst Fibers, Foss Manufacturing, Exxon Chemicals and Norfolk Southern, January.

12.7 Zimmerman, J. R. (1988). Norfolk-Southern 48-month field and laboratory geotextile tests. (personal communication.).

12.8 Chrismer, S. M., and Richardson, G. (1985). *In-track performance test of geotextiles at Caldwell, Texas.* Association of American Railroads Research Report No. R - 611, December.

12.9 Byrne, Brian J. (1989). *Evaluation of the ability of geotextiles to prevent pumping of fines into ballast.* Master's degree project report, Report No. AAR 89-367P, Department of Civil Engineering, University of Massachusetts, Amherst, Mass., October.

12.10 Salter, R. J. (1982). **An experimental comparison of the filtration characteristics of construction fabrics under dynamic loading.** *Geotechnique*, Vol. 32, No. 4, pp. 392-396.

12.11 Snaith, M. S. and Bell, A. L. (1978). **The filtration behaviour of construction fabrics under conditions of dynamic loading.** *Geotechnique,* Vol. 28, No. 4, p. 466-468.

12.12 Ayres, D. J., and McMorrow, J. C. (1980). **The filtration behaviour of construction fabrics under conditions of dynamic loading.** *Geotechnique*, Vol. 30, No. 1, pp. 87-88.

12.13 Hoare, D. J. (1982). **A laboratory study into pumping clay through geotextiles under dynamic loading.** *Proceedings*, 2nd International Conference of Geotextiles, Las Vegas, pp. 423-428.

12.14 Ayres, D. J. (1986). **Geotextiles or geomembranes in track? British Railways' Experience.** *Geotextiles and Geomembranes*, Vol. 3, pp. 129-142.

12.15 Glynn, D. T. and Cochrane, S. R. (1987). **The behaviour of geotextiles as separating membranes on glacial till subgrades.** *Proceedings*, Geosynthetic 1987 Conference, New Orleans, pp. 26-37.

12.16 ASTM D4751 (1990). **Standard test method for determining the apparent opening size of a geotextile.** *1990 Annual Book of Standards*, Vol. 4.08, Philadelphia, PA.

12.17 ASTM D4632 (1990). **Standard test method for breaking load and elongation of geotextiles (grab method).** *1990 Annual Book of Standards*, Vol. 4.08, Philadelphia, PA.

12.18 ASTM D4595 (1990). **Standard test method for determining the tensile properties of geotextiles by the wide-width strip method.** *1990 Annual Book of Standards,* Vol. 4.03, Philadelphia, PA.

12.19 ASTM D3786 (1990). **Standard test method for hydraulic bursting strength of knitted goods and nonwoven fabrics: Diaphragm bursting strength tester method.** *1990 Annual Book of Standards*, Vol. 7.01, Philadelphia, PA.

12.20 ASTM D4833 (1990). **Standard test method for index puncture resistance of geotextiles, geomembranes, and related products.** *1990 Annual Book of Standards*, Vol. 4.08, Philadelphia, PA.

12.21 ASTM D4533 (1990). **Standard test method for trapezoid tearing strength of geotextiles.** *1990 Annual Book of Standards*, Vol. 4.08, Philadelphia, PA.

12.22 Haliburton, T. Allan (1980). *Use of geotechnical fabric in railroad operations.* Department of Civil Engineering, Oklahoma State University, Report for Association of American Railroads, August.

12.23 Christopher, B. R. and Holtz, R. D. (1985). *Geotextile Engineering Manual.* Federal Highway Administration, National Highway Institute, Washington, D.C.

12.24 Coleman, D. M. and Taylor, H. M. (1991). *Engineering use of geotextiles in railroad track construction and rehabilitation.* Waterways Experiment Station, Corps of Engineers, Final Report, Technical Report No. GL-91-18, September.

12.25 Carroll, R. G., Jr. (1983). **Geotextile filter criteria**, TRR 916, *Engineering Fabrics in Transportation Construction.* Washington, D. C. pp. 46-53.

12.26 AREA. *American Railway Engineering Association Manual.*

12.27 Friedli, P., and Anderson, D. G. (1982). **Behaviour of woven fabrics under simulated railway loading.** *Proceedings*, 2nd International Conference on Geotextiles, Las Vegas.

12.28 Saxena, S. K. and Chin, D. (1982). **Evaluation of fabric performance in a rail-road system.** *Proceedings*, 2nd International Conference on Geotextiles, Las Vegas.

12.29 Raymond, G. P. (1986). **Performance assessment of a railway turnout geotextile.** *Canadian Geotechnical Journal*, Vol. 23, No. 4, pp. 472-480.

12.30 Selig, E. T. (1991). Field trip to Conrail site at Pottstown, PA (U.S.A.), June 5, 1990. Trip report, April.

12.31 Selig, E. T. (1986). Field trip to British Rail site at Fenny Compton, in Birmingham Division. Trip report, July 3.

12.32 Lacy, H. S., and Pannee, J. (1987). **Use of geosynthetics in the design of railroad tracks.** *Performance of aggregates in railroads and other track performance issues,* Transportation Research Record 1131, Transportation Research Board, pp. 99-106.

13. Track Drainage

13.1. Importance

The three sources of water entering the track substructure are indicated in Fig. 13.1. Because the ballast surface is open, precipitation (rain and snow) falling onto the track will enter the ballast rather than run off the surface as will occur for intact road pavements. Water flowing down adjacent slopes will also enter the ballast and underlying layers unless diverted. Finally, water can seep upward from the subsurface and enter the substructure zone which influences track performance.

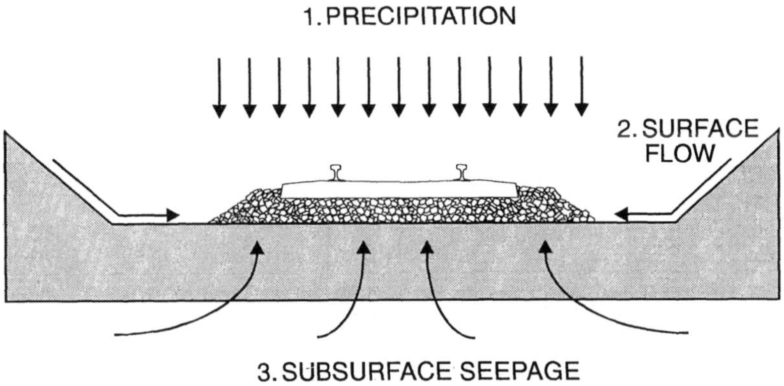

Fig. 13.1 Sources of water entering track substructure

Although at least portions of the ballast sometimes can be dry, generally ballast, and always subballast and subgrade contain some amount of moisture. In fact, subballast and subgrade perform best under repeated load when they are in a state which is intermediate to dry and saturated. Excess substructure water, particularly when it creates a saturated state, causes significant increases in track maintenance costs because of problems such as the following:

1) Pore pressure increase under cyclic load which causes increase in plastic strain accumulation, decrease in stiffness, and decrease in strength.

2) Loss of strength from water content increase.

3) Subgrade attrition and slurry formation from ballast action.

4) Hydraulic pumping of fine material.

5) Volume change from swelling.

6) Frost heave/thaw softening.

7) Ballast degradation from slurry abrasion, chemical action, and freezing of water.

8) Sleeper attrition from slurry abrasion.

Clearly, adequacy of drainage to prevent or minimize these problems has a major influence on maintenance costs. Because each source of water requires different drainage methods, the sources must be recognized in each case in order to determine the proper drainage solutions.

13.2. Principles of Flow

Flow through porous media such as ballast may take several forms. It may be steady or unsteady (variable). It may be laminar, if at low enough velocity, or turbulent, if at higher velocity. It may be saturated or unsaturated. However flow of water in soils can be considered to be incompressible.

Flow principles will be illustrated by assuming laminar flow under saturated conditions. In such cases the average fluid flow velocity is proportional to the hydraulic gradient, as expressed by

$$v = k i , \tag{13.1}$$

where

v = average flow velocity,

i = hydraulic gradient, and

k = proportionality constant.

Such flow is illustrated in Fig. 13.2. Here v is the average velocity defined as the quantity of flow divided by the soil cross sectional area. In reality the flow is through an irregular path through the soil pores at a higher velocity, v_s, which may be determined from

$$v = n v_s , \tag{13.2}$$

where n = porosity of the medium.

The proportionality constant, k, has traditionally been termed the coefficient of permeability, or permeability. However it is now preferred to call it hydraulic conductivity to recognize the distinction between the flow properties of the porous medium independent of the fluid properties and the flow properties with a particular fluid. Permeability, k_p, should be used to represent the properties of the porous medium only, while hydraulic conductivity, k_h , should be used to represent the combined effects of the fluid and the porous medium. These terms are related by:

$$k_h = k_p \left(\frac{\rho g}{\nu}\right) \quad , \tag{13.3}$$

where ρ = mass density of fluid, g/mm^3 ,

g = acceleration of gravity, mm/sec^2 ,

ν = viscosity of fluid, g/mm-sec.

k_p = permeability, mm^2 .

k_h = hydraulic conductivity, mm/sec.

The hydraulic gradient is the loss of total head per unit distance along the flow path. Total head, H_t , is given by

$$H_t = H_e + H_p \quad , \tag{13.4}$$

where H_e = elevation head, and

H_p = pressure head.

Elevation head is the height above the reference elevation. Pressure head is the height to which water will rise in a tube (with negligible capillary effects) for a water pressure, u, at the bottom. Thus

$$u = (\rho g)\, H_p \quad . \tag{13.5}$$

Flow is driven by total head differences and the movement is from the higher total head to the lower total head. This is shown in Fig. 13.2. The total head at point 1 is

$$H_{t1} = H_{e1} + H_{p1} = Z_1 + 0 \quad . \tag{13.6}$$

Since there is no head loss between points 1 and 3, then

$$H_{t3} = H_{t1} \quad . \tag{13.7}$$

Similarly

$$H_{t2} = Z_2 + 0 = H_{t4} \quad . \tag{13.8}$$

The total head loss through the porous medium is thus

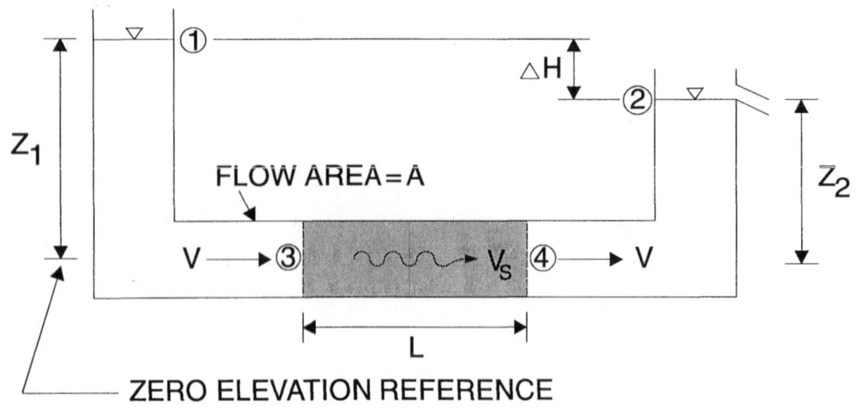

Fig. 13.2 Flow through porous medium

$$\Delta H = H_{t3} - H_{t4} = Z_1 - Z_2 \ . \tag{13.9}$$

The quantity of flow (in volume/unit time) through the porous medium is given by Darcy's law:

$$q = k\,i\,A \ , \tag{13.10}$$

where $i = \Delta H/L$.

13.3. Hydraulic Conductivity Values

Typical values of k_h for various soils are given in Table 13.1.

Table 13.1 Hydraulic conductivity values for soils

Soil Type	Representative k_h range (mm/sec)
Gravel	10^1 to 10^3
Sand	10^{-3} to 10^1
Silt	10^{-7} to 10^{-3}
Clay	10^{-9} to 10^{-4}

Parsons (Ref. 13.1) measured hydraulic conductivity of ballast specimens ranging in condition from clean to highly fouled. His gradations are shown in Fig. 7.6. The fouling index, F_I (Eq. 7.11), for these gradations range from 0 to 45. He found that as the voids in the ballast particles approached the condition of being filled with fine particles, the hydraulic conductivity for a given gradation became highly dependent on how the voids were filled. In track this is defined by the source of fouling material. Thus it matters whether the fouling material was infiltrated or was formed by ballast breakdown, and the extent to which the void material was compacted.

The results reported by Parsons (Ref. 13.1) are given in Table 13.2.

Table 13.2 Hydraulic conductivity values for ballast

Fouling Category	Fouling Index	Hydraulic Conductivity, k_h	
		(in./sec)	(mm/sec)
Clean	< 1	1 - 2	25 - 50
Moderately Clean	1 - 9	0.1 - 1	2.5 - 25
Moderately Fouled	10 - 19	0.06 - 0.1	1.5 - 2.5
Fouled	20 - 39	0.0002 - 0.06	0.005 - 1.5
Highly Fouled	> 39	< 0.0002	< 0.005

13.4. Modelling of Precipitation Drainage

Insights into factors influencing drainage of water from precipitation can be obtained from computer modelling of the process. Track substructure drainage is complicated by several factors including: partial saturation at least in some portions, turbulent flow in relatively clean ballast, non steady-state (transient) conditions, and highly distorted flow nets. However useful results can be obtained by simplifying the problem through assuming steady state, saturated, laminar flow as long as the boundary conditions of the layered system are still represented. Such an approach was taken by Cole (Ref. 13.2) using a finite element model.

The geometry for the model is illustrated in Fig. 13.3. The rainfall rates (mm/hr) required to saturate the ballast were determined for various material permeabilities and layer conditions. Two rainfall rates were defined: 1) R_{av} for the average flow over the entire ballast surface, and 2) R_{cr} for the flow entering that portion of the surface extending 300 mm (12 in.) each side of the track centerline. The second gives a much lower rainfall rate because the flow path lengths generally increase from the surface at the shoulder to the surface at the center. An

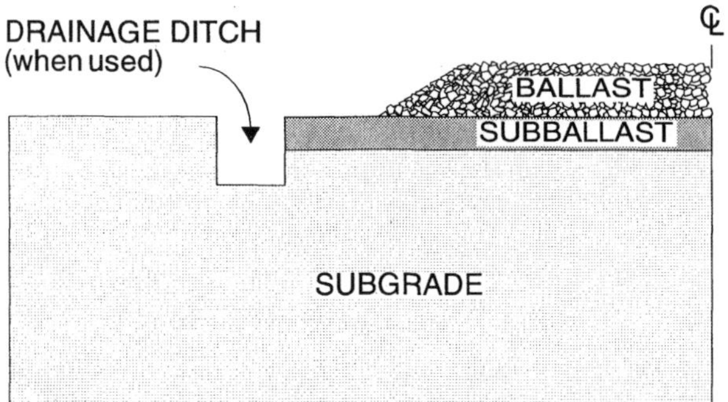

Fig. 13.3 Geometry assumed for drainage analysis

increase in flow path length decreases the quantity of flow needed to saturate the material. The simplifying assumption is also made that the crib ballast is sufficiently cleaner than the ballast below the sleeper to compensate for the surface flow area blocked by the sleeper. Thus the layered system was modelled for flow as though the sleepers were not present, and their space filled with ballast.

13.4.1. Low Permeability Subgrade

For cases in which the subgrade is of much lower permeability than that of the ballast and subballast, the model in Fig. 13.3 can be simplified to that in Fig. 13.4. A typical flow net for this geometry is shown in Fig. 13.5. The hydraulic conductivity of the ballast is designated k_b and that of the subballast is designated k_s. The direction of flow is predominantly lateral. Two-thirds (two of the three flow paths) of the water drained out of the ballast flows through the shoulder, yet this is only 10 percent of the horizontal ballast surface. This means that the water falling on the top of the ballast near the track center will have to travel some distance through the ballast and the subballast layers and discharge near the toe of the ballast layer. Hence, drainage will not occur as rapidly in the center of the ballast layer as for the side, and ponding of water will occur in the center much more quickly than at the sides. This is the reason that the critical steady state rainfall is defined considering the flow nearest the centerline. Of course once ponding begins near the track center the water will run over the surface to the adjacent flow channels. With a high enough rainfall rate the ballast will become saturated by this process.

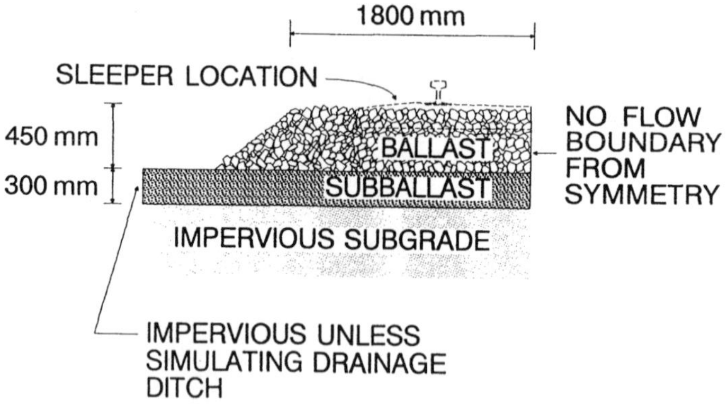

Fig. 13.4 Geometry for cases with low permeability subgrade

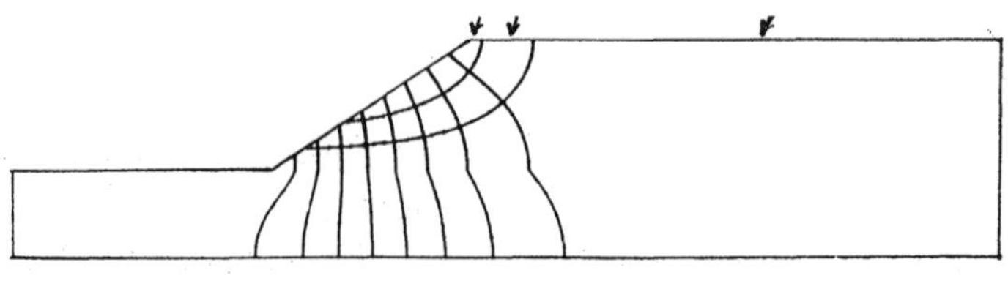

Fig. 13.5 Flow net for cases with low permeability subgrade (k_b/k_s = 1000)

For $k_b/k_s \geq 100$ the flow net is the same as shown in Fig. 13.5 and little drainage occurs through the subballast layer. The rainfall rates are

$$\frac{R_{av}}{k_b} = 0.067 \ , \tag{13.11a}$$

$$\frac{R_{cr}}{k_b} = 0.0016 \ . \tag{13.11b}$$

For $k_b/k_s = 1$ the flow net changes to that in Fig. 13.6, and

$$\frac{R_{av}}{k_b} = 0.11 \ , \tag{13.12a}$$

$$\frac{R_{cr}}{k_b} = 0.017 \ . \tag{13.12b}$$

The variation of R_{cr}/k_b with k_b/k_s is given by

$$\left(\frac{R_{cr}}{k_b}\right) = \left[1.6 + 8.4\left(\frac{k_b}{k_s}\right)^{-1} + 7.0\left(\frac{k_b}{k_s}\right)^{-2}\right] \text{x } 10^{-3} \ . \tag{13.13}$$

When $k_b/k_s >> 1$, then changing the subballast dimensions had no signficant effect on the rainfall rates. For $k_b/k_s = 1$, changing the distance by which the subballast extended beyond the ballast toe had only a small effect on the rainfall rates. Changing the subballast thickness mainly affected R_{cr}.

Adding the ditch at the edge of the subballast (Fig. 13.3) had no effect on either R_{av} or R_{cr} in most cases. Hence the primary function of the ditch was removal of surface water from the various sources rather than to help remove water from the ballast.

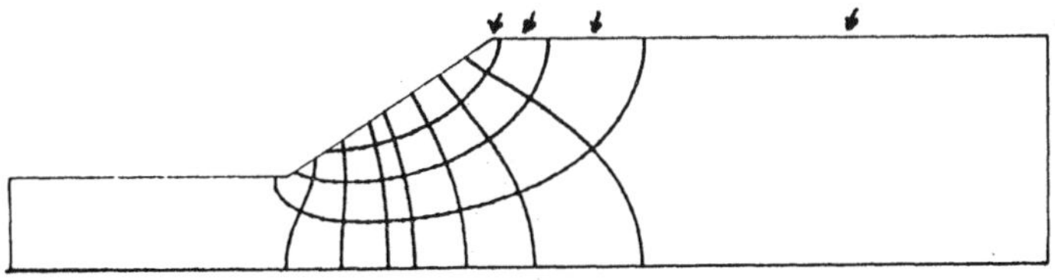

Fig. 13.6 Flow net with ballast and subballast of same permeability over low permeability subgrade

13.4.2. Homogeneous Substructure

When the hydraulic conductivities of the ballast, subballast and subgrade are the same, the substructure is homogeneous. In this case the subgrade conditions have a big effect on the drainage pattern and rates. The model in Fig. 13.3 is used for these cases, with variable boundary conditions. With the water table at the subgrade surface the rainfall rates are

$$\frac{R_{av}}{k_b} = 0.22 \ , \tag{13.14a}$$

$$\frac{R_{cr}}{k_b} = 0.15 \ . \tag{13.14b}$$

Note that the two ratios are similar in value.

Lowering the water table will substantially increase the rainfall rates needed to saturate the ballast. The drainage path will tend to become vertical, with the result that R_{cr} and R_{av} will become about the same. The drainage ditch has little effect on the rainfall rates for a homogeneous substructure.

13.4.3. Effects of Ballast Cleaning

Parsons (Ref. 13.1) used Cole's model to investigate the effects of both side (known as shoulder) cleaning and undercutting cleaning of the ballast. The cases compared are illustrated in Fig. 13.7. In every case the ballast layer was 300 mm (12 in.) thick beneath the sleeper. The ballast condition is designated by the fouling index. Where subballast is not shown it is assumed to have a low enough permeability that it does not significantly influence the rainfall rate. Where the subballast was included it was 150 mm (6 in.) thick with a hydraulic conductivity of about 0.00025 mm/sec. This is representative of a silty sand. The subgrade in all cases was considered to be relatively impermeable.

The critical rainfall results are given in Fig. 13.7. The cases are listed in order of decreasing critical rainfall rates, that is decreasing drainage ability. Note that the fouled ballast with cleaned shoulders (case b) has better drainage than the corresponding undercut ballast (case c). Both of these maintenance operations substantially improve drainage for the fouled ballast condition prior to cleaning (case e). For the highly fouled condition (case g), the undercutting (case d) improved drainage more than the shoulder cleaning (case f). The reason can be seen by comparing the two undercut cases (cases c and d) with the two shoulder cleaning cases (cases b and f). Increasing the degree of fouling beneath the sleepers has a much bigger effect on drainage for the shoulder cleaning cases because all the water near the center of the track must pass through this fouled ballast. However when the ballast is undercut, and the remaining ballast has a high degree of fouling, most of the drainage will be through the clean layer. Hence changing the degree of fouling of the remaining ballast will not significantly affect the drainage rate. Of course, full ballast layer cleaning will cause much greater drainage improvement than either partial undercutting with cleaning or shoulder cleaning for all degrees of fouling.

13.4.4. Drainage Criteria for Cleaning

Parsons (Ref. 13.1) determined the critical rainfall rates corresponding to a range of fouling index from 0 to 40. For F_I from 0 to 30 he used the geometry and boundary conditions represented by cases a and e in Fig. 13.7. For $F_I > 30$ he used the geometry and boundary conditions represented by case g in Fig. 13.7. The results are given in Fig. 13.8.

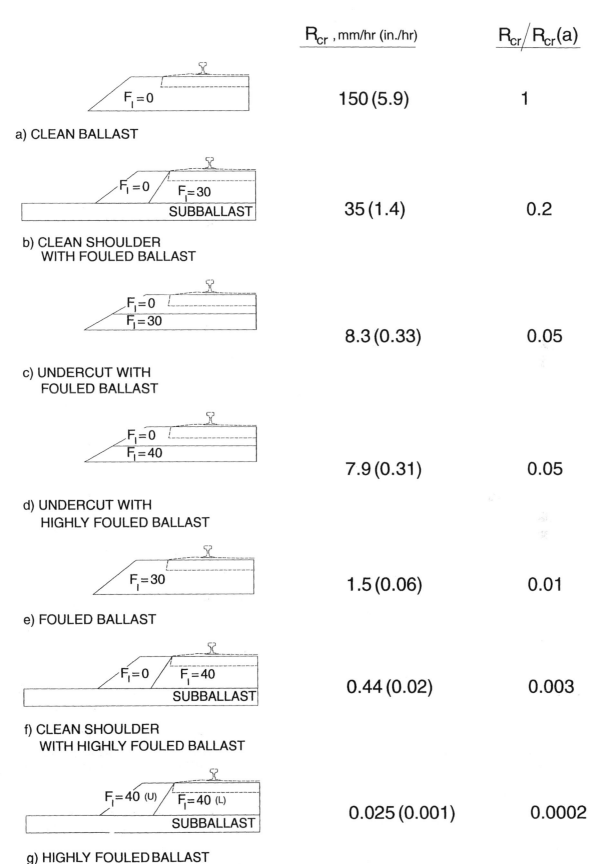

R_{cr}, mm/hr (in./hr) — $R_{cr}/R_{cr}(a)$

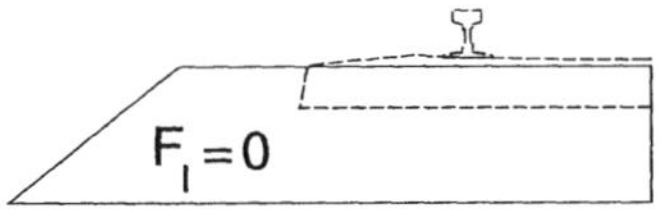

a) CLEAN BALLAST — 150 (5.9) — 1

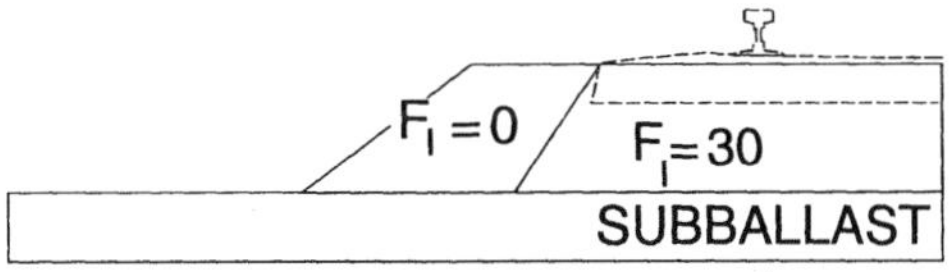

b) CLEAN SHOULDER WITH FOULED BALLAST — 35 (1.4) — 0.2

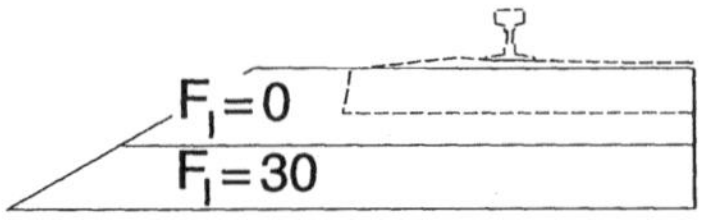

c) UNDERCUT WITH FOULED BALLAST — 8.3 (0.33) — 0.05

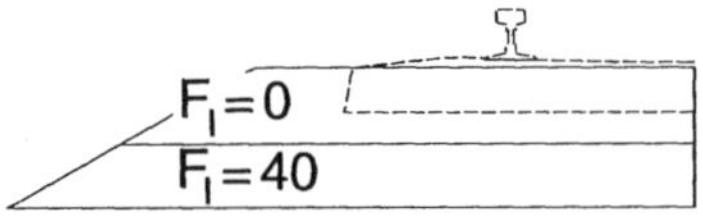

d) UNDERCUT WITH HIGHLY FOULED BALLAST — 7.9 (0.31) — 0.05

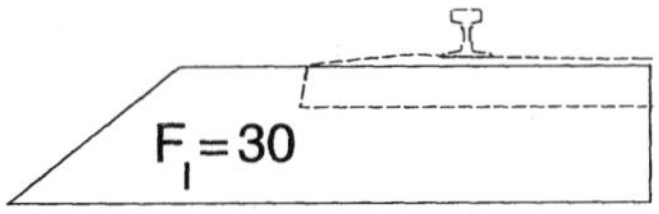

e) FOULED BALLAST — 1.5 (0.06) — 0.01

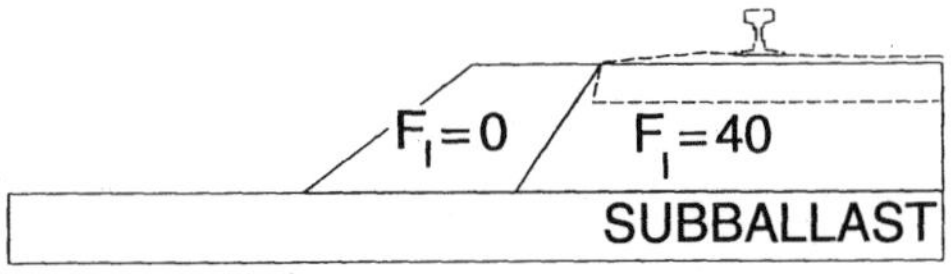

f) CLEAN SHOULDER WITH HIGHLY FOULED BALLAST — 0.44 (0.02) — 0.003

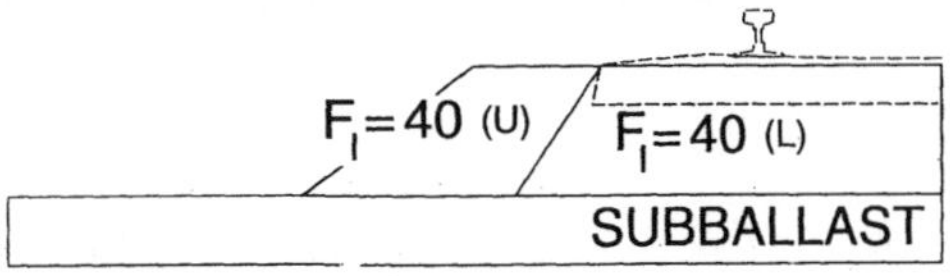

g) HIGHLY FOULED BALLAST — 0.025 (0.001) — 0.0002

Fig. 13.7 Drainage cases for maintenance evaluation

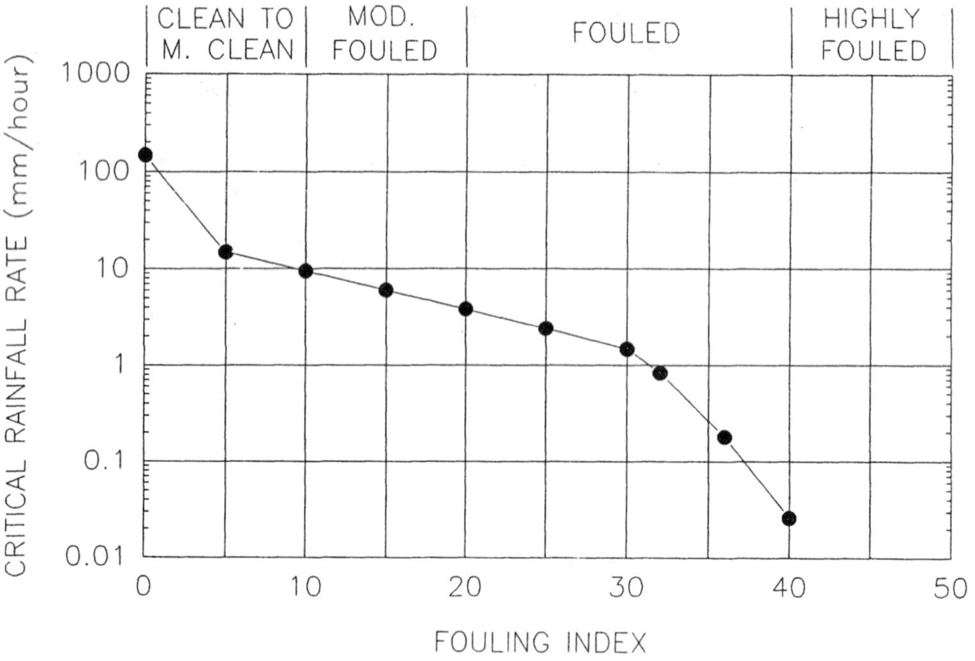

Fig. 13. 8 Variation of critical steady state rainfall intensity with degree of fouling

Expected maximum rainfall intensity during the ballast cleaning cycle can be estimated from available weather records. The allowable degree of fouling can then be determined from Fig. 13.8. The more extensive the ballast cleaning, the longer the time before cleaning is again needed. However the cost will be higher as well. A cost analysis is required to determine the optimum maintenance action at any point in time.

Parson's model assumes a single track with drainage symmetrical about the center line. For double track with drainage only to one side of each track the critical drainage rates can be expected to be about half that of single track. Adjustment for differences with each track layout will have to be considered in applying Fig. 13.8.

13.5. Modelling Seepage from Subgrade

Information is readily available on analysis of seepage of water through the ground. Applications have included seepage through dams to consider loss of water and structural stability, seepage through natural slopes to consider slope stability and erosion, seepage towards earth retaining structures to consider forces on the walls and drainage requirements, and upward seepage into pavement structures to consider effects on pavement performance and drainage requirements. Useful discussion of these subjects is given in Refs. 13.3 and 13.5.

Finite element techniques like that used to model ballast drainage are available for analyzing flow of water through the ground for applications like the above. However there are practical limitations in the accuracy of such analyses, especially when natural soil strata are involved rather than constructed earthworks. These limitations are a result of uncertainties about the groundwater conditions, and the soil permeabilities. It is often difficult to determine the ground water conditions because of the complex flow networks that are possible and because conditions vary over time. Furthermore soil deposits are not homogeneous nor isotropic in their hydraulic conductivity properties. Laboratory hydraulic conductivity tests on soil specimens often are not representative of field conditions. Nevertheless simplified analyses are useful in understanding potential ground water problems and how to handle them.

An example of a flow net for water seepage through and under a dam is shown in Fig. 13.9 Ref. 13.5). A flow net for upward seepage from a pervious aquifer into a horizontal surface drainage blanket is given in Fig. 13.10 (Ref. 13.4). Such flow nets have two primary purposes: 1) estimation of flow quantities, and 2) estimation of pore water pressure for determining soil behavior and for analysis of stability using effective stresses.

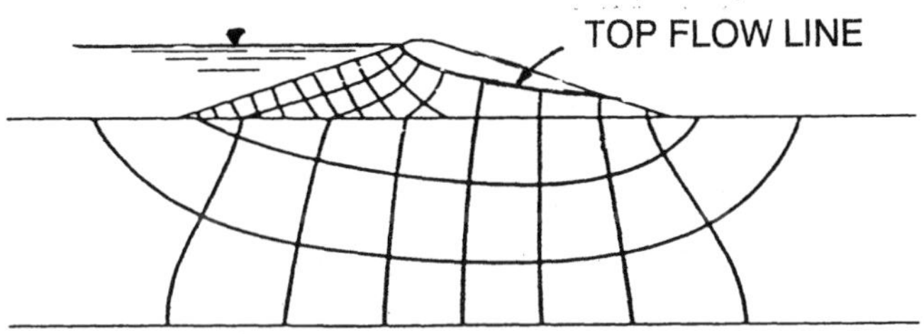

Fig. 13.9 Flow net for dam (reprinted from Ref. 13.5 with permission of the publisher)

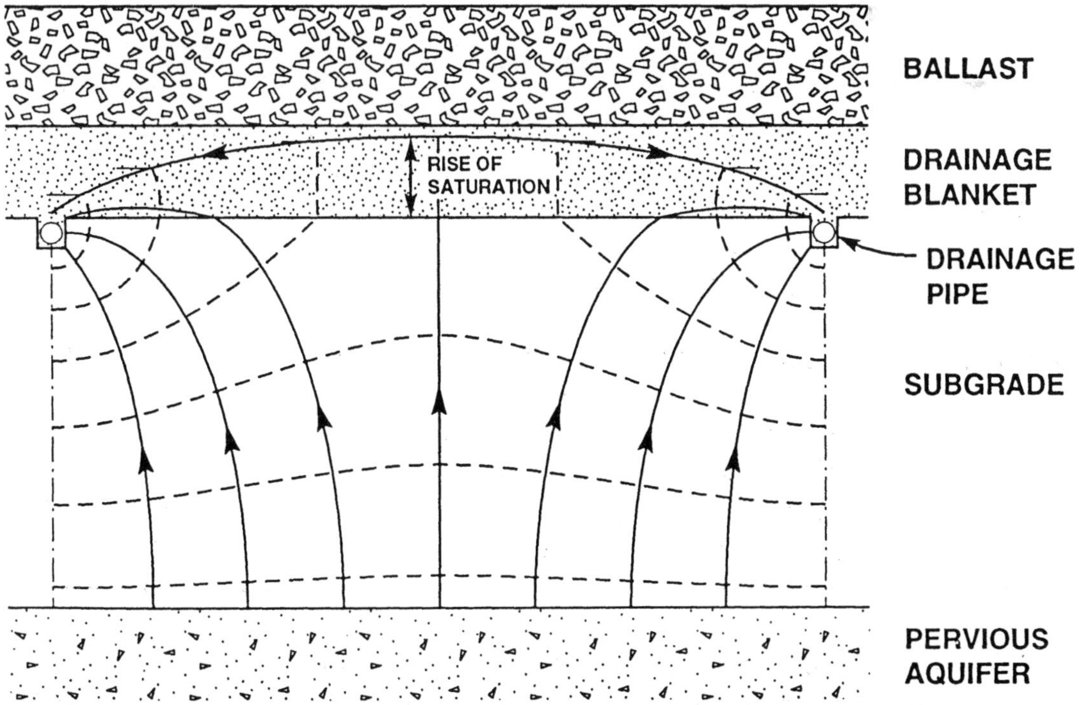

Fig. 13.10 Flow net for vertical seepage into horizontal drainage layer (adapted from Ref. 13.4 with permssion of the publisher)

The top flow line in Figs. 13.9 and 13.10, which designates the surface of zero (atmospheric) porewater pressure, is termed the phreatic surface. In the case of no water flow (seepage) the phreatic surface is known as the water table. Above the phreatic surface the pore water pressure will be negative, and the soil is generally unsaturated.

Seepage of ground water cannot occur into areas of the track substructure which are above the phreatic surface. Thus seepage of ground water will not be a problem with track structures on embankments. Likewise seepage of ground water will not be a problem for track structures on level ground if the water table is below the ground surface. Artesian conditions such as illustrated in Fig. 13.10 must be present to produce water from seepage in horizontal ground situations. These comments do not mean that ground water problems cannot exist for tracks on embankments or on level ground. Capillary water can produce a high enough degree of saturation that strength and stiffness loss can occur under repeated train loading. Capillary water can also feed ice lens formation, hence creating frost heave/thaw softening problems.

Ground water seepage is mainly a problem in cuts. The reason is that the natural ground water level prior to excavation is usually above the boundaries of the excavation (Fig. 13.11). If so, unless corrective action is taken after excavation, a flow pattern will develop in which water will seep out of the slope causing erosion and slope stability problems, and water will seep upward into the track substructure leading to problems which have previously been described.

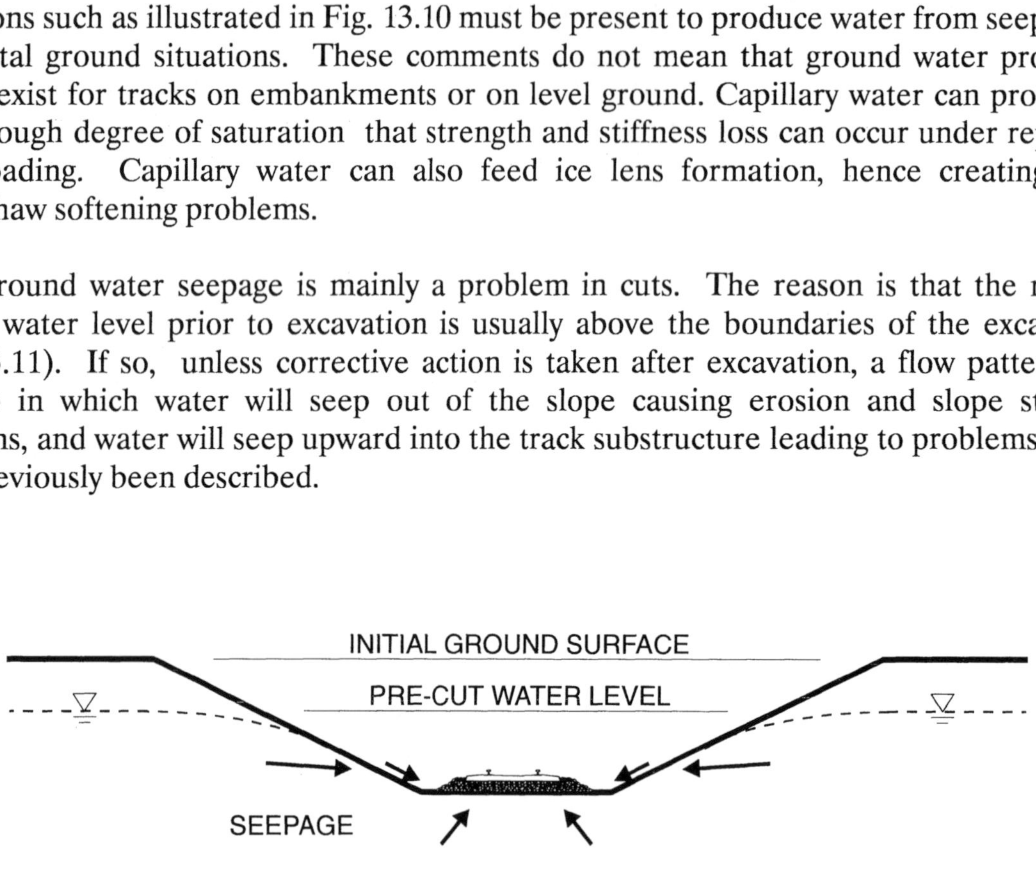

Fig. 13.11 Ground water seepage in cuts

13.6. Drainage Methods

Design of satisfactory drainage systems cannot be done without first properly examining existing conditions. All three potential sources of water shown in Fig. 13.1 must be considered.

Explorations prior to drainage system design should define the soil, groundwater and climatic conditions. Characteristics of subgrade soils must be known including type, layering, and thicknesses. In existing track, the explorations have to be more detailed close to the surface to assess local variations that may significantly affect the proposed drainage system. In particular, the lateral slope of the subgrade, the presence and depth of subgrade squeezes or ballast pockets, and the extent of ballast fouling will all have a marked effect on drainage system performance.

The depth to the groundwater table affects groundwater-lowering requirements, drainage measures for control of capillary rise, and the provisions of outlets for disposal of surface water. In assessing the groundwater conditions, the yearly high groundwater level is of concern. Explorations should be timed to coincide with this condition. Observation wells may be needed to monitor the variations in water levels over at least the critical period of time.

The drainage system capacity should be adequate to handle the highest expected inflow rate during the design life of the system. Designing for an average rate will not be adequate because then there will certainly be periods when water will accumulate.

Drainage system design is a normal part of geotechnical engineering practice. Consultation with experienced specialists in this field is usually desirable, when evaluating and implementing drainage systems, especially when dealing with ground water seepage problems.

To provide a background for more detailed drainage design and a basis for assessing drainage requirements, some examples of drainage control methods will be given. These examples have been obtained in part from Refs. 2.3, 9.4, 11.4, and 13.3.

13.6.1. Drainage from Track

13.6.1.1 Requirements

Water from precipitation falling onto the track and entering the ballast must be able to drain out of the substructure. As already shown, this normally means lateral flow out through the ballast and to some extent out through the subballast.

The **first requirement** to achieve substructure drainage is to keep the ballast clean enough to be able to drain the water as fast as it enters.

The **second requirement** for this drainage is to have the surface of the subballast and subgrade sloped towards the sides such as shown in Fig. 13.12. Progressive settlement of the subgrade can eventually (sometimes eventually is not far off) reverse the slope and cause depressions which trap water beneath the track (Fig. 13.13). Major depressions caused by soft subgrade are known as ballast pockets because the ballast will fill these depressions, with ballast added to the track to make up for the lost volume. The best solution to this problem when it is severe is to remove the track superstructure and reconstruct the substructure with an adequate granular layer depth and suitable subgrade properties as discussed in Chapters 10 and 11. Less severe conditions can perhaps be fixed just by

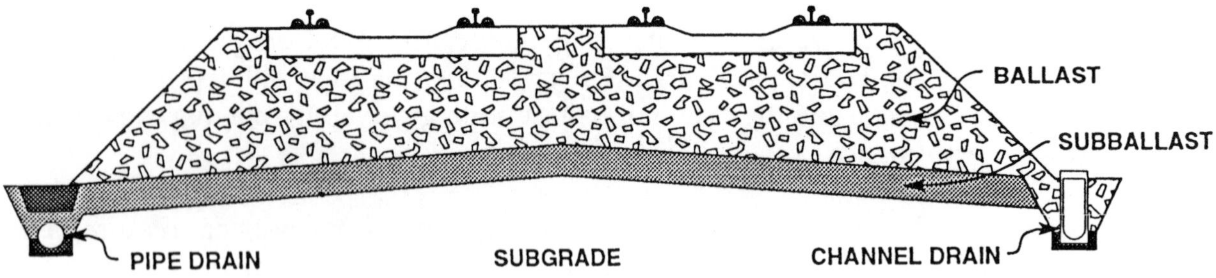

Fig. 13.12 Subballast and subgrade sloped for drainage

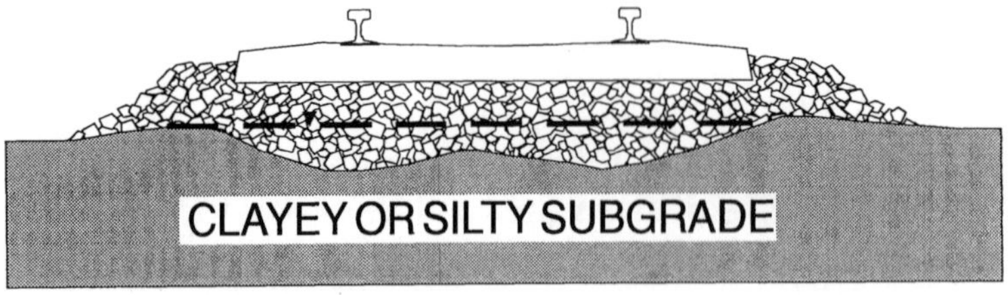

Fig.13.13 Water trapped in ballast pockets

re-shaping the subgrade surface and replacing the subballast and ballast to restore the condition shown in Fig. 13.12.

The **third requirement** is to provide a means of carrying away the water that comes out of the substructure. The bathtub analogy is often cited to describe this requirement. Water within a bathtub is free to move about, i.e. the permeability within the tub is infinite. However the water is confined by the impermeable tub sides and bottom as long as the plug is in place. There is little benefit in having a suitable ballast and subballast if they are confined on the sides and bottom by a low permeability soil, even with a ground water table well below the subgrade surface (Fig. 13.14). At the very least, a suitable edge drain will be required to unplug the bathtub.

The same bathtub effect will occur due to contamination of the shoulder ballast (Fig. 13.15). Track subgrade failure, described in Chapter 10 (Figs. 10.13-10.15), will also create a bathtub effect. This subgrade failure is related to the formation of ballast pockets illustrated in Fig. 13.13.

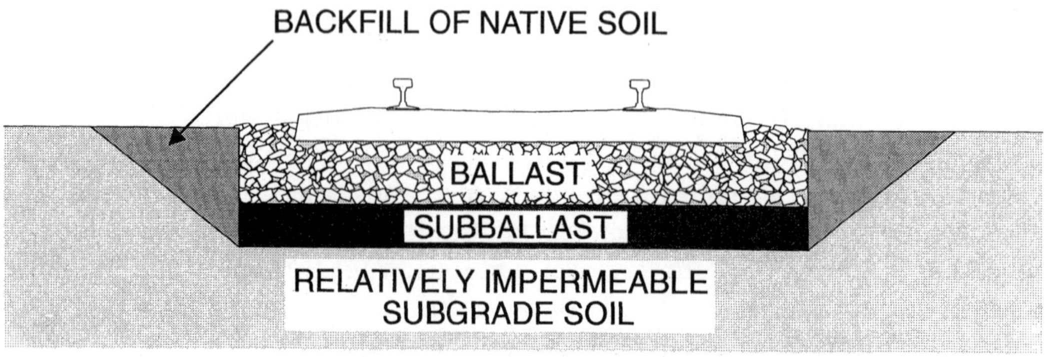

Fig. 13.14 Bathtub condition for ballast and subballast

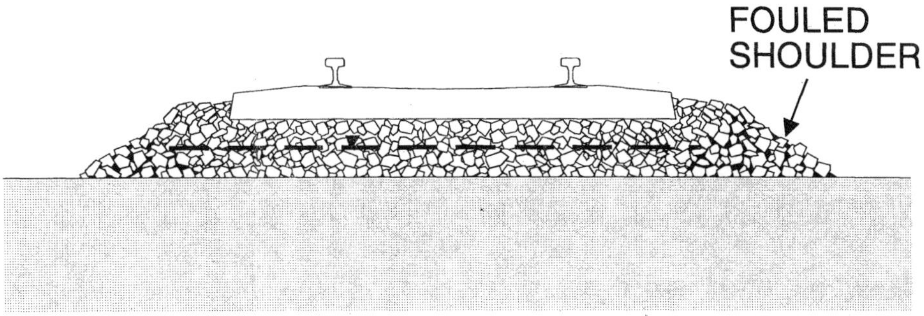

Fig. 13.15 Side ballast contamination effect

13.6.1.2 Side drains

Side drains are located on one or both sides of a track, paralleling its route until an outfall is reached. These are probably the most common of all railroad drain installations and can be designed to intercept and carry away surface water, as well as seepage from the ballast, subballast and subgrade. Offtake drains provide an intermediate outlet for the side drain systems to limit side drain length or to satisfy slope limitations. Offtake drains intersect side drains and provide a shorter distance to a natural drainage course than would be available parallel to the track.

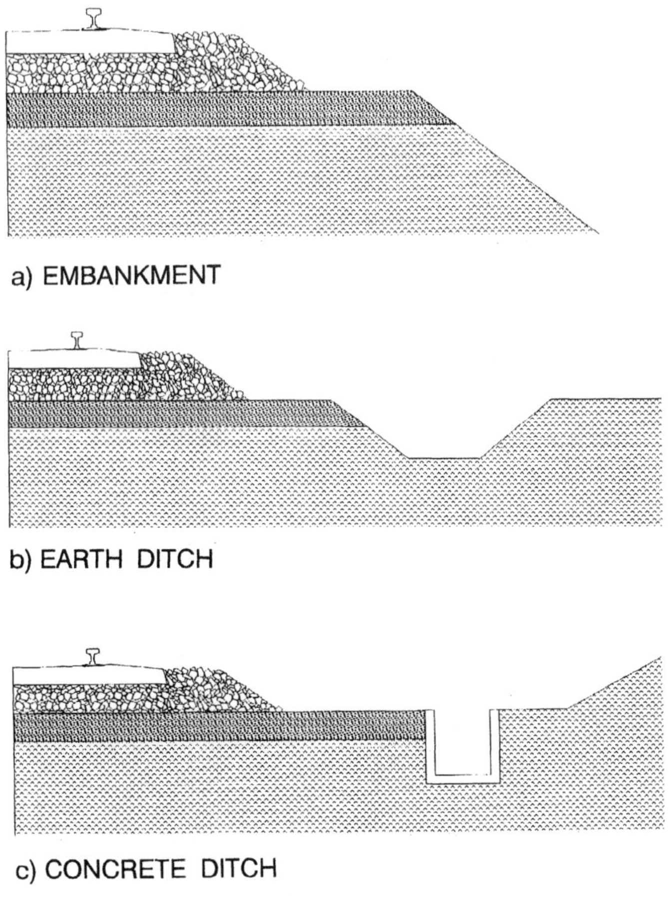

Fig. 13.16 Track side drainage

The most effective side drain is the side of the embankment if it is not too far away from the toe of the ballast and if the top of the embankment is sloped to shed water (Fig. 13.16a). However embankment shoulder protection is required to control erosion. Obviously embankments are selected for achieving the required track grade -- drainage is only an accompanying advantage. Keep in mind, though, that an embankment only satisfies the third drainage requirement. The first and second requirements, which are clean enough ballast and sloped subballast and subgrade surfaces, are also essential.

The next most effective drains are open ditches such as illustrated in Figs. 13.16b and 13.16c. They must have the capacity to carry away the water from the substructure as well as from adjacent surface runoff. Ditches must be sloped steeply enough to prevent sedimentation but not cause erosion of the ditches. If velocities higher than the soil erosion limits are anticipated, the ditch may be protected from erosion by lining. The principal advantages of ditch drains are that they are economical to construct and can handle large flows required for storm water control. However, ditch drain geometry is restricted by the geometry of the track and the topography of the surrounding ground.

Once installed the ditches must be maintained in the required condition. This requirement is obvious, but it is commonly neglected.

Ditch drains are not effective for removing subgrade water, either because they are not deep enough, or because they are lined. A deep drain is required to provide a sufficient hydraulic gradient to cause water movement, as well as to keep the phreatic surface well below the top of subgrade. One approach is a trench drain, examples of which are given in Figs. 13.17 and 13.18. These drains are buried, so they are less affected by track and site topography than surface drains, and can be installed beneath structures such as grade crossings, stations, and multi-track areas, and as cross drains beneath the track.

Trench drains would be used together with ditch drains if the latter were also required. One option is to place the trench drain directly below the ditch drain. Alternatively the ditch drain may be placed outside of the trench drain. This provides the option for locating the trench drain below the ballast slope. In this case the low permeability cap would be eliminated so that the trench drain would connect to the subballast and ballast at the top through a suitable filter material.

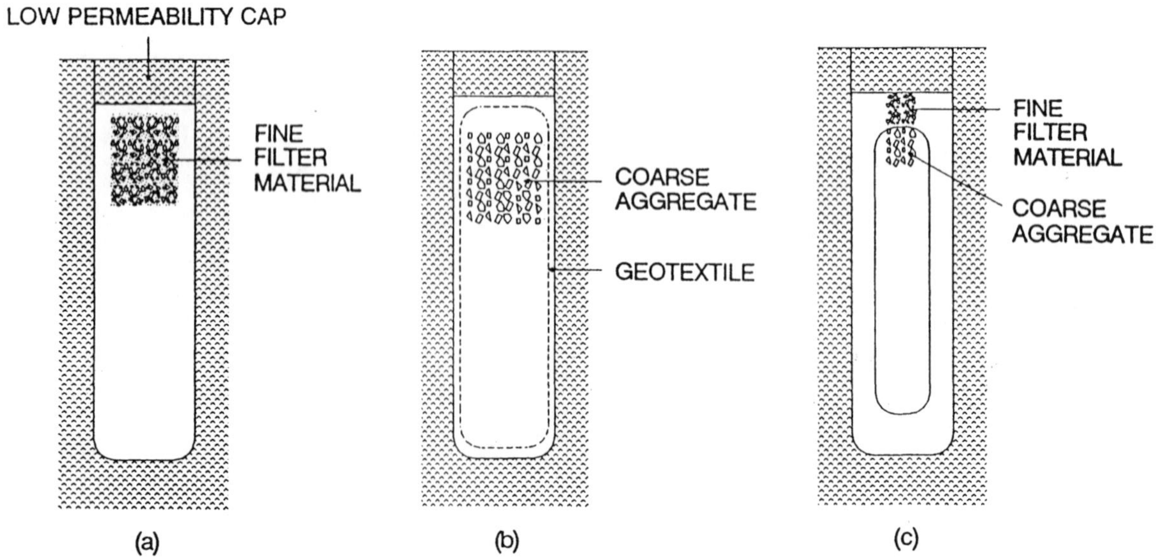

Fig. 13.17 French drain examples

French drains consist of a trench filled with a suitable granular material to collect water from the surrounding soil and also transmit it longitudinally. The permeability and separation criteria for granular filters in Chapter 9 must be satisfied by the selected trench fill. This will generally result in a relatively fine filter if the simplest version of french drain shown in Fig. 13.17a is used. Otherwise soil particles will enter the voids and inhibit the drainage. Hence the flow capacity will be relatively low for this type of drain.

A coarser granular material may be used together with a geotextile to improve the drain capacity (Fig. 13.17b). However the geotextile must satisfy the separation requirement that

$$AOS \leq D_{85} \text{ (protected soil)} \; , \qquad (13.15)$$

where AOS = apparent opening size of fabric (mm), and

D_{85} = diameter for which 85% of the soil particles are finer (mm).

As discussed in Chapter 12, this requirement poses a problem with fine-grained soils (silts and clays).

A third alternative for a french drain is a 2-zone granular filter (Fig. 13.17c). Better flow capacity is provided than with the single zone (Fig. 13.17a), but the 2-zone version is more difficult to construct.

Pipe drains are trench drains which use a perforated pipe (usually slots or circular holes) surrounded by a suitable filter material (Fig. 13.18) The filter material collects the water from the adjacent soil and transmits it to the pipe, which serves as the longitudinal conduit. This arrangement provides a better flow capacity than the french drain. As for the french drain the filter separation/permeability requirement of Chapter 9 must be met by the granular trench fill to prevent migration of adjacent soil particles into the drain.

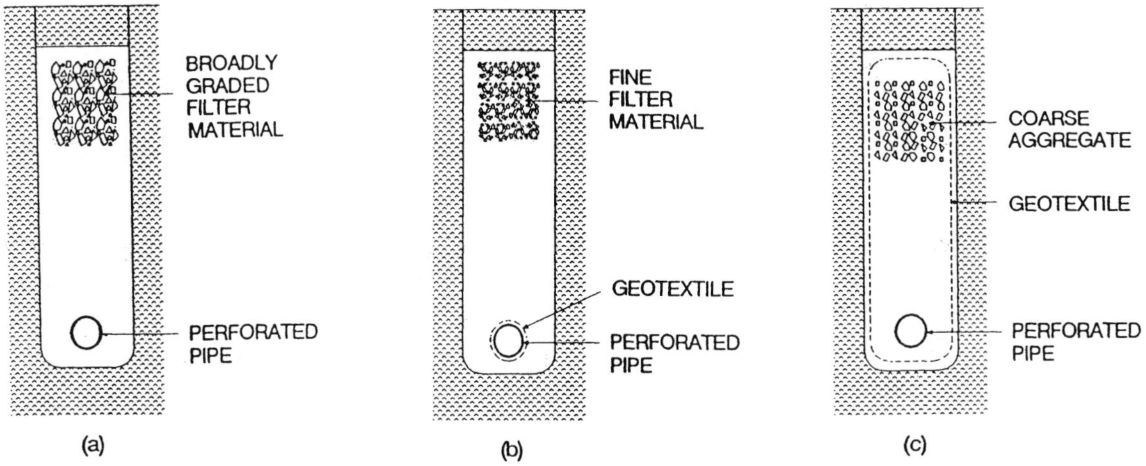

Fig. 13.18 Pipe drain examples

Alternatively the trench may be lined with a geotextile satisfying the criterion in Eq. 13.15. This permits a greater flexibility in the choice of granular fill material .

The gradation of granular envelope around the drain pipe must also bear a special relationship to the hole sizes to keep the granular particles from entering the pipe. One set of criteria is that (Ref. 13.5) for slots:

$$\text{D85 size of filter material} > 1.2 \text{ to } 2 \text{ slot width,} \tag{13.16a}$$

and for circular holes:

$$\text{D85 size of filter material} > 1.0 \text{ hole diameter.} \tag{13.16b}$$

Alternatively the pipe can be wrapped in a geotextile (Fig. 13.18b) to keep the granular particles from entering the pipe.

Geocomposite edge drains (Fig. 13.19) have been developed as an alternate to granular trench drains. They consist of a flat plastic core wrapped in a geotextile. These

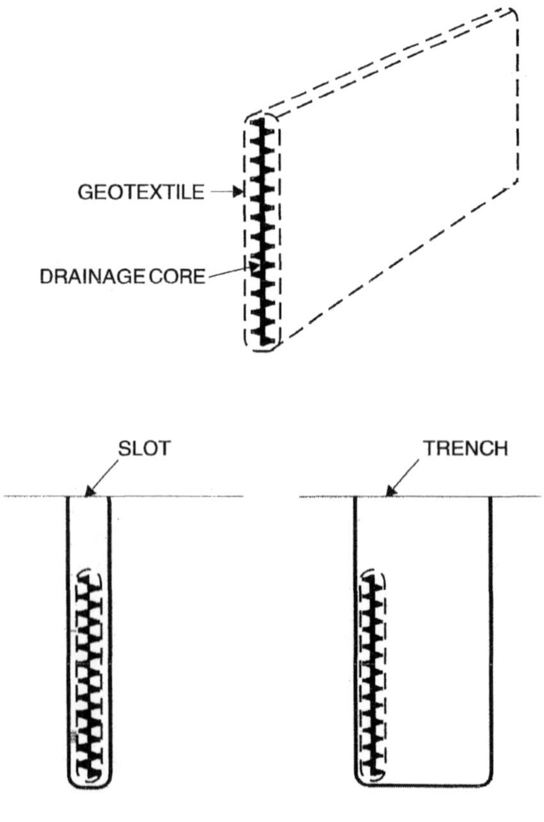

Fig. 13.19 Geocomposite edge drain

drains, also called fin drains, are designed to be inserted in a slot dug or formed in the soil along the edge of a track or pavement. This can be done in a continuous process without the need for backfill. Alternatively a narrow trench can be dug and backfilled with the excavated material. The criterion of Eq. 13.15 must be met by the geotextile.

Figure 13.20 is an example of the way in which a fin drain can be incorporated into a track drainage scheme. Fin drains act primarily as a rapid ground water collector. The water carrying capacity of such a drain is not intended to be equivalent to a long distance carrier drain. Where a large volume of water is involved, or the distance between water discharge points is great, the fin drain discharges into a lower carrier, which can be a closed jointed traditional carrier drain pipe.

In 1984, fin drains were installed at three locations on the Southern Railway at which wet subgrades were threatening to contaminate the ballast (Ref. 13.4). An inspection of the fin drains carried out two years after installation indicated that they were functioning satisfactorily.

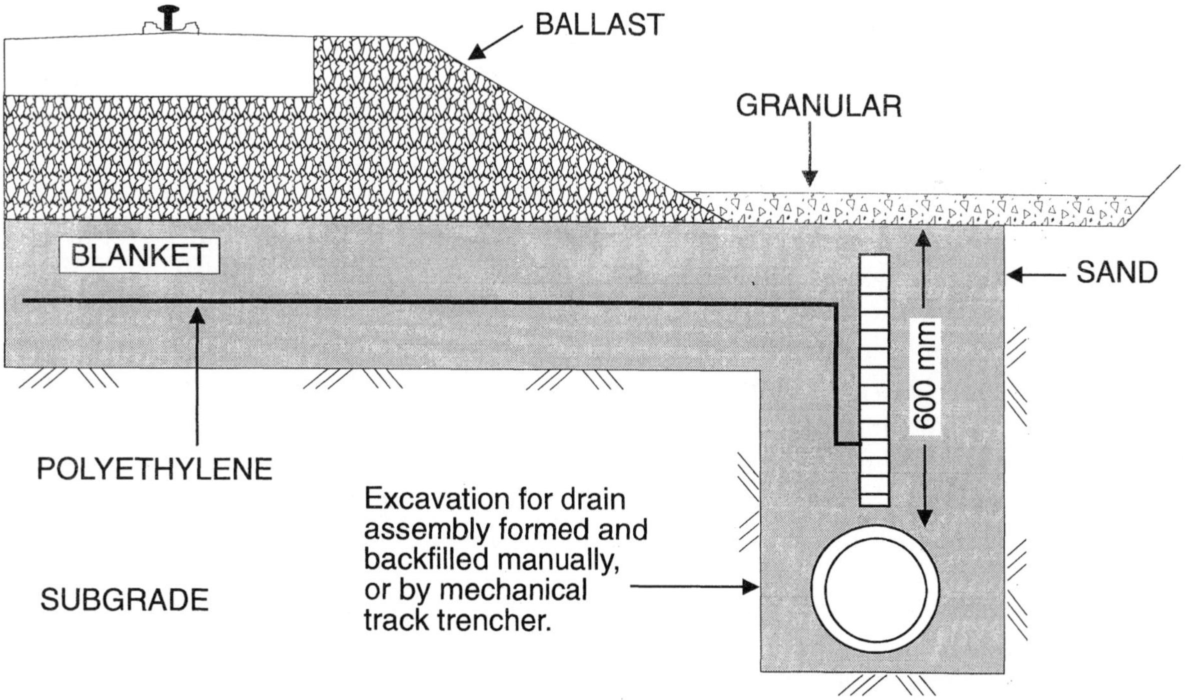

Fig. 13.20 Example of fin drain installation

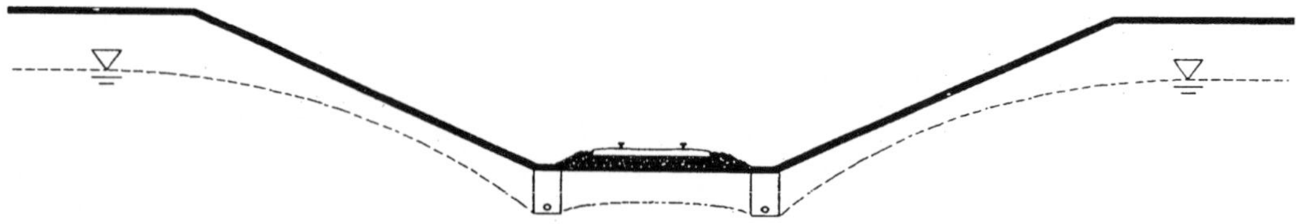

Fig. 13.21 Lowering ground water level with side drains

13.6.2. Drainage of Subgrade Seepage

Side drains may also be used to collect water seeping upward from the subgrade as indicated in Fig. 13.21. These drains must be deep enough to adequately lower the phreatic surface. For example, compare the phreatic surface in Fig. 13.20 with that in Fig. 13.11. The required drain depth depends on the permeability of the soil including the effects of fissures and pervious seams. Depths of at least 1 to 2 m may be required. Trench-type drains or other internal drains also may be installed on the slopes of cuts to help with the phreatic surface lowering.

Internal slope drains may also be required for maintaining slope stability. However this topic goes beyond the scope of this chapter. An experienced geotechnical engineer should be consulted for assistance with slope stability evaluation and means of handling potential slope stability problems.

Cross drains may be used to supplement side drains. As indicated in Ref. 13.4 cross drains are sometimes installed beneath track where lateral drains are insufficient to control groundwater beneath the centerline of the track. However, use of cross drains for general groundwater control is probably impractical. In high permeability soils, i.e. sands and gravels, cross drains are rarely required. In fine-grained soils, the permeability is so low that the cross drains must be closely spaced to be effective. This is uneconomical. A better solution is to raise the track and place it on a blanket of granular material or to excavate a portion of the natural subgrade and replace it with granular material. In these ways, the lateral drains can control the flow. In wide areas, cross drains, such as shown in Fig. 13.22, may be necessary.

13.6.3. Drainage from Surface Flow

Drainage of surface water from precipitation in the area of the track, or from adjacent slope runoff is handled by open ditch drains. The ditch drains at the side of the track will serve this purpose if they have adequate capacity.

Ditch drains also may be required on and at the top of slopes to intercept and divert surface flow from the area of the track. Observations during periods of rainfall, especially

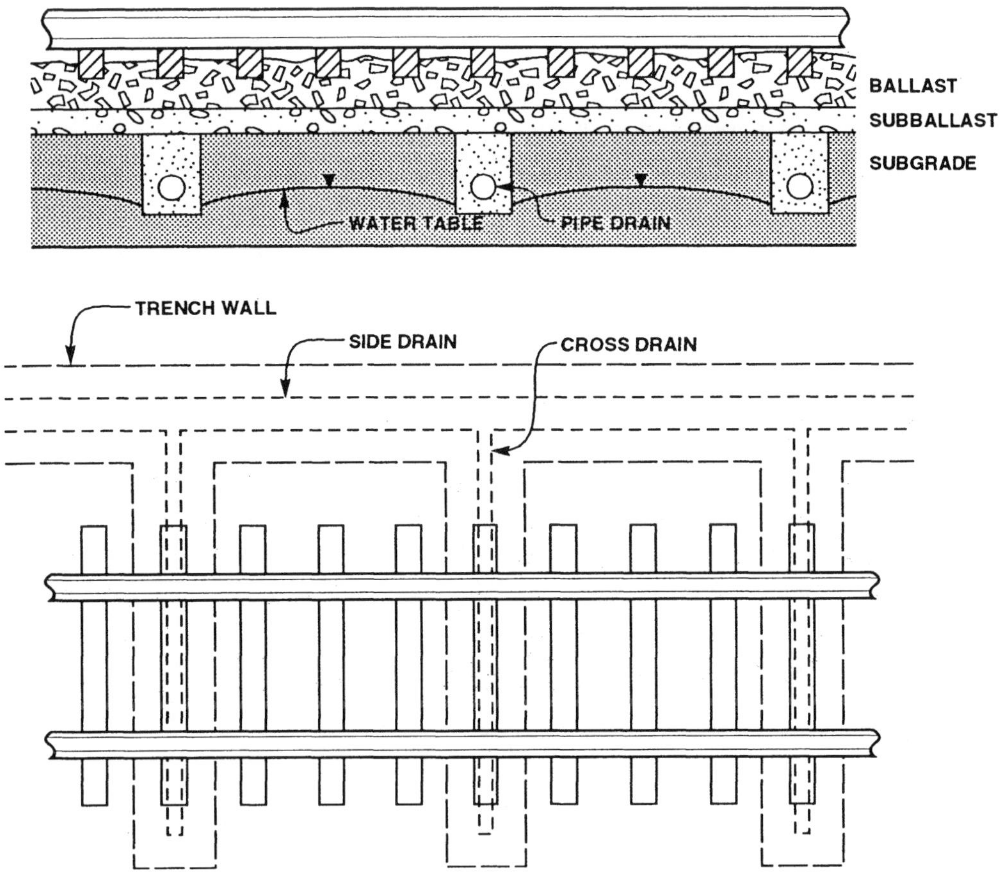

Fig. 13.22 Cross drain illustration

during rainy seasons, will be needed to determine the requirement for these supplemental drains.

When ditch drains are used on or at the top of slopes two potentially adverse effects on slope stability must be considered and avoided. First, the drains must not permit water to seep into the slope, causing weakening. Second, the ditch must not be cut into a critical slip surface, thus lowering the sliding resistance.

REFERENCES

13.1 Parsons, B. K. (1990). *Hydraulic conductivity of railroad ballast and track structure drainage.* Geotechnical Report No. AAR90-372P, Department of Civil Engineering, University of Massachusetts, Amherst, MA, February.

13.2 Cole, B. E. (1989). *Analysis of railroad track substructure drainage*. Geotechnical Report No. AAR89-365P, Department Civil Engineering, University of Massachusetts, Amherst, MA, June.

13.3 Cedergren, Harry R. (1974). *Drainage of highway and airfield pavements*. John Wiley and Sons.

13.4 **Southern tracking prefab drain's performance**. *Railway Track and Structures,* June 1985.

13.5 Cedergren, Harry R. (1977). *Seepage, drainage and flow nets*. John Wiley and Sons, New York.

14. Machines and Methods

A large number of machines are available to the track engineer which, either directly or indirectly, interact with the track foundations. These machines and the uses to which they may be put will be discussed in this chapter.

14.1. Track Level Adjustment

Machines designed to facilitate the adjustment of track level fall into two main categories:

- Tampers which remold the existing sleeper supporting ballast.
- Stoneblowers which add ballast to the surface of the existing ballast.

14.1.1. Tamping Machine

A typical tamping machine is shown in Fig. 14.1. It is self propelled and, by means of lifting and lining rollers that grip the head of the rail (Fig. 14.2), is able to lift the track to a pre-determined level, and move it sideways to a pre-determined alignment. Tamping tines, as shown in Fig. 14.3, are able to penetrate the ballast at the rail/sleeper interface, and then squeeze the ballast up and under the sleeper to retain it in its raised position.

Fig. 14.1 Self-propelled tamping machine

Fig. 14.2 Lifting and lining rollers

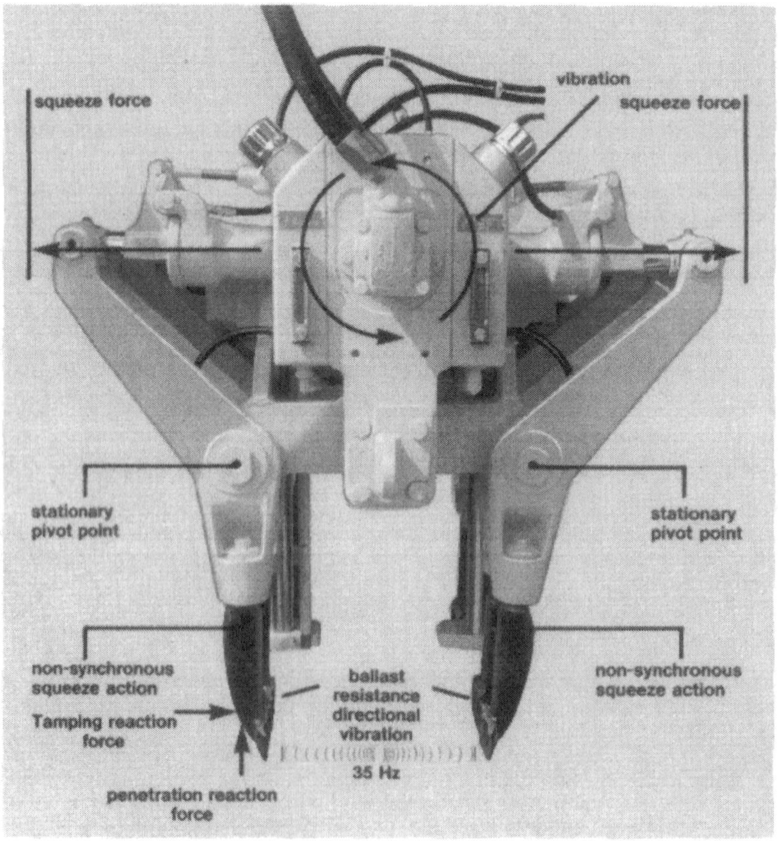

Fig. 14.3 Tamping tines attached to power unit

Figure 14.4 shows the following sequence of events involved in a sleeper tamping cycle:

A) The tamping machine positions itself over the sleeper to be tamped.

B) The lifting rollers raise the sleeper to be tamped to the target level and thereby create a void beneath the sleeper.

C) The tamping tines penetrate the ballast on either side of the sleeper.

D) The tamping tines squeeze ballast into the void beneath the sleeper, thereby retaining the sleeper in its raised position.

E) The tamping tines are withdrawn from the ballast, the lifting rollers lower the track, and the tamper moves forward to the next sleeper to be tamped.

14.1.1.1 Tamping and levelling principle

Tamping machines are operated in either a smoothing mode or a design mode. Operation of the tamping machine in a smoothing mode can be further sub-divided into automatic and controlled smoothing mode tamping.

The tamping machine shown in Fig. 14.5 is provided with a three point measuring system in which:

- The lower ends of three reference members AD, BE and CF are in contact with the rail head.
- The top ends of the front and rear reference members are connected at D and F, by a tensioned wire.
- Tamping is carried out in the direction AC.

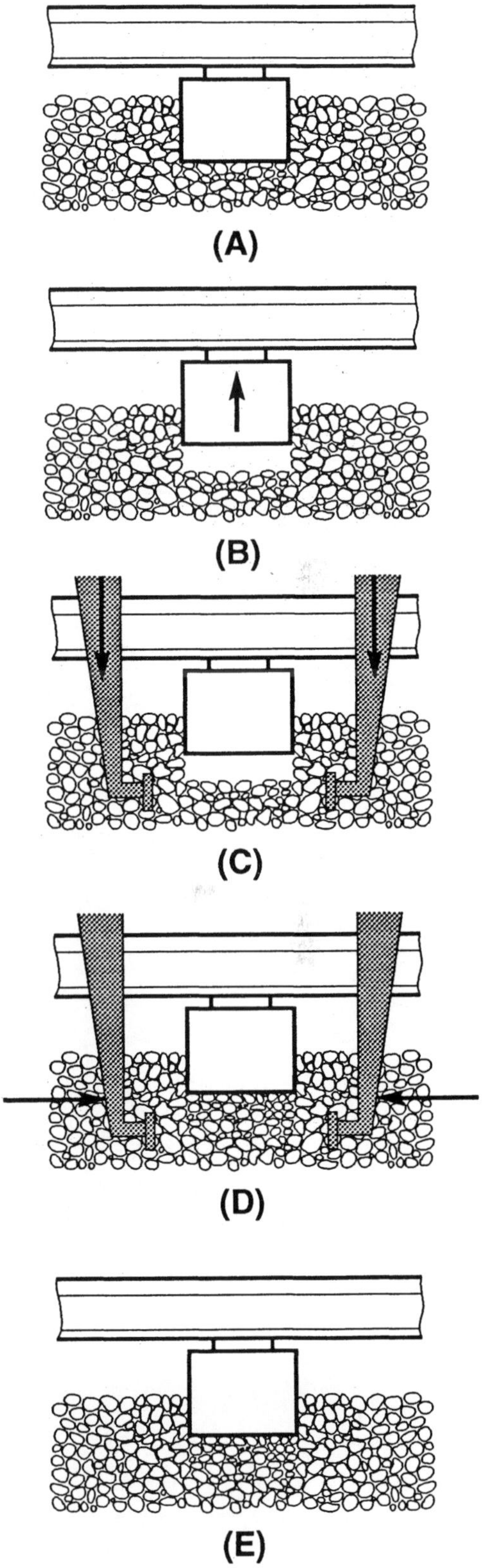

Fig. 14.4 Tamping sequence

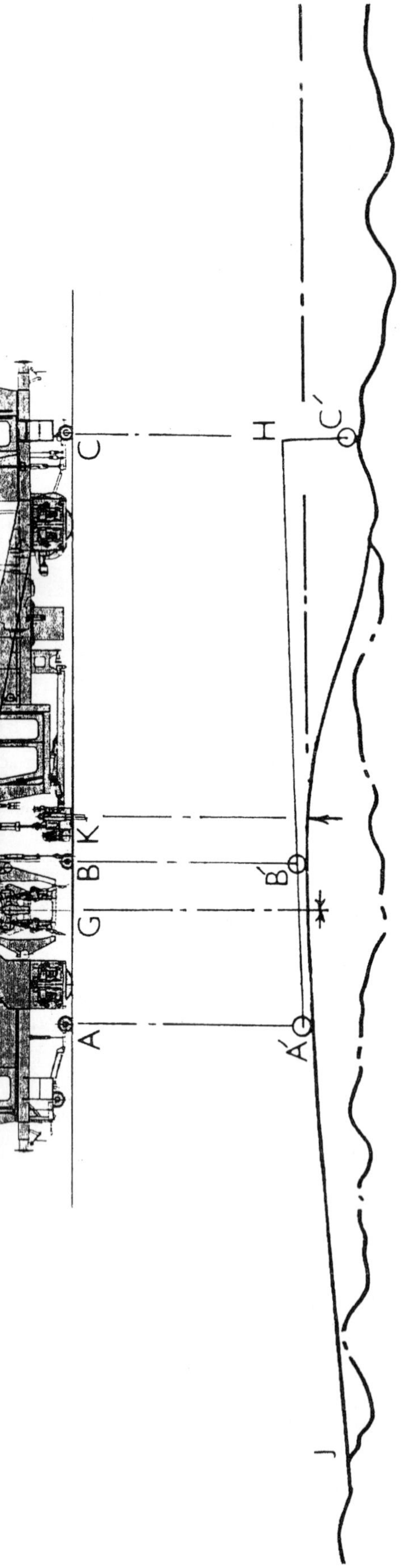

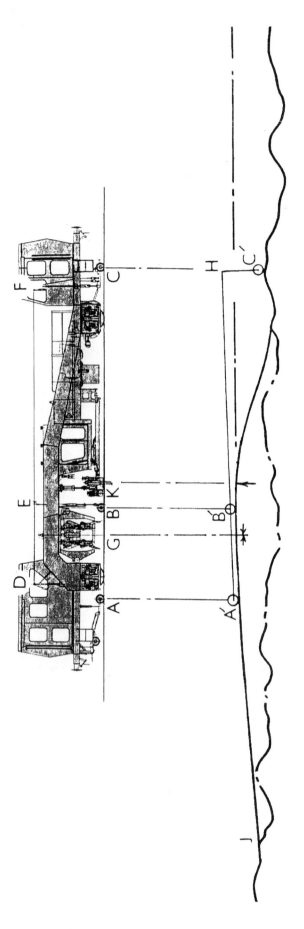

Fig. 14.5 Tamping machine with 3-point measuring system

- The rear reference member contacts the rail head of the tamped track at A.
- The intermediate reference member contacts the rail head of the untamped track at B, a point just in front of the tamping point G.
- The front reference member, often referred to as the front tower, contacts the rail head of the untamped track at a point C, which is beyond the front axle of the tamping machine.

Since the tamping process can lift, but not lower the track, the front reference member FC is extended downwards such that reference point C′ is set at a distance HC′ below the extrapolated line A′B′. This will ensure that the track is lifted at all points, during the automatic smoothing tamping operation. The location J marks the starting point of this operation.

14.1.1.2 Automatic smoothing tamping

The automatic smoothing mode operates as follows (Fig. 14.5):

- The track lifting equipment K, lifts the track at the tamping position G, until the top end E of reference member BE contacts the wire DF.
- The track lifting equipment retains the track in the lifted position while the tamping tines pack additional ballast beneath the sleeper to retain it in its lifted position.
- Once ballast packing is complete, the track lifting equipment releases the track and the tamper moves forward to the next tamping position.

It will be seen that a number of successive sleepers need to be tamped before the 'overall lift' CH is fully established, or 'run in'. For this reason, it is usual practice to commence tamping at a 'high spot' rather than at a 'low spot' to ensure that a track lift is always required.

The effectiveness of a tamping machine working in the automatic smoothing mode can best be expressed by means of its transfer function, which is the ratio of the amplitude of a track geometry fault after tamping to that before tamping, as a function of the wavelength of the fault being considered.

The transfer function or modulus, M, at any given wavelength is given by:

$$\frac{1}{M} = \frac{O}{R} = \left\{ \left[\frac{d+c}{c}\right]^2 + \left[\frac{d}{c}\right]^2 - 2\left[\frac{d+c}{c}\right]\frac{d}{c}\,(\cos\lambda\,c) \right\}^{0.5}, \qquad (14.1)$$

where M = modulus,

R = amplitude of fault remaining after tamping,

O = amplitude of fault existing before tamping,

d = length of the long chord (AC in Fig. 14.5),

c = length of the short chord (AB in Fig. 14.5),

λ = wavelength of fault being considered.

The transfer function considered above relates to a simple "three point" measuring system of the type shown in Fig. 14.5. In practice, the short wavelength performance of the tamping machine can be enhanced by providing a two point contact at the front reference point in which the front reference point, C, refers to the mid point of a 3 m trolley running on the rail head.

Figure 14.6 shows the modulus of the transfer function for three tamping machines in which the lengths of the long chord are 11 m, 20 m, and 50 m, respectively. For example, a tamping machine having a long chord length of 11 m will reduce the amplitude of a 40 m wavelength to 50% of its original amplitude. Note that a phase shift will occur in the wave pattern.

In general, it can be said that:

- For all wavelengths, the longer the long chord, the greater the smoothing effect.
- For any given long chord length, the shorter the wavelength of the fault, the greater the smoothing effect.

In addition to the foregoing, the track smoothing ability of the tamping machine is dependent upon the correct adjustment and functioning of the two chord measuring system. It is generally accepted that tamping machines operating in the smoothing mode are effective in reducing geometric faults adequately in the 5 m to 35 m waveband.

The effect of operating the tamping machine in an automatic smoothing mode is to reduce the amplitude of the faults present in the track by a factor that is proportional to the ratio AB/AC (Fig. 14.5). The tamped track will therefore contain a modulated, phase shifted version of the geometry of the track at C. The effect of this phase shift can result in the unwanted smoothing and translation along the track of fixed design points in a direction opposite to that in which the automatic smoothing operation is carried out. Such fixed design points could take the form of the start and finish points of vertical curves and transitions. These unwanted effects can be overcome by using the tamping machine in a design mode.

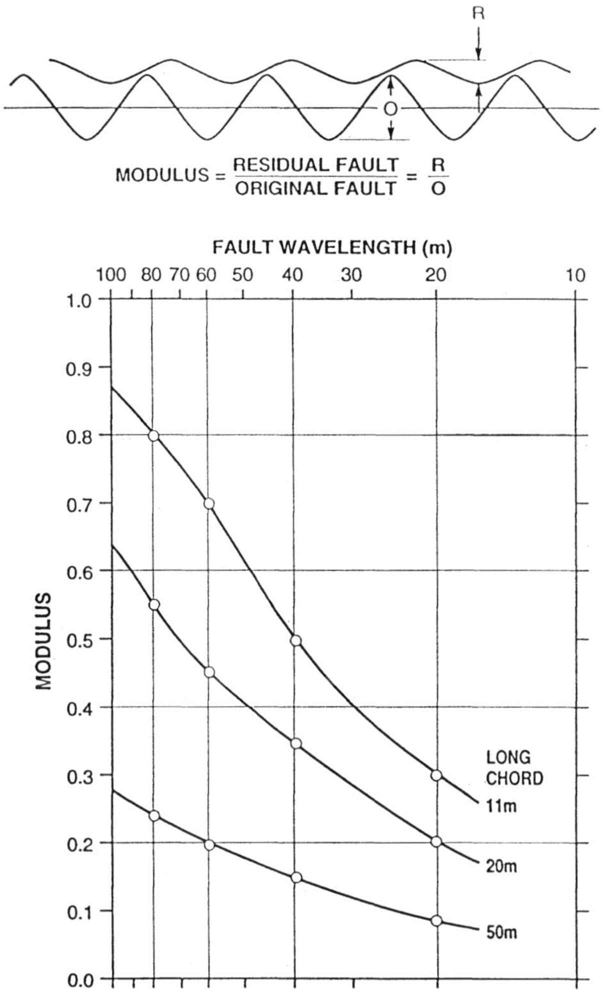

Fig. 14.6 Modulus variation with long chord length and wavelength

14.1.1.3 Controlled smoothing tamping

For high speed lines, the effects of long wavelength geometric faults on vehicle ride become very important. The higher the speed, the longer the wavelength of the fault that should be considered. For example, at train speeds of 200 km/h, which are modest by today's standards, wavelengths in excess of 50 m certainly have to be considered. In this connection, see Chapter 15.

It is the limitations in the automatic smoothing tamping process that have led to the development of 'controlled' smoothing tamping. The aim of controlled smoothing tamping is to achieve a smooth target geometry which, while conforming to prescribed track standards, corresponds as closely as possible to the existing track geometry. In this way, problems associated with phase shift are eliminated and the volumes of material tamped are kept to a minimum.

Adjustments in the height of point H (Fig. 14.5) above the rail, allow the lift applied to the sleeper at the tamping point G to be varied at will.

A suitable target geometry can be derived in a number of ways. The following are two examples.

The first uses the two chord measuring system of the tamping machine to collect existing track geometry data in the form of off-sets of the middle reference point, B, from the chord , AC, in Fig.14.5. An onboard computer is able to process these off-sets to give machine control data in the form of sleeper lifts required with respect to the existing track geometry, to achieve the required geometry.

The second uses an absolute geometry measuring approach. The existing track geometry is measured by reference to a datum laser, by traditional optical means, track survey cars, or as in the following example, by the FROG measuring trolley. These methods are described in Chapter 15. The data are processed as indicated in Fig. 14.7a to give:

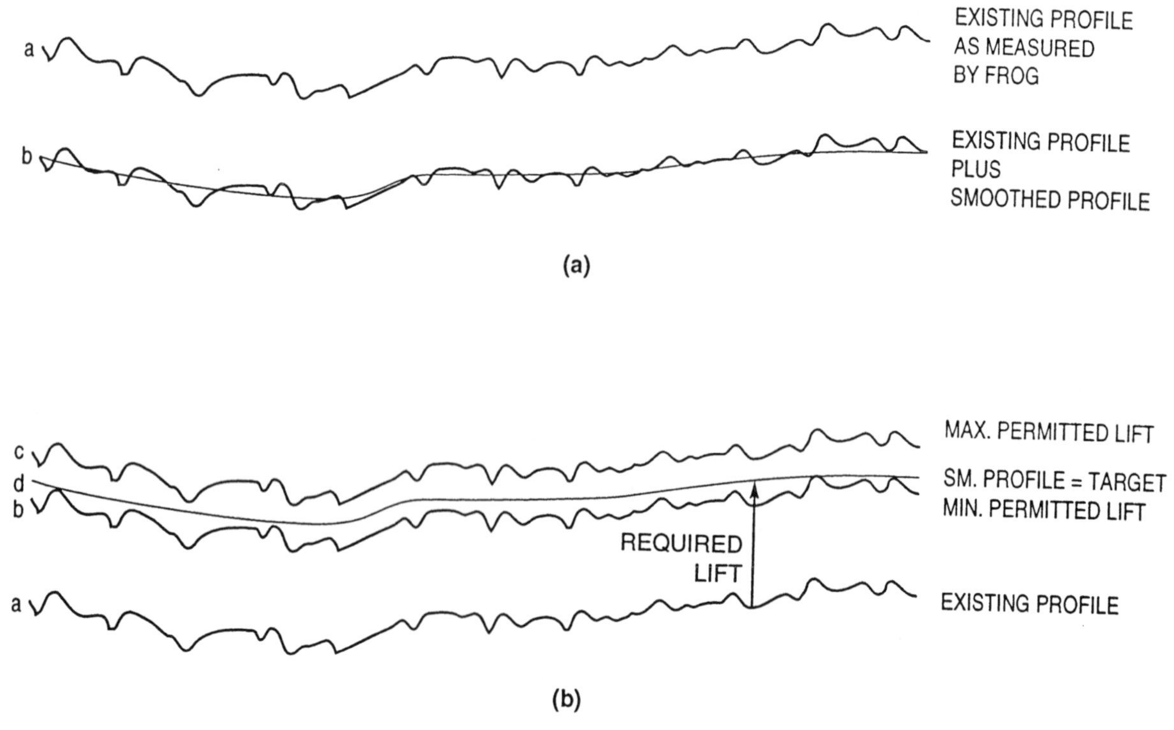

Fig. 14.7 Absolute geometry measuring approach to controlled smoothing tamping

1) The existing profile as measured by FROG of one rail (the datum rail) of the track which is to be tamped.

2) The existing profile of the datum rail and its associated smoothed mean profile, as described in Chapter 15.

Figure 14.7b shows the existing track profile translated vertically to indicate the minimum and maximum permitted track lifts. A minimum lift ensures that the permanence of the tamping effect is maximized. A maximum lift ensures that ballast consumption, associated with tamping, is not excessive. Fixed points, such as 'over-bridges' together with run in and run out distances also need to be taken into account.

The smoothed profile, as referred to in Fig. 14.7a, serves as the target profile and is placed at a height above the existing track equal to the required lift. At this stage, two options are available. The target geometry can be referred to a datum provided by the existing track profile as originally measured, or to an external datum such as a beacon laser. In either case, a listing needs to be prepared which indicates the lift to be applied by the tamper on a sleeper by sleeper basis. Such lifts are applied either manually by the tamper operator, or automatically by the track lift control system of the tamping machine.

14.1.1.4 Design tamping

When operating in the design mode, the target geometry need bear no relationship to the existing track geometry. The target profile is made up of a series of straights, curves and transitions, etc. The target track geometry is referred to fixed lineside objects, such as overhead wire supports, the positions of which are known. Automatic guidance systems are able to accept data relating to existing and target geometry and guide the tamping machine accordingly.

14.1.2. Stoneblower

The stoneblower is typical of the family of techniques that add stone to the surface of the existing ballast. Such techniques, which include trowelling and measured shovel packing (Ref. 14.1), are generally recognized as providing a durable result. The technique however has proved difficult to mechanize. The Stoneblowing system, which was developed by British Railways, has shown that such mechanization is possible.

The stoneblowing process operates in three stages:

1) The geometry of the existing track is measured.

2) The precise track lift required at each sleeper to restore it to an acceptable geometry is calculated.

3) The volume of stone that needs to be blown beneath the sleeper to achieve such a lift is deduced from the known relationship between volume of added stone and residual lift.

4) The track is stoneblown.

The various stages of the stoneblowing process are shown in Fig. 14.8. At (A) the sleeper rests in the ballast before adjustment. At (B) the sleeper is raised to create a void into which the stone can be blown. At (C) the stoneblowing tubes are driven down alongside the sleeper. At (D) a measured quantity of stone is blown by compressed air into the void. The quantity of stone blown is that which, when compacted by subsequent traffic, will raise the sleeper by the amount needed to achieve the required track geometry. At (E) the stoneblowing tubes are withdrawn, leaving the blown stone resting on the surface of the ballast that was originally supporting the sleeper. At (F) the sleeper is lowered onto the surface of the blown stone where it will be compacted by subsequent traffic.

The stone used in the stoneblowing process has to be of the highest possible quality since it will be positioned at the point of maximum stress, i.e. at the ballast/sleeper interface, immediately below the rail. The particles should be equidimensional and angular. The nominal particle size should be about 25 mm (1 in.).

A prototype Stoneblower (Fig. 14.9), manufactured for British Railways, has shown that a viable working speed can be achieved, and has confirmed the effectiveness of the system. Hand held stoneblowers, similar to that shown in Fig. 14.10, are available for spot maintenance purposes (Ref. 14.2).

By virtue of the size of stone used by the two systems, tamping and stoneblowing can be regarded as complementary systems. Tampers are suited to the relatively high lifts associated with the removal of long wavelength faults in track geometry, while stoneblowing is suited to the low lifts associated with the removal of short wavelength geometric faults.

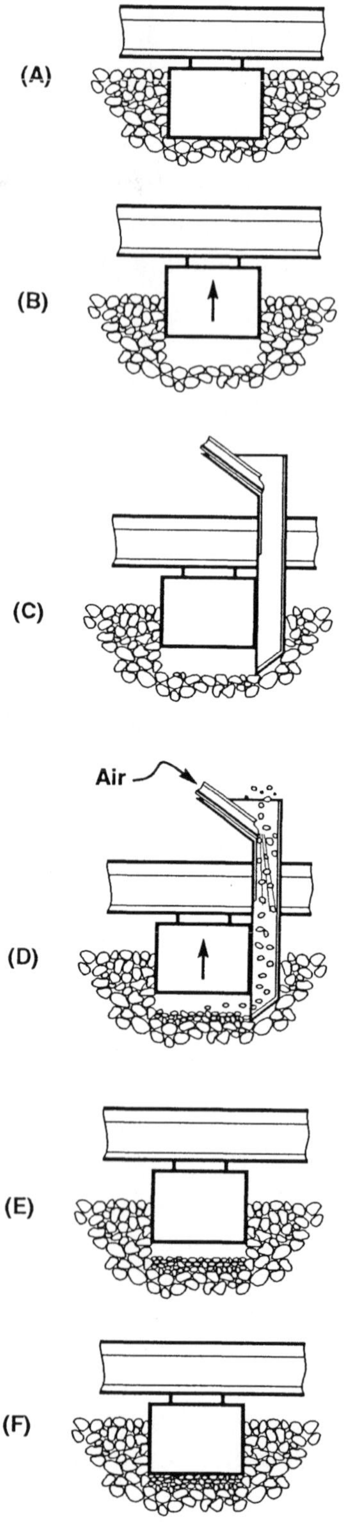

Fig. 14.8 The stoneblowing process

Fig. 14.9 Prototype stoneblowing machine

Fig. 14.10 Handheld stoneblower: left = installing tubes; right = blowing stones

14.2. Ballast Compaction

As demonstrated in Chapter 8, the process of tamping followed by refilling the cribs and shaping the shoulders loosens the top ballast, crib ballast and shoulder ballast. This causes a temporary loss of track lateral stability. Subsequent traffic will recompact the ballast, during which the lateral stability will be regained, accompanied by track settlement with a loss of proper geometry. Tamping will eventually be needed again.

Figure 14.11 shows the relationship between the resistance offered by the track to lateral movement of the track, and the lateral displacement of the track. The loss in lateral resistance caused by the disturbance associated with tamping can be seen, as can the regain in lateral resistance associated with the passage of the first 34 days of traffic which followed tamping. Similar reduction in lateral ballast resistance, and subsequent regains are associated with stoneblowing.

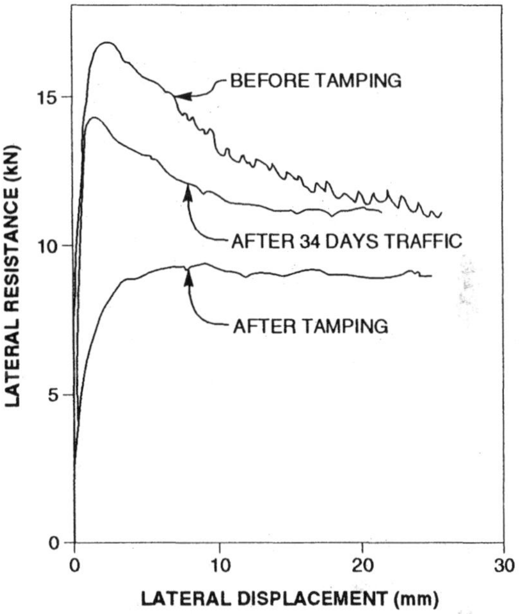

Fig. 14.11 Effect of tamping and traffic on sleeper lateral resistance

These adverse effects of tamping can be reduced by ballast compaction following tamping. Two different methods of compaction are available: 1) lateral track vibration using the Dynamic Track Stabilizer, and 2) crib and shoulder surface compaction. Both of these methods will be described.

14.2.1. Dynamic Track Stabilizer

The Dynamic Track Stabilizer, as shown in Fig. 14.12, applies a combination of horizontal vibration and static vertical load to the track which results in the compaction of loose ballast (Ref. 14.3).

The machine is shown diagrammatically in Fig. 14.13. Also shown are the two stabilizing units which run on the track beneath the main frame, and make contact with the rail heads through eight flanged roller discs which contact the outside of the rail. The vertical load is applied to the track by four hydraulic cylinders which react against the machine frame. Two synchronized vibration units cause the track to vibrate horizontally across the track. The vibrating frequency is adjustable from 0 to 45 Hz. The resulting centrifugal force is up to a maximum of 320 kN.

Fig. 14.12 Dynamic Track Stabilizer

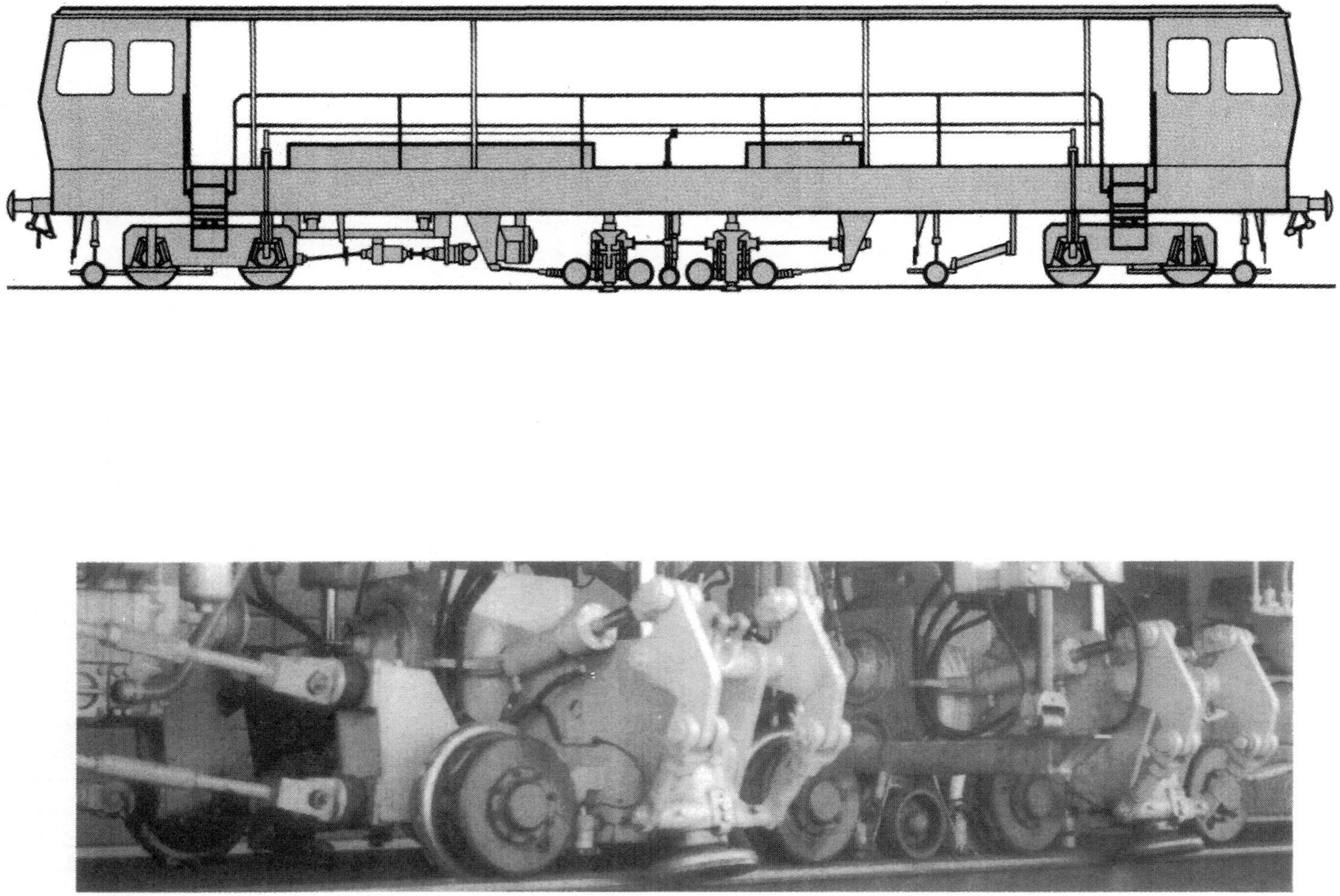

Fig. 14.13 Diagram of Dynamic Track Stabilizer (top); stabilizing units with rollers (bottom)

When operating continuously at full load, following a tamping operation, the effect on the restoration of lateral stability of one pass of the Dynamic Track Stabilizer is regarded as being equivalent to the passage of 100,000 tonnes of normal traffic. Thus the Dynamic Track Stabilizer can be used as an alternative to normal traffic, to induce a regain in lateral resistance following tamping. Figure 14.14 shows that about half the loss in lateral resistance from tamping is regained by the application of the stabilizer. This effect is of particular value following ballast cleaning, ballast renewal and tamping operation, since it allows speed restrictions associated with low lateral resistance to be dispensed with.

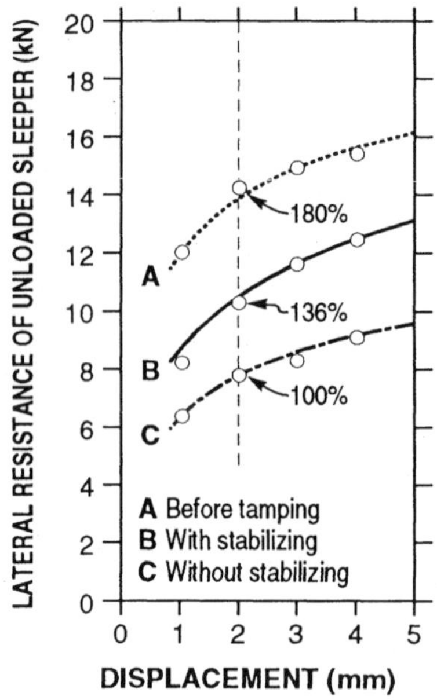

Fig. 14.14 Effect of tamping and stabilizing on lateral resistance

As far as vertical stability is concerned, one pass of the Dynamic Track Stabilizer is equivalent to the trafficking effects of betwen 100,000 and 700,000 tonnes of normal traffic. This effect is of particular value following ballast cleaning, ballast renewal and tamping operations (see Section 15.8.2).

14.2.2. Surface Compactors

Examples of ballast surface compaction machines are shown in Fig. 14.15. The ballast is compacted by application of a vertical vibratory force to pads pressed down onto the ballast surface by a vertical static force. The vibratory force is generated by either a rotating eccentric mass producing a constant dynamic force amplitude for a given frequency or by an eccentric shaft drive producing a constant dynamic displacement amplitude. The cribs are compacted near the rails in the zone penetrated by the tamping tools. The shoulders may also be compacted on the top next to the sleeper ends and on the upper portion of the side slopes. Pressure plates are often used on the side slopes to prevent lateral flow of the ballast during vibration.

The main compactor parameters are static down pressure, generated dynamic force, vibration frequency, and duration. The values of these parameters on available machines vary significantly. The static force ranges from about 7.6 to 9.8 kN (1700 to 2200 lb), the rated dynamic force from approximately 4.9 to 30.7 kN (1100 to 6900 lb) and the vibration frequency from 25 to 75 Hz. Most of the machines are equipped to vary the duration of vibration. These parameters all have important roles in determining the effectiveness of

Fig. 14.15 Crib and shoulder surface compactor

ballast compaction. Other factors influencing compaction are the general characteristics of the compactor, the conditions of the ballast being compacted, and the foundation supporting the ballast. Ballast compaction under vibrating machines is a complex problem. The ballast and the compactor interact during vibration in such a way that a particular compactor will react differently with different ballast and machine conditions, and hence result in a different degree of compaction.

A summary of reported results with vibratory surface compactors prior to 1980 is given in Ref. 14.3. Very little information was available to assess the effects of the individual parameters, and their interaction. However the following trends are indicated:

1) A minimum static pressure is needed to transmit the vibratory force to the ballast, but too large a static force can restrict densification or cause excessive pad penetration. High pressure on the ballast particles also increases the extent of ballast degradation, during compaction.

2) The rate of ballast density increase is most rapid at the start of vibration and diminishes quickly. Therefore a long duration of compaction, which adversely affects productivity of the compactor, is not warranted. A time of 2 to 4 seconds seems to be reasonable with the available ballast compactors.

3) The effectiveness of ballast compaction is very much dependent on the track conditions prior to compaction, such as quality of tamping, the amount and quality of ballast in the crib and shoulder area, and ballast support conditions.

4) Ballast density is significantly increased in the cribs and shoulders beneath and near the areas of pad application.

5) The surface compactor immediately restores a substantial amount of the loss of ballast resistance to sleeper lateral displacement caused by tamping. An example is given in Fig. 14.16 (Ref. 14.4). Other examples are given in Chapter 8.

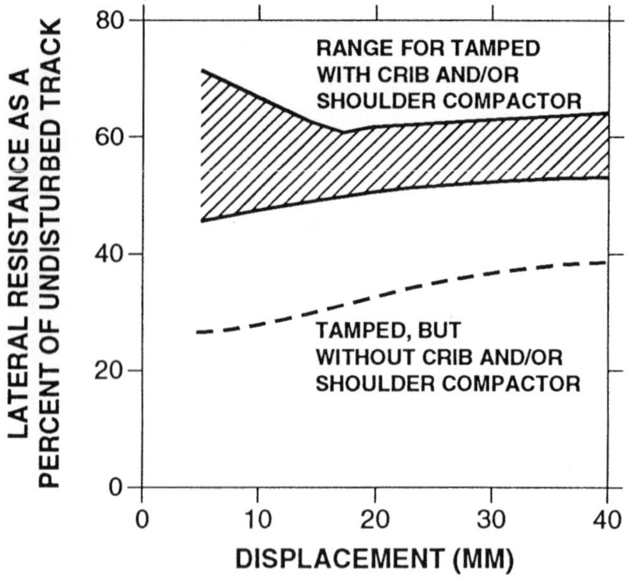

Fig. 14.16 Lateral track resistance with and without surface compaction

6) Surface compaction helps reduce settlement from traffic and retains the tamped track geometry longer than without the compaction. An example is given in Fig. 14.17 from experiments at the British Rail research test track (Ref. 14.5). Also the values of standard

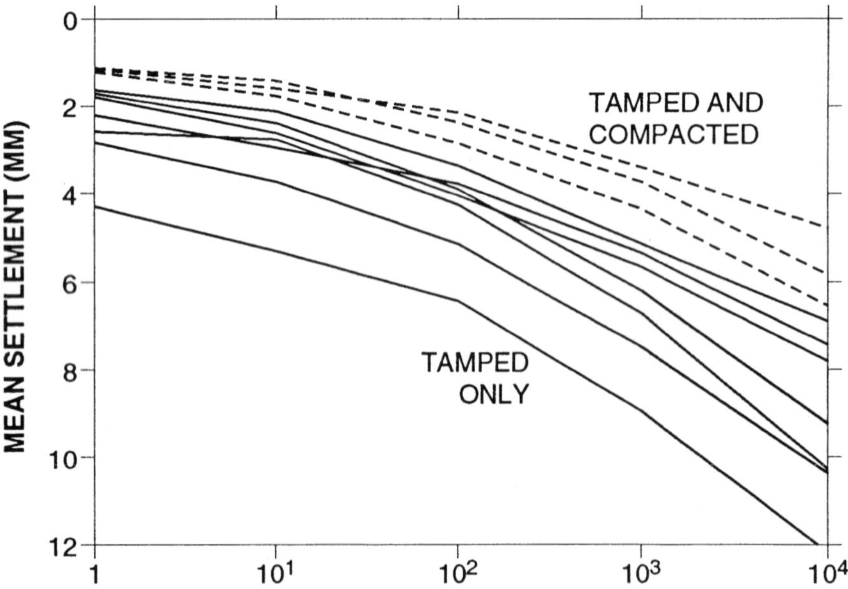

Fig. 14.17 Effect of surface compaction on track settlement

deviation of settlement readings for the tamped-only track were 2 to 4 times larger than those for the tamped and compacted track. Other examples are described in Ref. 14.3.

7) Resistance to longitudinal track forces is improved by surface compaction.

14.2.3. Regain of Resistance to Lateral Displacement

Figure 14.18 compares the relative effects of surface compaction and dynamic track stabilization upon the restoration of the resistance of the track to lateral displacement following tamping (Ref. 14.7). If the resistance of the track to lateral displacement immediately following tamping is assumed to be 100% then, for a lateral displacement of 2 mm, the resistance to lateral displacement following crib consolidation is 115%, and following dynamic stabilization is 134%.

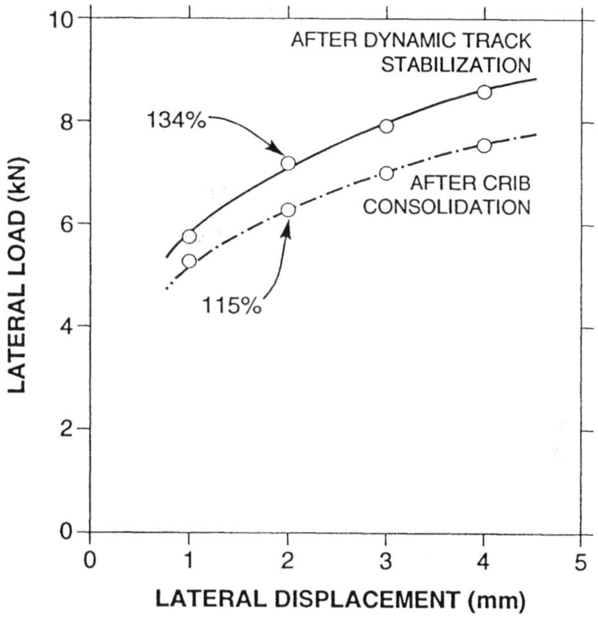

Fig. 14.18 Restoration of track resistance to lateral displacement following tamping

14.3. Ballast Cleaning and Ballast Renewal

Ballast cleaning and ballast renewal are costly and time consuming operations which are highly disruptive to traffic. Therefore they are not to be undertaken casually. The need to ballast clean or renew the track foundation is not easy to establish. As has been shown in Chapter 8, the existence of fine material on the surface of the ballast can arise from many causes. Examples are:

- Ballast fouled by material pumping from a local subgrade failure which would call for track foundation renewal incorporating a blanket layer.
- Worn ballast which would call for ballast renewal.
- Ballast fouled by wagon spillage which would call for ballast cleaning.

A decision regarding the correct measure to employ can only be safely made on the basis of an examination of the track foundation materials down to, and including the subgrade. Traditionally, such an examination has been carried out from a series of cross track trenches. The sinking of bore holes provides an alternative method, although the need to avoid interruption of traffic by the boring rig limits its use.

14.3.1. Rail Mounted Boring Machine

A recent innovation has been the on-track boring rig shown in Fig.14.19. Three boring units are installed on the machine which allow mechanized boring to be carried out through the ballast into the subgrade. The middle boring unit is fixed and bores in the center of the track. As is shown in Fig. 14.20, two units are mounted on side panels that allow them to be swung outwards from the side of the vehicle up to a distance of 3.3 m from the center line of the track. The maximum probing depth is normally 1.9 m below rail head level. If required, the boring depth of the outer units can be extended to 5.0 m below rail head level by fitting tube extension units.

Following boring, the boring tube can be withdrawn together with a core of the materials encountered. The core sample can be inspected through a slot in the side of the tube which is covered during boring (Fig. 14.21).

Figure 14.21 shows that an intact blanketing layer exists and that no fouling of the ballast by subgrade materials is occurring. Such a site would probably respond well to ballast cleaning.

Fig. 14.19 Boring machine for subsurface examination

Fig. 14.20 Moveable boring tube on side panel

14.3.2. Ballast Cleaner

In time, the ballast supporting the track will become fouled with dirt and the ballast will have to be cleaned, or possibly replaced. The ballast cleaner (Fig. 14.22) is a track mounted machine for this purpose. It is equipped with an endless excavating chain which passes beneath the track. As the ballast cleaner moves forward, the excavating chain removes the ballast from beneath the track and conveys it up to vibrating screens which separate the dirt from the ballast (Fig. 14.23). The dirt is conveyed away to lineside or to spoil wagons for subsequent disposal. The cleaned ballast is returned to the track for re-use.

Care must be taken during ballast cleaning to make sure that any existing blanket layer is not inadvertently removed or damaged. Care must also be taken to ensure that the width and inclination of the cut surface are such that the water collecting in the cleaned ballast is able to flow freely across the cut surface and into the track side drainage system.

Ideally, the function of a ballast cleaner is to provide a uniform depth of clean ballast, in a state of uniform compaction, resting on the geometrically smooth cut surface of a compact sub-ballast layer. Such a ballast bed gives the very best chance of compacting uniformly under subsequent traffic loading, to give a stable track of high geometric quality requiring a minimum of future maintenance.

Fig. 14.21 Core inspection through side of boring tube

Unfortunately, the cutter bar of a ballast cleaner used in a traditional way is not able to cut the required geometrically smooth surface in the compact subballast layer. Since the cutter bar is, in general, maintained at a constant depth below the chassis of the machine at a point between the front and rear bogies, the depth of cut is affected by vertical movements of the front and rear bogies. Vertical movements of the rear bogie occur as a result of variations in track geometry behind the cutter bar, resulting from variations in the volume and distribution of the returned clean ballast, which supports the track beneath the rear bogie. Unevenness of the cutting depth can also be caused by the machine operator over-correcting the cutting depth which results from a lack of a reference to which the operator can refer. This tendency is well illustrated in Fig. 14.27 in the vicinity of sleeper number 260.

Fig. 14.22 Ballast cleaner

Fig. 14.23 Excavating chain for ballast cleaner

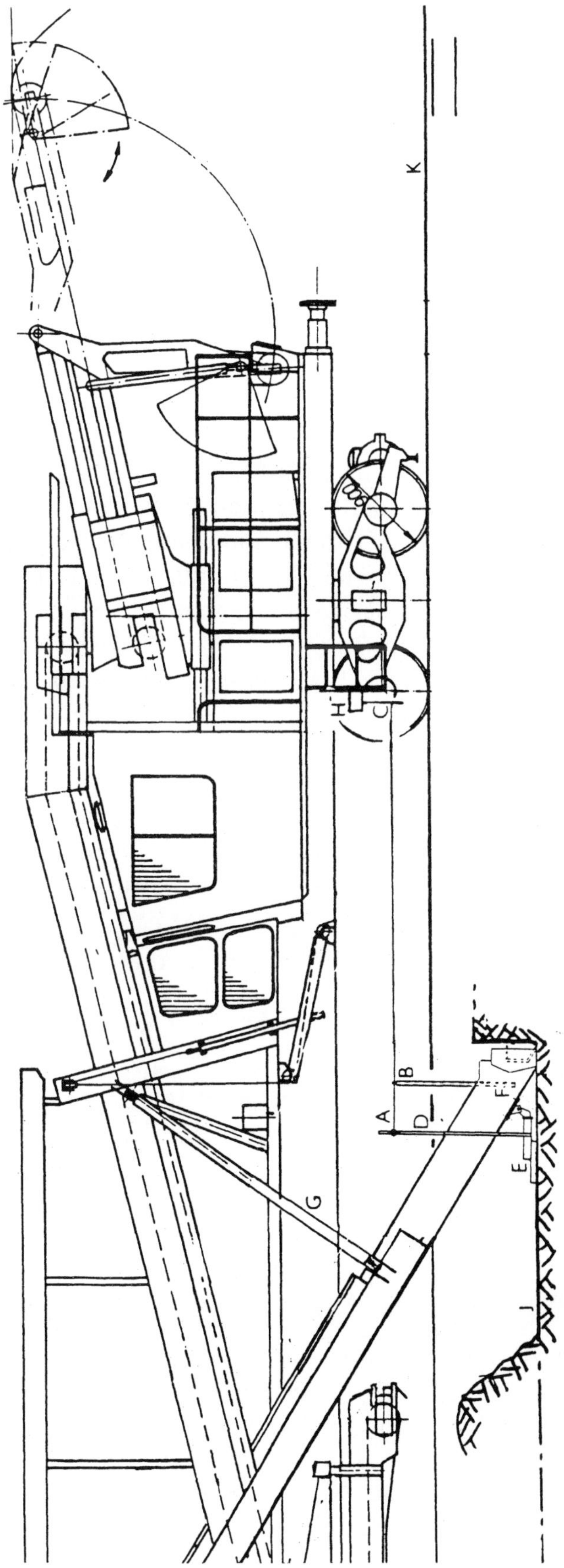

Fig. 14.24 Rear reference depth control for ballast cleaner

In view of the great importance of achieving a geometrically smooth cut surface, it is necessary to provide the cutter bar with an alternative reference that will overcome the undesirable feedback from the rear bogie of the ballast cleaner.

14.3.2.1 Depth control by rear reference

A simple method of creating an alternative reference, which is still in a development stage, is shown in Fig. 14.24. The ballast cleaner is provided with a rear reference which allows it to be used in a design or automatic smoothing mode that is analogous to that of a tamping machine (Ref. 14.8). The system relies for its operation on a reference plate which slides on the surface of the freshly cut sub-ballast surface immediately behind the cutter bar. The cutter bar is related to a line joining this rear reference to a front reference running on the undisturbed track ahead of the cutter bar.

With the system shown in Fig. 14.24 a wire (AC) is contacted by a detector (B) which is attached to the cutter bar. The rear wire attachment point (A) is connected via the support (D) to the reference plate (E) which slides on the cut ballast surface (J). The reference plate is connected to the cutter bar at the pivot point (F). The front wire attachment point (C) is fixed to the axle box of the rear axle of the front bogie. The height of the front wire attachment point can be adjusted by operating the actuator (H). Should the detector lose contact with the wire, a 'LIFT' signal is sent to the cutter bar height control ram (G) which lifts the cutter bar until contact with the wire is re-established.

Used in the automatic smoothing mode, no pre-measurement of the track is necessary. The cutter bar depth is adjusted automatically and no intervention is required on the part of the ballast cleaner operator. The height of the front reference point above the track remains constant and the profile cut is a smoothed, phase shifted copy of the original track geometry. Since the cutter bar follows a path that is parallel to the mean path of the track, the volumes of dirty material to be disposed of, and the volume of make-up ballast to be imported, are both kept to a minimum.

Used in a design mode, the cutter bar cuts a surface having a geometry that is pre-determined by the engineer. In this case the geometry of the track to be ballast cleaned is pre-measured and a target geometry for the cut surface is determined. During the ballast cleaning operation, the height of the front reference point is adjusted either with respect to the track, or to a datum laser plane, such that the path of the cutter bar is that of the required geometry.

14.3.2.2 Hydraulic 'U' tube depth control

The system makes use of a number of inter-connected, water-filled 'U' tubes to establish a horizontal plane to which the height of the track, before and after ballast cleaning, and the height of the cutter bar can be referred. Floats connected to the potentiometers measure water levels within the 'U' tubes and transmit the level data to a central processing unit. Four instruments near the operator indicate the depth and cross level of the cutter bar and ballast grading unit.

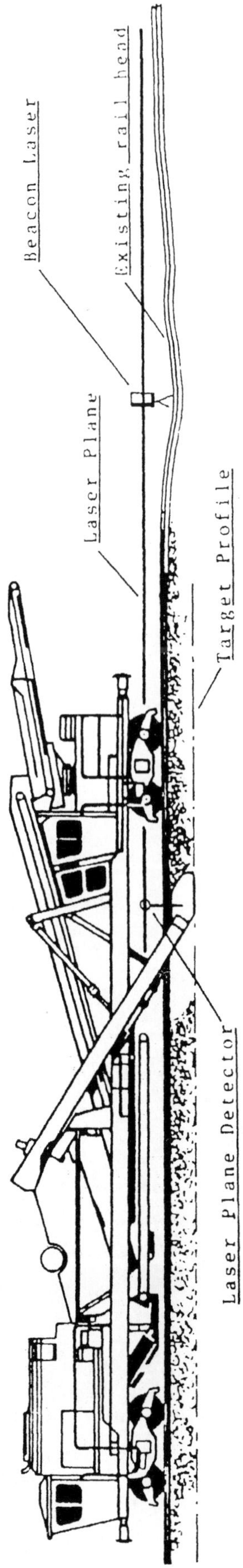

14.25 Laser based depth control for ballast cleaner

The system can be used in a design or an automatic mode. In the automatic mode, the system produces a phase shifted copy of the 'pre-ballast clean' track geometry at the cut surface. Initial problems associated with settling time following a change in track level, and with oscillations of the water resulting from horizontal accelerations of the ballast cleaner, have now been overcome by the use of pressure transducers.

14.3.2.3 Laser datum based depth control

Rotating lasers, as used for surveying purposes (Chapter 15), can also be used for machine control. In general, so called 'grade' lasers are used for this purpose as these allow the laser to be set with either a single or compound grade instead of the normal horizontal plane. Lasers used for machine control are normally rugged versions of those used for surveying purposes.

Figure 14.25 shows how the laser plane is set up parallel to the target profile. Shown also is a ballast cleaner equipped with laser plane detectors that can detect the beam of laser light, and which either operate a simple indicator to which the machine operator responds, or operate the cutter bar height control system of the machine directly.

Fig. 14.26 Ballast cleaner operating under laser control

Figure 14.26 shows a ballast cleaner operating under laser control. In this case the laser is attached to the signal gantry in the foreground.

Figure 14.27 compares the smoothness of the surfaces cut by a ballast cleaner working under laser guidance to that of the same ballast cleaner working under manual guidance. The advantages to be gained from laser guidance are clear.

Ballast cleaners and tampers are usually equipped with two systems: one system for the control of vertical level, the other for the control of cross-level.

A further development of the automatic machine control system allows the machine itself to survey the track on which it is to work, and subsequently calculate on a computer an optimum design for the surface to be formed.

A limitation of laser based systems in the past has been the inability to cope with vertical curvature. A recent development (Ref. 14.9) allows a small computer to be pre-programmed with curvature details which, when connected to a distance wheel and an electric telescopic mast upon which the laser detector is mounted, allows the laser datum to be offset according to distance travelled. With this device, vertical curves of any required geometry can be formed.

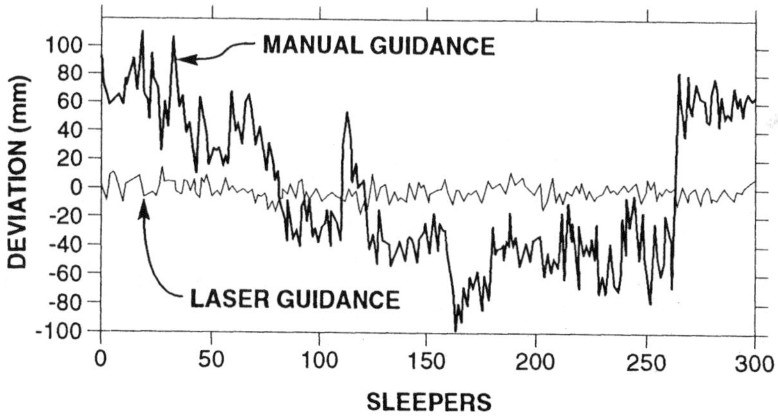

Fig. 14.27 Comparison of cut surface roughness with manual and laser guidance

14.3.3. Off-Track Ballast Cleaning

It is occasionally necessary to totally remove the ballast from beneath the track, for example when the ballast is excessively dirty, or as part of a blanket placing operation. As a result of a world shortage of economic supplies of new ballast, there is a growing interest, in such cases, in recycling such ballast which would normally have been discarded and replaced by new ballast. In such an operation, the ballast cleaner cuts the ballast and places it into a train of wagons, for example of the type shown in Fig. 14.28 in which the floor is in the form

Fig. 14.28 Conveyor/hopper wagons

of a conveyor belt which interconnects with an elevating conveyor. It is thus possible to transfer material along a train of such wagons.

Once ballast cutting has been completed, the conveyor/hopper wagons are moved to a siding where they discharge the cut material into a ballast cleaning unit. The cleaned ballast and the dirt are deposited into trains of wagons of similar type for reuse, and disposal, respectively.

14.3.4. Partial Ballast Cleaning

Cleaning the shoulder ballast will improve the drainage of the ballast supporting the track (see Chapter 13, Section 13.4.3). To be effective the shoulder must be cleaned to the bottom of the ballast layer being drained. Also the cleaning must extend all the way to the ballast slope to avoid leaving a dam of uncleaned ballast. Cleaning the shoulder ballast periodically may prolong ballast life by allowing fines migration from the ballast voids in the cribs and under the sleepers, thus lengthening the time until full undercutting/cleaning is required. Cleaning the shoulder ballast and leaving the ballast beneath the track uncleaned is only recommended when the drainage and elastic properties of the ballast supporting the track are adequate.

14.3.5. Placing a Blanket Layer in an Existing Track Sub-structure

A ballast cleaner can be used to place a blanket layer of up to 150 mm (6 in.) in thickness, in an existing track substructure, as shown in Fig. 14.29. The ballast is first cleaned in the normal way. Blanketing material is then deposited along the track in sufficient quantities to produce a blanketing layer of the required thickness. Prior to making a second pass, the screens of the ballast cleaner are modified. A fine mesh screen is fitted into the top screening unit, and the center screen is covered with a rubber plate. The ballast cleaner then makes a second pass.

The "previously cleaned" ballast/blanketing material mix is cut and conveyed to the modified screening unit, the effect of the modification being for the fine screen to retain the ballast but pass the blanketing material. A conveyor returns the blanketing material to the ballast surface cut by the cutter bar at the time of the second pass. A plough and vibrator unit form the blanketing material into a compact layer of the desired thickness. A second conveyor returns the ballast to the surface of the blanket layer.

Where a greater depth of blanketing material is required, e.g. 50-60 cm, the following technique can be employed:

- The ballast cleaner makes two passes.

- In the first pass, the ballast cleaner cuts to a depth of 25-30 cm, returning the dirt to the track and conveying the cleaned ballast to ballast wagons, e.g. conveyor/hopper wagons as shown in Fig. 14.28.

- In the second pass, the ballast cleaner cuts the lower layer of ballast, together with the dirt previously deposited on the track during the first pass of the ballast cleaner, and conveys it to waste.

- The sand blanket is placed, profiled and compacted by conventional means.

- The ballast previously saved in the conveyor/hopper wagons is returned to the surface of the blanketing material, together with any additional ballast that is found to be necessary.

In connection with the foregoing, it is worth noting that a blanket layer does not necessarily have to be placed immediately above the subgrade to be effective. It may still be effective if it is placed between the bottom ballast layer and the underlying granular material. This assumes that there is sufficient depth of clean ballast above the blanket layer. The layer of granular material, including uncleaned ballast, between the underside of the blanket layer and the top of the subgrade will act as part of the subballast layer. Any products of subgrade attrition may well enter this subballast layer, but their upward migration will be arrested by the blanket layer.

14.4. Ballast Regulators and Grading Machines

The stability of the track in the horizontal plane depends to a large extent upon the spaces between the sleepers being filled with ballast to top of sleeper level, and the shoulder ballast being correctly formed.

Formation and subsequent maintenance of correct track ballasting is best carried out by ballast distributing and grading machines, an example of which is shown in Fig. 14.30. These machines are able to 1) move existing ballast to any part of the track cross section, 2) fill the spaces between the sleepers with ballast up to top of sleeper level, 3) brush surplus ballast away from the top surface of the sleeper, and 4) form the shoulder ballast to any desired cross sectional profile. Some models of ballast regulator are also able to collect surplus ballast, store it in a hopper, discharge and re-distribute it at some other point along the track. Use of such machines may increase as supplies of acceptable ballast become more scarce.

14.5. Summary of Problems and Solutions

The application of available techniques to the remedying of track foundation failure conditions is costly, time consuming and disruptive. Therefore, it is important to insure that:

1) the cause of the failure condition has been correctly identified, and

2) an appropriate remedial measure is applied to the failure condition being considered.

Table 14.1 lists some of the more general track foundation failure conditions, and the solutions that are appropriate to their cure.

Table 14.1 Failure condition and appropriate remedial action

CONDITION	SOLUTION				
	Ballast Clean	Ballast Clean + Blanket	Ballast Renewal	Ballast Renewal + Blanket	Increase Ballast Depth
Ballast/Sleeper Attrition	Short Term	N.A.	Long Term	N.A.	N.A.
Ballast Wear	Short Term	N.A.	Long Term	N.A.	N.A.
Fouled Ballast	Long Term	Long Term (a)	N.A.	N.A.	N.A.
Subgrade Attrition and Pumping	Short Term	Long Term	N. A.	Long Term (b)	N.A.
Progressive Shear Failure	N.A.	N.A.	N.A.	N.A.	Long Term

a- Advisable if no blanket (subballast) present; b-Advisable if ballast requires renewal.

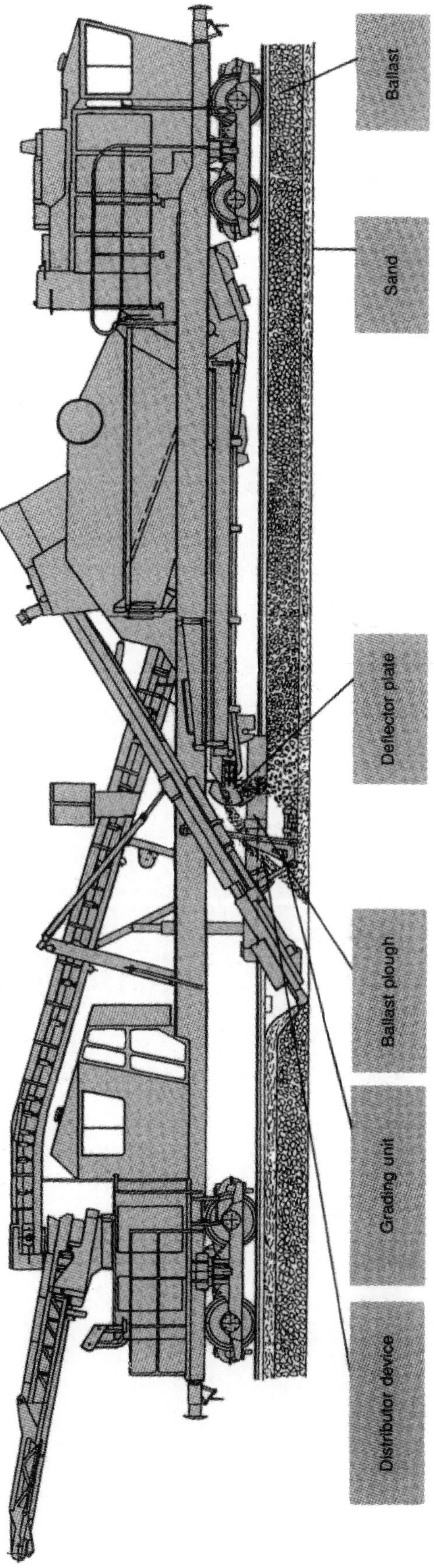

Fig. 14.29 Method for placing blanket layer with a ballast cleaner

Fig. 14.30 Ballast regulator

REFERENCES

14.1 Waters, J. M. (1981). **Pneumatic Stone Injection as a Means of Adjusting Track Level.** *Proceedings*, 4th International Rail and Sleeper Conference, Adelaide, Australia.

14.2 **Efficient track ballasting.** Journal of Applied Pneumatics, November 1988.

14.3 Selig, E. T., Yoo, T. S. and Panuccio, C. M. (1982). *Mechanics of ballast compaction, Vol. 1: technical review of ballast compaction and related topics.* Final Report, Department of Civil Engineering, State University of New York at Buffalo, for US Department of Transportation, Report Nos. FRA/ORD- 81/16.1 and DOT-TSC-FRA-81-3, I, March.

14.4 Riessberger, K. (1972). *The effect of various track maintenance procedures, especially of ballast compaction on lateral resistance of track.* A lecture given in Freilassing on June 15.

14.5 Powell, Michael C. (1972). *The effect of consolidation by the VDM 800 consolidator on the vertical deformation of track under subsequent traffic.* Interim Report, British Railways, Research Department Track Group, Technical Note S.M. 100, January.

14.6 Schubert, E. (1976). Report to OVG session. Bad Gastain, Austria.

14.7 Waters, J. M. (1989). **Improving the Geometric Quality of the Surface Cut by a Ballast Cleaner.** *Proceedings*, The Fourth International Heavy Haul Railway Conference, Brisbane, Australia.

14.8 Johnson, D. M. (1991). **The Use of Lasers for Track Surfacing.** Transportation Research Board Annual Meeting, Washington, D.C., U.S.A.

15. Track Geometry and Track Quality

15.1. Measurement of Track Geometry

It is often necessary to know the vertical profile of the existing track with respect to ground datum so that target profiles for the new track can be established. In addition to traditional systems for measuring track geometry based upon the use of optical levels, a number of alternative systems are now available. Examples of such systems are given below.

15.1.1. Datum Lasers

Laser surveying follows the same general principles as conventional optical surveying. In laser surveying, however, a laser provides a datum to which levels are referred. Readings are recorded at the staff end, and it is thus possible to survey using a single operator.

With this method a low power laser (generally about 2 mW) is set up to provide a collimated rotating beam of light which defines a horizontal plane. The beam may be either visible red or invisible infra-red. The range of the laser beam varies but is generally between 300 m and 600 m.

A measuring staff is equipped with a laser light detector. The operator locates the laser beam by extending the measuring staff until an audible warning indicates that the center of the laser beam has been located. The staff reading is then logged in the same manner as with conventional optical surveying.

A number of different types of staff exist, some of which have integral detectors. An early type of staff had a motorised detector which travelled up and down until it located the laser beam. Upon locating the beam, the staff stopped moving and indicated to the operator, via an audible warning, that the reading should be logged. A modern variant is a completely solid state device which continuously locates the laser beam and displays the staff reading on a liquid crystal display. With some staffs, the extension is logged directly to a microchip when a button is pressed. Using this type of staff in connection with a hand-held computer significantly reduces data processing time.

The accuracy of laser surveying is approximately ± 1 mm.

15.1.2. FROG Trolley

The FROG system (Ref. 15.1) shown in Fig. 15.1 was developed by British Railways. FROG measures the absolute, unloaded, longitudinal vertical profile of the track. The system comprises three sub-systems:

Fig. 15.1 The FROG system: a) measuring track geometry, and b) mini-computer

1) A trolley which runs on the rails (Fig. 15.1a).

2) A pair of highly accurate inclinometers, one measuring longitudinal slope, the other measuring cross level.

3) A portable mini-computer (Fig. 15.1b).

With the trolley stationary on the rails, the output of the inclinometer measuring longitudinal slope is fed to the computer and stored. At the same time, the output of the inclinometer measuring cross level is also fed to the computer and stored. FROG is then moved forward to the next sleeper and the process repeated.

On completion of the measuring run, the stored data is processed by the portable computer. The profile of one rail is calculated by placing end to end, the sequence of longitudinal slope measurements. The profile of the other rail is calculated by adding the cross level measurements to the longitudinal slope.

The results of the FROG survey can be plotted, together with such information as the standard deviation of the left and right hand rails as shown in Fig.15.2. Alternatively, the data can be transferred to a standard spread sheet for further processing, e.g. calculation of the mean line and its standard deviation, and design calculations (see Chapter 14).

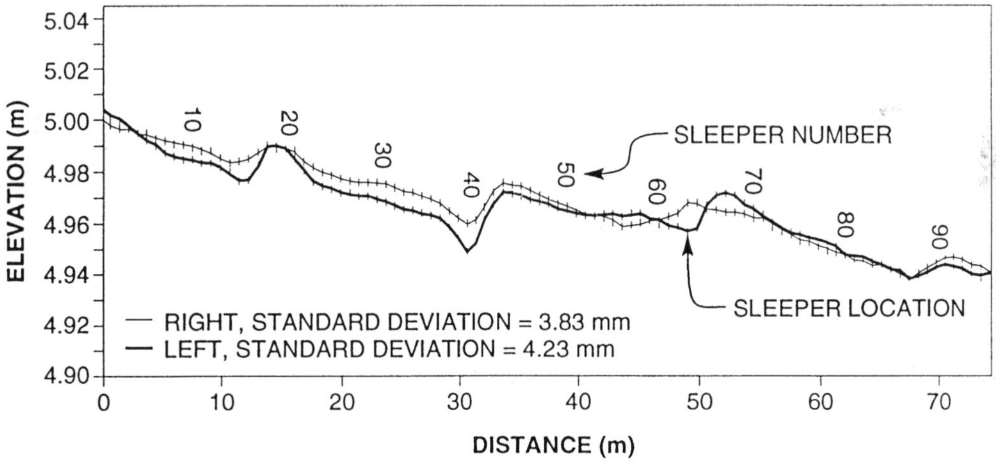

Fig. 15.2 Example plots of FROG survey data

15.1.3. Track Survey Cars

Track survey cars of the type illustrated in Fig. 15.3 permit the accurate and absolute simultaneous measurement of track level and alignment using laser reference chords. The laser reference chords are set up from fixed point to fixed point (e.g. mast). The spacing of the fixed points is normally 40 m to 80 m. Using this technique, both short (up to 20 m) and long wavelength faults are detected and measured.

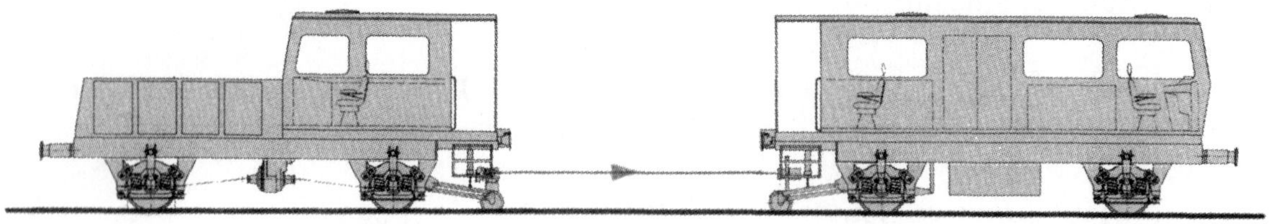

Fig. 15.3 Track survey cars using laser reference chords to measure level and alignment

The target track data can be entered via floppy disc, or keyed manually into the computer of the measuring car. By comparing the target data with the actual data, the required correction values for line and level can be calculated. This data can then be passed to the tamping machine in the form of a list or a floppy disc.

As shown in Fig. 15.3, the survey car consists of a transmitter section (A) and a receiver section (B). The transmitter section contains a laser, the emitting direction of which can be controlled remotely from the cabin of the receiver section. The receiver section also contains a laser detector unit. Existing external reference points (e.g. overhead electrification masts) are usually placed along the track at spacings of between 40 m and 80 m.

The laser detector within the receiver section is positioned adjacent to the first reference point, and its lateral displacement from the reference point is measured. The laser emitter, which is housed within the transmitter section, is positioned beyond the second reference point.

A video camera, mounted adjacent to the laser in the transmitter section, and a TV monitor mounted in the cab of the receiver section, allows the laser beam to be directed towards the laser detector by remote control from the cab of the receiver section. Once the laser beam has been directionally adjusted such that it strikes the detector, its direction is fixed. The position of the laser beam with respect to the laser detector is logged, and then re-logged, at every 250 mm forward movement of the receiver unit along the track.

When the receiver section reaches the second reference point, the transmitter section is moved to a point beyond the third reference point. Once the laser beam is re-positioned with respect to the laser detector and directionally fixed, surveying of the track between the second and third reference point can commence.

15.2. Track Settlement

While traffic can tolerate a certain degree of unevenness in the track, a point is eventually reached at which the track geometry has to be improved. The minimum geometric quality to which the track has to be maintained is a function of the speed and type of traffic that is being carried. Clearly, tracks carrying high speed and/or passenger traffic will need to be maintained to a higher geometric quality than will tracks dedicated to low speed and/or freight traffic.

The level of geometric quality that can be achieved above the minimum required is indicative of the time that it will elapse before the track will require resurfacing. Thus, no matter what the type of traffic being carried, there is always an incentive to achieve the highest geometric quality possible at the time of resurfacing.

Routine maintenance by most railway administrations to correct geometry errors is carried out by combined tamping and lining machines. After the rectification of the geometry of the track by these machines, the loading from the traffic causes the ballast to compact and settle. Since the settlement is not uniform, faults in track geometry develop.

Figure 15.4a shows the longitudinal rail head profile of one rail of a 100 m length of track. The vertical scale is as shown. The profiles refer to:

- PRE TAMP immediately prior to tamping.
- POST TAMP immediately following tamping.
- 1 MONTH 1 month following tamping.
- 6 MONTHS 6 months following tamping.
- 33 MONTHS 33 months following tamping.

A comparison between the PRE TAMP and the POST TAMP profiles shows that the tamping machine was very successful in removing the short wavelength faults in vertical geometry.

A comparison between the POST TAMP, 1 MONTH, 6 MONTH, and 33 MONTH profiles, shows the deterioration in the quality of vertical geometry that occurred under subsequent traffic.

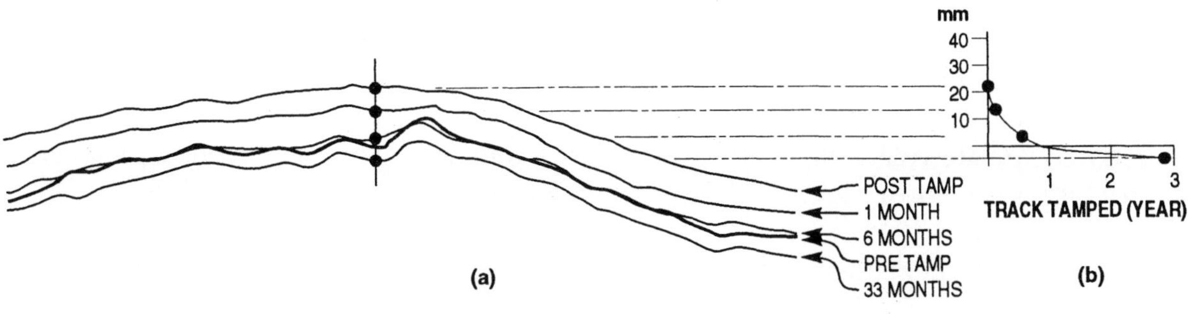

Fig. 15.4 Longitudinal railhead profile (a) and the associated track settlement at one location (b)

Figure 15.4b shows the relationship between track settlement and time, for the typical section indicated. It can be seen that the rate of settlement is initially rapid but ultimately reaches a near stable condition.

Of particular interest is the comparison between the PRE TAMP and the 33 MONTH profiles. It will be seen that these profiles are virtually identical. The track has clearly inherited its 33 MONTH geometry from the long term geometry associated with the previous maintenance cycle. The reasons for the persistence of this long term geometry are discussed in greater detail in a later section.

Figure 15.5 shows similar longitudinal rail head profile data superimposed in a three dimensional form to separate the time intervals.

Clearly, it is differential track settlement and not total track settlement that is of prime interest to the track engineer, since it is differential settlement that gives rise to faults in vertical track geometry.

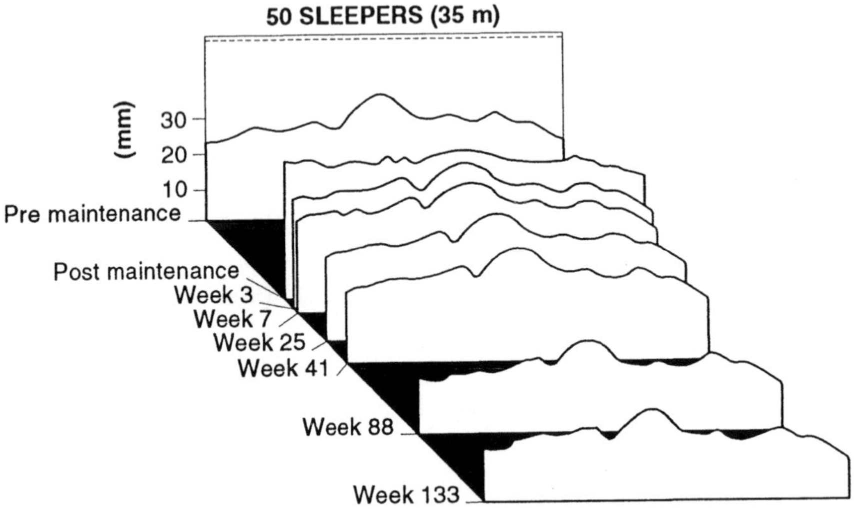

Fig. 15.5 Three dimensional presentation of rail head profile data

15.3. Track Geometry

For any given train speed, undulations in the track induce vertical vibrations. Depending upon their frequency, these vibrations have a deleterious effect on passenger comfort and vehicle ride which can result in track damage. The longer the wavelength of the undulation, the lower the frequency of the induced vibration.

Very long wavelength undulations in the track induce very low frequency vibrations in the vehicles passing over them. In the vertical plane, these low frequency vibrations can be largely ignored. For example, at train speeds of 100 km/h, it is unlikely that undulations in the track having a wavelength in excess of 30 m will have a deleterious effect on either passenger comfort or vehicle ride. Similarly, at 200 km/h, it is unlikely that wavelengths in excess of 50 m will be significant. Since there is no advantage to be gained from removing these longer wavelength undulations from the track, they can be removed from the measured track profile by filtering. Attention can then be concentrated on the remaining shorter wavelengths. In the horizontal plane, however, because of the lower natural frequency of typical rolling stock in the lateral plane, these low frequency vibrations, and the longer wavelength track faults giving rise to them, cannot be ignored.

15.3.1. Filtering of Measured Track Geometry

Two types of filters are available for removing long wavelength components, Box Car and Triangular. A "Box Car" filter ascribes to each point being considered (e.g. sleeper location), an elevation equal in value to the average of all the elevation values being considered by the filter. A triangular filter ascribes to each point being considered, an elevation equal in value to the average of all the elevation values being considered by the filter, such elevation values having been weighted in proportion to their distance from the point being considered. For example, the elevation values derived from the application of a 5 point triangular filter would be as shown in Table 15.1.

Table 15.1 Example of triangular filter

Position	1	2	3	4	5	6
True Elevation (mm)	13	16	29	42	41	32
Weighting	1/9	2/9	3/9	2/9	1/9	0
Weighted Elevation	1.4	3.6	9.7	9.3	4.6	0
Average elevation ascribed to position 3: 1.4 + 3.6 + 9.7 + 9.3 + 4.6 = 28.6 mm						
Position	1	2	3	4	5	6
True Elevation (mm)	13	16	29	42	41	32
Weighting	0	1/9	2/9	3/9	2/9	1/9
Weighted Elevation	0	1.8	6.4	14.0	9.1	3.6
Average elevation ascribed to position 4: 1.8 + 6.4 + 14.0 + 9.1 + 3.6 = 34.9 mm						

All positions within the section of track being considered are processed in a similar manner. The line passing through the average elevations so determined, represents the filtered profile.

The larger the number of points considered by a filter, the greater will be the smoothing effect of the filter, i.e. the greater will be the rejection of the short wavelength component of the profile being filtered.

15.3.2. Standard Deviation

A convenient way of quantifying the geometric shape of a section of track is by its standard deviation with reference to a smoothed line from which the short wavelength components of the track geometry have been removed by filtering.

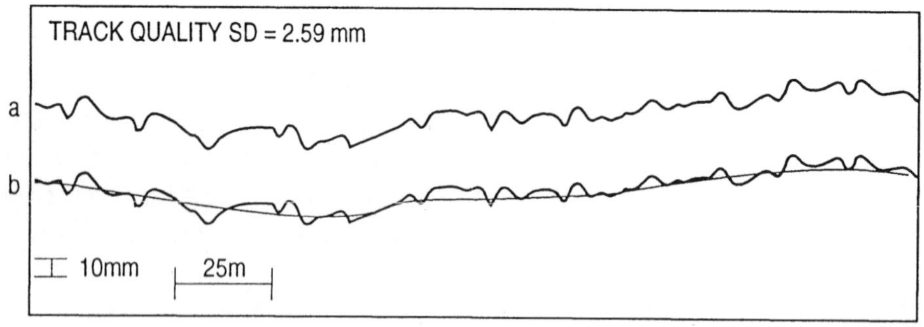

Fig. 15.6 Geometric shape of a track section with reference to a smoothed line

Figure 15.6a shows the longitudinal vertical rail head profile of one rail of a length of track. Figure 15.6b shows the rail head profile with a superimposed smoothed line obtained by removing the short wavelength components of the track geometry by the application of 51 point triangular filter, each point corresponding to a sleeper position. In this example the standard deviation is 2.59 mm. The standard deviation in this case was determined over successive 200 m lengths of track.

15.4. Track Quality

The standard deviation of a section of track provides a convenient way of quantifying its geometric quality. The higher the standard deviation, the poorer the quality of the track. Figure 15.7 is an example of the way in which the quality of the track, in terms of standard deviation, changes during a maintenance cycle. In this case, the tamping machine has improved the standard deviation of the track from 2.9 mm to 1.2 mm. During the month following tamping however, there has been a relatively rapid deterioration in track geometry from a standard deviation of 1.2 mm to 2.2 mm. This has been followed by a slower rate of deterioration, which is approximately linear with time. After one year, the standard deviation has risen to 3.2 mm.

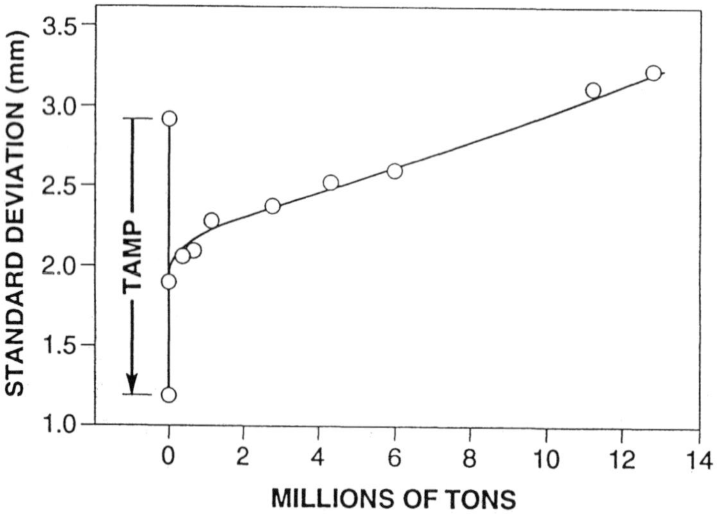

Fig. 15.7 Track quality changes during maintenance cycle

15.4.1. Track Quality Standards

The minimum quality to which track geometry may be allowed to deteriorate ('Minimum Permissible Track Standard' or 'Intervention Level') will depend, among other things, upon the minimum geometric quality of track that can be tolerated. This in turn will depend upon the type of traffic being considered, i.e., freight or passenger.

For passenger trains, travelling at 200 km/hour, the limits in vertical smoothness given in Table 15.2 would be considered reasonable. In the case of lateral irregularlities for any given speed, the longer wavelengths are of more importance.

Table 15.2 Smoothness limits at 200 km/h

Dominant Component	Wavelength	Desirable Maximum Peak-to-Peak Amplitude	Parameter Affected
	m	mm	
Ballast	50	16	Vehicle Ride (Comfort)
	20	9	
	10	5	
	5	2.5	
Rail	2	0.6	Dynamic Track Forces (Track deterioration)
	1	0.3	
	0.5	0.1	
	0.05	0.005	

15.4.2. Track Recording Cars

Following track maintenance, there is a need to regularly monitor the geometric quality of the track to ensure that it has not fallen below the minimum required standard. Track recording cars, such as that shown in Fig. 15.8, are well suited to this task, since they are able to make the required qualitative assessment of the geometric quality of track at a speed that is compatible with that of normal traffic. Recording cars are manufactured in the form of self-propelled vehicles, or coaches which can form part of a high speed train formation.

Successive track recording cars run over the same section of track allow track geometry deterioration trends to be established. Once this has been done, the probable future date at which the quality will have deteriorated to the minimum permissible, can be calculated and the required maintenance resources allocated. An acceleration in the rate of deterioration of a section of track from one maintenance cycle to the next can be identified and the reason sought, e.g. the possible need for ballast cleaning.

Fig. 15.8 Self propelled track recording car

15.4.3. Inherent Track Quality

Figure 15.9 shows two sections of track, 1 km apart, both carrying the same traffic (Ref. 15.2). The downward pointing arrows indicate tamping operations. The track shown in the upper part of the drawing has a standard deviation of 1.5 mm and has required 2 tamping operations in 5 years to maintain the track quality at that level. Such a track can be regarded as having a good inherent quality. The track shown in the lower part of the drawing however, has a standard deviation of 3.2 mm and has required 6 tamping operations in the same 5 year period, to maintain the track quality at that level and can thus be regarded as having a relatively poor inherent quality.

Since the magnitude of the faults giving rise to subsequent track quality deterioration, as indicated by initial track quality, are less in the case of the 'Good' track than in the case of the 'Poor' track, it would seem reasonable to expect that the rate at which the quality of the 'Good' track subsequently deteriorated would also be reduced, as was the case.

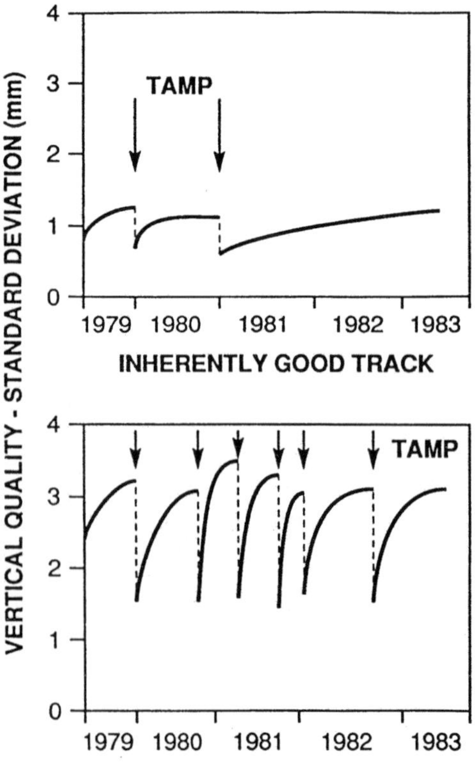

Fig. 15.9 Tamping requirements as a function of track quality

Since the two track sections are of the same age, have the same construction and carry the same traffic, the differences in inherent quality can probably, to a large extent be attributed to differences in the inherent shapes as discussed below.

15.4.4. Conclusions Regarding Track Quality

The conclusions that have been reached with regard to track quality can be summarized as follows:

1) Track has an inherent quality which is determined during the early part of its life, and which is a function of the quality of the components from which the track was constructed, as well as the smoothness and compactness of the supporting ballast bed.

2) Track having a good inherent quality gives a good ride to traffic and requires little maintenance.

3) Track having a poor inherent quality gives a poor ride to traffic and requires much maintenance.

4) The advantages to be gained from good inherent quality are of value, no matter what the speed or type of traffic being carried.

15.5. Inherent Track Shape

It has been observed over the years that track appears to have an inherent shape which remains with it throughout its life. This inherent shape appears to be introduced into the track at the time of its original construction. Achieving subsequent changes in the inherent track shape is very difficult.

There can be little doubt that, to a large extent, Inherent Track Quality is a function of Inherent Track Shape.

15.5.1. Persistence of Inherent Track Shape

An example of the way in which inherent track shape persists through a tamping cycle is shown in Fig. 15.10. This figure shows the longitudinal rail head profile of one rail of a length of track following tamping. It can be seen that although the roughness of the track is increased by subsequent trafficking, the inherent shape of the track remains unaltered.

Figure 15.11 is an example of the way in which inherent track shape persists from tamping cycle to tamping cycle. The upper part of the figure shows the way in which the track quality of a section of track has changed over a number of years. The improvements in track quality correspond to mechanical tamping operations.

The moments in time at which track quality determinations were made are numbered, and the corresponding track profiles are shown in the lower part of the drawing. The

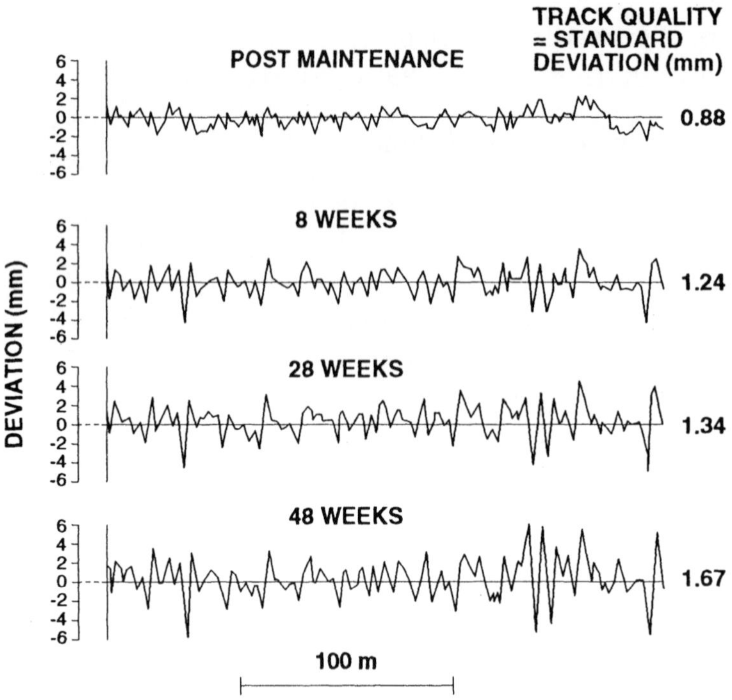

Fig. 15.10 Persistance of inherent track shape through a tamping cycle

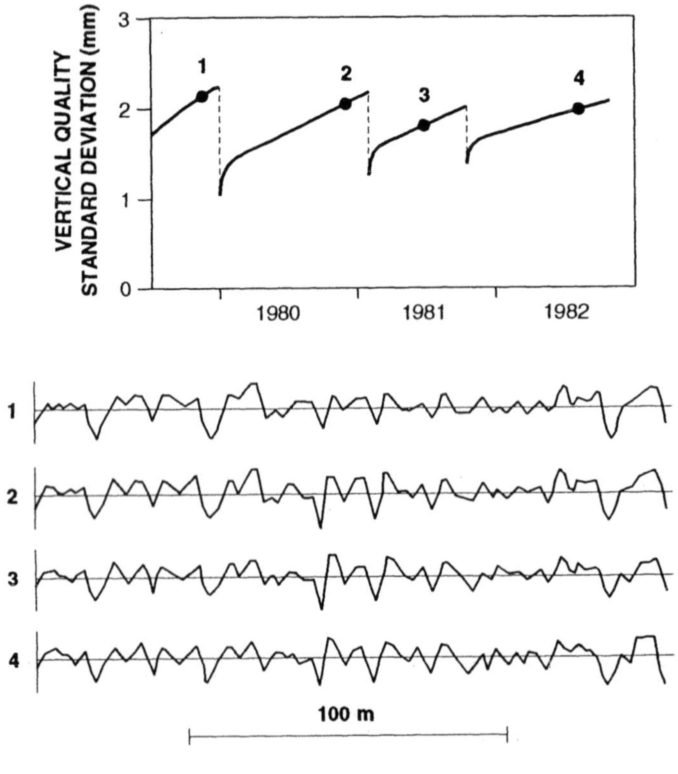

Fig. 15.11 Persistence of inherent track shape from tamping cycle to tamping cycle

drawing shows that no matter how many times the track is tamped, it always deteriorates towards the same inherent shape.

15.5.2. Influence of Rail Shape and Ballast Surface Profile

The reasons for the existence of an inherent track shape can be largely attributed to the shape of the rail and the profile of the supporting ballast bed as indicated diagrammatically in Fig. 15.12.

Figure 15.12a indicates that for short wavelengths of less than approximately 5 m, the bending stiffness of the rail is high, compared with the resistance to deformation offered by the supporting ballast bed. Thus, the rail imprints, via the sleepers, the short wavelength component of its shape into the surface of the ballast bed.

Conversely, Fig. 15.12b indicates that for longer wavelengths, the bending stiffness of the rail is low compared with the resistance to deformation offered by the supporting ballast bed, with the result that the track conforms to the long wavelength component of the surface profile of the ballast bed.

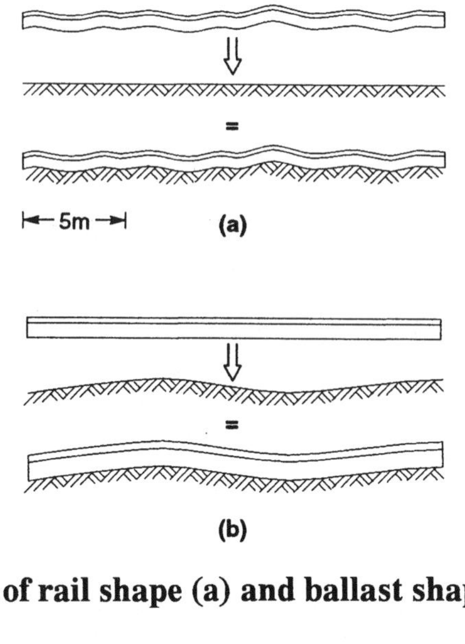

Fig. 15.12 Influence of rail shape (a) and ballast shape (b) on track shape

15.5.2.1 Rail shape

Examples illustrating the persistence of rail shape are shown in Figs. 15.13 and 15.14.

Figure 15.13 shows the longitudinal head profile of rail through the various stages of tamping following a ballast renewal. It can be seen from a comparison of rail head profiles corresponding to days 0 and 313, that despite ballast renewal and four tamping operations, the short wavelength features that existed in the rail prior to ballast renewal are still present 313 days after ballast renewal. Clearly, the original rail is once again imprinting its shape into the surface of the new ballast bed.

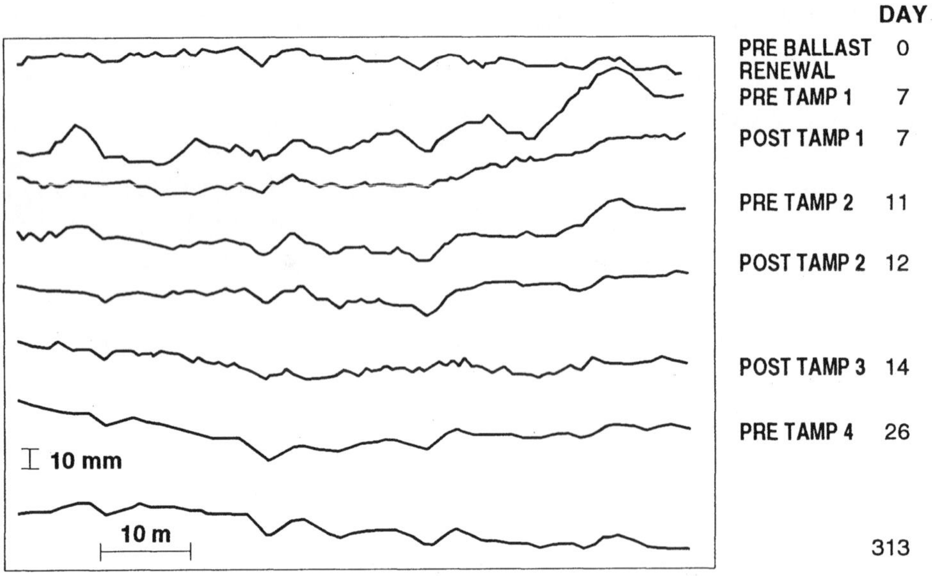

Fig. 15.13 Rail shape persisting following ballast renewal

Figure 15.14 is an example of the way in which rail shape persists through a tamping operation. The figure shows the longitudinal rail head profile of both rails of a 170 m long section of track from which the long wavelength component of the track geometry has been removed by filtering.

A comparison of the shapes shows that the short wavelength component of the track geometry was temporarily removed by tamping, but had, to a large extent re-established

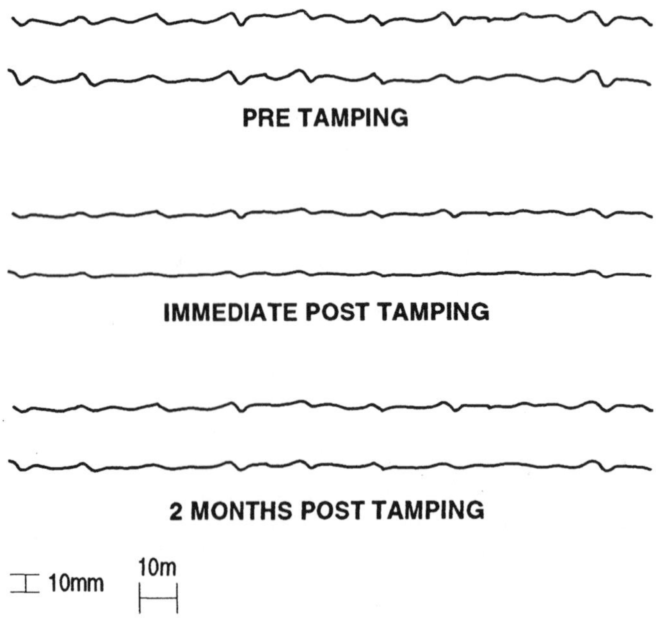

Fig. 15.14 Rail shape persisting following tamping

itself during the two months which followed tamping. The original rail shape is once again imprinting its shape into the surface of the newly tamped ballast bed.

Clearly, rails should be as straight as possible and to this end, great care must be taken to ensure that new rails are not bent during handling. Bends in existing rails should be removed by in situ rail straightening.

15.5.2.2 Top ballast surface profile

Figure 15.15 is an example of the way in which the geometric shape of the track is influenced by the surface profile of the ballast bed upon which it is laid. This figure also shows the way in which the 'as laid' surface profile of the ballast persists in spite of subsequent trafficking, and many tamping operations. The figure relates to the first three years in the life of a section of a newly constructed railway track.

Trace 15.15A shows the uncompacted surface profile of the ballast layer upon which the track was ultimately laid. Trace 15.15B shows the longitudinal rail head profile of one rail of the track immediately following laying. The long wavelength component of the track geometry has clearly adopted the long wavelength component of the ballast bed upon which it has been laid. Traces 15.15C through F show the longitudinal rail head profiles that were achieved by the first six smoothing tamping operations which were carried out in the 4th, 5th, 16th, and, 28th, weeks following track laying, respectively.

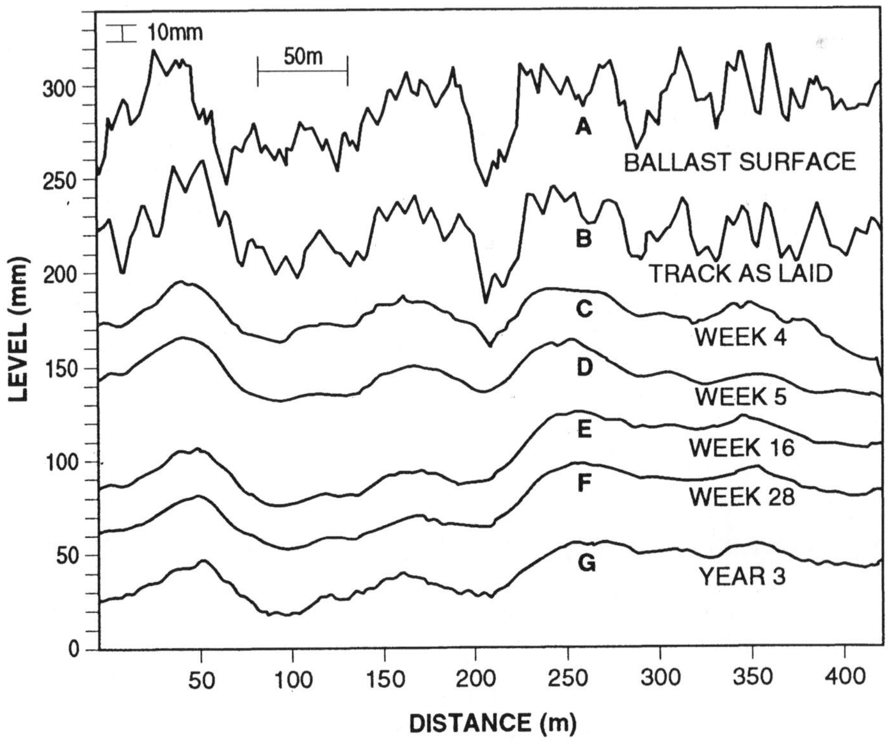

Fig. 15.15 Influence of ballast surface profile on track geometry

It can be seen from the longitudinal rail head profile in trace 15.15G, that long wavelength geometric faults present in the surface profile of the ballast bed, as originally laid, are still present in the longitudinal profile of the rail despite 3 years of trafficking and 6 tamping operations.

In the case of new construction, ballast renewal or ballast cleaning, it is clear that every effort should be made to ensure that the ballast surface upon which the track is laid is compact and has a surface that is free from longitudinal and cross level faults. Techniques described in a later section allow such a ballast bed to be achieved.

15.6. Improving the Inherent Shape of Track

It will be clear from the foregoing, that permanent improvements to fundamental track shape can only be achieved by improving the shape of the rail and/or improving the surface profile of the supporting ballast bed.

15.6.1. Rail Straightening

Improvements in rail shape can be achieved by rail straightening which falls outside the scope of this book. Improvements in ballast bed profile can be achieved by high lift tamping which is discussed in a later section.

15.6.2. Tamping

15.6.2.1 Lift/settlement relationship

While improving track shape by rail straightening is effective, attempts to improve the surface profile of the supporting ballast bed by smoothing tamping is not usually found to be effective, there being a tendency following tamping, for the track to revert to its inherent shape as shown in Fig. 15.11. The reason for this lack of success can largely be attributed to the phenomenon known as the lift/settlement relationship.

In Fig.15.16, each point corresponds to a sleeper end. The lifts given by a tamping machine to the sleeper at the time of tamping, are plotted against the settlements that occurred in the subsequent 66 weeks of trafficking. The scatter that is apparent in the lift/settlement relationship is no doubt in some measure, due to the influence of rail shape.

It can be seen that for relatively low lifts, the lift given by the tamping machine is approximately equal to the settlement that occurs in the subsequent 66 weeks of traffic. Thus, no lasting change in the inherent track shape has been achieved. This effect is often referred to as "ballast memory". For higher lifts however (i.e. greater than 25 mm in Fig. 15.16) there is a residual lift, and a lasting improvement in the inherent shape of the track has been achieved. The track settlement of 4 mm corresponding to a zero tamping lift, is associated with ballast wear and compaction resulting from the 66 days of normal trafficking.

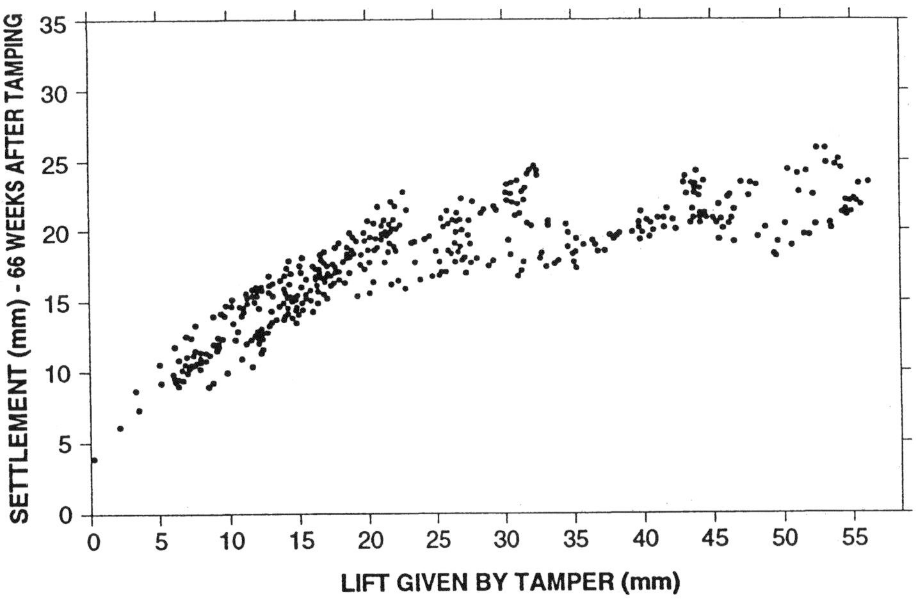

Fig. 15.16 Sleeper settlement as a function of tamping lift

In general, a high lift can be regarded as a lift which is in excess of the D50 size of the ballast, i.e. the sieve size that will retain 50% of a representative sample of the ballast being tamped.

A possible explanation of this phenomenon is that for low tamping lifts, the tamping tines laterally squeeze the vertically compacted ballast which dilates, and expands upwards into the void between the ballast surface and the underside of the sleeper. The ballast skeleton deforms, but re-arrangement of the particles does not take place.

Upon contacting the underside of the sleeper, further deformation of the ballast skeleton is not possible. Since the arrangement of the ballast particles within the ballast skeleton has remained unchanged, re-imposition of the vertical traffic loading will re-compact the ballast and the particles will adopt their original positions with respect to each other. The track will thus revert to its original geometry.

Where high tamping lifts are concerned, the tamping tines again laterally squeeze the vertically compacted ballast which dilates, and expands upwards. In this case however there is sufficient room for maximum ballast dilation to take place. Further tamping will result in additional ballast particles being absorbed by the ballast skeleton. Since in this case, the arrangement of the ballast particles within the ballast skeleton has changed, re-imposition of the vertical traffic loading will re-compact the ballast to a new skeleton and the track will adopt a new geometry.

Limited head room and/or a shortage of crib ballast, could preclude the use of high lift tamping as a means of improving the inherent quality of the track.

Results similar to those shown in Fig. 15.16 have been observed in South Africa (Ref. 15.3), and can be derived from data reported in America (Ref. 15.4).

15.6.2.2 High lift design tamping

Once the lift/settlement relationship for a site has been established, a high lift design tamp can be undertaken in which the DESIGN component ensures that the required track geometry is obtained, and the HIGH LIFT component ensures that the geometry achieved is long lasting.

Figure 15.17 shows the data contained within Fig. 15.16 re-plotted in the form of a lift/residual lift relationship. The relationship between lift and residual lift contained in Fig. 15.17 was used to determine the lifts that needed to be applied by the tamper to the sleepers to achieve the 52 week profile shown in Fig. 15.18. Also shown are the pre-tamp profile, and the profiles corresponding to immediate post tamp and week 16. It can be seen

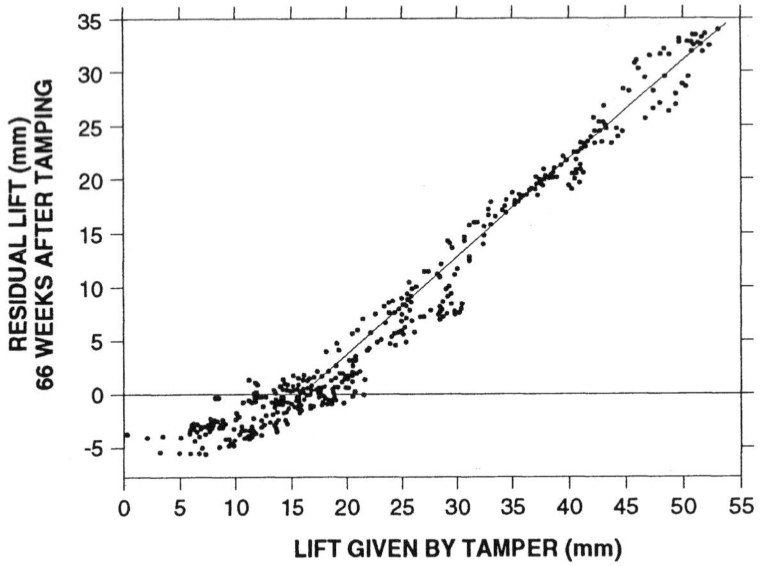

Fig. 15.17 Residual sleeper lift as a function of tamping lift

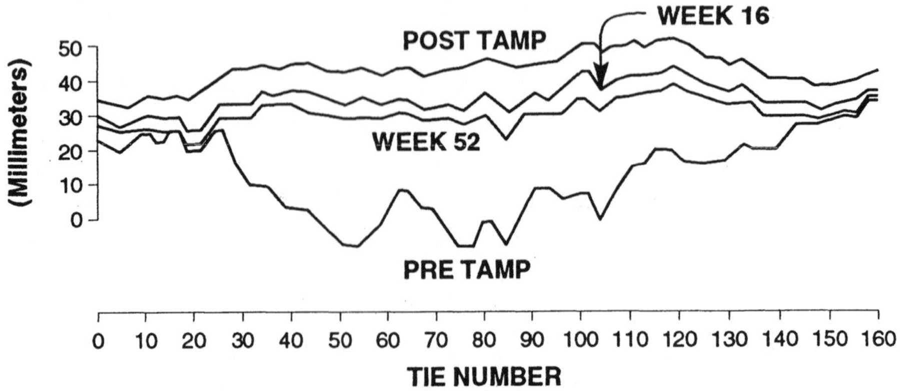

Fig. 15.18 Effect of design lift tamping on track geometry

that under traffic, the track settled from its immediate post-tamp profile, to a very acceptable '52 week' profile as was intended.

The short wavelength geometric faults in the vicinity of sleeper numbers 80 to 100 in the pre-tamp, post-tamp, week 16 and week 52 profiles, are undoubtedly associated with inherent rail shape.

15.7. Benefits from Good Inherent Quality Track

15.7.1. General Considerations

The benefits to be gained from good inherent quality track are always of value since the maintenance interval is extended, and the vehicle ride is improved, no matter what the speed or type of traffic being carried. Speed and type of traffic determine the level to which the geometric quality of the track may be allowed to deteriorate before re-surfacing is required. 'Goodness' or 'Poorness' of the inherent quality determines the rate at which the track geometry deteriorates to the level at which re-surfacing is required.

Consider a passenger train running on an inherently poor quality track (Fig. 15.19). Surfacing will be required when the intervention level is reached. Consider the same passenger train running on an inherently good quality track. The need to surface will not be reached so quickly, i.e. the maintenance interval is extended. A similar argument can be applied to freight traffic.

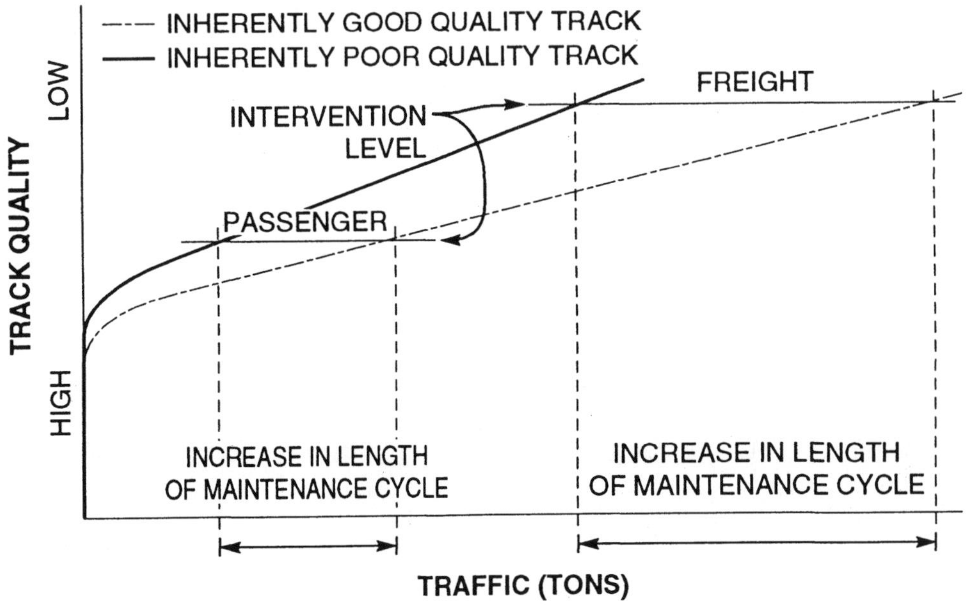

Fig.15.19 Track quality effect on maintenance cycle length

Further considerations with regard to good inherent quality track are the associated low levels of tamping maintenance which result in low levels of ballast damage. These low levels of ballast damage and associated production of fine material, in turn, result in a long ballast life, and a long ballast cleaning cycle.

Clearly, it costs money to achieve a good inherent track quality. However, it also costs money to live with the consequences of poor inherent track quality. In the case of freight traffic, the consequences of poor track geometry include: increased maintenance frequency, and/or poor track geometry resulting in excessive wear to track and train components, and poor ride resulting in damage to materials being carried. In the case of passenger traffic, a reduction in passenger comfort results ultimately in the loss of fare paying passengers who may well opt for alternative forms of transport.

15.7.2. Conclusions Regarding Inherent Track Quality

Good inherent quality track utilizes straight rails attached to sleepers resting on a geometrically smooth bed of compact ballast of uniform thickness which is in turn resting on a geometrically smooth and compact sub-ballast layer. Such a track foundation gives the very best chance of compacting uniformly under traffic to give a stable track of inherently good quality, requiring minimum future maintenance. The track construction techniques described in the next section are compatible with these requirements.

15.8. Achieving Good Inherent Quality Track

Ballast cleaning and ballast renewal exercises provide ideal opportunities to improve the inherent quality of existing track since they provide access to all the components which dictate inherent quality. For similar reasons, ballast cleaning and ballast renewal can, if not carried out with great care, result in the formation of inherently poor quality track.

For purposes of illustration, a ballast renewal operation will be considered. However, similar principles apply both to the construction of new track and to ballast cleaning.

The key to the success of the method adopted is that the ballast is placed in layers (Ref. 15.3). Each layer is compacted by the Dynamic Track Stabilizer, geometric faults associated with differential compaction being buried by the subsequently placed ballast layer. Geometric faults associated with compaction by the Dynamic Track Stabilizer of the final ballast layer are removed by the tamping machine which will still be on site.

Such procedure is clearly preferable to compaction being achieved by normal traffic, and the importation of tamping machines to deal with the geometric faults that will have resulted from differential compaction.

Consider Fig. 15.20, in which the track has been omitted for clarity:

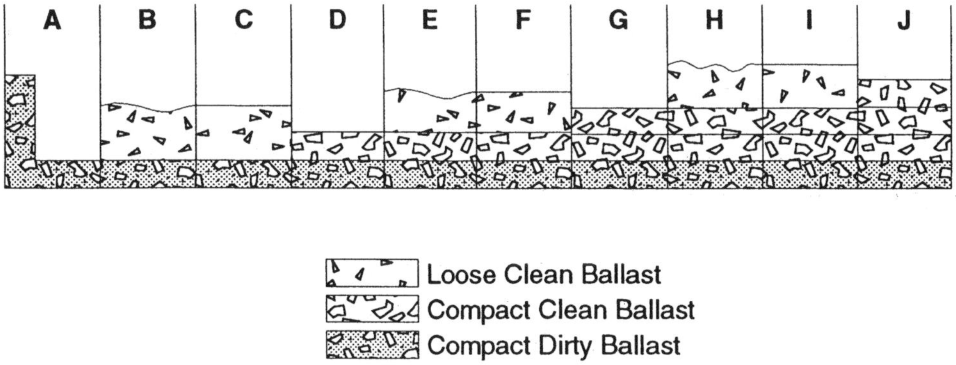

Fig. 15.20 Illustration of construction sequence with Dynamic Track Stabilizer

1) A shows the freshly cut surface of the original, highly compacted ballast layer. A typical depth of cut would be 400 mm.

The cutter bar of the ballast cleaner is operating under some form of 'depth of cut' control, e.g. a 'laser datum' as described in Chapter 14. Thus, the surface profile of the original, compact ballast layer will be free of longitudinal and cross level faults, and will have a cross fall that will ensure satisfactory cross-track drainage.

2) B shows the new loose ballast placed on the cut surface. C shows this layer following regulating and tamping, the 'track lift' control system of the tamping machine being referred to the same datum as was the cutter bar of the ballast cleaner.

3) D shows the ballast surface profile following one or more passes of the Dynamic Track Stabilizer.

The ballast compaction induced by these passes of the Dynamic Track Stabilizer will have resulted in the formation of minor geometric faults in the surface profile of the ballast layer. Clearly, the better the quality of the tamping operation referred to in 2) above, the fewer will be the geometric faults revealed by the Dynamic Track Stabilizer.

4) E shows the minor geometric faults referred to in 3) eradicated by the placing of the next ballast layer.

5) F and G show regulating, tamping and compacting operations identical to those described in 2) and 3) above.

Operations H, I, and J are repeated as described in 4) and 5), respectively, until the time that the ballast bed reaches the required level, at which point the rails are de-stressed and the track is re-opened to traffic.

The final pass of the Dynamic Track Stabilizer imparts a measure of lateral stability to the track. Geometric faults associated with this pass cannot be removed since they are the last operation prior to re-opening the track to traffic. For this reason, the compaction control system of the Dynamic Track Stabilizer is activated, and the loss of track geometry associated with this last pass is reduced. In this way, loss of track geometry associated with the dynamic loading resulting from the initial passes of normal traffic, travelling at normal speed, is minimized.

It will be seen from Fig. 15.20 that the track foundation which has resulted from the operation described has all the properties required of a track having a good inherent track quality as specified in Section 15.7.2. Specifically, the sleepers are resting on the geometrically smooth surface of a uniform depth of clean ballast, in a state of uniform compaction, which is in turn resting on the geometrically smooth surface of a highly compacted, and stable sub-ballast layer.

The procedures outlined above are now widely practiced and, when applied to ballast renewal exercises, have resulted in tracks having an initial geometric quality that allows them to be re-opened to traffic at full line speed immediately after ballast renewal. In addition, the inherent quality of the tracks are such that the maintenance cycle is extended well beyond that which would be expected, had the ballast been renewed by more conventional means.

On British Railways, the use of these techniques has allowed the old track to be removed, the ballast to be cleaned, new track to be replaced and reopened to traffic at the full line speed of 200 km/hr, within a 48 hour track possession.

15.9. Conclusions

The following conclusions can be drawn from the discussion in this chapter:

1) Tamping should only be undertaken when necessary. In this context, skipping tamping is a good thing and tamping on a routine basis is a bad thing.

2) Low lift tamping is unlikely to achieve a permanent improvement in track quality.

3) The inherent quality of existing track can be improved.

4) The maintenance interval can usually be lengthened.

5) Track can be constructed such that it has a good inherent quality.

6) It is worth striving for a good inherent quality track no matter what the train speed and type of traffic.

REFERENCES

15.1 Waters, J. M. (1984). **The FROG Track Geometry Measuring System.** *International Railway Journal*, December.

15.2 Shenton, M. J. (1984). **Ballast Deformation and Track Deterioration.** *Track Technology for the Next Decade*, Thomas Telford Ltd., London, pp. 253-265.

15.3 Kearsley, E. P. and Van As, S. C. (1993). **Effect of Heavy Haul Trafffic on Track Geometry Deterioration.** *Proceedings,* Fifth International Heavy Haul Railway Conference, Beijing, China, June, pp. 369-378.

15.4 Chrismer, S. M. (1990). *Track surfacing with conventional tamping and stone injection.* Association of American Railroads Research and Test Department, Report No. R-719.

15.5 **British Rail Looks to Business Motive.** *New Civil Engineer,* 8 October 1987.

16. Mechanics-Based Maintenance Model

16.1. Introduction

The decision about what type of ballast maintenance to use and when to use it is generally determined by precedent and by the need to work within existing maintenance budgets. To provide an alternative, a preliminary model has been developed which predicts the timing of, and least cost alternative among various ballast maintenance methods. The model combines research results on substructure deformation with economic analysis.

The model can be used by the maintenance planning engineer for a region or for an entire track system. The goal of the model is to allow the engineer to arrive at the most economical ballast choice and maintenance solution. In addition, the process of model development is beneficial in understanding the fundamentals of system interaction, and track performance.

The following model description is a modification of Ref. 16.1 to incorporate some of the subsequent changes.

16.2. Model Components

Figure 16.1 shows the construction of the model and the sequence in which the program interactively asks the user to respond to choices or provide the value of certain inputs. The main segments, shown in rectangular boxes are: 1) the track data input, 2) the ballast settlement and remaining life calculations, and 3) the maintenance costing calculation.

The analysis begins with a designation of track conditions. These include: initial ballast gradation and Abrasion Number, yearly traffic (MGT/yr), and an estimate of wheel load distribution. The user then selects one of four maintenance options. The track profile can be displayed to assist in selecting these options. The analysis for the desired planning period should start with tamping, undercutting/cleaning or plowing. However, it can start with no maintenance by designating the type of ballast maintenance last performed and the degree of ballast fouling at that time. In these cases where maintenance is selected, the track conditions are updated. For example with tamping, the required information is the number of tamps and the height of raise. With undercutting/cleaning the required information is depth of undercut, height of raise and type of replacement ballast. The input track information will provide a basis for estimating the ballast cleanability and expected recovery (amount of ballast returned to the track). Plowing of course requires specifying the depth of ballast to be removed, the replacement ballast thickness, and the type of replacement ballast.

The program then calculates the settlement for the next interval (nominally one year, but longer or shorter periods if conditions warrant), and the remaining ballast life. If the end of the planning period has been reached the cost of the plan will be determined. If not, another calculation cycle will be carried out. The option of changing certain conditions will be given. These conditions include traffic level, labor and equipment costs, and ballast material.

The end of the planning period is based on either a specified number of years or reaching the end of the ballast life. After the cost of the selected plan is determined, alternative plans can be analyzed for comparison.

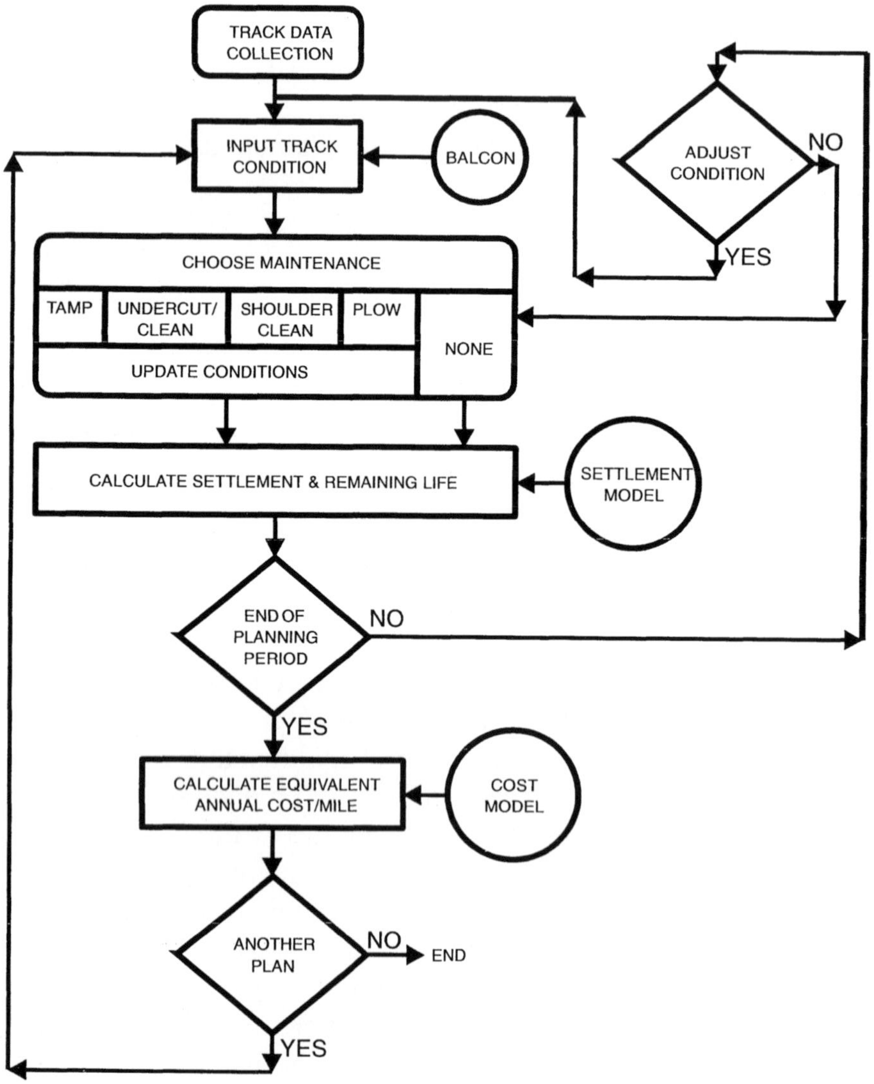

Fig. 16.1 Flow chart of ballast maintenance model

16.3. Ballast Condition Inspection

A means of determining values for the controlling system parameters is necessary for quantitative analysis. Some of these are known, such as traffic and track superstructure. However others, such as ballast, subballast and subgrade conditions, are usually not available.

Currently, ballast condition is most likely to be judged by its outward appearance. This can be highly deceiving because the clean ballast may only be of one particle thickness on the surface. The ability of a ballast to maintain a durable smoothing lift or to allow fast drainage is determined by the layers under the surface.

An efficient means of inspecting and documenting ballast conditions can be achieved using a portable computer with a program known as BALCON (Ref. 16.2). BALCON has three subsections: 1) site description, 2) subsurface layer delineation, and 3) sample description. The information needed by BALCON is readily obtained by one or two field personnel with digging tools or backhoe. A few sites judged to be representative of a "track segment" would be selected and excavations dug into the ballast to a depth of 300 to 600 mm (one to two feet) or more under the sleeper. The layers are usually easily distinguished by their different coloration and/or gradation.

With a small sample collected from each layer, the process of sample identification may proceed in the field, or later in the laboratory. The ASTM Visual-Manual method (Ref. 3.7) is used to identify and describe samples collected from each layer. Because BALCON was designed to allow rapid sample collection and identification, many sites can be inspected in a day. Judgment on rail or sleeper replacement would not be made without at least some kind of inspection to determine the condition of the materials as they exist in the field. Inspection of ballast is equally important. Considering the multi-million dollar

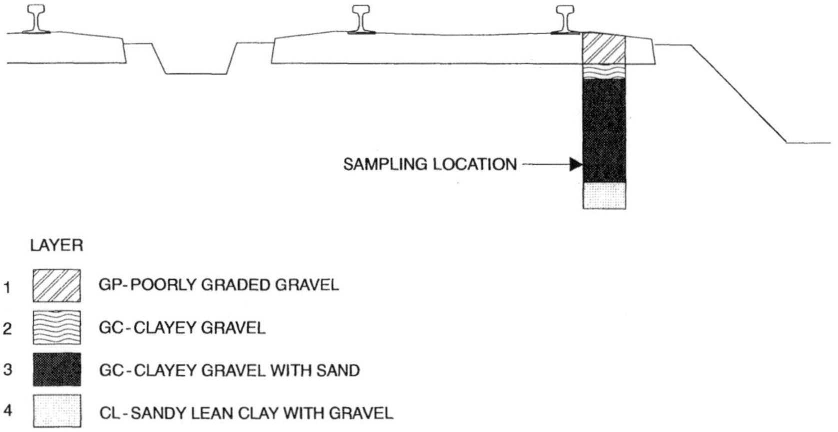

Fig. 16.2 Example of track cross section generated from BALCON database

investment in ballast that a railroad typically makes in any year, this minimum level of ballast inspection is economically justified.

At the user's discretion information in the BALCON database can be converted into track cross-sections by a plotting program as shown in Fig. 16.2. Note the indicated soil classification of each layer, and the layer depths. Although not shown, these track cross-sections can also reveal gradations of any of the layers, and the amount of reclaimable ballast. A gradation plot can provide convincing evidence of the degree to which a ballast section is at, or near, the end of its life.

The potential amount of recoverable ballast may be used to judge the benefit of undercutting/cleaning as opposed to wasting the entire section (Fig. 16.3). The graph of recoverable ballast versus depth of undercutting/cleaning is obtained when the user specifies the undercutter sieve size and depth of undercutting. The gradation of the ballast layers above the undercutting depth are used to calculate the percent of ballast which is bigger than the screen opening size. This amount is the estimated percent return of coarse ballast. Note the discontinuous slope of the plot in Fig. 16.3. This is due to the often distinct gradations of the ballast layers encountered with increasing depth. The layering is generally horizontal, which allows the estimation of percent of recoverable ballast from just one excavation below the sleeper as shown in Fig. 16.2.

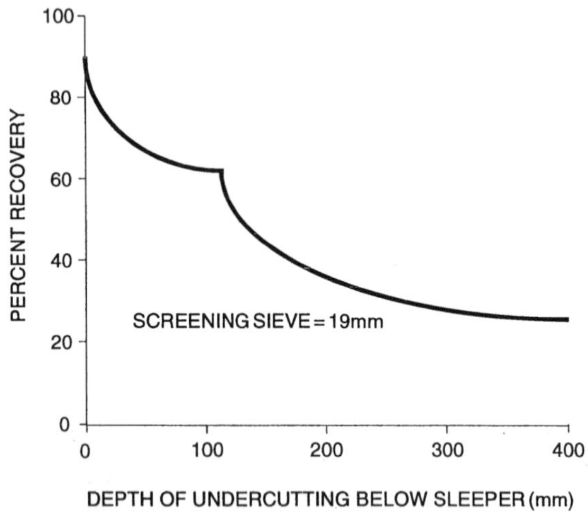

Fig. 16.3 Example of relationship between ballast recovery and depth of undercutting

16.4. Selection of Ballast Maintenance Option

At the end of any given year of the planning period the user must choose one of the four ballast maintenance options (or the no maintenance option) shown in Fig. 16.1. In the first few years of the analysis the BALCON information just described can be used to aid the maintenance technique selection. For example, tamping with a small amount of new ballast placed under the sleeper may be the preferred choice because tamping usually has a lower initial cost compared to ballast renewal. However, if the existing ballast is highly fouled, this tamping would only temporarily delay the need for ballast cleaning and/or replacing. The total cost would then be based on tamping followed shortly by ballast replacement.

For each maintenance option the user would define or update the labor, equipment, and ballast information in the "Update Conditions" segment of the flow chart. If, for example, the undercutting option were chosen, the equipment and labor cost per hour, and productivity would be entered.

The current basis for determining which option to choose is as follows:

1) Tamping

Use the following criteria, whichever comes first:

a) After completion of rail and/or sleeper renewal.

b) After undercutting/cleaning.

c) After ballast replacement (undercut, sled or plow).

d) After sufficient geometry loss as determined by

i) Roadmaster judgment or ride quality.

ii) Geometry car measurements.

iii) Tolerances allowed by safety standards.

e) To reduce fuel consumption.

f) To reduce cargo damage.

The geometry loss will be examined after each year of the analysis in order to decide whether tamping is necessary. The analysis will indicate when the specified settlement limit is reached during the year so that tamping can be selected immediately if desired, instead of imposing a speed restriction.

2) Undercut/clean

Use the following criteria:

a) Fouling index reaches 40 and ballast can be cleaned.

b) When judged necessary by the engineer.

3) Plow/sled

Use the following criteria:

a) Ballast life is so short that a better ballast is needed.

b) Ballast is fouled with wet and/or plastic fines and will not clean.

4) Shoulder clean

Criteria need to be developed for this option.

16.5. Ballast Life Calculation

Ballast life is defined as the amount of traffic (in MGT) that results in fouling of the ballast to the extent that voids are completely filled. The ballast life model used in the ballast settlement subprogram is based on a slightly modified form of that used by CP Rail (Fig. 8.56). One field study that seems to confirm the CP Rail model is given in Ref. 16.3. Although research to improve the prediction of ballast life is continuing, the Abrasion Number concept provides the best available estimate of ballast life when fouling is mainly a result of ballast breakdown with only a minor contribution from other sources. Adjustment to the model is needed to account for external sources of fouling (Table 8.2). In the CP ballast life model, the Abrasion Number is the measure of ballast material quality. This and the gradation of the ballast are the only two inputs needed to determine the ballast life. A difference from the CP model is in the relationship between ballast void volume and gradation. AAR research (Ref. 16.4) found a somewhat smaller effect of void volume on ballast life than that used by CP Rail.

The model requires that the Abrasion Number based ballast life subprogram be used. Because of the interaction of the value of Abrasion Number with the workings of the model, it is not possible to allow the user to specify ballast life (in MGT) without using the Abrasion Number based life. One of these interactive features is the relation between number of tamping applications and the resulting reduction of ballast life. With every tamping cycle the model decreases the remaining ballast life due to tamping damage by an amount dependent upon the Abrasion Number, and number of squeezes. This is used because some experimental evidence exists to suggest the relationship between Abrasion Number and degradation of ballast due to tamping (Ref. 16.6). The model simultaneously increases the ballast life by the amount of clean ballast under the sleeper.

16.6. Ballast Settlement Calculation

The remaining life of the ballast and the cumulative settlement are shown by the model for each year (or calculation interval) of the analysis. While it is the amount of differential track settlement which largely determines the need for track surfacing, the model currently uses average track settlement as an indication. Work is progressing toward developing the relationship between average and differential ballast settlement as discussed below.

Average settlement is determined by multiplying the calculated ballast vertical strain by the ballast layer thickness. Field and laboratory data for ballast layer deformations under various loading conditions were used to develop the ballast strain predictive equations in the model. Three factors were thought to have greatest influence on strain: 1) wheel load magnitude, 2) ballast material quality, and 3) amount of ballast fouling. The ballast strain is thus represented by the equation:

$$\varepsilon_N = \varepsilon_1 \kappa_{wi} \kappa_A \kappa_F [N_{ei} + N_i]^b \quad , \qquad (16.1)$$

where ε_N is the strain accumulated after N load applications,

ε_1 is the first load cycle strain for the base condition,

κ_{wi} is the factor which adjusts for changes in wheel load from the base condition,

κ_A is the factor which adjusts for changes in abrasion number of the ballast from the base condition,

κ_F is the factor which adjusts for changes in fouling index from the base condition,

N_{ei} pertains to the equivalent number of axle loads of magnitude i in yielding a strain equivalent to all of the applications of lesser loads, and

N_i represents the number of axle loads which occur for wheel load i.

This model represents the trend indicated in Fig. 8.6.

The base condition has a wheel load of 33 kips, an Abrasion Number of 45, and a zero initial fouling index. For these initial conditions the first cycle strain ε_1 has a value of 1.9%, the base case. For conditions which are different from this base case value, the κ terms account for differences between the base case and the current values of wheel load distribution, Abrasion Number, and ballast fouling index.

The settlement model calculates accumulated strain as shown in Fig. 16.4. Each line corresponds to the strain that would occur for a given number of load cycles at a certain wheel load. For greater wheel loads the curves shift upward as shown. As track roughness increases with settlement the induced dynamic component shifts the distribution of wheel loads upward from those for "smooth" track (left graph) to those for rougher track (right graph of Fig.16.4). The wheel loads are automatically updated after every calculation time period and then are used to find the settlement over the next time period. If maintenance is specified, the program assumes "smooth" track once again as a result of tamping. The wheel load mix returns to the original distribution unless the traffic conditions change.

Based on the trends for ballast strain increase with number of cycles described in Chapter 8, the settlement model in Eq. 16.1 is being replaced with a model of the form of Eq. 8.5. The ballast layer will be divided into two sublayers, one to represent the top ballast which is disturbed by tamping, and the other to represent the bottom ballast which is not disturbed. In addition settlement contributions from the subballast and subgrade are being added based on information in Chapter 8 to represent the total settlement situation shown in Fig. 8.4.

The strain magnitude increases at a diminishing rate until the ballast nears the end of its life, whereafter its behavior is dominated by the fines in the voids and the rate of settlement rapidly increases (Fig. 8.5). Laboratory data provide evidence of this rapid settlement at the end of ballast life. Unfortunately, information on the mechanical behavior of ballast near the end of its life is lacking.

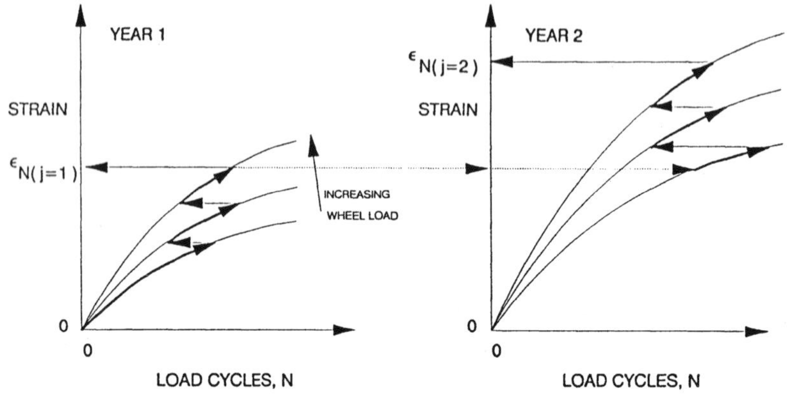

Fig. 16.4 Dynamic wheel load effect on ballast strain accumulation

16.7. Ballast Settlement and Track Roughness

In order to relate ballast strain to maintenance requirements, there is a need to establish the link between track average settlement and the resulting differential settlement or roughness. Some data showing this correlation are given in Fig. 16.5 (Ref. 16.5). One convenient measure of roughness is the standard deviation of rail profile as determined by discrete measurements (such as at every sleeper). Evidence has shown that, excluding the initial settlement just after maintenance, the trend of standard deviation with traffic tonnage is approximately linear (Refs. 16.6 and 16.7) over at least a moderate range of ballast life. This fact is helpful for the prediction of maintenance timing.

The problem, however, is that this relation is dependent not only on the characteristics of the ballast but also upon the inherent shape of the track as discussed in Chapter 15. Esveld writes that because of the apparent randomness and complexity of the interaction between ballast settlement, increasing rail profile roughness and dynamic loads, each track "section" should be monitored to determine its own characteristic differential settlement behavior (Ref. 16.7) and to base future maintenance requirements on this. However, a more powerful predictive approach would be to incorporate the mechanisms causing the differential settlement, as attempted here for ballast. Admittedly, this task is greatly complicated by the interaction of increasing dynamic wheel load and track profile roughness. Nonetheless the problem must be approached by initially considering each part of the problem by itself as is done here for ballast.

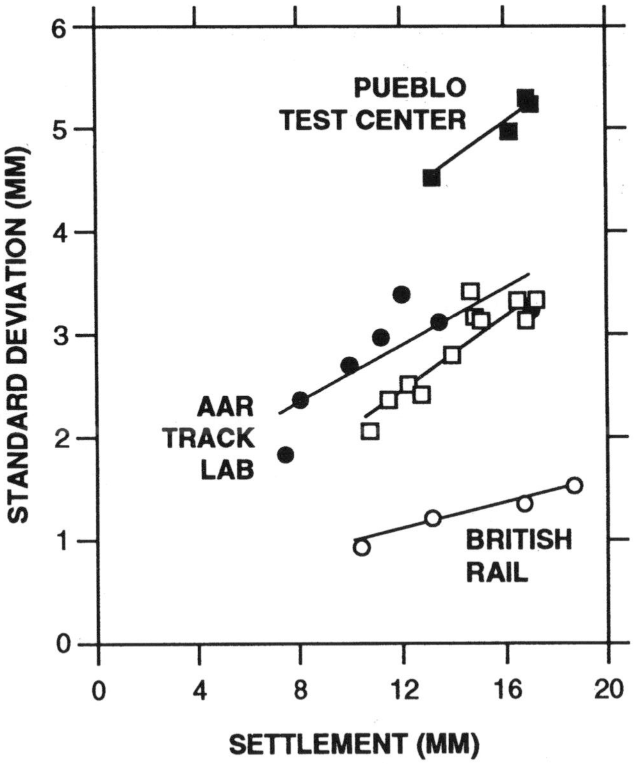

Fig. 16.5 Relationship between track roughness and average settlement

16.8. Cost Analysis

If the user decides to stop the analysis after so many years and find the cost of the maintenance sequence employed over the planning period, the cost model is invoked. The costing model (Ref. 16.4) assigns an equivalent annual cost per mile of track based on such considerations as the selected maintenance sequence and ballast choice. The program uses standard accounting equations to bring each cost outlay back to the present (year 0) and then finds the annual cost over the number of years considered in the analysis.

The equivalent annual cost per mile for a particular planning period may be noted, and another comparative run made to see if there is a scenario with a lower overall cost. Several such calculations can be made and the "least cost" maintenance plan determined. To compare the economics of one run with another it is not necessary to end on the same year.

But it is generally necessary for proper cost comparison that the user continues the analysis until the end of the life of the current ballast being used.

The user may wish to consider the effects of changing ballast material or changing the acceptable amount of track settlement which determines when maintenance is required. Many combinations of parameter may be analyzed in the pursuit of the least cost ballast maintenance program.

16.9. Summary and Conclusions

A ballast maintenance model has been described which has three main components: 1) gathering ballast and track information, 2) predicting ballast life and settlement, and 3) determining the least cost maintenance option. Research into settlement of ballast of various material qualities and gradations, subject to a range of load magnitudes, has allowed development of the ballast settlement-based maintenance model. The settlement model is mechanistic with its settlement behavior largely determined by ballast characteristics and load magnitudes.

With a greater understanding of the mechanisms of track deterioration, a more complete track maintenance model can be developed. Future work will include attempting to relate uniform settlement to roughness using field data, and adjusting ballast life for elastic deformation of track and for ballast contact pressure effects.

REFERENCES

16.1 Chrismer, S. M. and Selig, E. T. (1991). **Mechanics-based model to predict ballast-related maintenance timing and costs.** *Proceedings,* International Heavy Haul Railway Conference, Vancouver, B. C., Canada, June.

16.2 Selig, E. T. and Selig T. (1988). *BALCON - menu driven data collection system for ballast and subgrade condition,* November.

16.3 Chrismer, S. M., Selig, E. T., Laine, K., and DelloRusso, V. (1991). *Ballast durability test at Sibley, Missouri.* Association of American Railroads, Report No. R-801, December.

16.4 Chrismer, S. M. (1988). *Recent developments in predicting ballast life and economics.* Association of American Railroads, Research Report No. R-700, Chicago, IL.

16.5 Johnson, D. M. (1983). *Development of a track deterioration index.* AAR internal report (not published).

16.6 Chrismer, S. M. (1989). *Track surfacing with conventional tamping and stone injection.* Association of American Railroads Research Report No. R-719, Chicago, Illinois, March.

16.7 Esveld, Coenraad (1989). *Modern railway track.* MRT Productions, Germany.

Index

A

Author Index

B

C

D

End of index